MOS DEVICES FOR LOW-VOLTAGE AND LOW-ENERGY APPLICATIONS

MOS DEVICES FOR LOW-VOLTAGE AND LOW-ENERGY APPLICATIONS

Yasuhisa Omura
Kansai University, Japan

Abhijit Mallik
University of Calcutta, India

Naoto Matsuo
Hyogo Pref. University, Japan

Registered Office
John Wiley & Sons Singapore Pte. Ltd., 1 Fusionopolis Walk, #07-01 Solaris South Tower, Singapore 138628

For details of our global editorial offices, for customer services and for information about how to apply for permission to reuse the copyright material in this book please see our website at www.wiley.com.

Library of Congress Cataloging-in-Publication data applied for

ISBN: 9781119107354

Set in 10/12pt Times by SPi Global, Pondicherry, India
Printed and bound in Singapore by Markono Print Media Pte Ltd

10 9 8 7 6 5 4 3 2 1

Contents

Preface

The twin goals of low-voltage operation and low power consumption have been pursued for 40 years, since the adoption of the scaling law. In the 1970s, however, the electronics industry did not pay much attention to the scaling law, despite its advantages, because nobody imagined at that time that the industry would grow so rapidly or that the global social infrastructure would be so thoroughly altered by the Internet.

Since the final decade of the twentieth century, the limitations of petroleum-related chemicals and the threat of global warming have compelled industry leaders and scientists to make serious efforts to find solutions. Population pressure is negatively impacting energy resources and global warming, giving the situation particular urgency.

Against this background, many kinds of monitoring technologies using sensors are being developed. However, the power consumption of such products is very high and high-power batteries are needed as a result.

We, the authors, are interested in the development of low-voltage and low-power semiconductor devices to contribute to the energy efficiency of electronic products. In this book we introduce the concept of "low energy" in discussing the above issues. The meaning of "low energy" is described in the following chapters in detail.

We hope that this book will be helpful in developing low-energy device technologies for the electronics industry and the world.

Yasuhisa Omura, Osaka, Japan
Abhijit Mallik, Kolkata, India
Naoto Matsuo, Himeji, Japan

Acknowledgments

The first editor, Professor Yasuhisa Omura, wishes to express his sincere thanks to Dr. Katsutoshi Izumi (the former NTT laboratory director and a professor at Osaka Prefecture University) for his warm support for this study when Dr. Izumi was with NTT Laboratories. He also thanks Ms. Yu Azuma Yoshioka and Mr. Yoshimasa Yoshioka (Graduate School of Engineering, Kansai University; presently with Panasonic, Japan), Mr. Yoshio Mimura (Graduate School of Engineering, Kansai University; presently with Renesas Electronics, Japan), Mr. Daiki Sato (Graduate School of Engineering, Kansai University; presently with Panasonic, Japan), Mr. Shunsuke Nakano (Graduate School of Engineering, Kansai University; presently with Renesas Electronics, Japan), Mr. Osanori Hayashi (Graduate School of Engineering, Kansai University; presently with Roam, Japan), Mr. Hidehiko Nakajima (Graduate School of Engineering, Kansai University; presently with Canon, Japan), Dr. Kenji Komiya (Graduate School of Engineering, Kansai University; presently with Denso, Japan), and Mr. Daishi Ino (Graduate School of Engineering, Kansai University; presently Alps Tech., Japan) for their documentation of experiments, device simulations, and circuit simulations. He gives his special thanks to Dr. Hirobumi Watanabe and Mr. Hidenori Kato (Ricoh, Japan) for the fabrication of XCT SOI devices.

Professor Omura expresses his deep appreciation to Mr. Mike Blackburn (TECH-WRITE, Japan) for his continued guidance in English communication over the past 25 years. He also expresses his gratitude to his family members – wife Kikuyo, son's family, and daughter's family – for their various metal supports.

The second editor, Abhijit Mallik, would like to acknowledge his research collaborator and the first editor, Professor Yasuhisa Omura, who encouraged and convinced him to be a coeditor of this book. Without his tireless efforts, this book would not exist. He would also like to acknowledge the benefit he has gained from the years of collaboration and interaction with Professor V. Ramgopal Rao (Indian Institute of Technology, Mumbai, India), Prof. Chandan Sarkar (Jadavpur University, Kolkata, India), and Prof. Anupam Karmakar (University of Calcutta, Kolkata, India). He expresses his thanks to his graduate students, Dr. Saurav Chakraborty (presently with Tyfone Communications Development (India) Pvt. Ltd., Bangalore, India), Professor Srimanta Daishya (currently with the National Institute of

Technology, Silchar, India), Dr. Avik Chattopadhyay (currently with the Birla Institute of Technology and Science, Pilani, India), Ms. Shilpi Guin (presently a research scholar with the University of Calcutta, Kolkata, India), and Mr. Asish Debnath, who helped to bring this work to its present form over the years. He is also grateful to his family members – mother Muktakeshi Mallik, wife Arpita Mallik, son Abhishek, and daughter Annika – for their constant encouragement and patience during the entire process. Finally, he expresses his gratitude to his late father Raghunath Mallik, whose memory helped him through the process of writing.

The third editor of this book, Naoto Matsuo, wishes to thank sincerely Yasuhisa Omura, and also Dr. Shin Yokoyama, Hiroshima University, and Dr. Hiroki Hamada, Director of the Research Department of Sanyo Electric Co., presently at the Kinki University, for their useful discussions. Measurement of the electrical characteristics of the tunneling dielectric thin-film transistor (TDTFT) and conventional thin-film transistor (TFT) at low and high temperatures was performed at the Omura Laboratory, Kansai University. The TDTFT and the conventional TFT were fabricated at Research Institute for Nanodevices and Bio Systems, Hiroshima University, supported by the Nanotechnology Support Project of the Ministry of Education, Culture, Sports, Science and Technology (MEXT), Japan. A part of this study was supported by *Kakenhi* (grants-in-aid) MEXT and by the Ozawa and Yoshikawa Memorial Electronics Research Foundation. Naoto Matsuo also thanks Dr. Akira Heya, associate professor at the University of Hyogo, Dr. Takahiro Kobayashi, Yumex Inc., Japan, Dr. Naoya Kawamoto, Yamaguchi University, Dr. Kensaku Ohkura, Hiroshima University, and the many master's course students of Yamaguchi University and the University of Hyogo, for carrying out experiments and simulations.

He is deeply grateful to Dr. Akio Sasaki, an emeritus professor of Kyoto University and Osaka Electrical and Communication University, and Dr. Hiroyuki Matsunami, an emeritus professor of Kyoto University, for their useful comments and encouragement during the earlier period of this study.

On behalf of the editors, Professor Omura expresses his thanks to Mr. James W. Murphy for his assistance because he warmly accepted the task of examining the book proposal, and to Ms Maggie Zhang, Ms Victoria Taylor and Ms Nivedhitha Elavarasan for their guidance.

Finally, the authors would all like to express their appreciation to the Japan Society of Applied Physics (JSAP), the Institute of Electronics, Information, and Communication Engineers (IEICE), the Electrochemical Society (ECS), the Institute of Electrical and Electronics Engineers (IEEE), Elsevier Limited, and Springer for their cooperation in granting permission to reproduce original papers or figures.

Part I

INTRODUCTION TO LOW-VOLTAGE AND LOW-ENERGY DEVICES

1

Why Are Low-Voltage and Low-Energy Devices Desired?

The original scaling rule [1] indicated that the dissipation power density (W/cm^2) of an integrated circuit is not changed by scaling [1, 2]. However, this guideline is only really applicable to DRAM devices. In most integrated circuits, the supply voltage is not scaled according to the designer's intent, and devices have faced negative impacts, such as hot-carrier phenomena [3] and negative bias temperature instability (NBTI) phenomena [4], due to high supply voltages [5]. In the twentieth century, central processing unit (CPU) revealed dramatic advances in device performance (high-speed signal processing with increases in data bit length), and the guideline seemed to be ignored (see http://www.depi.itch.edu.mx/apacheco/asm/Intel_cpus.htm, accessed May 18, 2016).

However, in the 1990s, CPU designers noticed the limitations of CPU cooling efficiency, which triggered an urgent and ongoing discussion of low-power device technology. The following possible solutions have been proposed:

- the introduction of a silicon-on-insulator integrated circuit (SOI IC) strategy based on advanced substrate technology [6];
- a multicore strategy [7];
- a low-voltage strategy [8].

These major strategic proposals have led the worldwide electronics industry to the Internet of Things.

Business opportunities based on information technology have increased in the real world without taking account of the issues raised by technologies such as cloud computing [9] and datacenter construction (see http://www.datacenterknowledge.com/,

MOS Devices for Low-Voltage and Low-Energy Applications, First Edition.
Yasuhisa Omura, Abhijit Mallik, and Naoto Matsuo.
© 2017 John Wiley & Sons Singapore Pte. Ltd. Published 2017 by John Wiley & Sons Singapore Pte. Ltd.

accessed May 18, 2016), which have rapidly increased global energy consumption [10]. We must propose viable and innovative ideas on semiconductor device technologies to suppress such global-warming factors.

Information technology has widened perspectives on improving the quality of our future social life. Many companies are creating highly desirable products in the field of sensing technology, such as house monitoring (temperature, humidity, air pollution, fire, human health, security), office monitoring (temperature, humidity, air pollution, fire, security), traffic monitoring (for aspects such as car speed, traffic jams, and accidents), agriculture monitoring (for temperature, humidity, air pollution, rain, wind, lighting, storms), space monitoring (moon, sun, stars, meteorites, and other astronomical phenomena), defense monitoring, and so on. Many of these products use dry batteries or solar power/batteries for 24-hour monitoring. In the case of portable equipment, large batteries are impractical, which has triggered the development of small batteries with high energy density. This is also applicable to cellular phones and smart phones [11].

In battery-powered sensing devices, the battery volume must be small. This may be achieved by lowering the supply voltage, which in turn reduces the battery energy as it is proportional to the square of the voltage. Hence, it is more important to reduce the dissipation energy than the dissipation power for sensing devices. This will be addressed again later. We must, therefore, contribute to the solution of urgent social problems by proposing low-energy devices and integrated circuits.

References

[1] R. H. Dennard, F. H. Gaensslen, H.-N. Yu, V. L. Rideout, E. Bassous, and A. R. LeBlanc, "Design of ion implanted MOSFETs with very small physical dimensions," *IEEE J. Solid-State Circuits*, vol. SC-9, pp. 256–268, 1974.

[2] P. K. Chatterjee, W. R. Hunter, T. C. Holloway, and Y. T. Lin, "The impact of scaling laws on the choice of N-channel or P-channel for MOS VLSI," *IEEE Electron Device Lett.*, vol. EDL-1, pp. 220–223, 1980.

[3] T. H. Ning, "Hot-electron emission from silicon into silicon dioxide," *Solid-State Electron.*, vol. 21, pp. 273–282, 1978.

[4] D. K. Schroder and J. A. Babcock, "Negative bias temperature instability: road to cross in deep submicron silicon semiconductor manufacturing," *J. Appl. Phys.*, vol. 94, pp. 1–18, 2003.

[5] B. Kaczer, R. Degraeve, M. Rasras, K. Van de Mieroop, P. J. Roussel, and G. Groeseneken, "Impact of MOSFET gate oxide breakdown on digital circuit operation and reliability," *IEEE Trans. Electron Devices*, vol. 49, pp. 500–506, 2002.

[6] M. Canada, C. Akroul, D. Cawlthron, J. Corr, S. Geissler, R. Houle, P. Kartschoke, D. Kramer, P. McCormick, N. Rohrer, G. Salem, and L. Warriner, "Impact of MOSFET Gate Oxide Breakdown on Digital Circuit Operation and Reliability," IEEE Int. Solid State Circ. Conf. (ISSCC), Dig. Tech. Papers, pp. 430–431, 1999.

[7] T. Chen, R. Raghavan, J. N. Dale, and E. Iwata, "Cell broadband engine architecture and its first implementation – A performance view," *IBM J. Res. Dev.*, vol. 51, pp. 559–570, 2007.

[8] S. Hanson, B. Zhai, K. Bernstein, D. Blaauw, A. Bryant, L. Chang, K. K. Das, W. Haensch, E. J. Nowak, and D. M. Sylvester, "Ultralow-voltage minimum-energy CMOS," *IBM J. Res. Dev.*, vol. 50, pp. 469–490, 2006.

[9] Meil, P., and Grance, T., "The NIST Definition of Cloud Computing," NIST, 2011.

[10] G. Fettweis and E. Zimmermann, "ICT Energy Consumption – Trends and Challenges," The 11th Int. Symp. on Wireless Personal Multimedia Communications (WPMC 2008) (Finland, Sept., 2008), Session WG1: W-GREEN 2008 (I), 2008.

[11] Y. Orikasa, T. Masese, Y. Koyama, T. Mori, M. Hattori, K. Yamamoto, T. Okado, Z.-D. Huang, T. Minato, C. Tassel, J. Kim, Y. Kobayashi, T. Abe, H. Kageyama, and Y. Uchimoto, "High energy density rechargeable magnesium battery using earth-abundant and non-toxic elements," *Sci. Rep.*, vol. 4, Article number: 5622, 2014.

2

History of Low-Voltage and Low-Power Devices

2.1 Scaling Scheme and Low-Voltage Requests

Dennard *et al.* [1] considered the impact of device scaling on device performance. They assumed the device parameters and voltage parameters shown in Table 2.1 when the so-called "constant-field scaling" method was proposed. They assumed the following expression for the metal oxide semiconductor field-effect transistor (MOSFET) drain current:

$$I_D = \left(\frac{W_G}{L_G}\right) C_{ox} \mu_{eff} \left(V_G - V_{TH} - V_D / 2\right) V_D \tag{2.1}$$

where W_G is the gate width, L_G is the gate length, C_{ox} is the gate capacitance per unit area, μ_{eff} is the carrier mobility, V_G is the gate voltage, V_{TH} is the threshold voltage, and V_D is the drain voltage. Calculation results for the scaling factor (k) are summarized in Table 2.2. They reveal the following features of scaled parameters:

1. The electric field in the device and the averaged carrier velocity are static.
2. Capacitance components, including the depletion layer capacitance, shrink at the rate of $1/k$.
3. The carrier density of the inversion layer in the "ON" state does not change.
4. Drain current (drift current) decreases at the rate of $1/k$.
5. Channel resistance does not change.
6. Switching delay time/device (intrinsic delay time) decreases at the rate of $1/k$.
7. The dissipation power/device decreases at the rate of $1/k^2$.
8. The power-delay product/device shrinks at the rate of $1/k^3$.
9. The dissipation power of devices/unit area does not change.

MOS Devices for Low-Voltage and Low-Energy Applications, First Edition.
Yasuhisa Omura, Abhijit Mallik, and Naoto Matsuo.
© 2017 John Wiley & Sons Singapore Pte. Ltd. Published 2017 by John Wiley & Sons Singapore Pte. Ltd.

Table 2.1   Physical parameters of MOSFET [1].

Parameters	Initial value
t_{ox}	100 nm
N_A	5×10^{15} cm^{-3}
L_G	5 μm
W_G	5 μm
V_D, V_G	20 V

Table 2.2   Constant field scaling and results [1].

Parameters	Scaling factor
Circuit performance	
Device dimension t_{ox}, L_G, W_G	1/k
Doping concentration N_A	k
Voltage V	1/k
Current I	1/k
Capacitance C	1/k
Delay time/circuit VC/I	1/k
Power dissipation/circuit VI	1/k^2
Power density VI/A	1
Interconnection lines	
Line resistance R_L	k
Normalized voltage drop IR_L/V	k
Line response time R_L/C	1
Line current density I/A	k

One important feature that attracted attention in the 1970s was 6. but the concept of "constant-field scaling" was not adopted in industry until after the 1980s [2, 3]. As a result, feature 6. was successful, but features 7.–9. were lost.

From the latter half of the 1990s, the semiconductor industry was able to employ lower supply voltages. The popularity of cellular phones created new demands on the semiconductor industry. As cellular phones must be extremely portable, they need a small battery; this strong demand resulted in low-voltage IC designs. In addition, many battery vendors contributed to battery downsizing and battery energy density enhancement.

Lowering the supply voltage accelerates device scaling. However, the scaling trend ignored the original concept proposed by Dennard *et al.* [1]; the dissipation power of integrated circuits (ICs) per unit area was already much higher than the value estimated by the original scaling concept even though the supply voltage was successfully lowered [4]. The surface temperatures of integrated processor circuits are reaching dangerous values (the intrinsic temperature) [5]. Moreover, we now face the negative influence of several physical device parameters on IC performance because they do not follow any scaling rule. The following are typical of the parameters showing such undesirable behavior:

1. Electric-field induced mobility degradation.
2. Depletion capacitance of poly Si gate and inversion layer capacitance.

3. Subthreshold swing.
4. Parasitic resistance of devices (gate electrode, source diffusion, drain diffusion, contacts).
5. Leakage current (pn junctions, gate insulator).
6. Threshold voltage.

Issue 1. can basically be overcome using strain technology [6]. Issue 2. can be addressed by using high-κ dielectrics [7], and issue 4. can be improved drastically by the gate-last process [8] and the silicidation process [9]. Issue 6. is aided somewhat by adopting the metal-gate electrode. Issue 5. can be improved by using hetero-junctions and introducing high-κ dielectrics. However, issue 3. remains unresolved because there is no agreement on the design methodology that will allow use of subthreshold operation, or on how to design steep swing devices. Subthreshold logic circuits were proposed to realize the ultimate in low-energy operation [10]. In contrast, steep-swing devices, such as tunnel field-effect transistors (TFETs), were proposed in order to lower the supply voltage and to advance the radio frequency performance of MOSFETs [11]. The TFET device technology is still under investigation, and doesn't yet appear to be a reliable solution for future electronics. It is, however, a leading candidate. This book will discuss the above two concerns; that is, useful applications of subthreshold characteristics and the potential of steep-swing devices.

Before discussing individual solutions, we estimate the power consumption and dissipation energy of conventional complementary metal oxide semiconductor (CMOS) devices. Here we assume the following two equations:

$$Dissipation\ power\ density\ \left(\text{W}\ /\ \text{cm}^2\right) = P_{ON} + P_{OFF}$$

$$= r_{ON} f_{clock} \frac{1}{2} C V_D^2 N_{device} + r_{OFF} I_{leak} V_D N_{devices}$$

(2.2)

$$Dissipation\ energy\ density\ \left(\text{J}\ /\ \text{cm}^2\right) = E_{ON} + E_{OFF}$$

$$= T_{one-sec} \left(r_{ON} f_{clock} \frac{1}{2} C V_D^2 N_{device} + r_{OFF} I_{leak} V_D N_{devices} \right)$$

(2.3)

where P_{ON} denotes the dissipation power density in the "ON" state, P_{OFF} denotes the dissipation power density in the "OFF" state, N_{device} denotes the number of devices per unit area, r_{ON} denotes the fraction of devices working, r_{OFF} denotes the fraction of devices in standby, f_{clock} denotes the clock frequency, I_{leak} denotes the leakage current of the device, V_D denotes the supply voltage, C denotes the gate capacitance of single CMOS, and $T_{one-sec}$ denotes the one-second period.

We calculated the dissipation energy of ICs using the equations as shown in Figure 2.1, where we assumed post-1980 device technology. Device parameters assumed in the calculations are summarized in Table 2.3.

Figure 2.1 raises the following key points:

1. E_{OFF} has significantly increased this century;
2. V_{TH} is approaching the thermal voltage;
3. the subthreshold swing should be steep.

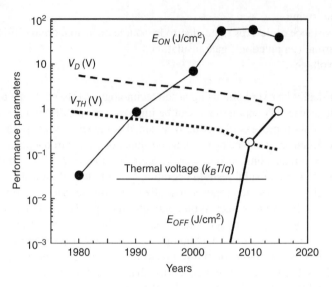

Figure 2.1 Evolution of performance parameters. It is assumed that $r_{ON} = 0.5$, subthreshold swing = 70 mV/dec, and $W_G/L_G = 5$.

Table 2.3 Parameters in simulations.

Years	L_G (nm)	EOT (nm)	V_D (V)	V_{TH} (V)	f_{clock} (Hz)	N_{device} (cm^{-2})
1980	2000	50	5.0	1.0	5×10^6	7×10^4
1990	600	20	3.3	0.7	1×10^8	1×10^6
2000	200	8.0	2.5	0.5	1×10^9	5×10^6
2005	55	2.2	2.0	0.4	3×10^9	8×10^7
2010	30	1.4	1.5	0.2	3×10^9	3×10^8
2015	24	1.1	1.0	0.15	3×10^9	5×10^8

EOT: equivalent oxide thickness.

The calculation results suggest that we must continuously lower the supply voltage, but also that we will reach its floor (the thermal voltage, ~26 mV at 300 K) [12].

2.2   Silicon-on-Insulator Devices and Real History

In the 1980s, silicon-on-insulator (SOI) substrate technology emerged and was commercialized. Before the 1980s, silicon-on-sapphire (SOS) substrate technology was the major material for SOI device applications [13]. The SOI substrate was studied in order to develop high-speed switching devices because the silicon-on-sapphire metal oxide semiconductor field-effect transistor (SOS MOSFET) had a much smaller parasitic capacitance beneath the source and drain diffusions than the bulk MOSFET. However, the SOS material had some serious issues, such as the autodoping of Al from the sapphire substrate and high-density defects near the Si/sapphire interface. The Si-based SOI substrate technology basically replaced the SOS substrate technology because the latter's shortcomings could not be ignored by 1990.

Before the 1990s, the electronics industry focused on the signal processing speed of ICs. After the 1990s, however, the industry had to pay attention to IC dissipation power because energy concerns were becoming more pronounced. Accordingly, leading scientists and engineers studied the low-power performance of SOI ICs with low-power dissipation closely [14], and many companies began studies to develop high-performance ICs [15]. Many scientists and engineers, however, remained skeptical about whether sub-100 nm gate silicon-on-insulator complementary metal oxide semiconductor (SOI CMOS) technology was actually suitable for creating the ICs needed in the twenty-first century because no aggressive technical vision had been published. Such negative perceptions were familiar in the electronics industry because nobody predicted the dramatic expansion of worldwide use of mobile phones and tablet PCs. Before specific ICs for mobile phones were developed, the power supply voltage was still high (~3 V). The voltage level was too high to yield a 100 nm gate SOI CMOS with fully depleted mode operation; the parasitic bipolar phenomenon induces the single-transistor latch [16], and many engineers were very skeptical as to the reality of scaled fully depleted silicon-on-insulator metal oxide semiconductor field-effect transistors (SOI MOSFETs). Fortunately, in the latter half of the 1990s, many businessmen began to use cellular phones and notebook PCs, forcing the electronics industry to overcome the historical barrier of power-supply voltage. This breakthrough gave the fully depleted SOI devices a good business opportunity with the result that they are used in many commercial electronics products.

Dr. K. Izumi and his group developed the ITOX-SIMOX (Internal Thermal OXidation-Separation by IMplanted OXygen) substrate [17], which can be applicable to scaled CMOS devices, and one of the editors (Omura) and his colleagues in NTT Laboratories demonstrated the potential of the 100 nm gate SOI CMOS IC with fully depleted components [18]. With the development of the oxygen ion-implanter, commercially viable SIMOX substrates became possible [19], SIMOX substrates were used to fabricate SOI devices and circuits in the laboratories of many companies. After the 1980s, other SOI substrate fabrication technologies, such as ELTRAN® and UNIBOND®, were also proposed [20–23], and have now entered commercialization. In the twenty-first century, companies that could catch this technology trend are doing well, while those that could not faltered.

Let us consider how we can use the low-power performance of SOI MOSFETs. Intrinsic switching delay time (τ) and dissipation power (P_{ON}) of SOI MOSFET are given by

$$Switching\ delay\ time\ (\tau\ /\ Tr) = \left(L_G W_G C_{ox} + L_G W_G C_{S,D}\right)\frac{V_D}{I_D} + C_P \frac{V_D}{I_D} \tag{2.4}$$

$$Dissipation\ power\ (P_{ON}\ /\ Tr) = V_D I_D \tag{2.5}$$

where C_{ox} denotes the gate capacitance per unit area, $C_{S,D}$ denotes the source and drain capacitance per unit area, and C_P denotes the parasitic capacitance except $C_{S,D}$. Energy dissipated by a single switching event (τP_{ON}) can be estimated as

$$Power \cdot Delay\ product\ (P_{ON} \cdot \tau\ /\ Tr) = \left\{\left(L_G W_G C_{ox} + L_G W_G C_{S,D}\right) + C_P\right\} V_D^2. \tag{2.6}$$

Estimated dissipation power of IC (P_{ON}) is given by using the clock frequency (f_{clock}):

$$P_{ON} = N_{device} f_{clock} \left\{\left(L_G W_G C_{ox} + L_G W_G C_{S,D}\right) + C_P\right\} V_D^2 \tag{2.7}$$

We must also take account of the standby power of the IC (P_{OFF}), given by

$$P_{OFF} = N_{device} I_{leak} V_D \tag{2.8}$$

As a result, the overall dissipation power (P_{total}) of the LSI is given by

$$P_{total} = P_{ON} + P_{OFF}$$
$$= N_{device} f_{clock} \left\{ \left(L_G W_G C_{ox} + L_G W_G C_{S,D} \right) + C_P \right\} V_D^2 + N_{devices} I_{leak} V_D \tag{2.9}$$

In the above expressions, the terms that are significantly different from the estimated values of the bulk MOSFET are those that include $C_{S,D}$ and C_p. In SOI CMOS devices, we can predict a reduction in I_{leak} due to a reduction in effective junction area, suppression of I_{leak} increases based on better subthreshold swing values, and C_p reductions due to shrinkage of device isolation area, resulting in the suppression of increases in P_{total}. However, as the IC designer is always seeking to enhance signal-processing performance, P_{ON} will increase due to the increase in the value of $N_{device} f_{clock}$. When the IC designer lowers the MOSFET threshold voltage to raise the switching speed, I_{leak} inevitably increases.

Recent P_{OFF} values are comparable to P_{ON} due to I_{leak}. IC designers are suffering from this conundrum. Therefore, one of the primary concerns of this book is how to suppress P_{OFF} as well as whether we can find applications that do not need high-speed performance.

In Part II, we review the fundamentals of conventional MOSFETs, SOI MOSFETs, and TFETs. This part will help readers to follow easily the discussions in the other parts. In Part III, we discuss how we can use the low-energy performance of bulk MOSFETs and how we can apply such devices to low-energy circuits. In Part IV, we review the low-energy performance potential of fully depleted SOI MOSFETs and introduce examples. In Part V, we address the low-energy performance potential of cross-current tetrode (XCT) SOI MOSFETs, which were proposed to realize extremely low energy circuits. In Part VI, the low-energy potential of quantum-effect devices, proposed by taking account of geometrical aspects, is considered. In Part VII, we discuss comprehensively how we can suppress the energy dissipation by using TFETs, one of the more recent quantum-effect devices. In Part VIII, finally, we briefly compare the performance of various devices and review considerations described in Part III to Part VII. We also address the latest low-energy devices, circuit applications, and future perspectives.

References

[1] R. H. Dennard, F. H. Gaensslen, H.-N. Yu, V. L. Rideout, E. Bassous, and A. R. LeBlanc, "Design of ion implanted MOSFETs with very small physical dimensions," *IEEE J. Solid-State Circuits*, vol. SC-9, pp. 256–268, 1974.

[2] P. K. Chatterjee, W. R. Hunter, T. C. Holloway, and Y. T. Lin, "The impact of scaling laws on the choice of N-channel or P-channel for MOS VLSI," *IEEE Electron Device Lett.*, vol. EDL-1, pp. 220–223, 1980.

[3] T. H. Ning, "Hot-electron emission from silicon into silicon dioxide," *Solid-State Electron.*, vol. 21, pp. 273–282, 1978.

[4] C. Fiegna, H. Iwai, T. Kimura, S. Nakamura, E. Sangiorgi, and B. Riccò, "Monte Carlo Analysis of Hot Carrier Effects in Ultra Small Geometry MOSFETs," Int. Workshop on VLSI Process and Device Modeling: VPAD (Nara, 1993), pp. 102–103, 1993.

[5] S. M. Sze and K. K. Ng, "Physics of Semiconductor Devices," 3rd ed, (John Wiley & Sons, Inc, 2007), p. 20 and p. 26.

[6] Z. Y. Cheng, M. T. Currie, C. W. Leitz, G. Taraschi, E. A. Fitzgerald, J. L. Hoyt, and D. A. Antoniadis, "Electron mobility enhancement in strained-Si n-MOSFETs fabricated on SiGe-on-insulator (SGOI) substrates," *IEEE Electron Device Lett.*, vol. 22, pp. 321–323, 2001.

[7] V. Misra, G. Lucovsky, and G. Parsons, "Issues in high-κ gate stack interfaces," *MRS Bull.*, vol. 27, pp. 212–216, 2001.

[8] K. Mistry, C. Allen, C. Auth, B. Beattie, D. Bergstrom, M. Bost, M. Brazier, M. Buehler, A. Cappellani, R. Chau, C.-H. Choi, G. Ding, K. Fischer, T. Ghani, R. Grover, W. Han, D. Hanken, M. Hattendorf, J. He, J. Hicks, R. Huessner, D. Ingerly, P. Jain, R. James, L. Jong, S. Joshi, C. Kenyon, K. Kuhn, K. Lee, H. Liu, J. Maiz, B. McIntyre, P. Moon, J. Neirynck, S. Pae, C. Parker, D. Parsons, C. Prasad, L. Pipes, M. Prince, P. Ranade, T. Reynolds, J. Sandford, L. Shifren, J. Sebastian, J. Seiple, D. Simon, S. Sivakumar, P. Smith, C. Thomas, T. Troeger, P. Vandervoorn, S. Williams, and K. Zawadzki, "A 45 nm Logic Technology with High-κ+Metal Gate Transistors, Strained Silicon, 9 Cu Interconnect Layers, 193 nm Dry Patterning, and 100% Pb-free Packaging," IEEE IEDM Tech. Dig. (Washington DC, 2007), pp. 247–250, 2007.

[9] J. Kedzierski, P. Xuan, E. H. Andersonf, J. Bokor, T.-J. King, and C. Hu, "Complementary Silicide Source/Drain Thin-Body MOSFETs for the 20 nm Gate Length Regime," Tech. Dig. IEEE IEDM (San Francisco, 2000) pp. 57–60, 2000.

[10] S. A. Vitale, P. W. Wyatt, N. Checka, J. Kedzierski, and C. L. Keast, "FESOI process technology for sub-threshold-operation ultralow-power electronics," *Proc. IEEE*, vol. 98, pp. 333–342, 2010.

[11] K. K. Bhuwalka, J. Schulze, and I. Eisele, "Performance enhancement of vertical tunnel field-effect transistor with SiGe in the layer," *Jpn. J. Appl. Phys.*, vol. 43, no. 7A, pp. 4073–4078, 2004.

[12] R. Landauer, "Irreversibility and heat generation in the computing process," *IBM J. Res. Dev.*, vol. 5, pp. 183–191, 1961.

[13] J.-P. Colinge, "Silicon-on-Insulator Technology: Materials to VLSI," 3rd ed. (Kluwer Academic Publishers, 2004), pp. 12–13.

[14] M Itoh, Y. Kawai, S. Ito, K. Yokomizo, Y. Katakura, Y. Fukuda, F. Ichikawa, "Fully Depleted SIMOX SOI Process Technology for Low-Power Digital and RF Device," Proc. 10th Int. Symp. Silicon on Insulator Technology and Devices (The Electrochem. Soc.) (Washington DC, 2001) Vol. 2001-3, pp. 331–336, 2001.

[15] J.-P. Colinge, "Silicon-on-Insulator Technology: Materials to VLSI," 3rd ed. (Kluwer Academic Publishers, 2004), Chapter 8.

[16] C.-E. D. Chen, M. Matloubian, R. Sundaresan, B. Mao, C. C. Wei, and G. P. Pollack, "Single-transistor Latch in SOI MOSFETs," *IEEE Electron Device Lett.*, vol. 9, pp. 636–638, 1988.

[17] S. Nakashima, Y. Omura, and K. Izumi, "A High-Quality SIMOX Wafer and Its Application to Ultrathin-Film MOSFETs," Proc. 5th Int. Symp. on SOI Technol. Dev. (The Electrochem. Soc., 1992) Vol. 92–13, pp. 358–367, 1992.

[18] Y. Omura, S. Nakashima, K. Izumi, and T. Ishii, "0.1-μm-Gate, Ultrathin-Film CMOS Devices Using SIMOX Substrate with 80-nm-Thick Buried Oxide Layer," 1991 IEEE Int. Electron Devices Meeting, Tech. Dig., pp. 675–678, 1991.

[19] K. Izumi, Y. Omura, and S. Nakashima, "Promotion of practical SIMOX technology by the development of a 100 mA-class high-current oxygen implanter," *Electron. Lett*, vol. 22, no. 15, pp. 775–777, 1986.

[20] T. Yonehara, K. Sakaguchi, and N. Sato, "Epitaxial layer transfer by bond and etch back of porous Si," *Appl. Phys. Lett.*, vol. 64, pp. 2108–2110, 1994.

[21] T. Yonehara and K. Sakaguchi, "ELTRAN®; novel SOI wafer technology," *Jpn. Soc. Appl. Phys. Int.*, no. 4, pp. 10–16, 2001.

[22] M. Bruel, "Application of hydrogen ion beams to silicon on insulator. Material technology," *Nucl. Instrum. Methods Phys. Res., Sect. B*, vol. 108, pp. 313–319, 1996.

[23] G., Celler "Smart Cut®, A guide to the technology, the process, the products," SOITEC, Press Release, July 1, 2003.

3

Performance Prospects of Subthreshold Logic Circuits

3.1 Introduction

The possibility of lowering the dissipation energy is addressed for the following conventional device technologies:

- multithreshold circuits [1];
- power management [2];
- subthreshold logic [3].

In a multithreshold circuit, designers can assume that different circuit blocks can have different threshold voltages. They can set low threshold voltages for high-speed circuit blocks and high threshold voltages for low-power circuit blocks. This technique is widely applied to commercial integrated circuits (ICs). In the case of power management, the power supply is switched in individual circuit blocks; this is a more recent technique, used most often in mobile electronics. In some cases the signal-processing speed might be degraded. In the case of subthreshold logic, conventional device technology is assumed but the supply voltage is lowered to the threshold voltage or much less. The logic circuit works in the subthreshold current range. In the last century this design approach was applied to control circuits in wrist watches [4].

3.2 Subthreshold Logic and its Issues

As described in the previous section, circuit designers initially used multithreshold devices and power management to reduce the dissipation power of ICs. These ideas are not innovative, as they are mere extensions of conventional techniques. Reducing dissipation power by lowering

MOS Devices for Low-Voltage and Low-Energy Applications, First Edition.
Yasuhisa Omura, Abhijit Mallik, and Naoto Matsuo.
© 2017 John Wiley & Sons Singapore Pte. Ltd. Published 2017 by John Wiley & Sons Singapore Pte. Ltd.

the supply voltage is also a conventional design guideline, where metal oxide semiconductor field-effect transistors (MOSFETs) are used in the conventional manner in the circuit.

On the other hand, the subthreshold logic and near-subthreshold logic approaches are quite different from the design idea of focusing on "ON" state performance. Circuit designers are now paying close attention to the quasi-"OFF" state. Rather than using the high impedance of the device, they target extremely low currents for logic circuits.

When we apply the subthreshold current of the MOSFET to low-power circuits, we must be careful of the noise issue, the variability of characteristics, and nonlinearity of *I-V* characteristics. Verma *et al.* recently addressed these points in detail [5].

3.3   Is Subthreshold Logic the Best Solution?

Low-power logic circuits, like the subthreshold logic circuits described in the previous section, are now commercialized as ICs for wrist watches [6]; such circuits were the concern of a small group of engineers and the design methodology was not familiar to most engineers. There was skepticism regarding whether a design technique based on subthreshold logic circuits would be utilized frequently.

In order to change these negative impressions, one of the editors (Omura) proposed the cross-current tetrode-silicon-on-insulator metal oxide semiconductor field-effect transistor (XCT-SOI MOSFET) [7, 8]. One of its advantages is the fact that we can basically design the required circuits by applying conventional design methodology in terms of supply voltage and device layout patterns. We can reduce the dissipation power by two orders when the "source potential floating effect" is significant. The most important point in the device design is that we can simply assume the stable saturation region of the drain current despite a very low drain current comparable to the subthreshold level. Details of device characteristics and aspects are described in Part V. We therefore still have many choices for reducing dissipation energy in various circuit applications.

References

[1] T. Douseki, J. Yamada, and H. Kyuragi, "Ultra Low-Power CMOS/SOI LSI Design for Future Mobile Systems," Dig. of Technical Papers, Int. Symp. on VLSI Circuits, June 2002, pp. 6–9, 2002.

[2] S. Mutoh, T. Douseki, Y. Matsuya, T. Aoki, S. Shigematsu, and J. Yamada, "1-V power supply high-speed digital circuit technology with multithreshold-voltage CMOS," *IEEE J. Solid-State Circuits*, vol. 30, pp. 847–854, 1995.

[3] S. A. Vitale, P. W. Wyatt, N. Checka, J. Kedzierski, and C. L. Keast, "FESOI process technology for subthreshold-operation ultralow-power electronics," *Proc. IEEE*, vol. 98, pp. 333–342, 2010.

[4] E. Vittoz, B. Gerber, and F. Leuenberger, "Silicon-gate CMOS frequency divider for the electronic wrist watch," *IEEE J. Solid-State Circuits*, vol. 7, pp. 100–104, 1972.

[5] N. Verma, J. Kwong, and A. P. Chandrakasan, "Nanometer MOSFET variation in minimum energy subthreshold circuits," *IEEE Trans. Electron Devices*, vol. 55, pp. 163–174, 2008.

[6] C. Piguet, "Low-Power CMOS Circuits –Technology, Logic Design and CAD Tools," (CRC Press, 2006), Chapter 1.

[7] Y. Omura, Y. Azuma, Y. Yoshioka, K. Fukuchi, and D. Ino, "Proposal of preliminary device model and scaling scheme of cross-current tetrode silicon-on-insulator metal-oxide-semiconductor field-effect transistor aiming at low-energy circuit applications," *Solid-State Electron.*, vol. 64, pp. 18–27, 2011.

[8] Y. Omura and D. Sato, "Mechanisms of low-energy operation of XCT-SOI CMOS devices – Prospect of sub-20-nm regime," *J. Low Power Electron. Appl.*, vol. 4, pp. 14–25, 2014.

Part II

SUMMARY OF PHYSICS OF MODERN SEMICONDUCTOR DEVICES

4

Overview

This part describes the fundamentals of metal oxide semiconductor (MOS) device physics. First, we describe the physics of bulk metal oxide semiconductor field effect transistors (MOSFETs). Then, some important physics related to silicon-on-insulator metal oxide semiconductor field-effect transistors (SOI MOSFETs) (partially depleted SOI MOSFET, fully depleted SOI MOSFET, FinFET, Triple-gate FET, gate-all-around MOSFET) are introduced, followed by the theoretical basis of tunnel field-effect transistors (TFETs).

The accumulation condition, the depletion condition, and the inversion condition in bulk MOSFET are explained. Threshold voltage and subthreshold swing are also explained as important parameters of MOSFET [1, 2]. In SOI MOSFETs, aspects of partially depleted (PD) SOI MOSFETs and fully depleted (FD) SOI MOSFETs are outlined [3, 4]. The reasons why FinFET, Triple-gate FET, and Gate-All-Around (GAA) MOSFET are superior to the PD and FD SOI MOSFETs are discussed [5, 6], and their details are described in the following parts. Finally, we discuss the physics of TFETs [7].

It is obvious that we must minimize the subthreshold swing (SS) value of the MOS device in designing device parameters for low-power applications. However, there is still some controversy regarding the issue of whether the subthreshold issue is the substantial problem in overcoming the stand-by power issue. In many MOS devices we can face a simultaneous increase in the band-to-band tunneling (BTBT) current around the source and drain junctions because there must be shallow junctions in bulk devices or an extremely thin semiconductor layer in silicon-on-insulator (SOI) devices in order to suppress short-channel effects. Readers will find that MOS devices with steep swing values and low BTBT current values often suffer from the low drivability. We must therefore take account of such tradeoff issues in optimizing the device's performance even when we discuss the low-standby energy concept and device applications. The following chapters review theoretical models of various MOSFETs.

MOS Devices for Low-Voltage and Low-Energy Applications, First Edition.
Yasuhisa Omura, Abhijit Mallik, and Naoto Matsuo.
© 2017 John Wiley & Sons Singapore Pte. Ltd. Published 2017 by John Wiley & Sons Singapore Pte. Ltd.

References

[1] S. M. Sze and K. K. Ng, "Physics of Semiconductor Devices," 3rd ed. (John Wiley & Sons, Inc., 2007), Chapter 6.

[2] Y. Tsividis, "Operation and Modeling of the MOS Transistor," 2nd ed. (Oxford University Press, 1999).

[3] J.-P. Colinge, "Silicon-on-Insulator Technology: Materials to VLSI," 3rd ed. (Kluwer Academic Pubishers, 2004).

[4] T. Sakurai, A. Matsuzawa, T. Douseki (eds.) "Fully-Depleted SOI CMOS Circuits and Technology for Ultralow-Power Applications," (Springer, 2006), pp. 23–58.

[5] J.-P. Colinge (ed.) "FinFETs and Other Multi-Gate Transistors" (Springer, 2008), Chapters 1 and 2.

[6] S. Deleonibus (ed.) "Electronic Device Architectures for Nano-CMOS Era –from Ultimate CMOS Scaling to beyond CMOS Devices" (Pan Stanford Publishing, 2009).

[7] A. C. Seabaugh, "Low-voltage tunnel transistors for beyond CMOS logic," *Proc. IEEE*, vol. 98, pp. 2095–2110, 2010.

5

Bulk MOSFET

5.1 Theoretical Basis of Bulk MOSFET Operation

Figure 5.1 shows a schematic cross section of a bulk n-channel metal oxide semiconductor field-effect transistor (MOSFET). In the following, we assume an n-channel MOSFET. As there is a single channel inside the body, there are three current flows at the onset of inversion: the front-channel current, the band-to-band tunneling (BTBT) currents at the front interface of the drain junction, and double injection currents at the source junction. In the subthreshold region, the subthreshold current is at the front interface. This chapter focuses on the DC characteristics of bulk MOSFET, and presents a theoretical analysis of the subthreshold and post-threshold current characteristics as an aid to low-energy device design.

5.2 Subthreshold Characteristics: "OFF State"

5.2.1 Fundamental Theory

Many papers have described the basic features of the subthreshold characteristics of bulk MOSFETs [1, 2]. The input capacitance consists of four components in series: (i) the capacitance of the gate insulator; (ii) the capacitance of the channel depletion layer; (iii) the capacitance of the depletion layer of the source junction (this is available when there exists potential difference between the source terminal and the substrate), and (iv) the capacitance of the depletion layer of the drain junction [2].

In the subthreshold regime, the drain current (Figure 5.2) is given by

$$I_{D(sub)} = I_{D(sub),front} + I_{D(sub),other} \tag{5.1}$$

MOS Devices for Low-Voltage and Low-Energy Applications, First Edition.
Yasuhisa Omura, Abhijit Mallik, and Naoto Matsuo.
© 2017 John Wiley & Sons Singapore Pte. Ltd. Published 2017 by John Wiley & Sons Singapore Pte. Ltd.

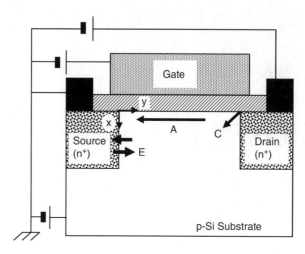

Figure 5.1 Bulk MOSFET – cross-section and operation. A: normal front channel, C: avalanche and band-to-band tunnel current (front side), and E: double injection on parasitic bipolar action.

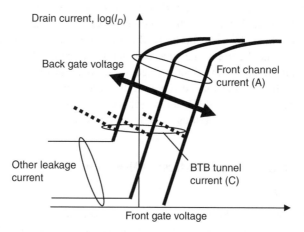

Figure 5.2 Subthreshold and off-state current characteristics of bulk MOSFET. (With kind permission from Springer Science + Business Media: Fully-Depleted SOI CMOS Circuits and Technology for Ultralow-Power Applications, 2006, pp. 48–58, Yasuhisa Omura (edited by T. Sakurai, A. Matsuzawa, and T. Douseki),© 2006 Springer.)

where $I_{D(sub),front}$ is the subthreshold current near the front interface and $I_{D(sub),other}$ is the combined current attributable to parasitic phenomena, such as a simple avalanche at the drain junction [3, 4], BTBT at the drain junction [5, 6] and the generation-recombination (GR) process [7]. For simplicity, this discussion will focus on $I_{D(sub),front}$. For an n-channel MOSFET, it is

$$I_{D(sub),front} = qAD_n \frac{dn}{dy} - qAD_n \frac{n(0) - n(L_{eff})}{L_{eff}} \qquad (5.2)$$

where A is the cross-sectional area of the channel, q is the elementary charge, D_n is the diffusion constant of electrons, L_{eff} is the effective channel length, and $n(0)$ and $n(L_{eff})$ are the electron concentrations at the edges of the source and drain, respectively. They are given by

$$n(0) = n_i \exp\left[\frac{\left(E_F - E_i + q\varphi_{s,front}\left(x = 0, y = 0\right)\right)}{k_B T}\right] \tag{5.3}$$

and

$$n\left(L_{eff}\right) = n_i \exp\left[\frac{\left(E_F - E_i + q\varphi_{s,front}\left(x = 0, y = 0\right) - qV_D\right)}{k_B T}\right] \tag{5.4}$$

where $\varphi_{s,front}$ is the surface potential at the top surface, n_i is the intrinsic carrier density, E_F is the Fermi level, E_i is the intrinsic Fermi level, k_B is the Boltzmann's constant, and T is the temperature in K. If we assume the effective channel depth to be $k_B T/qE_s$ [8], where E_s is the surface electric field [9], the subthreshold current can be written as

$$I_{D(sub),front} = \mu_n \left(\frac{W_{eff}}{L_{eff}}\right) q \left(\frac{k_B T}{q}\right)^2 \left(\frac{n_i^2}{N_A}\right) \left\{1 - \exp\left[\frac{-qV_D}{k_B T}\right]\right\} \frac{\exp\left[\dfrac{q\varphi_{s,front}}{k_B T}\right]}{E_s} \tag{5.5}$$

where μ_n is the electron mobility calculated from the Einstein relation $D_n = \mu_n k_B T/q$, N_A is the doping concentration of substrate, W_{eff} is the channel width, and V_D is the drain voltage. We can derive the following basic expression for the subthreshold swing (SS) from Eq. (5.5) if we neglect the influence of interface states for simplicity.

$$SS = \left(\frac{kT}{q}\right) \ln\left(10\right) \left\{1 + \frac{C_s}{C_{ox}}\right\} \tag{5.6}$$

where

$$C_s = \frac{\varepsilon_s}{W_D} \tag{5.7}$$

$C_{ox} \left(= \varepsilon_{ox}/t_{ox}\right)$ is the capacitance of the gate oxide layer, C_s is the depletion capacitance of the substrate, t_{ox} is the thickness of the gate oxide layer, ε_s is the permittivity of semiconductor, and W_D is the depletion layer width. As seen in Eq. (5.6), the basic expression for SS depends on doping concentration of the substrate, gate oxide layer thickness, permittivity of the gate oxide layer, and permittivity of the substrate.

The main conclusions that can be drawn from Eq. (5.6) are:

- Influence of gate oxide thickness (t_{ox}) on SS. One of the most common ways to suppress short-channel effects is to reduce t_{ox}, which makes C_s/C_{ox} small, as shown by Eq. (5.6). This means that the t_{ox} leads to the straightforward reduction of SS.

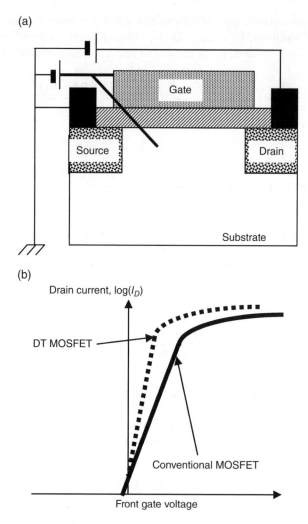

Figure 5.3   Dynamic-threshold MOSFET. (a) Cross section and (b) characteristics.

- Influence of doping concentration (N_A) on SS. Another common way to suppress short-channel effects is to increase N_A. This often causes SS to increase. To prevent that, t_{ox} should also be reduced when N_A is increased.
- Influence of substrate potential on SS: dynamic-threshold (DT) MOSFET operation (Figure 5.3). The substrate potential is often set to a certain value. In that case, SS is almost independent of the front gate voltage, but depends on the device parameters. In a dynamic-threshold MOSFET [10], the substrate terminal is connected to the front gate electrode; so, the substrate potential increases with the front gate voltage. This increase reduces the built-in potential of the source junction, thereby increasing the channel current. So, a simplified expression for the subthreshold current of a dynamic threshold MOSFET is

$$I_{D(sub),DT} = I'_{D(sub),front} \, C_u \exp\left[\frac{qV_{FG}}{mk_BT}\right] \tag{5.8}$$

where

$$I'_{D(sub),front} = \mu_n \left(\frac{W_{eff}}{L_{eff}} \right) q \left(\frac{k_B T}{q} \right)^2 \left(\frac{n_i^2}{N_A} \right) \left\{ 1 - \exp\left[\frac{-qV_D}{k_B T} \right] \right\} \frac{\exp\left[\dfrac{q\varphi_{s,front}\left(x = 0, y = 0, V_{FG} \right)}{k_B T} \right]}{E_S}$$

(5.9)

C_0 is a constant, m is the ideality factor ($m > 1$), and $\varphi_s(x = 0, y = 0, V_{FG})$ is the surface potential modified by V_{FG}. It should be noted that the subthreshold channel region depends on the initial surface condition of the device because the front gate voltage does not change the band structure near the top surface drastically. This is an important aspect of the dynamic threshold MOSFET. An approximate expression for the SS of a dynamic threshold MOSFET is

$$SS = \eta \left(\frac{k_B T}{q} \right) \ln(10) \left[1 + \frac{C_s}{C_{ox}} \right]$$

(5.10)

where $\eta < 1$ and C_s is given by Eq. (5.7).

- Drain-induced barrier lowering (DIBL) (front interface) (Figure 5.4). In a short-channel bulk MOSFET, DIBL [11] near the source junction at the front interface significantly degrades the SS. The reason for this is that the drain-induced lateral electric field extends quite far into the channel region. There are two ways to suppress the DIBL: one is to make N_A higher, and the other is to replace the gate oxide with a high-κ material. However, the former degrades the SS because C_s increases in Eq. (5.6). Therefore, the pocket ion-implantation technique [12] is frequently introduced in order to suppress the extension of the drain-induced lateral electric field.

5.2.2 Influence of BTBT Current

Band-to-band tunneling often occurs in the gate-drain overlap region [13]. It is sometimes called the gate-induced drain leakage (GIDL) current. For MOSFETs with a thin gate oxide, it has a significant influence on DC and AC operation [14]. An empirical expression for this current is [15]

$$I_{D(sub),others} = A_0 E_{SS} \exp\left[\frac{-B_0}{E_{SS}} \right]$$

(5.11)

where A_0 and B_0 are constants that depend on the device parameters and E_{SS} is the surface electric field of the gate overlap region. As the gate oxide of recent MOSFETs tends to be very thin, the electric fields of the gate-drain and drain-source overlap regions are higher than those of past devices. As a result, the contribution of the BTBT current to the subthreshold characteristics is apt to be significant [16].

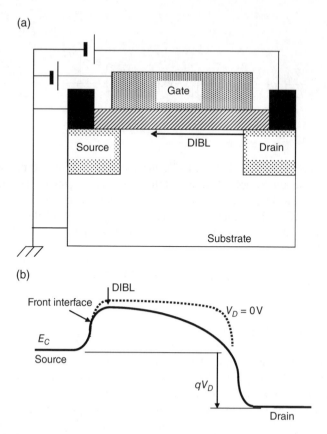

Figure 5.4 Drain-induced barrier lowering phenomenon in the bulk MOSFET. (a) Schematic view of DIBL and (b) impact of DIBL on the surface potential profile.

5.2.3 Points to Be Remarked

The "OFF state" current in the subthreshold region limits the standby power of the MOSFET even when the threshold voltage is sufficiently high. In the case of the bulk MOSFET, the reverse-biased leakage current generated at a large metallurgical junction area, the weak avalanche-based current, and the BTBT current around the gate-overlapped region of the drain diffusion region shares a large part of the total leakage current. These are apt to increase when the short-channel effects are suppressed because the suppression of the short-channel effects raises the local electric field around the drain diffusion. Therefore the lowering of the drain voltage is requested for the purpose of the suppression of those leakage current components. These points will be discussed again later.

5.3 Post-Threshold Characteristics: "ON State"

5.3.1 Fundamental Theory

In the post-threshold regime, the drain current is given by

$$I_{D(post)} = I_{D(post),front} + I_{D(post),other} \tag{5.12}$$

where $I_{D(post),front}$ is the inversion channel current along the front interface and $I_{D(post),other}$ is the parasitic current, which consists of the leakage current of the reverse-biased drain junction [17], a parasitic bipolar current [18], and an avalanche current [19]. The discussion below focuses on $I_{D(post),front}$.

In practice, the post-threshold channel current consists mainly of the drift current, where carriers are accelerated by the drain-to-source electric field. It is analyzed in the conventional gradual channel approximation. The expression for the drain current of the bulk MOSFET is derived below.

We start from the Ohmic relation

$$I_{D(post),front} = -W_{eff}\mu_n Q_n(V)\frac{dV(y)}{dy} \tag{5.13}$$

where $V(y)$ is the local channel potential and $Q_n(V)$ is the local electronic charge density. This is the expression for the local current inside the device, if we neglect the diffusion and other currents like the generation-recombination current.

We obtain the following relation for $Q_n(V)$ from Poisson's equation for the channel region [20]:

Q_n = total induced charge density of semiconductor – depletion charge density of substrate,

$$= -C_{ox}\left(V_{FG} - V_{FB,front} - \varphi(0) - V\right) + qN_A W_D\left(\varphi(0) + V\right) \tag{5.14}$$

Integrating it from source to drain based on the current continuity, we obtain the following expression for $I_{D(post),front}$:

$$I_{D(post),front} = \left(\frac{W_{eff}}{L_{eff}}\right)\mu_n C_{ox}\left[V_{FG} - V_{TH}^* - \frac{V_D}{2}\right]V_D + \left(optional - term\right) \tag{5.15}$$

where

$$V_{TH}^* = V_{FB,front} + \varphi_{s,front}(x=0, y=0) + \frac{\sqrt{2q\varepsilon_s N_A \varphi_{s,front}(x=0, y=0)}}{C_{ox}} + \left(optional - term\right) \tag{5.16}$$

When the doping level near the surface is not so high, substituting the new expression for $Q_n(V)$ into the Ohmic relation (5.13), and assuming current continuity, yield the following expression for $I_{D(post)}$, front:

$$I_{D(post),front} = \left(\frac{W_{eff}}{L_{eff}}\right)\mu_n C_{ox}\left[V_{FG} - V_{TH,front} - \frac{V_D}{2}\right]V_D \tag{5.17}$$

where the threshold voltage of a MOSFET ($V_{TH,front}$) is

$$V_{TH,front} = V_{FB,front} + \varphi_{s,front}(x=0, y=0) + \frac{qN_A W_D\left[\varphi_{s,front}(x=0, y=0)\right]}{C_{ox}} \tag{5.18}$$

The saturated drain current ($I_{D(post)sat}$) is given by:

$$I_{D(post)sat} = \left(\frac{W_{eff}}{2(1+\alpha)L_{eff}} \right) \mu_n C_{ox} \left[V_{FG} - V_{TH,front} \right]^2 \qquad (5.19)$$

where α is the factor depending on the substrate doping concentration profile. When the substrate has a uniform doping profile and its doping concentration is not so high, $\alpha \to 0$. The drain current of the bulk MOSFET on the saturation condition is basically independent of the drain voltage (V_D) as far as short-channel effects are not seen. Other important factors are discussed below.

5.3.2 Self-Heating Effects

Powered devices suffer from self-heating effects as shown in Figure 5.5. In the bulk MOSFET, the thermal conductivity of the substrate is not so high [21]. According to simulations [22], the temperature near the drain junction quickly rises to 100 °C. Since self-heating effects degrade the surface mobility of carriers, they must be taken into account. In electrostatic-discharge (ESD) protection circuits, self-heating effects reduce the temperature margin for the second breakdown. So, adequate thermal paths must be added during metallization.

If we need to take into account the influence of self-heating effects on drain current, we have to use the following expression for the mobility [23]:

$$\mu_{n(self-heating)} = C_1 \left[T_0 + \theta I_{D(post),front} V_D \right]^{-\delta} \qquad (5.20)$$

where C_1 is a constant, T_0 is room temperature, and θ and δ are fitting parameters.

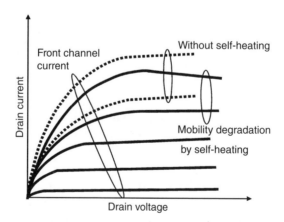

Figure 5.5 I_D-V_D characteristics of bulk MOSFET.

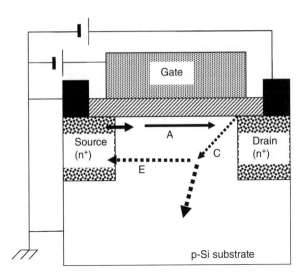

Figure 5.6 Parasitic bipolar action in the n-channel bulk MOSFET. Avalanche-induced holes (C) are injected into the source (E), and electrons are injected into the body (A). This positive feedback loop turns on parasitic bipolar action.

5.3.3 *Parasitic Bipolar Effects*

When the drain-to-source voltage is high, the electric field around the drain pn junction takes a high value. In the depletion region around the drain pn junction, a weak avalanche multiplication of carriers takes place. In that case, as shown in Figure 5.6, most holes generated near the drain are absorbed to the bulk substrate ("C" in Figure 5.6) and they are collected at the substrate terminal; this is the so-called "substrate current." However, some holes flow into the source junction ("E" in Figure 5.6). When the substrate resistivity is high, the body potential automatically rises by the substrate current. When the "forward-biased" current is made at the source junction, it works as the "base current" of the "NPN" parasitic bipolar transistor. This frequently results in a fatal "burnout" phenomenon.

5.4 **Comprehensive Summary of Short-Channel Effects**

Considering a simple phenomenalistic model of short-channel effects, we obtain

$$V_{TH} = V_{FB,front} + f_{DIBL}\varphi_{s,front}\left(x=0,y=0\right) + f_{CS}\frac{qN_A W_D\left[\varphi_{s,front}\left(x=0,y=0\right)\right]}{C_{ox}} - \eta V'_{BG} \tag{5.21}$$

where f_{DIBL} is the DIBL factor, f_{CS} is the charge-sharing factor, and η is the factor expressing the impact of the substrate bias effect.

References

[1] S. M. Sze and K. K. Ng, "*Physics of Semiconductor Devices*," 3rd ed. (John Wiley & Son, Inc., 2007), Chapter 6.

[2] Y. Tsividis, "*Operation and Modeling of the MOS Transistor*," 2nd ed. (Oxford University Press, 1999).

[3] S. M. Sze and G. Gibbons, "Avalanche breakdown voltages of abrupt and linearly graded p-n junctions in Ge, Si, GaAs, and GaP," *Appl. Phys. Lett.*, vol. 8, pp. 111–113, 1966.

[4] G. Merckel, F. Van de Wiele, W. Engl, and P. Jespers, "*NATO Course on Process and Device Modeling for Integrated Circuit Design*," (Noordhoff, 1975), p. 725.

[5] L. Esaki, "New phenomenon in narrow germanium p-n junctions," *Phys. Rev.*, vol. 109, p. 603, 1958.

[6] E. O. Kane, "Theory of tunneling," *J. Appl. Phys.*, vol. 32, pp. 83–91, 1961.

[7] W. Shockley and W. T. Read, Jr., "Statistics of the recombination of holes and electrons," *Phys. Rev.*, vol. 87, pp. 835–842, 1952.

[8] S. M. Sze and K. K. Ng, "*Physics of Semiconductor Devices*," 3rd ed. (John Wiley & Sons, Inc., 2007), p. 314.

[9] S. M. Sze and K. K. Ng, "*Physics of Semiconductor Devices*," 3rd ed. (John Wiley & Sons, Inc., 2007), p. 301.

[10] F Assaderaghi, D. Sinitsky, S. A. Parke, J. Boker, P. K. Ko, and C. Hu, "A Dynamic Threshold Voltage MOSFET (DTMOS) for Ultra-Low Voltage Operation," Tech. Dig. Int. Electron Devices Meeting (San Francisco, Dec., 1994) pp. 809–812, 1994.

[11] G. W. Taylor, "Subthreshold conduction in MOSFET's," *IEEE Trans. Electron Devices*, vol. 25, pp. 337–350, 1978.

[12] T. Hori, "A 0.1 μm CMOS Technology with Tilt-Implanted Punchthrough Stopper (TIPS)," Tech. Dig., IEEE IEDM (San Francisco, 1994) pp. 75–78, 1994.

[13] E. Takeda, H. Matsuoka, and S. Asai, "A Band to Band Tunneling MOS Device (B²T-MOSFET) – A Kind of 'Si quantum Device'," Tech. Dig., IEEE IEDM (San Francisco, 1988) pp. 402–405, 1988.

[14] M. Chan, J. Lin, S. N. Shih, T.-C. Wu, B. Huang, J. Yang, and P.-I. Lee, "Impact of gate-induced drain leakage on retention time distribution of 256 Mbit DRAM with negative wordline bias," *IEEE Trans. Electron Devices*, vol. 50, pp. 1036–1041, 2003.

[15] H.-J. Wann, P.-K. Ko, and C. Hu, "Gate-Induced Band-to-Band Tunneling Leakage Current in LDD MOSFETs," Tech. Dig. IEEE IEDM (San Francisco, 1992) pp. 147–150, 1992.

[16] J. Chen, F. Assaderaghi, P.-K. Ko, and C. Hu, "The enhancement of Gate-Induced-Drain-Leakage (GIDL) current in short-channel MOSFET and its application in measuring lateral bipolar current gain," *IEEE Electron Device Lett.*, vol. 11, pp. 572–574, 1992.

[17] S. M. Sze and K. K. Ng, "*Physics of Semiconductor Devices*," 3rd ed. (John Wiley & Sons, Inc., 2007), pp. 90–114.

[18] L. Wakeman, "Silicon-gate CMOS chips gain immunity to SCR latchup," *Electronics*, vol. 56, pp. 136–140, 1983.

[19] S. M. Sze and K. K. Ng, "*Physics of Semiconductor Devices*," 3rd ed. (John Wiley & Sons, Inc., 2007), pp. 104–106.

[20] S. M. Sze and K. K. Ng, "*Physics of Semiconductor Devices*," 3rd ed. (John Wiley & Sons, Inc., 2007), pp. 298–303.

[21] D. Sharma, J. Gautier, and C. Merckel, "Negative resistance in MOS devices," *IEEE J. Solid-State Circuits*, vol. SC-13, pp. 378–380, 1978.

[22] D. Sharma and K. Ramanathan, "Modeling thermal effects on MOS I-V characteristics," *IEEE Electron Device Lett.*, vol. EDL-4, p. 362, 1983.

[23] R. van Langevelde and F. M. Klaassen, "Accurate Drain Conductance Modeling for Distortion Analysis in MOSFETs," Tech. Dig. IEEE IEDM (Washington, D. C., 1997) pp. 313–316, 1997.

6

SOI MOSFET

6.1 Partially Depleted Silicon-on-Insulator Metal Oxide Semiconductor Field-Effect Transistors

The partially depleted (PD) n-channel silicon-on-insulator metal oxide semiconductor field-effect transistor (SOI MOSFET) has a quasineutral region in the silicon-on-insulator (SOI) layer as shown in Figure 6.1. This configuration is very similar to that of the bulk metal oxide semiconductor field-effect transistor (MOSFET) except for the buried oxide layer beneath the device region. Therefore, most operation characteristics are similar to those of the bulk MOSFET. However, the device has a few different characteristics from the bulk MOSFET, such as the kink in I_D versus V_D characteristics [1–4] and the body-floating characteristics [1–4]; that is, the potential of the quasineutral region is not fixed to a specific potential. When the potential of the quasineutral region floats, the potential of the quasineutral region of the SOI MOSFET cannot respond quickly to changes in the gate electrode and the drain electrode, which slows the operation of the partially depleted SOI MOSFET [5].

When a positive bias is applied to the substrate, the depletion layer expands from the bottom surface of the silicon layer, which yields an optional leakage current path on the bottom surface of the silicon layer. We have to increase the body doping concentration near the bottom surface of the silicon layer in order to suppress this optional leakage current [6]. When a high positive bias is applied to the substrate, the silicon body can be fully depleted [6]. When the full depletion of the silicon body is held, the SOI MOSFET exhibits the specific characteristics described in the following section.

MOS Devices for Low-Voltage and Low-Energy Applications, First Edition.
Yasuhisa Omura, Abhijit Mallik, and Naoto Matsuo.
© 2017 John Wiley & Sons Singapore Pte. Ltd. Published 2017 by John Wiley & Sons Singapore Pte. Ltd.

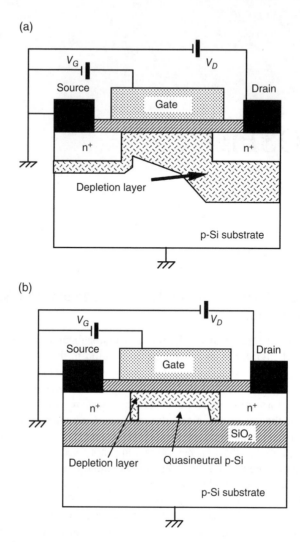

Figure 6.1   Schematic structures of bulk MOSFET and partially depleted MOSFET. (a) Bulk MOSFET and (b) partially depleted SOI MOSFET without substrate bias.

6.2   Fully Depleted (FD) SOI MOSFET

6.2.1   Subthreshold Characteristics

Figure 6.2 shows a schematic cross section of an SOI MOSFET. In the following, we assume an n-channel SOI MOSFET. As there are two channels inside the body, there are six current flows at the onset of inversion: the front-channel current, the back-channel current, the band-to-band tunneling (BTBT) currents at the front and back interfaces of the drain junction, and double injection currents at the source junction. On the other hand, in the subthreshold region, there are subthreshold currents at the front and back interfaces. This section focuses on the DC characteristics of fully depleted silicon-on-insulator metal oxide semiconductor field-effect

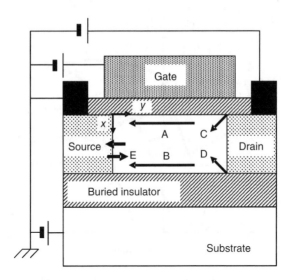

Figure 6.2 Fully depleted SOI MOSFET: Cross section and operation. A: Normal front channel, B: normal back channel, C: avalanche and band-to-band tunneling current (front side), D: avalanche and band-to-band tunneling current (back side), and E: double injection due to parasitic bipolar action. With kind permission from Springer Science + Business Media: T. Sakurai, A. Matsuzawa, and T. Douseki (eds.), "Fully-Depleted SOI CMOS Circuits and Technology for Ultralow-Power Applications," 2006, pp. 48–58. Copyright © 2006, Springer.

transistors (FD-SOI MOSFETs), and presents a theoretical analysis of the subthreshold and post-threshold current characteristics as an aid to device design.

Many papers have described the basic features of the subthreshold characteristics of FD-SOI MOSFETs [7]. The input capacitance consists of four components in series: (i) the capacitance of the gate insulator, (ii) the capacitance of the SOI layer (which is equivalent to the depletion layer of a bulk MOSFET), (iii) the capacitance of the buried insulator, and (iv) the capacitance of the depletion layer beneath the buried insulator, if the buried insulator is thin. As a result, the effective input capacitance in the subthreshold region is quite small, which makes the subthreshold swing sharp [7].

In the subthreshold regime, the drain current (Figure 6.3) is given by

$$I_{D(sub)} = I_{D(sub),front} + I_{D(sub),back} + I_{D(sub),other} \tag{6.1}$$

where $I_{D(sub),front}$ and $I_{D(sub),back}$ are the subthreshold currents near the front and back interfaces, respectively, and $I_{D(sub),other}$ is the combined current attributable to parasitic phenomena, such as a simple avalanche at the drain junction [8], BTBT at the drain junction [9, 10], and the generation-recombination (GR) process [11]. For simplicity, this discussion will focus on $I_{D(sub),front}$. For an n-channel SOI MOSFET, it is

$$I_{D(sub),front} = -qAD_n \frac{dn}{dy} = qAD_n \frac{n(0) - n(L_{eff})}{L_{eff}} \tag{6.2}$$

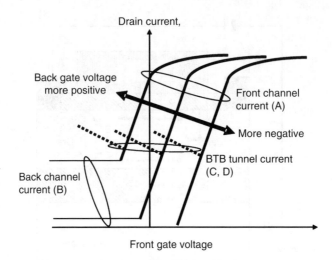

Figure 6.3 Subthreshold and off-state current characteristics of FD-SOI MOSFET. With kind permission from Springer Science+Business Media: T. Sakurai, A. Matsuzawa, and T. Douseki (eds.), "Fully-Depleted SOI CMOS Circuits and Technology for Ultralow-Power Applications," 2006, pp. 48–58. Copyright © 2006, Springer.

where A is the cross-sectional area of the channel; and $n(0)$ and $n(L)$ are the electron concentrations at the edges of the source and drain, respectively. They are given by

$$n(0) = n_i \exp\left[\frac{\left(E_F - E_i + q\varphi_{s,front}\right)}{k_B T}\right] \text{ and} \tag{6.3}$$

$$n\left(L_{eff}\right) = n_i \exp\left[\frac{\left(E_F - E_i + q\varphi_{s,front} - qV_D\right)}{k_B T}\right] \tag{6.4}$$

where φ_s, front is the surface potential at the top of the SOI layer. If we assume the effective channel depth to be $k_B T/qE_s$, where E_s is the surface electric field [12], the subthreshold current can be written as

$$I_{D(sub),front} = \mu_n \left(\frac{W_{eff}}{L_{eff}}\right) q \left(\frac{k_B T}{q}\right)^2 \left(\frac{n_i^2}{N_A}\right) \left\{1 - \exp\left[\frac{-qV_D}{k_B T}\right]\right\} \frac{\exp\left[\dfrac{q\varphi_{s,front}}{k_B T}\right]}{E_S} \tag{6.5}$$

where W_{eff} is the channel width and μ_n is the electron mobility calculated from the Einstein relation, $D_n = \mu_n k_B T/q$. The theoretical expression for the E_s of an FD-SOI MOSFET is different from that for a bulk MOSFET because the depth profile of the internal potential is different [7]. Nevertheless, we can derive the following basic expression for the subthreshold swing (SS) from Eq. (6.5) if we neglect the influence of interface states for simplicity.

$$SS = \left(\frac{k_B T}{q}\right) \ln(10)\left\{1 + \frac{C_s}{C_{ox}}\right\} \tag{6.6}$$

where

$$C_s = \frac{C_{SOI} C_{BOX} C_{sub}}{\left(C_{SOI} C_{BOX} + C_{BOX} C_{sub} + C_{sub} C_{SOI}\right)} \tag{6.7}$$

$C_{ox} (= \varepsilon_{ox}/t_{ox})$ is the capacitance of the gate oxide layer, C_s is the total capacitance of the substrate, $C_{SOI} (= \varepsilon_{si}/t_s)$ is the equivalent capacitance of the SOI layer, $C_{BOX} (= \varepsilon_{ox}/t_{BOX})$ is the capacitance of the buried-oxide (BOX) layer, C_{sub} is the capacitance of the depletion layer beneath the BOX when the substrate is p-type, t_{ox} is the thickness of the gate oxide layer, t_s is the thickness of the SOI layer, and t_{BOX} is the thickness of the BOX. As seen in Eq. (6.6), the basic expression for SS is the same for fully depleted silicon-on-insulator (FD-SOI) and bulk MOSFETs, although the complete expression for C_s is quite different. SS is usually smaller for an FD-SOI MOSFET than for a bulk MOSFET because C_s is smaller than the depletion layer capacitance of a bulk MOSFET.

When $I_{D(sub), back}$ exists, we obtain the following expression for SS:

$$SS = \frac{dV_{FG}}{d\left[\log\left(I_{D(sub),front} + I_{D(sub),back}\right)\right]} \tag{6.8}$$

This means that the subthreshold current near the back interface degrades the subthreshold swing of the front gate.

The main conclusions that can be drawn from Eq. (6.6) are listed below.

- Influence of t_s on SS. The most common way to suppress short-channel effects is to reduce t_s [7], which makes C_s approach $C_{BOX} (= \varepsilon_{ox}/t_{BOX})$, as shown by Eq. (6.7). This means that t_{BOX} limits the straightforward reduction of SS. In other words, t_{BOX} should be large in consideration of SS.
- Influence of t_{BOX} on SS. A less common way to suppress short-channel effects is to reduce t_{BOX} [13, 14], which makes C_s approach $C_{SOI} (= \varepsilon_{si}/t_s)$, as shown by Eq. (6.7). This often causes SS to increase. To prevent that, t_{ox} should also be reduced when t_{BOX} is reduced.
- Influence of body potential on SS: dynamic threshold MOSFET operation (Figure 6.4). The body potential is often set to a certain value. In that case, SS is almost independent of the front gate voltage, but depends on the device parameters. In a dynamic-threshold MOSFET [15], the body terminal is connected to the front gate electrode; so, the body potential increases with the front gate voltage. This increase reduces the built-in potential of the source junction, thereby increasing the channel current. So, a simplified expression for the subthreshold current of a dynamic threshold MOSFET is

$$I_{D(sub),DT} = I'_{D(sub),front} C \exp\left[\frac{qV_{FG}}{mk_B T}\right] \tag{6.9}$$

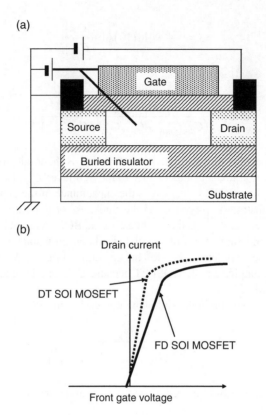

Figure 6.4 Dynamic-threshold FD-SOI MOSFET. (a) Device cross section and (b) schematic characteristic.

where

$$
I'_{D(sub),front} = \mu_n \left(\frac{W_{eff}}{L_{eff}} \right) q \left(\frac{k_B T}{q} \right) t_{SOI} \left(\frac{n_i^2}{N_A} \right)
$$
$$
\left\{ 1 - \exp\left[\frac{-qV_D}{k_B T} \right] \right\} \exp\left[\frac{q\varphi_s (V_{FG})}{m k_B T} \right]
\tag{6.10}
$$

C is a constant, m is the ideality factor ($m > 1$), and $\varphi_s(V_{FG})$ is the surface potential modified by V_{FG}. It should be noted that the effective depth of the subthreshold channel is t_s in the formula for the subthreshold current (Eq. (6.5)) because there is almost no surface depletion in a dynamic threshold MOSFET. A simple algebraic manipulation yields $d\varphi_s(V_{FG})/dV_{FG} = 1$, which is different from the value for a bulk MOSFET. An approximate expression for the SS of a dynamic threshold MOSFET is

$$
SS = \eta \left(\frac{k_B T}{q} \right) \ln(10) \left[1 + \frac{C_s}{C_{ox}} \right]
\tag{6.11}
$$

where $\eta < 1$ and C_s is given by Eq. (6.7).

- Drain-induced barrier lowering (DIBL) (front interface) and buried-insulator-induced barrier lowering (BIIBL) (back interface) (see Figure 6.5). In a short-channel bulk MOSFET, DIBL [16] near the source junction at the front interface significantly degrades the subthreshold swing. In an FD-SOI MOSFET, BIIBL [13] near the source junction at the back interface also significantly influences the subthreshold swing. The reason for this is that the lateral electric field extends quite far into the buried insulator because the insulator has a much lower permittivity than Si. There are two ways to suppress this extension: one is to make the buried insulator (e.g., SiO_2) thinner, and the other is to replace it with a high-κ material. However, the latter degrades the subthreshold swing because reducing the equivalent oxide thickness (EOT) pulls down the body potential at the back interface of the SOI layer.

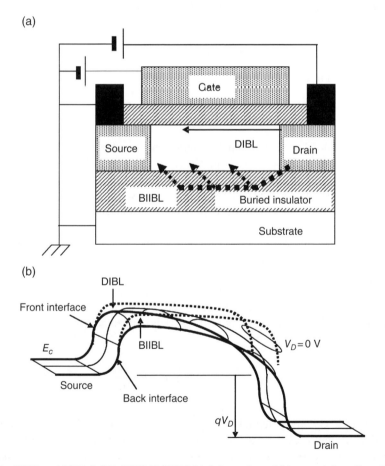

Figure 6.5 DIBL and BIBL in FD-SOI MOSFET. (a) Schematic and (b) potential profiles of DIBL and BIBL. With kind permission from Springer Science + Business Media: T. Sakurai, A. Matsuzawa, and T. Douseki (eds.), "Fully-Depleted SOI CMOS Circuits and Technology for Ultralow-Power Applications," 2006, pp. 48–58. Copyright © 2006, Springer.

- Influence of band-to-band tunneling current. Band-to-band Tunneling current often occurs in the gate-drain overlap region [9]. It is sometimes called the gate-induced drain leakage (GIDL) current. For a thin SOI MOSFET, it also has a significant influence on DC and AC operation [17, 18]. An empirical expression for this current is [9]

$$I_{D(sub),others} = A_0 E_{SS} \exp\left[\frac{-B_0}{E_{SS}}\right] \tag{6.12}$$

where A_0 and B_0 are constants that depend on the device parameters and E_{SS} is the surface electric field of the overlap region. Since the gate oxide and buried oxide layers of recent ultrathin SOI MOSFETs tend to be very thin, the electric fields of the gate-drain and drain-substrate overlap regions are higher than those of previous devices. As a result, the contribution of the BTBT current to the subthreshold characteristics is apt to be significant [18].

6.2.2   Post-Threshold Characteristics

In the post-threshold regime, the drain current is given by

$$I_{D(post)} = I_{D(post),front} + I_{D(post),back} + I_{D(post),other} \tag{6.13}$$

where $I_{D(post),front}$ is the inversion channel current along the front interface, $I_{D(post),back}$ is the inversion channel current along the back interface, and $I_{D(post),other}$ is the parasitic current, which consists of an additional floating-body-induced current [1], a parasitic bipolar current [19], and an avalanche current [20]. The discussion below focuses on $I_{D(post),front}$.

In practice, the post-threshold channel current consists mainly of the drift current, which is accelerated by the drain-to-source electric field. It is analyzed in a way similar to the conventional gradual channel approximation. The key difference comes from the limited depletion region beneath the channel, as shown in Figure 6.2. When t_s is larger than the depth of the depletion layer, an SOI MOSFET operates just like a bulk one under a DC bias, and is called a partially depleted silicon-on-insulator metal oxide semiconductor field-effect transistor (PD-SOI MOSFET). So, it should have almost the same drain current as a bulk MOSFET. The expression for the drain current of an SOI MOSFET is derived below.

We start from the Ohmic relation

$$I_{D(post),front} = -W_{eff}\mu_n Q_n(V)\frac{dV}{dy} \tag{6.14}$$

where V is the local channel potential and $Q_n(V)$ is the local electronic charge density. This is the expression for the local current inside the device, if we neglect the diffusion and other currents like the GR current. As an SOI MOSFET has both front and back channels, the discussion will focus on the front one ($I_{D(post),front}$) for simplicity.

We obtain the following relation for $Q_n(V)$ from Poisson's equation for the SOI layer [21, 22]:

Q_n = total induced charge density of semiconductor – depletion charge density of SOI layer – total induced charge density of substrate beneath buried insulator,

$$= -C_{ox}\left(V_{FG} - V_{FB,front} - \varphi(0) - V\right) + qN_A t_S + \varepsilon_s \left.\frac{d\varphi(x)}{dx}\right|_{x=t_S}$$

(6.15)

where the potential gradient at the back interface $(d\varphi(x)/dx|_{x=t_S})$ is [21]

$$\left.\frac{d\varphi(x)}{dx}\right|_{x=t_S} = \frac{qN_A}{2C_{SOI}} + \frac{\left(\varphi(t_S) - \varphi(0) - V\right)}{t_S}$$

(6.16)

This yields the approximation [13]

$$\varphi(t_S) = \frac{\left\{-\left(\frac{1}{2}\right)qN_A t_S + C_{BOX}V_{BG}' + C_{SOI}\left[\varphi(0) + V\right]\right\}}{\left(C_{SOI} + C_{BOX}\right)}$$

(6.17)

where V'_{BG} is the effective back gate bias. Substituting Eqs. (6.15)–(6.17)) into Eq. (6.14) and integrating it from source to drain, we obtain the following expression for $I_{D(post),front}$:

$$I_{D(post),front} = \left(\frac{W_{eff}}{L_{eff}}\right)\mu_n C_{ox}\left[V_{FG} - V_{TH}^* - \frac{(1+\alpha)V_D}{2}\right]V_D$$

(6.18)

where

$$V_{TH}^* = V_{FB,front} + (1+\alpha)\varphi_{s,front} + \left(\frac{3}{2}\right)(1+\gamma)\frac{qN_A t_S}{C_{ox}} - \alpha V'_{BG}$$

(6.19)

and α and γ are the geometrical parameters

$$\alpha = \frac{C_{SOI}C_{BOX}}{C_{ox}\left(C_{SOI} + C_{BOX}\right)}$$

(6.20)

$$\gamma = \frac{C_{SOI}}{3\left(C_{SOI} + C_{BOX}\right)}$$

(6.21)

α is not zero in either an FD-SOI MOSFET or a bulk MOSFET. When we reduce t_{BOX} in order to suppress short-channel effects, we must also reduce t_{ox} and/or t_S to suppress the increase in α.

In many cases, we have the condition $t_{BOX} \gg t_S$ and $t_{BOX} \gg t_{ox}$. When C_{BOX} approaches 0 $(t_{BOX} \rightarrow$ infinity), Eq. (6.18) can be simplified. That is, if we assume that the buried oxide layer is very thick, the electric field across it is quite low; so there is almost no depletion or accumulation of charge beneath it. Thus, $Q_n(V)$ can be simplified to

Q_n = total induced charge density of semiconductor – depletion charge density of SOI layer

$$= -C_{ox}\left(V_{FG} - V_{FB,front} - \varphi_{s,front} - V\right) + qN_A W_D(V)$$

$$(6.22)$$

For a PD-SOI MOSFET, this expression holds because $W_D(V) < t_S$. This suggests that $I_{D(post)}$ is almost the same as that for a bulk MOSFET. However, for an FD-SOI MOSFET, $W_D(V) > t_S$. So, we have to replace the above expression with

$$Q_n = -C_{ox}\left(V_{FG} - V_{FB,front} - \varphi_{s,front} - V\right) + qN_A t_S \qquad (6.23)$$

As this is an approximate expression, the quantitative agreement with experimental results is not discussed. Substituting the new expression for $Q_n(V)$ into the Ohmic relation (6.14) and assuming current continuity yields the following expression for $I_{D(post),front}$:

$$I_{D(post),front} = \left(\frac{W_{eff}}{L_{eff}}\right)\mu_n C_{ox}\left[V_{FG} - V_{TH,front} - \frac{V_D}{2}\right]V_D \qquad (6.24)$$

where the threshold voltage of an FD-SOI MOSFET ($V_{TH,front}$) is

$$V_{TH,front} = V_{FB,front} + \varphi_{s,front} + \frac{qN_A t_S}{C_{ox}} \qquad (6.25)$$

This expression was derived in order to explain the threshold voltage of an FD silicon-on-sapphire (SOS) MOSFET. It means that the threshold voltage of an FD-SOI MOSFET is usually smaller than that of a bulk one with the same device parameters. As the depth of the depletion layer is limited to t_S, the drop in the gate oxide voltage is also smaller, which often results in a lower effective surface electric field, accompanied by the apparent enhancement of the electron mobility [23]. However, it should be noted that there are some conditions for t_S and N_A that produce a lower surface electric field [24]. The saturated drain current ($I_{D(post)sat}$) is given by

$$I_{D(post)sat} = \left(\frac{W_{eff}}{2(1+\alpha)L_{eff}}\right)\mu_n C_{ox}\left[V_{FG} - V_{TH,front}\right]^2 \qquad (6.26)$$

This expression means that an FD-SOI MOSFET has a larger drivability than either a PD-SOI MOSFET or a bulk MOSFET because of the low threshold voltage and high carrier mobility. This is a significant advantage.

Finally we address the issue of volume inversion. When t_S is very small, the potential profile in the depth direction is flat at the onset of surface inversion, and the inversion layer also spreads over the body [25]. Simulations show that the drivability is doubled [26]. However, we can really only expect such good performance in long-channel devices [26] because the transverse field is effectively reduced when the body is extremely thin [27].

Other important factors are discussed below.

• *Self-heating effects* (Figure 6.6). Powered devices suffer from self-heating effects. Since an SOI MOSFET has a buried insulator (often with a low thermal conductivity) beneath the body, thermal paths to carry away the Joule heat are needed [28]. Otherwise, according to simulations, the body temperature quickly rises to $100\,°C$ [29]. As self-heating effects degrade the surface mobility of carriers, they must be taken into account. In electrostatic-discharge (ESD) protection circuits, self-heating effects reduce the temperature margin for the second breakdown. So, adequate thermal paths must be added during metallization. If we need to take into account the influence of self-heating effects on drain current, we have to use the following expression for the mobility [30]:

$$\mu_{n(self-heating)} = C\left[T_0 + \theta I_{D(post),front} V_D\right]^{-\delta} \tag{6.27}$$

where C is a constant, T_0 is room temperature, and θ and δ are fitting parameters.

• *Influence of body potential* (Figure 6.7). Body potential is basically isolated from the substrate potential, so the influence of the substrate potential is quite limited, except when the body is very thin. On the other hand, avalanche-induced holes easily reach the source junction because the effective barrier height for holes is quite low when the body is fully depleted. This is why parasitic bipolar action (PBA) readily occurs in an FD-SOI MOSFET. To suppress parasitic bipolar action, we have to use a body terminal [31] or perform silicide source diffusion [32].
• *Influence of short-channel effects.* Charge sharing is automatically suppressed in an FD-SOI MOSFET, as expected from the expression for the threshold voltage, Eq. (6.19). The most serious short-channel effect in such a device is BIIBL (Figure 6.5). BIIBL arises from the fact that the permittivity of the buried oxide layer ($\varepsilon_{ox} = 3.8\varepsilon_0$) is much lower than that of the Si substrate ($\varepsilon_{Si} = 11.7\varepsilon_0$). As a result, the electric field is higher inside the buried oxide layer

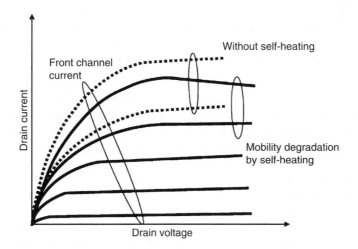

Figure 6.6 I_D–V_D characteristics of FD-SOI MOSFET.

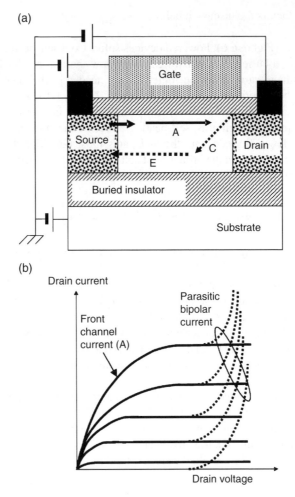

Figure 6.7 Parasitic bipolar action in FD-SOI MOSFET. (a) Current components. Avalanche-induced holes (C) are injected into the source (E), and electrons are injected into the body (A). This positive feed-back loop turns on parasitic bipolar action. (b) Parasitic bipolar action.

than inside the Si layer. So, to suppress BIIBL, two-dimensional (2-D) or three-dimensional (3-D) electric-charge sharing is required; in other words, the drain-to-source field flux through the buried oxide layer should be less than the drain-to-substrate field flux. It is possible to do this by making the buried oxide layer thinner [13]. The possible disadvantages of thinning the buried oxide layer are that the threshold voltage is higher, SS is larger, and the delay time is longer.

• *Velocity saturation and velocity overshoot.* Since carrier velocity is a key factor in obtaining a high drivability, most engineers try to obtain high carrier mobility. In an actual SOI MOSFET, the saturation velocity is about 70% that of a bulk one [33]. However, this has little effect on overall device performance because the inversion charge density and the low field mobility are both higher at a lower supply voltage

6.2.3   Comprehensive Summary of Short-Channel Effects

First, we derive the theoretical expression for the threshold voltage of an FD-SOI MOSFET [13, 21]. The depletion approximation and Poisson's equation yield the following relation:

$$\varphi(x) = \varphi_b + \left(\frac{qN_A}{2\varepsilon_{si}}\right)(t_s - x)^2 - (t_s - x)\frac{d\varphi(x)}{dx}\Bigg|_{x=t_S} \tag{6.28}$$

where

$$\varphi_b = \frac{\left[-\left(\frac{1}{2}\right)qN_A t_s + C_{BOX}V'_{BG} + C_s\varphi_{s,front}\right]}{(C_s + C_{BOX})} \tag{6.29}$$

This in turn yields

$$V_{TH} = V_{FB,front} + \varphi_{s,front} + \frac{qN_A t_S}{C_{ox}} + \left(\frac{\varepsilon_{si}}{C_{ox}}\right)\frac{d\varphi(x)}{dx}\Bigg|_{x=t_S}$$

$$= V_{FB,front} + (1+\alpha)\varphi_{s,front} + \left(\frac{3}{2}\right)(1+\gamma)\frac{qN_A t_S}{C_{ox}} - \alpha V'_{BG} \tag{6.30}$$

Considering a simple phenomenalistic model of short-channel effects, we obtain

$$V_{TH} = V_{FB,front} + f_{DIBL}(1+\alpha)\varphi_{s,front} + \left(\frac{3}{2}\right)f_{CS}(1+\gamma)\frac{qN_A t_S}{C_{ox}} - \alpha V'_{BG} \tag{6.31}$$

where f_{DIBL} is the DIBL factor and f_{CS} is the charge-sharing factor. Generally speaking, an FD-SOI MOSFET is not very susceptible to DIBL because the effective surface potential is deeper than for a bulk MOSFET, as can be seen in Eq. (6.30). It can also be seen that there is comparatively little charge sharing in an FD-SOI MOSFET because the depletion charge density is lower than in a bulk MOSFET.

In FD-SOI MOSFETs, the body is fully depleted, which means that the entire SOI layer functions as a charged insulator. The drain-induced potential easily penetrates the SOI and buried oxide layers, which results in BIIBL, as shown in Figure 6.5. To suppress it, we have to make the buried oxide layer thinner, as explained in the previous section. However, that degrades the subthreshold swing because the potential at the SOI-layer/buried oxide layer interface is easily pinned by the substrate potential. Thus, it is necessary to optimize the thickness of the SOI and buried oxide layers.

6.3   Accumulation-Mode (AM) SOI MOSFET

6.3.1   Aspects of Device Structure

Generally speaking, most MOSFETs have pn junctions that work as terminals. However, we can realize an SOI MOSFET without any pn junctions. The accumulation mode silicon-on-insulator

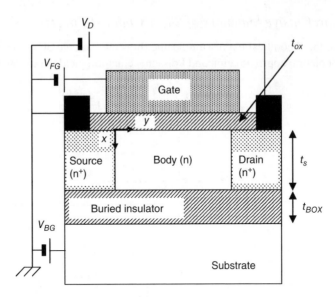

Figure 6.8   Schematic structure of n-channel accumulation mode SOI MOSFET.

metal oxide semiconductor field-effect transistor (AM SOI MOSFET) shown in Figure 6.8 is one such device; its source and drain have the same polarity as the body. Notations for bias parameters and dimensions are shown in Figure 6.8. Its electrical characteristics are very similar to those of conventional SOI MOSFETs having pn junctions [34]; the difference from the conventional MOSFET with pn junction is that the channel carriers are of the same type as the majority carriers of the body material.

6.3.2   Subthreshold Characteristics

Potential distribution in the silicon layer of the SOI device is complex. It is a function of the front gate voltage (V_{FG}), the back gate voltage (V_{BG}), and the drain voltage (V_D). Lim and Fossum analyzed the potential distribution and categorized the operation mode into several aspects [6]. They started their calculation of subthreshold characteristics with the following relations [6, 35]:

$$V_{FG} = V_{FB,front} + \varphi_{S,front}\left(1 + \frac{C_{SOI}}{C_{OX}}\right) - \varphi_{S,back}\frac{C_{SOI}}{C_{OX}} + \frac{qN_D t_s}{2C_{OX}} \qquad (6.32)$$

$$V_{BG} = V_{FB,back} + \varphi_{S,back}\left(1 + \frac{C_{SOI}}{C_{BOX}}\right) - \varphi_{S,front}\frac{C_{SOI}}{C_{BOX}} + \frac{qN_D t_s}{2C_{BOX}} \qquad (6.33)$$

The contributions of interface traps were ignored for simplicity. We first calculate the electron density distribution ($n(y)$) from the source to the drain in the silicon layer. It is given by the following calculation:

$$n(y) = N_D \int_0^{t_S} \exp\left(\frac{q\varphi(x)}{k_B T}\right) dx \qquad (6.34)$$

where $\varphi(x)$ denotes the electrostatic potential of the silicon layer. Although a possible expression for $\varphi(x)$ can be analytically modeled, the electron density has to be numerically calculated. Thus we simply calculate $n(y)$ as

$$n(y) = N_D \exp\left(\frac{q\varphi_{S,front}}{k_B T}\right) F_n\left(\varphi_{S,front}, \varphi_{S,back}\right) \qquad (6.35)$$

where function F_n is the numerical calculation result of the integration shown in Eq. (6.34). After some algebra, shown in Chapter 5, the approximated subthreshold swing SS is given by [35]:

$$SS = \frac{k_B T}{q} \ln(10)\left(1 + \frac{C_{SOI}}{C_{OX}} - \frac{\dfrac{C_{SOI}^2}{C_{OX} C_{BOX}}}{1 + \dfrac{C_{SOI}}{C_{BOX}}}\right) \qquad (6.36)$$

Here, the contributions of interface traps are not taken into account. This theoretical expression for the subthreshold swing shows that the SS value is always larger than $SS = \frac{k_B T}{q} \ln(10)$. Therefore, we don't address the details of the subthreshold current characteristics here.

6.3.3   Drain Current Component (I) – Body Current ($I_{D,body}$)

The physics of operation modes are roughly the same as those of junction field-effect transistors (FETs) except for the fact that the primary current component consists of carriers in the surface accumulation layer [36]. The total drain current is expressed as

$$I_D = I_{D,acc} + I_{D,body} \qquad (6.37)$$

where $I_{D,acc}$ denotes the drain current component, basically composed of the surface accumulation layer, and $I_{D,body}$ denotes the drain current component due to the majority carrier flow in the nondepleted region.

We next calculate the body current $I_{D,body}$. When the device has depletion regions near the top surface of the body silicon layer and near the bottom surface of the body silicon layer, the theoretical expression for $I_{D,body}$ is complicated and there are six function categories depending on gate biases (V_{FG} and V_{BG}) and drain bias (V_D). The body current component ($I_{D,body}$) is expressed as [34, 37, 38]:

$$I_{D,body} = \left(\frac{W_{eff}}{L_{eff}}\right) q N_D \mu_{n,b} K_F\left(V_{FG}, V_{BG}, V_D, t_S, t_{ox}, t_{BOX}, N_D\right) \qquad (6.38)$$

where N_D is the doping concentration of the body silicon layer, $\mu_{n,b}$ is the bulk electron mobility, and K_F is the function of biases and device parameters. Here we take the depletion approximation for simplicity. Function K_F is categorized as follows:

- The silicon layer is fully depleted. There is no body current given the depletion approximation.

$$K_F\left(V_{FG}, V_{BG}, V_D, t_S, t_{ox}, t_{BOX}, N_D\right) = 0 \tag{6.39}$$

- The front surface of the silicon layer is in accumulation.

$$K_F\left(V_{FG}, V_{BG}, V_D, t_S, t_{ox}, t_{BOX}, N_D\right) = t_{S,eff} V_D \tag{6.40}$$

where $t_{s,eff}$ is the thickness of the quasineutral silicon layer and is expressed as

$$t_{S,eff} = t_S + \frac{\varepsilon_S}{C_{OX}} - \left[\left(\frac{\varepsilon_S}{C_{BOX}}\right)^2 + \frac{2\varepsilon_S}{qN_D}\left(V_{BG} - V_{FB,back} - V(y)\right)\right]^{1/2} \tag{6.41}$$

where $V_{FB,back}$ denotes the flat-band voltage at the bottom surface of the silicon layer.

- Neither the surface accumulation channel nor the quasineutral silicon layer is pinched off.

$$K_F\left(V_{FG}, V_{BG}, V_D, t_S, t_{ox}, t_{BOX}, N_D\right) = \left(t_{S,eff} + \frac{\varepsilon_S}{C_{OX}}\right)V_D$$
$$+ \left(\frac{qN_D}{3\varepsilon_S}\right)\left[\left(\frac{\varepsilon_S}{C_{OX}}\right)^2 + \left(\frac{2\varepsilon_S}{qN_D}\right)\left(V_{FG} - V_{FB,front} - V_D\right)\right]^{3/2}$$
$$- \left(\frac{qN_D}{3\varepsilon_S}\right)\left[\left(\frac{\varepsilon_S}{C_{OX}}\right)^2 + \left(\frac{2\varepsilon_S}{qN_D}\right)\left(V_{FG} - V_{FB,front}\right)\right]^{3/2} \tag{6.42}$$

- The silicon layer is pinched off by the depletion region, but not fully depleted. The accumulation layer is not near the source terminal.

$$K_F\left(V_{FG}, V_{BG}, V_D, t_S, t_{ox}, t_{BOX}, N_D\right) = \left(t_{S,eff} + \frac{\varepsilon_S}{C_{OX}}\right)\left(V_{FG} - V_{FB,front} - V_{dep}\right)$$
$$+ \left(\frac{qN_D}{3\varepsilon_S}\right)\left[\left(\frac{\varepsilon_S}{C_{OX}}\right)^2 + \left(\frac{2\varepsilon_S}{qN_D}\right)V_{dep}\right]^{3/2}$$
$$- \left(\frac{qN_D}{3\varepsilon_S}\right)\left[\left(\frac{\varepsilon_S}{C_{OX}}\right)^2 + \left(\frac{2\varepsilon_S}{qN_D}\right)\left(V_{FG} - V_{FB,front}\right)\right]^{3/2} \tag{6.43}$$

$$V_{dep} = \frac{qN_D t_{S,eff}^2}{2\varepsilon_S} + \frac{qN_D t_{S,eff}}{C_{OX}} \tag{6.44}$$

- The silicon layer is not pinched off. The surface accumulation layer is near the source terminal.

$$K_F\left(V_{FG},V_{BG},V_D,t_S,t_{ox},t_{BOX},N_D\right)=t_{S,eff}\left(V_{FG}-V_{FB,front}\right)+\left(t_{S,eff}+\frac{\varepsilon_S}{C_{OX}}\right)\left(V_D-V_{FG}+V_{FB,front}\right)$$

$$+\left(\frac{qN_D}{3\varepsilon_S}\right)\left[\left(\frac{\varepsilon_S}{C_{OX}}\right)^2+\left(\frac{2\varepsilon_S}{qN_D}\right)\left(V_{FG}-V_{FB,front}-V_D\right)\right]^{3/2}-\left(\frac{qN_D}{3\varepsilon_S}\right)\left(\frac{\varepsilon_S}{C_{OX}}\right)^3$$

(6.45)

- The accumulation layer is near the source terminal, but the silicon layer is pinched off.

$$K_F\left(V_{FG},V_{BG},V_D,t_S,t_{ox},t_{BOX},N_D\right)=t_{S,eff}\left(V_{FG}-V_{FB,front}\right)+\left(t_{S,eff}+\frac{\varepsilon_S}{C_{OX}}\right)V_{dep}$$

$$+\left(\frac{qN_D}{3\varepsilon_S}\right)\left[\left(\frac{\varepsilon_S}{C_{OX}}\right)^2+\left(\frac{2\varepsilon_S}{qN_D}\right)V_{dep}\right]^{3/2}\cdot-\left(\frac{qN_D}{3\varepsilon_S}\right)\left(\frac{\varepsilon_S}{C_{OX}}\right)^3$$

(6.46)

6.3.4 Drain Current Component (II)–Surface Accumulation Layer Current ($I_{D,acc}$)

The surface accumulation layer works as a conductive channel, like the inversion layer, so the drain current is calculated by assuming the drift current of surface electrons. The resistance of an elementary resistor in the accumulation layer channel ($dR(y)$) is given by

$$dR(y)=\frac{dy}{W_{eff}\mu_{n,acc}Q_{acc}}$$

(6.47)

where $\mu_{e,acc}$ denotes the mobility of the electron accumulation layer and Q_{acc} denotes the charge density of the accumulation layer. Based on Eq. (6.42) and the discussion in Section 5.3, we have the following expression for $I_{D,acc}$ in the nonsaturation condition:

$$I_{D,acc}=\frac{W_{eff}}{L_{eff}}\mu_{n,acc}C_{OX}\left(V_{FG}-V_{FB,front}-\frac{V_D}{2}\right)V_D$$

(6.48)

For the postsaturation condition, we have

$$I_{D,acc}=\frac{W_{eff}}{2L_{eff}}\mu_{n,acc}C_{OX}\left(V_{FG}-V_{FB,front}\right)^2$$

(6.49)

6.3.5 Optional Discussions on the Accumulation Mode SOI MOSFET

In section 5.2.2 and section 6.2.1 we addressed the BTBT current around the drain terminal for SOI devices in the OFF state. Readers may think that the accumulation mode SOI MOSFET is free from the BTBT current based on its pn-junction-free structure. However, it is well known that the buried-channel SOI MOSFET and the accumulation model SOI MOSFET do exhibit PBA [39]. This is due to the fact that the BTBT frequently forms around the drain terminal [10]. Therefore, we must address this undesirable phenomenon.

6.4 FinFET and Triple-Gate FET

6.4.1 Introduction

It is well known that a simple FinFET shows the characteristics of the double-gate MOSFET or the triple-gate metal oxide semiconductor field-effect transistor (TG MOSFET), depending on device geometry and device parameters. In this section, we reveal how the FinFET structure changes drivability with the aid of three-dimensional (3D) device simulations. A discussion of device geometry to achieve low-energy operation is given in Chapter 18.

As has been noted frequently, 3D-FETs like FinFETs and TG MOSFETs are being studied extensively to realize high-performance circuits and to suppress short-channel effects (SCEs) [40, 41]. These devices have peculiar issues, such as corner effects, due to their 3D structure. Many people feel that the conventional FinFET is inferior to other MOSFETs because corner effects are not easily controlled. In many cases, the corner effects yield a hump in the subthreshold region, which results in an extra leakage current and undesirable low threshold voltages. Such complaints stem from the uncontrollability of the corner shape. However, recent process technology has overcome this issue [42]. So, it is worth considering how we can use corner effects.

As the top surface of the fin of the TG MOSFET works as an additional channel, drivability is enhanced by the contribution of the top surface to the channel current. However, it is well known that the effective channel width (W_{eff}) of such 3D-FETs is not a simple multiple of W_{eff} of the planar-single-gate SOI MOSFET because of the complicated 3D effects [43]. These complications have prevented the development of a current model equation for 3D-FETs.

This section compares the simulation results of planar double-gate silicon-on-insulator metal oxide semiconductor field-effect transistors (p-DG SOI MOSFETs) to those of SOI FinFETs that work as triple-gate FETs. We propose an empirical quantitative model of W_{eff} that includes 3D-effects for FinFETs working like Triple-gate FETs.

6.4.2 Device Structures and Simulations

Device structures assumed in these simulations are illustrated in Figure 6.9; Figure 6.9a shows the cross-sectional view of the double-gate SOI MOSFET and Figure 6.9b shows that of the FinFET and TG FET. Device parameters are summarized in Table 6.1. Si-fin width (w) and height (h) of the FinFET represent the SOI layer thickness (t_S) and channel width (W_{eff}) of the p-DG SOI MOSFET, respectively. The channel length is assumed to be 100 nm to suppress the SCEs because we focus only on 3D effects. Device characteristics were derived at the drain voltage (V_D) of 1 V using a 2D or 3D device simulator, Synopsys Inc.-DESSIS (Device Simulation for Smart Integrated Systems) [44], and the hydrodynamic transport model. Drive current (I_{on}) is defined as drain current (I_D) at $V_G = V_{TH} + 0.75$ [V] = 1.0 [V], where V_G is the gate voltage and V_{TH} is the threshold voltage.

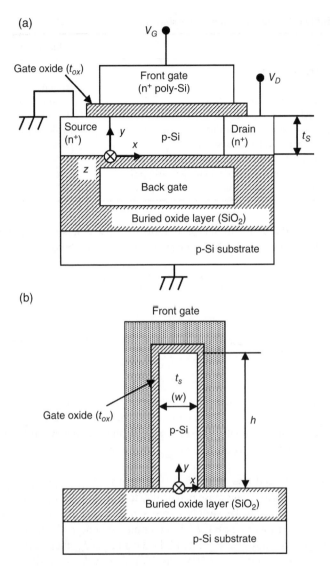

Figure 6.9 Device structure assumed in simulations. (a) Cross sectional view of double gate structure and (b) cross sectional view of FinFET and TG FET.

6.4.3 Results and Discussion

We varied four parameters in the simulations: gate oxide thickness (t_{ox}), Si-fin width (SOI layer thickness) w (t_s), Si-fin height (channel width) h (W_{eff}), and channel doping concentration (N_A). Here, we introduce the drain current ratio, α_{sim} ($= I_{on(FinFET)}/I_{on(p\text{-}DG\ FET)}$), which is automatically calculated from simulation results, and try to express it by empirical equations. For this

Table 6.1   Device parameters assumed.

Parameters	Values (unit)
Channel length, L_{eff}	100 (nm)
Top gate oxide thickness	50 (nm)
Side gate oxide thickness, t_{ox}	1.5–4.0 (nm)
Buried oxide thickness	100 (nm)
Si-fin width (SOI layer thickness), w (t_S)	10–40 (nm)
Si-fin height (channel width), h (W_{eff})	10–40 (nm)
Channel doping concentration, N_A (p-type)	10^{15}–10^{17} (cm^{-3})
Source/drain doping concentration (n-type)	10^{20} (cm^{-3})
Si-substrate doping concentration (p-type)	10^{15} (cm^{-3})

goal, we propose a model of the drain current ratio, α_{model}, which is a function of four parameters (t_{ox}, $w(t_S)$, $h(W_{eff})$, and N_A). We show our expression of α_{model} below:

$$\alpha_{model} = \beta \cdot \exp\left(\frac{t_{ox}/A - 2}{18}\right) \cdot \exp\left(\frac{w/A - 10}{300}\right) \tag{6.50}$$

$$\beta = -257 \times 10^{-2} \ln\left(\frac{N_A}{B}\right) - 0.02 \times 10^{-3} \cdot \frac{(w \cdot h \cdot t_{ox})}{A^3} + 1.24 \cdot \left(\frac{h}{4}\right)^{-0.255} + 0.85 \cdot \left(\frac{t_{ox}}{A}\right)^{-0.053} + \partial \tag{6.51}$$

where $A = 1$ nm, $B = 1$ cm^{-3}, $\delta = 0.365$. The effective channel width of a FinFET ($W_{eff(FinFET)}$) is given by $W_{eff(FinFET)} = \alpha_{model} \times W_{eff(p\text{-}DG\ FET)}$. Calculated α_{sim} dependence on α_{model} is shown in Figure 6.10. We can see that the simulation results α_{sim} adequately match the model (α_{model}). This means that, by the above empirical model, (α_{model}) can identify the device parameters that yield superior drivability.

In Figure 6.10, the dashed line indicates $\alpha_{sim} = 1$, that is, $I_{ON(FinFET)} = I_{ON(p\text{-}DG\ FET)}$. It should be noted that α_{sim} decreases with an increase in h or W_{eff}. In a p-DG FET, I_{ON} increases in proportion to W_{eff} because $W_{eff(p\text{-}DG\ FET)}$ can be simply expressed as twice W_{eff}. On the other hand, in a FinFET working like a triple-gate FET, I_{ON} is a complex function of h, and $W_{eff(FinFET)}$ is at most twice h. Accordingly, in many cases, p-DG FETs have larger I_{ON} than FinFETs.

Note that α_{sim} exceeds unity when $h(W_{eff})$ is 10 nm. In order to verify the cause for this, Figure 6.11 plots the electron density profiles from BOX to the top interface along the z-axis near the source junction for 10 nm $h(W_{eff})$ devices. We can see that, for FinFETs, the electron density increases at the corner because of 2-D field enhancement. It may also be seen that this increase in electron density spreads out about 3 nm as shown in Figure 6.11.

Figure 6.12 shows the product of electron density (n) and velocity (v) as a function of distance measured from the center of the Si-fin (along the y-axis). The nv product of the FinFET on the 5 nm deep plane from the BOX interface, where the corner effect is almost insignificant, is smaller than that of the p-DG FET. On the other hand, the nv products of the FinFET at both BOX and the top interfaces are larger than those of the p-DG FET. These two facts indicate that the superior drivability of FinFETs with 10 nm height and width is due to the corner effect. Consequently, low-fin devices retain their superior drivability even with a high population of inversion carriers because of the corner effect.

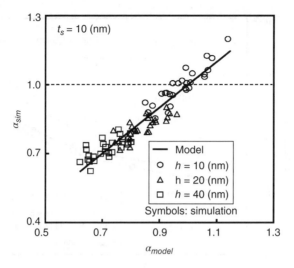

Figure 6.10 α_{sim} dependence on α_{model} at V_D of 1 V.

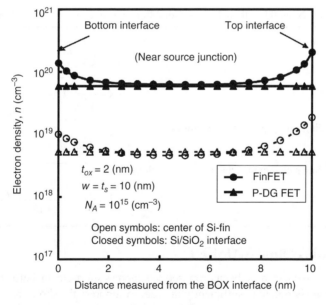

Figure 6.11 Electron density profile across the silicon layer.

6.4.4   Summary

In this chapter we considered the impact of geometrical effects on the drivability of FinFET. It was demonstrated that the triple-gate SOI MOSFET has superior drivability when the corner current component is well managed by proper design.

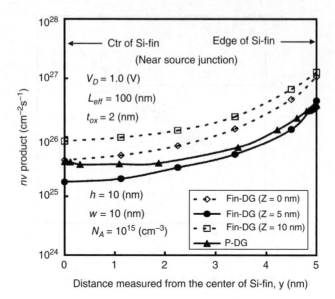

Figure 6.12   *nv* product along the *y*-axis at $V_G = V_{TH} + 0.75$ [V].

Table 6.2   Natural length in devices with different geometries (square section) [43].

Single gate	$\lambda = \sqrt{\dfrac{\varepsilon_s}{\varepsilon_{ox}} t_s t_{ox}}$
Double gate	$\lambda = \sqrt{\dfrac{\varepsilon_s}{2\varepsilon_{ox}} t_s t_{ox}}$
Surrounding gate (square channel cross section)	$\lambda = \sqrt{\dfrac{\varepsilon_s}{4\varepsilon_{ox}} t_s t_{ox}}$
Surrounding gate (circular channel cross section)	$\lambda = \sqrt{\dfrac{2\varepsilon_s t_S^2 \ln\left(1 + \dfrac{2t_{ox}}{t_S}\right) + \varepsilon_{ox} t_S^2}{16\varepsilon_{ox}}}$

6.5   Gate-all-Around MOSFET

Although various devices have been proposed to suppress short-channel effects [45], we must still control the body potential sufficiently in order to suppress the short-channel effects. This consideration is based on the solution of the following Poisson equation for a fully depleted multigate MOSFET:

$$\frac{\partial^2 \Phi(x,y,z)}{\partial x^2} + \frac{\partial^2 \Phi(x,y,z)}{\partial y^2} + \frac{\partial^2 \Phi(x,y,z)}{\partial z^2} = \frac{qN_A}{\varepsilon_s} \tag{6.52}$$

where a p type body silicon is assumed and $\Psi(x,y,z)$ denotes the electrostatic potential of the body silicon. Solutions of Eq. (6.52) for various MOSFETs give natural lengths that characterize the device geometry.

The natural lengths derived theoretically are summarized in Table 6.2. It is theoretically obvious that multigate structures are inevitable in designing scaled devices. It might be expected that the gate-all-around MOSFET, or gate-all-around wire MOSFET, would be the ultimate geometry for the scaled MOSFET. However, such devices still face various issues, one of which is high dissipation energy due to various leakage currents [17]. We address performance aspects in later parts.

References

[1] J. Tihanyi and H. Schlotterer, "Influence of the floating substrate potential on the characteristics of ESFI MOS transistors," *Solid-State Electron.*, vol. 18, pp. 309–314, 1975.

[2] J. Tihanyi and H. Schlotterer, "Properties of ESFI MOS transistors due to the floating substrate and the finite volume," *IEEE Trans. Electron Devices*, vol. ED-22, pp. 1017–1023, 1975.

[3] R. T. Jerdonek and W. R. Brandy, "Modeling the operation of the n-channel deep-depletion SOS/MOSFET," *IEEE Trans. Electron Devices*, vol. 25, pp. 899–907, 1978.

[4] S. S. Eaton and B. Lalevic, "The effect of a floating substrate on the operation of silicon-on-sapphire transistors," *IEEE Trans. Electron Devices*, vol. 25, pp. 907–912, 1978.

[5] P. Su, S. K. H. Fung, S. Tang, F. Assaderaghi, and C. Hu, "BSIMPD: A Partial-Depletion SOI MOSFET Model for Deep-Submicron CMOS Designs," IEEE Custom Integrated Circ. Conf., pp. 197–200, 2000.

[6] H. K. Lim and J. G. Fossum, "Threshold voltage of thin-film Silicon-on-insulator (SOI) MOSFET's," *IEEE Trans. Electron Devices*, vol. 30, no. 10, pp. 1244–1251, 1983.

[7] "Silicon-on-Insulator Technology: Materials to VLSI," ed. By J.-P. Colinge, 3rd ed. (Kluwer Academic Publishers, 2004) Chapter 5.

[8] G. Merckel, F. Van de Wiele, W. Engle, and P. Jespers, "NATO Course on Process and Device Modeling for Integrated Circuit Design," (Noordhoff, 1975), p. 725.

[9] J. Chen, T. Y. Chan, I. C. Chen, P. K. Ko, and C. Hu, "Subbreakdown drain leakage current in MOSFET," *IEEE Electron Device Lett.*, vol. EDL-8, pp. 515–518, 1987.

[10] T. Ishiyama and Y. Omura, "Influence of superficial Si layer thickness on band-to-band tunneling current characteristics in ultra-thin n-channel metal-oxide-semiconductor field-effect-transistor by separation by implanted oxygen (nMOSFET/SIMOX)," *Jpn. J. Appl. Phys.*, vol. 36, pp. L264–L267, 1997.

[11] W. Shockley and W. T. Read, Jr., "Statistics of the recombination of holes and electrons," *Phys. Rev.*, vol. 87, pp. 835–842, 1952.

[12] J. R. Brews, "A charge-sheet model of the MOSFET," *Solid State Electron.*, vol. 21, pp. 345–355, 1978.

[13] Y. Omura, S. Nakashima, K. Izumi, and T. Ishii, "0.1-μm-Gate, Ultrathin-Film CMOS Devices Using SIMOX Substrate with 80-nm-Thick Buried Oxide Layer," Tech. Dig. Int. Electron Devices Meeting (Washington DC, Dec., 1991) pp. 675–678, 1991.

[14] L. T. Su, J. B. Jacobs, J. E. Chung, and D. A. Antoniadis, "Deep-submicrometer channel design in silicon-on-insulator (SOI) MOSFET's," *IEEE Electron Device Lett.*, vol. 15, pp. 366–368, 1994.

[15] F Assaderaghi, D. Sinitsky, S. A. Parke, J. Boker, P. K. Ko, and C. Hu, "A Dynamic Threshold Voltage MOSFET (DTMOS) for Ultra-Low Voltage Operation," Tech. Dig. Int. Electron Devices Meeting (San Francisco, Dec., 1994) pp. 809–812, 1994.

[16] R. R. Troutman, "Subthreshold design considerations for IGFET's," *IEEE J. Solid-State Circuits*, vol. SC-9, pp. 55–60, 1974.

[17] Y. Omura, T. Ishiyama, M. Shoji, and K. Izumi, "Quantum Mechanical Transport Characteristics in Ultimately Miniaturized MOSFETs/SIMOX," Proc. 7th Int. Symp. Silicon-on-Insulator Technology and Devices (The Electrochem. Soc., Los Angeles, 1996) PV-96-3, pp. 199–211, 1996.

[18] H. Nakajima, S. Yanagi, K. Kenji, and Y. Omura, "Off-leakage and drive current characteristics of sub-100-nm SOI MOSFETs and impact of quantum tunnel current," *IEEE Trans. Electron Devices*, vol. 49, pp. 1775–1782, 2002.

[19] C.-E. D. Chen, M. Matloubian, R. Sundaresan, B.-Y. Mao, C. C. Wei, and G. P. Polack, "Single-transistor latch in SOI MOSFET's," *IEEE Electron Device Lett.*, vol. 9, pp. 636–638, 1988.

[20] Y. Omura and K. Izumi, "Physical background of substrate current characteristics and hot-carrier immunity in short-channel ultrathin-film MOSFET's/SIMOX," *IEEE Trans. Electron Devices*, vol. 41, pp. 352–358, 1994.

[21] Y. Omura, S. Nakashima, and K. Izumi, "Theoretical analysis on threshold characteristics of surface-channel MOSFET's fabricated on a buried oxide," *IEEE Trans. Electron Devices*, vol. ED-30, pp. 1656–1662, 1983.

[22] Y. Omura, S. Horiguchi, M. Tabe, and K. Kishi, "Quantum-mechanical effects on the threshold voltage of ultra-thin-SOI nMOSFET's," *IEEE Electron Device Lett.*, vol. 14, pp. 569–571, 1993.

[23] M. Yoshimi, H. Hazama, M. Takahashi, S. Kambayashi, T. Wada, K. Kato, and H. Tango, "Two-dimensional simulation and measurement of high-performance MOSFET's made on a very thin SOI film," *IEEE Trans. Electron Devices*, vol. 36, pp. 493–503, 1989.

[24] M. Shoji, Y. Omura, and M. Tomizawa, "Physical basis and limitation of universal mobility behavior in fully depleted silicon-on-insulator Si inversion layers," *J. Appl. Phys.*, vol. 81, pp. 786–794, 1997.

[25] F. Balestra, S. Cristoloveanu, M. Benachir, J. Brini, and T. Elewa, "Double-gate silicon-on-insulator with volume inversion: a new device with greatly enhanced performance," *IEEE Electron Device Lett.*, vol. EDL-8, pp. 410–412, 1987.

[26] J.-P. Colinge, M. H. Gao, A. Romano, H. Maes, and C. Claeys, "Silicon-on-Insulator Gate-All-Around Device," Tech. Dig. IEEE Int. Electron Devices Meeting (San Francisco, Dec., 1990) pp. 595–599, 1990.

[27] S. Yanagi and Y. Omura, "Consideration of performance limitation of sub-100-nm double-gate silicon-on-insulator (SOI) metal-oxide-semiconductor field-effect transistors (MOSFET's)," *Jpn. J. Appl. Phys.*, vol. 41, pp. L1096–L1098, 2002.

[28] K. Oshima, S. Cristoloveanu, B. Guillaumot, G. L. Carval, H. Iwai, C. Mazure, M. S. Kang, Y. H. Bae, J. W. Kwou, S. Deleonibus, and. J. H. Lee, "Replacing the BOX with Buried Alumina: Improved Thermal Dissipation in SOI MOSFETs," Proc. 11th Int. Symp. Silicon-on-Insulator Technology and Devices (The Electrochem. Soc., Paris, May, 2003) pp. 45–50, 2003.

[29] J. Jomaah, G. Ghibaudo, F. Balestra, and J. L. Pelloie, "Impact of Self-Heating Effects on the Design of SOI Devices versus Temperature," Proc. 1995 IEEE Int. SOI Conf. (Arizona, Oct., 1995) pp. 114–115, 1995.

[30] N. D. Arora, L. T. Su, B. S. Doyle, and D. A. Antoniadis, "Modeling the I-V Characteristics of Fully-Depleted SOI MOSFETs Including Self-Heating," Proc. 1994 IEEE Int. SOI Conf. (Massachusetts, Oct., 1994) pp. 19–20, 1994.

[31] M. Matloubian, "Smart Body Contact for SOI MOSFETs," Proc. 1989 IEEE SOS/SOI Technology Conf. (Nevada, Oct., 1989) pp. 128–129, 1989.

[32] L. J. McDaid, S. Hall, W. Eccleston, and J. C. Alderman, "Suppression of latch in SOI MOSFETs by silicidation of source," *Electron. Lett*, vol. 27, pp. 1003–1005, 1991.

[33] F. Assaderaghi, J. Chen, P. Ko, and C. Hu, "Measurement of Electron and Hole Saturation Velocities in Silicon Inversion Layers Using SOI MOSFETs," Proc. 1992 IEEE Int. SOI Conf. (Florida, Oct., 1992) pp. 112–113, 1992.

[34] J.-P. Colinge (ed.),"Silicon-on-Insulator Technology: Materials to VLSI," 3rd ed. (Kluwer Academic Pub., 2004), p. 217.

[35] D. J. Wouters, J.-P. Colinge, and H. E. Maes, "Subthreshold slope in thin-film SOI MOSFET's," *IEEE Trans. Electron Devices*, vol. ED-37, no. 9, pp. 2022–2033, 1990.

[36] S. M. Sze and K. K. Ng, "Physics of Semiconductor Devices," 3rd ed. (John Wiley & Sons, Inc., 2007), Chapter 7.

[37] J.-P. Colinge, "Conduction mechanisms in thin-film accumulation-mode SOI p-channel MOSFETs," *IEEE Trans. Electron Devices*, vol. 37, pp. 718–723, 1990.

[38] K. W. Su and J. B. Kuo, "Analysis of current conduction in short-channel accumulation-mode SOI pMOS devices," *IEEE Trans. Electron Devices*, vol. 44, pp. 832–840, 1997.

[39] J. Gautier and A.-J. Auberton-Herve, "A latch phenomenon in buried N-body SOI nMOSFET's," *IEEE Electron Device Lett.*, vol. 12, pp. 372–374, 1991.

[40] B. Doyle, B. Boyanov, S. Datta, M. Doczy, S. Hareland, B. Jin, J. Kavalieros, T. Linton, R. Rios, and R. Chau, "Tri-gate Fully-Depleted CMOS Transistors," Tech. Dig. Int. Symp. VLSI. Tech. Dig. (Kyoto, 2003) pp. 133–134, 2003.

[41] S.-H. Kim, J. G. Fossum, and V. P. Trivedi, "Bulk Inversion in FinFETs and the Implied Insignificance of the Effective Gate Width", Proc. IEEE Int. SOI Conf., pp. 145–147, 2004.

[42] Y. Liu, K. Ishii, T. Tsutsumi, M. Masahara, and E. Suzuki, "Ideal rectangular cross-section Si-fin channel double-gate MOSFETs fabricated using orientation-dependent wet etching," *IEEE Trans. Electron Device Lett.*, vol. 24, pp. 484–486, 2003.

[43] J.-P. Colinge, "FinFETs and Other Multi-Gate Transistors" (Springer, 2008), Chapter 1.

[44] Synopsys Inc., "Sentaurus Device User Guide," Version A-2007, 2007, ftp://147.46.117.90/synopsys/manuals/PDFManual/data/sdevice_ug.pdf (accessed June 10, 2016.)

[45] R. H. Yan, A. Ourmazd, and K. F. Lee, "Scaling the Si MOSFET: from bulk to SOI to bulk," *IEEE Trans. Electron Devices*, vol. 39, pp. 1704–1710, 1992.

7

Tunnel Field-Effect Transistors (TFETs)

7.1 Overview

This chapter introduces physics-based analytical device models for the lateral tunnel field-effect transistor (TFET) and the vertical TFET because they are primary device structures that have been studied widely. The models reproduce important characteristics such as OFF-to-ON *I-V* characteristics and subthreshold swing characteristics. In addition, such models predict technical issues.

Recent studies suggest that vertical TFETs composed of Si and Ge might be promising because high drivability and a steep swing are easily produced according to device simulations. On the vertical TFET, a theoretical consideration that addresses the impact of tunnel dimensionality on drivability is provided because drivability is still an issue with TFETs.

7.2 Model of Double-Gate Lateral Tunnel FET and Device Performance Perspective

7.2.1 Introduction

Metal oxide semiconductor field-effect transistor (MOSFET) devices have contributed significantly to the design of high-speed and low-power ICs that now power the worldwide Internet, resulting in more reliable social infrastructures. The poor energy efficiency of conventional MOSFET devices, including the silicon-on-insulator metal oxide semiconductor field-effect transistor (SOI MOSFET), is now impacting circuit designs because the energy loss in the OFF state approaches the energy dissipated in the "ON" state [1]. This is obviously seen as a serious problem.

MOS Devices for Low-Voltage and Low-Energy Applications, First Edition.
Yasuhisa Omura, Abhijit Mallik, and Naoto Matsuo.
© 2017 John Wiley & Sons Singapore Pte. Ltd. Published 2017 by John Wiley & Sons Singapore Pte. Ltd.

Discussions on "More-than-Moore" devices may contribute to some proposals for new devices that can overcome this issue but we propose promising near-term devices that can directly contribute to the energy issue described above. In this sense, the TFET is attracting much attention [2–9] because of its very sharp swing.

Some studies suggest that the sharp swing of the TFET is due to the transport yielded by the tunneling phenomenon [2–9]. In contrast, another study does not concur with the performance promises due to unknown issues with device materials and fabrication processes [10]. In addition to such difficulties, TFET drivability is still low in comparison to conventional MOSFET devices.

Device simulations will contribute to the elucidation of issues and proposals for new device structures. As the key step prior to simulation, analytical device modeling is essential in considering phenomena observed in the testing of prototype devices. Many device models have already been proposed for the purpose of increasing our insight [11–15] but their capability is still limited and we must improve them.

In this chapter, we propose analytical models that can reproduce the I_D-V_G and I_D-V_D characteristics of the lateral TFET. Here we focus on the double-gate lateral TFET. We introduce device simulations [16] in order to examine model quality.

7.2.2   Device Modeling

7.2.2.1   Device Structure and Potential Model

The schematic device structure for device modeling is shown in Figure 7.1a, where the x-axis parallels the source-to-drain direction with its origin at the source junction. The y-axis parallels the source junction surface. Device parameters are summarized in Table 7.1. Here we assume that the doping levels of the p-type source and n-type drain regions are very high ($\sim 10^{21}\,\mathrm{cm}^{-3}$).

Figure 7.1b is a simplified band diagram around the source junction, where the solid lines show the band edges near the gate-oxide/Si interface and the broken lines show those around the middle of the Si body. We will propose a critical field-line model to take account of the nonlocality of tunneling around the source junction. Accordingly, we show a schematic image of the critical field-line model in Figure 7.2, where F_J denotes the source-junction-induced field, F_{FG} denotes the gate-induced field, and $a(V_G)$ is the distance measured from the source junction surface. We assume that $a(V_G)$ is the tunneling-path length.

When the electrostatic potential of the body at $y=0$ is labeled φ_0, we have

- *Partial depletion of the body.* Assuming t_d is the depletion width, the potential profile function is given by

$$\varphi(y) = \varphi_S \left\{ \frac{y}{\dfrac{t_S}{2} - t_d} - 1 \right\}^2 \tag{7.1}$$

$$t_d = \left[\frac{2\varepsilon_S \varphi_S}{e N_A} \right]^{1/2} \tag{7.2}$$

$$\varphi\left(\pm \frac{t_s}{2} \right) = \varphi_S \tag{7.3}$$

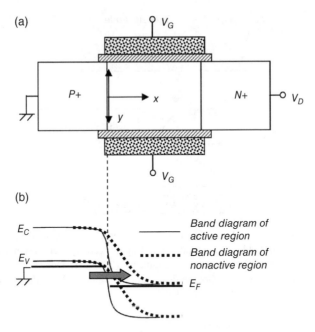

Figure 7.1 Schematic of double-gate lateral TFET and energy band diagram. (a) Device structure and (b) energy band diagram. Reproduced by permission of the Institute of Electronics, Information, and Communication Engineers (2013). Y. Omura, D. Sato, S. Sato, and A.Mallik, "Theoretical Modeling of Double-Gate Lateral Tunnel FET," IEICE Technical Report, SDM2013-109, pp. 55–60. Copyright © 2013 IEICE.

Table 7.1 Device parameters assumed here.

Parameters	Values
Effective gate oxide thickness (EOT)	0.5–2 nm
Si layer thickness (t_s)	5–20 nm
Body doping concentration (N_A)	10^{15}–10^{18} cm^{-3}
Electron mobility (μ_n)	500 cm^2/V s
Channel length (L_{eff})	40 nm
Channel width (W_{eff})	1 µm

Reproduced by permission of the Institute of Electronics, Information, and Communication Engineers (2013). Y. Omura, D. Sato, S. Sato, and A.Mallik, "Theoretical Modeling of Double-Gate Lateral Tunnel FET," IEICE Technical Report, SDM2013-109, pp. 55–60. Copyright © 2013 IEICE.

where φ_s denotes the surface potential at $y=\pm t_s/2$.

Following Figure 7.1b, we can estimate the gate-induced field as

$$\frac{d\varphi}{dy} \equiv \frac{E_G}{ea(V_G)} \tag{7.4}$$

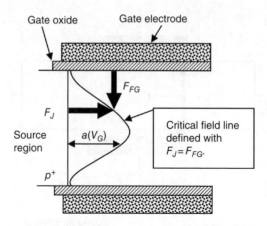

Figure 7.2 Model for the critical line defined by the condition of "lateral field strength equals vertical field strength." F_J: normal field. F_{FG}: lateral field. Reproduced by permission of the Institute of Electronics, Information, and Communication Engineers (2013). Y. Omura, D. Sato, S. Sato, and A.Mallik, "Theoretical Modeling of Double-Gate Lateral Tunnel FET," IEICE Technical Report, SDM2013-109, pp. 55–60. Copyright © 2013 IEICE.

where E_G is the bandgap energy, and e is the elementary charge. Parameter $a(y, V_G)$ is given as

$$a(y,V_G) = \frac{E_G}{4e\varphi_s} \frac{\left(\frac{t_S}{2} - t_n\right)^2}{\left\{y - \left(\frac{t_S}{2} - t_n\right)\right\}} \tag{7.5}$$

for positive values of y.

- *Full depletion of the body*. Equation (7.4) yields the following form

$$\varphi(y) = by^2 + \varphi_0 \tag{7.6}$$

where b is the parameter. Equation (7.6) is rewritten as

$$\varphi(y) = (\varphi_s - \varphi_0)\frac{4y^2}{t_S^2} + \varphi_0 \tag{7.7}$$

Similarly, following Eq. (7.4), we have

$$a(y,V_G) = \frac{t_S^2 E_G}{8ey(\varphi_s - \varphi_0)} = \frac{t_S^2 E_G}{8ey\left\{\frac{eN_A t_S^2}{8\sigma_S} + \frac{eN_S(V_G)}{C_{ox}}\right\}} \tag{7.8}$$

where $N_s(V_G)$ denotes the surface electron density, which depends on gate voltage (V_G), ε_s the permittivity of Si, and C_{ox} the gate oxide capacitance per unit area.

7.2.2.2 Tunnel Probability Model

In order to simplify the calculation of tunnel current, we use the following single-band model for the tunnel probability $(T_t(y, E))$ [17, 18].

$$T_t(y,E) = \left[1 + \frac{E_G^2/4}{4E(E_G/2 - E)}\sinh^2(k_t a(y,V_G))\right]^{-1} \tag{7.9}$$

$$k_t = \left[\frac{2m_x^*(E_G/2 - E)}{\hbar^2}\right]^{1/2} \tag{7.10}$$

where m_x^* is the effective mass of electrons tunneling from the source to the channel. In a later section, Eq. (7.10) will be reconsidered with a view to calculating the tunnel current integral.

7.2.2.3 Theoretical Formulation of Tunnel Current

When we assume that the effective tunnel resistance is larger than the MOSFET channel resistance, the total tunnel current across the source junction I_t of the device with the channel width of W_{eff} is given as

$$I_T = \frac{2eW_{eff}}{(2\pi)^3}\int_{-\frac{t_S}{2}}^{\frac{t_S}{2}} dy \int d^2k_\perp \int dk_x T_t(y,k_x,V_G,V_D)\frac{\partial E}{\hbar \partial k_x}\left[f(E) - f(E + eV_D)\right]$$

$$= \frac{2eW_{eff}}{\hbar(2\pi)^3}\int_{-\frac{t_S}{2}}^{\frac{t_S}{2}} dy \int d^2k_\perp \int dE_x T_t(y,k_x,V_G,V_D)\left[f(E) - f(E + eV_D)\right] \tag{7.11}$$

where $f(E)$ denotes the Fermi–Dirac function. In this chapter, as we don't take account of the impact of energy quantization in the Si body on the tunnel current, the minimal value of Si layer thickness (t_s) would be 10 nm [19]. In addition, we introduce the following assumptions to simplify the integration.

- The inversion layer is metallic and degenerate.
- $f(E)$ is replaced with a step function based on assumption 1.
- The energy range contributing to electron tunneling extends from the Fermi level of the source to the conduction-band bottom of the body region (see Figure 7.3).

We provide the integration of Eq. (7.11) as

$$I_T = \frac{2eW_{eff}\rho_t}{h}\left[\int_{-\frac{t_S}{2}}^{\frac{t_S}{2}} dy \int_0^{E_{FS}-E_x} dE_t \int_{E_{FS}-eV_D+\delta}^{E_{FS}} dE_x T_t(y,k_x,V_G,V_D)\right]$$

$$+ \frac{2eW_{eff}\rho_t}{h}\left[\int_{-\frac{t_S}{2}}^{\frac{t_S}{2}} dy \int_{E_{FS}-E_x-eV_D+\delta}^{E_{FS}-E_x} dE_t \int_0^{E_{FS}-E_x} dE_x T_t(y,k_x,V_G,V_D)\right] \tag{7.12}$$

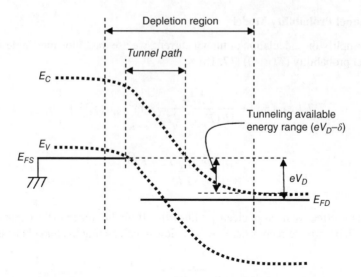

Figure 7.3 Model of energy band diagram at the "ON" state of lateral TFET. Parameters used in the theoretical model are shown. Reproduced by permission of the Institute of Electronics, Information, and Communication Engineers (2013). Y. Omura, D. Sato, S. Sato, and A.Mallik, "Theoretical Modeling of Double-Gate Lateral Tunnel FET," IEICE Technical Report, SDM2013-109, pp. 55–60. Copyright © 2013 IEICE.

where ρ_t denotes the density of states of the two-dimensional inversion layer electrons, E_{FS} denotes the Fermi level of the source region, and δ denotes the difference between the conduction band bottom and the Fermi level of the body region (see Figure 7.3). For the purpose of separation of the integration, Eq. (7.12) is rewritten as

$$I_T = \frac{2eW_{eff}\rho_t}{h}\left[\int_{-\frac{t_S}{2}}^{\frac{t_S}{2}} dy\left(I_1(y)-I_2(y)+I_3(y)\right)\right] \tag{7.13}$$

where

$$I_1(y)=E_{FS}\int_{E_{FS}-eV_D+\delta}^{E_{FS}}T_t(y,E_x,V_G,V_D)dE_x \tag{7.14}$$

$$I_2(y)=\int_{E_{FS}-eV_D+\delta}^{E_{FS}}E_xT_t(y,E_x,V_G,V_D)dE_x \tag{7.15}$$

$$I_3(y)=(eV_D-\delta)\int_0^{E_{FS}-eV_D+\delta}T_t(y,E_x,V_G,V_D)dE_x \tag{7.16}$$

Details of the derivation of Eq. (7.13) are given in section 7.4. After finalizing the integration, we have

$$I_T = \frac{16\eta e W_{eff} \rho_t t_S}{\hbar E_G} f_{TFET}(V_D, V_G) \tag{7.17}$$

$$f_{TFET}(V_D, V_G) = -\frac{2}{B^3} \exp(-E_G B) + (eV_D - \delta)\frac{1}{B^2}\exp(-E_G B)$$
$$+ \left[\frac{1}{B^2}(eV_D - \delta) + \frac{2}{B^3}\right]\exp\left[-(E_G + eV_D - \delta)B\right] \tag{7.18}$$

where

$$B = \frac{4K(V_G)}{\hbar t_S}\sqrt{\frac{m_x^*}{E_G}} \tag{7.19}$$

$$K(V_G) = \frac{t_S^2 E_G}{8e\left\{\dfrac{eN_A t_S^2}{8\varepsilon_S} + \dfrac{eN_S(V_G)}{C_{ox}}\right\}} \tag{7.20}$$

- *Subthreshold condition:*

$$N_S(V_G) = \frac{n_i^2}{N_A} \frac{k_B T}{e\left(\dfrac{eN_A t_S}{4\varepsilon_S} + \dfrac{k_B T}{eL_D}\right)} \exp\left\{\frac{e\varphi_S}{k_B T}\right\} \tag{7.21}$$

$$\sqrt{\varphi_S} = \frac{-\left[\dfrac{\varepsilon_S e N_A}{2C_{OX}^2}\right]^{1/2} + \sqrt{\dfrac{\varepsilon_S e N_A}{2C_{OX}^2} + 4(V_G - V_{FB})}}{2} \tag{7.22}$$

$$\delta = \frac{E_G}{2} + k_B T \ln\left[\frac{N_A}{n_i}\right] - e\varphi_S \tag{7.23}$$

where n_i denotes the intrinsic carrier density and V_{FB} is the flat-band voltage.

- *Inversion condition:*

$$N_S(V_G) = \frac{C_{ox}}{e}(V_G - V_{TH}) + N_A L_D \tag{7.24}$$

$$V_{TH} = V_{FB} + 2\varphi_F + \frac{eN_A t_S}{C_{ox}} \tag{7.25}$$

$$\delta = \frac{E_G}{2} - kT \ln\left[\frac{N_S(V_G)}{n_i L_D}\right] \tag{7.26}$$

7.2.2.4 Drain Current Saturation Condition

Tunneling electrons are supplied in the source region and electrons flow along the surface of the body region as inversion-layer electrons. Finally, they reach the drain junction. The behavior of electrons near the drain junction mirrors that seen in the MOSFET. Therefore, we can assume that the saturation condition of the TFET is just the same as that of the MOSFET for simplicity. It may be anticipated that such a simplified model would not be available for the n-type body [20]. Applying the gradual-channel approximation, we find that the drain voltage at saturation (V_{Dsat}) is given by V_G-V_{TH}, where V_{TH} is the threshold voltage for MOSFET operation.

Therefore, we have the following expression for TFET drain current (I_{TSAT}) at saturation:

$$I_{TSAT} = \frac{16\eta e W_{eff} P_t t_S}{\hbar E_G} f_{TFET-SAT}(V_D, V_G) \tag{7.27}$$

$$f_{TFET-SAT}(V_D, V_G) = -\left[\frac{2}{B^3} + (eV_{Dsat} - \delta_{Sat})\frac{1}{B^2}\right]\exp(-E_G B)$$
$$+ \left[\frac{1}{B^2}(eV_{Dsat} - \delta_{sat}) + \frac{2}{B^3}\right]\exp\left[-(E_G + eV_{Dsat} - \delta_{sat})B\right] \tag{7.28}$$

$$\delta_{sat} = \frac{E_G}{2} - kT \ln\left[\frac{N_{S2}(V_G)}{n_i L_D}\right] \tag{7.29}$$

$$N_{S2}(V_G) = \frac{C_{ox}}{2e}(V_G - V_{TH}) + N_A L_D \tag{7.30}$$

7.2.2.5 Enhancement of the Model

In this subsection, we assume, for simplicity, that the effective tunnel resistance is larger than the MOSFET channel resistance. In scaled lateral TFETs, however, it is expected that the above assumption will fail if the lateral TFET doesn't suffer from short-channel effects. In that case, we can change the above model by replacing drain voltage V_D with the following new notation V_{DM}.

$$V_{DM}(V_G, V_D) = V_D - I_{T,R} R_{ch}(V_G, V_D) \tag{7.31}$$

where R_{ch} denotes the inversion channel resistance and $I_{T,R}$ is the effective tunnel current. Using Eq. (7.31), we have the following set of equations for the drain current in the ON state:

- *Linear region:*

$$I_{T,R} = \frac{16\eta e W_{eff} P_t t_S}{\hbar E_G} f_{TFET,R}(V_{DM}, V_G) \tag{7.32}$$

$$f_{TFET,R}\left(V_{DM},V_G\right) = -\frac{2}{B^3}\exp\left(-E_G B\right) + \left(eV_{DM} - \delta\right)\frac{1}{B^2}\exp\left(-E_G B\right)$$
$$+ \left[\frac{1}{B^2}\left(eV_{DM} - \delta\right) + \frac{2}{B^3}\right]\exp\left[-\left(E_G + eV_{DM} - \delta\right)B\right] \tag{7.33}$$

$$R_{ch}\left(V_G,V_D\right) = \frac{L_{eff}}{W_{eff}\mu_n C_{ox}\left(V_G - V_{TH} - V_D/2\right)} \tag{7.34}$$

where μ_n is electron mobility and L_{eff} is channel length.

- *Saturation region:*

$$I_{TSAT,R} = \frac{16\eta e W_{eff}\rho_t t_S}{\hbar E_G} f_{TFET-SAT,R}\left(V_{DMsat},V_G\right) \tag{7.35}$$

$$f_{TFET-SAT,R}\left(V_{DMsat},V_G\right) = -\left[\frac{2}{B^3} + \left(eV_{DMsat} - \delta_{Sat}\right)\frac{1}{B^2}\right]\exp\left(-E_G B\right)$$
$$+ \left[\frac{1}{B^2}\left(eV_{DMsat} - \delta_{sat}\right) + \frac{2}{B^3}\right]\exp\left[-\left(E_G + eV_{DMsat} - \delta_{sat}\right)B\right] \tag{7.36}$$

$$V_{DMsat}\left(V_G,V_D\right) = V_D - I_{TSAT,R}R_{chsat}\left(V_G,V_D\right) \tag{7.37}$$

$$R_{chsat}\left(V_G,V_D\right) = \frac{2L_{eff}}{W_{eff}\mu_n C_{ox}\left(V_G - V_{TH}\right)} \tag{7.38}$$

where $R_{ch,sat}$ is the effective channel resistance at drain current saturation. It is anticipated that the channel resistance will reduce the drain current somewhat but its impact on drivability is not significant.

7.2.3   *Numerical Calculation Results and Discussion*

7.2.3.1   **Drain Current *versus* Gate Voltage**

Assuming the material is Si, we calculate the drain current (I_D) as a function of the gate voltage where $I_D = I_T$, I_{TSAT}. Calculated I_D-V_G characteristics are shown in Figure 7.4, where the device simulation results by technology computer-aided design (TCAD) [16] are also shown for comparison. In the calculations based on the model, we assumed Eqs. (7.17)–(7.26); the impact of channel resistance is discarded for simplicity. It is assumed in TCAD that $V_D = 1\,V$, effective gate oxide thickness (EOT) = 2 nm, $W_{eff} = 1\,\mu m$, and $L_{eff} = 40\,nm$. For Figure 7.4, we took the value of 10^{-3} for η.

The model basically reproduces the I_D-V_G characteristics obtained with TCAD except at low gate voltages. Thinning the SOI layer promises a great drivability advantage, as is well known

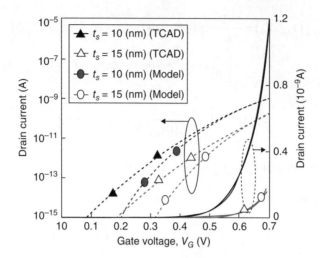

Figure 7.4 Calculated I_D vs. V_G characteristics for various silicon layer thicknesses. The model is compared with TCAD results. We assume $V_D = 1.0$ V and EOX $= 2$ nm. Reproduced by permission of the Institute of Electronics, Information, and Communication Engineers (2013). Y. Omura, D. Sato, S. Sato, and A.Mallik, "Theoretical Modeling of Double-Gate Lateral Tunnel FET," IEICE Technical Report, SDM2013-109, pp. 55–60. Copyright © 2013 IEICE.

[21, 22]. It is anticipated that the model's underestimation of the drain current is due to its failure to consider trap-assisted tunneling (TAT) current.

Another calculated plot of I_D-V_G characteristics with the EOT parameter is shown in Figure 7.5, where EOT ranges from 2.0 to 0.5 nm. Other parameters are the same as those in Figure 7.4. I_D-V_G characteristics and drivability are not sensitive to EOT.

7.2.3.2 Subthreshold Swing

Subthreshold swing values are extracted from Figures 7.4 and 7.5 and plotted in Figure 7.6 and Figure 7.7, respectively. As expected, TCAD shows larger swing values than the model. The swing values in the sub-60 mV/dec regime are not easily realized in the lateral TFET. Technology computer-aided design also suggests that the drain current calculation should consider both the TAT current and the junction leakage current. Figure 7.7 also suggests to us that EOT does not play an important role in improving drivability.

7.2.3.3 Drain Current *versus* Drain Voltage

Using Eqs. (7.27)–(7.30), we calculated the I_D-V_D characteristics; the results are plotted in Figure 7.8. As expected from Figure 7.4, the drain current is a logarithmic function of the gate voltage. Hence the vertical axis takes a logarithmic scale. Although the TCAD data is not plotted in Figure 7.8, we think that the model should basically predict the drain current behavior because Figure 7.4 reproduces the I_D-V_D characteristics of TCAD.

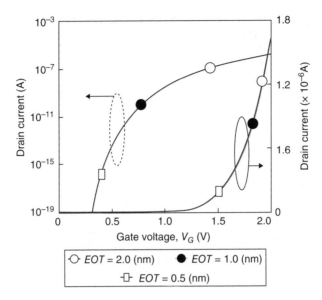

Figure 7.5 Calculated I_D vs. V_G characteristics for various EOT values. We assume $V_D = 1.0V$ and $t_s = 10$ nm. Reproduced by permission of the Institute of Electronics, Information, and Communication Engineers (2013). Y. Omura, D. Sato, S. Sato, and A. Mallik, "Theoretical Modeling of Double-Gate Lateral Tunnel FET," IEICE Technical Report, SDM2013-109, pp. 55–60. Copyright © 2013 IEICE.

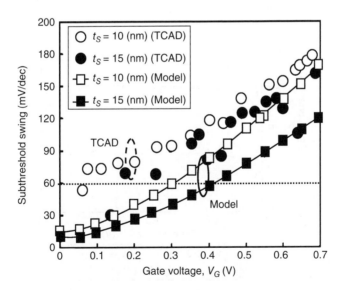

Figure 7.6 Calculated subthreshold swing versus V_G characteristics for various silicon layer thicknesses. The model is compared with the TCAD results. We assume $V_D = 1.0V$ and EOX = 2 nm. Reproduced by permission of the Institute of Electronics, Information, and Communication Engineers (2013). Y. Omura, D. Sato, S. Sato, and A. Mallik, "Theoretical Modeling of Double-Gate Lateral Tunnel FET," IEICE Technical Report, SDM2013-109, pp. 55–60. Copyright © 2013 IEICE.

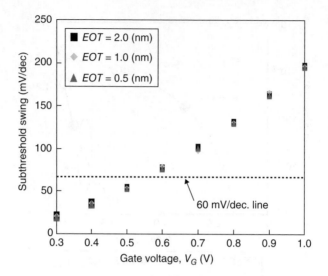

Figure 7.7 Calculated subthreshold swing vs. V_G characteristics for various EOT values. We assume $V_D = 1.0\,$V and $t_s = 10\,$nm. Reproduced by permission of the Institute of Electronics, Information, and Communication Engineers (2013). Y. Omura, D. Sato, S. Sato, and A. Mallik, "Theoretical Modeling of Double-Gate Lateral Tunnel FET," IEICE Technical Report, SDM2013-109, pp. 55–60, Copyright © 2013 IEICE.

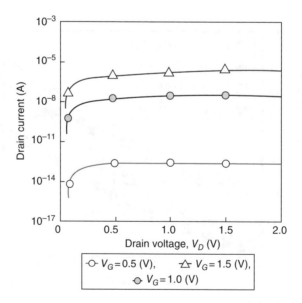

Figure 7.8 Calculated I_D vs. V_D characteristics for various gate voltages. We assume $t_s = 10\,$nm and EOX = 2 nm. Reproduced by permission of the Institute of Electronics, Information, and Communication Engineers (2013). Y. Omura, D. Sato, S. Sato, and A. Mallik, "Theoretical Modeling of Double-Gate Lateral Tunnel FET," IEICE Technical Report, SDM2013-109, pp. 55–60. Copyright © 2013 IEICE.

7.2.4　Summary

We derived a closed and analytical model to predict drain current characteristics based on quantum mechanics. Although the model underestimates the tunnel current, it reproduces important aspects well, such as I_D-V_G characteristics and subthreshold swing of the lateral TFET by introducing a fitting parameter. We also proposed a lateral TFET model for operation in the saturation region. These models are useful for consideration of aspects of the lateral TFET that have been studied recently.

7.3　Model of Vertical Tunnel FET and Aspects of its Characteristics

7.3.1　Introduction

Integrated circuits dissipate large amounts of energy because their primary components, MOSFETs, consume large amounts of energy in both the "ON" state and the "OFF" state [1]. As altering the threshold voltage of MOSFETs does not substantially resolve this issue, it is expected that subthreshold control approaches, such as steep swing, will provide useful solutions. One such technology, the planar lateral tunnel field-effect transistor (LTFET) was proposed some time ago [2–9].

Tunnel field-effect transistors use quantum mechanics to establish tunneling and so promise the important benefit of a steep swing [2–9]. However, the steep swing expected has yet to be realized because tunneling is sensitive to material quality and device geometry, optimization of which remains unresolved at the laboratory level [10]. Moreover, the low drivability imposed by the tunnel phenomena hinders their application to commercial ICs.

In order to overcome such issues, vertical tunnel field-effect transistors (VTFETs) are now attracting attention; they offer high drivability because the tunnel area is easily enlarged [23–26]. Various device simulations are, and will be, performed to advance the performance and to optimize the device architecture. However, we also need physics-based models because we must consider what the substantial issues are that limit the performance of TFETs; design guidelines are necessary to improve performance. Although some analytical LTFET models have already been proposed [11–15], they assume very limited conditions, and this degrades their usefulness. Very few models have been proposed for VTFETs.

In this section, we propose a physics-based model based on a theoretical consideration of VTFET operation. The proposed model is examined with the aid of TCAD simulations [27].

7.3.2　Device Structure and Model Concept

7.3.2.1　Device Structure Assumed

A schematic of the n-channel VTFET assumed here is shown in Figure 7.9a. Device parameters are summarized in Table 7.2. The thickness of the top n-type layer is denoted by t_n, the doping concentration of the n-type layer by N_D, the gate oxide layer thickness by t_{ox}, and the doping concentration of the p-type substrate by N_A ($\sim 10^{19}\,\text{cm}^{-3}$).

In this chapter, we assume that the substrate material is monocrystalline Si or Ge given the goal of high drivability. Although the doping level of the substrate is high, we take Boltzmann's approximation because our discussion is limited to room temperature.

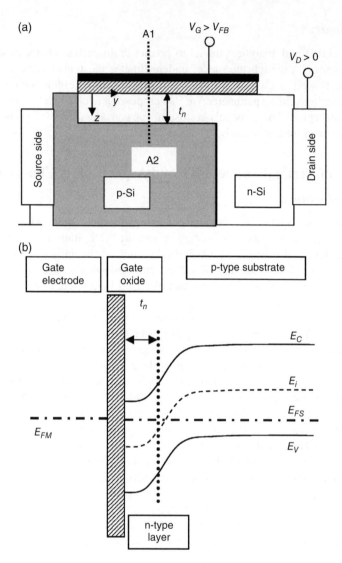

Figure 7.9 Schematics of vertical tunnel FET and its energy band diagram. (a) Device structure and (b) energy band diagram along the A1-to-A2 cut line in the flat-band condition. Reproduced by permission of the Institute of Electronics, Information, and Communication Engineers (2014). Y. Omura, S. Sato, and A. Mallik, "Physics-Based Analytical Model for Gate-on-Germanium Source (GoGeS) TFET," IEICE Technical Report, SDM2014-97, pp. 7–12. Copyright © 2014 IEICE.

In this chapter, we derive a comprehensive device model that addresses four key characteristics: the impact of n-type layer thickness (t_n) on tunnel current; the impact of EOT on drivability; the impact of the doping concentration of the n-type layer and the p-type substrate on tunnel attributes, and the impact of semiconductor material on drivability. We will make the theoretical expression of the tunnel current as simple as possible to provide a clear perspective on TFET performance.

Table 7.2 Device parameters for Si and Ge.

Parameters	Values	Units
Effective gate oxide thickness (EOT)	0.4–2.0	nm
Thickness of n-Si layer (t_n)	1.0–4.0	nm
Doping concentration of n-Si layer (N_D)	10^{16}–10^{18}	cm^{-3}
Channel width (W_{eff})	1.0	μm
Channel length (L_{eff})	30	nm
Source doping concentration (N_A)	10^{19}	cm^{-3}

Reproduced by permission of the Institute of Electronics, Information, and Communication Engineers (2014). Y. Omura, S. Sato, and A. Mallik, "Physics-Based Analytical Model for Gate-on-Germanium Source (GoGeS) TFET," IEICE Technical Report, SDM2014-97, pp. 7–12. Copyright © 2014 IEICE.

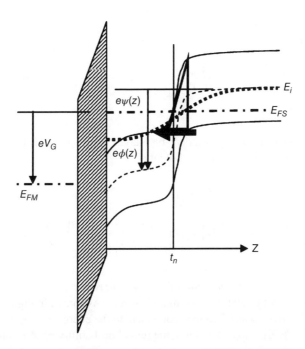

Figure 7.10 Expected energy band diagram of vertical tunnel FET in the "ON" state. The red triangle suggests the energy barrier for the tunnel electrons. Electrostatic potential is defined from the base line of E_i of the p-type substrate. The bold dotted line suggests the E_i level in the flat-band condition. Reproduced by permission of the Institute of Electronics, Information, and Communication Engineers (2014). Y. Omura, S. Sato, and A. Mallik, "Physics-Based Analytical Model for Gate-on-Germanium Source (GoGeS) TFET," IEICE Technical Report, SDM2014-97, pp. 7–12. Copyright © 2014 IEICE.

7.3.2.1.1 *Potential Model and Electron Density Function*

When the VTFET uses a positive gate bias to drive the tunnel current, we can assume the energy band diagram shown in Figure 7.10.

Device simulations demonstrate that the electrostatic potential of the semiconductor near the top surface is a roughly linear function of the distance measured from the top surface. Thus we can assume the following expression for the electrostatic potential near the top surface:

$$\varphi(z,y) = \varphi_{S0}\left(1 - 4z/3t_n\right) + \left(y/L_{eff}\right)V_D \tag{7.39}$$

where ϕ_{s0} denotes the surface potential, V_D denotes the drain voltage, the positive z-axis parallels the depth direction, and the positive y-axis parallels the gate oxide/n-type layer interface. Equation (7.39) yields the following expression for the electron volume density:

$$n(z,y) = N_D \exp\left[\frac{e\varphi_{S0}\left(1 - 4z/3t_n\right) - e\left(y/L\right)V_D}{k_BT}\right] \tag{7.40}$$

Using Eq. (7.40), we can calculate the surface electric field as

$$\left.\frac{d\varphi(z,y)}{dz}\right|_{z=0} \cong -\frac{e}{\varepsilon_S}\int_0^{t_n}\left[n(z,y) - N_D\right]dz$$

$$= -\left(\frac{3t_n k_B T N_D}{4\varepsilon_S\varphi_{S0}}\right)\exp\left(\frac{e\varphi_{S0}}{k_BT}\right) \times \left[1 - \exp\left(-\frac{4e\varphi_{S0}}{3k_BT}\right)\right]\exp\left(-\frac{e\left(y/L\right)V_D}{k_BT}\right) + \frac{eN_Dt_n}{\varepsilon_S} \tag{7.41}$$

The boundary condition at the gate oxide/n-type layer interface yields the following relation between the gate voltage and the surface potential.

$$\frac{C_{ox}}{e}\left(V_G - V_{FB} - \varphi_{S0} - \left(y/L_{eff}\right)V_D\right)$$

$$= \left(\frac{3t_n k_B T N_D}{4e\varphi_{S0}}\right)\left[\exp\left(\frac{e\varphi_{S0}}{k_BT}\right) - \exp\left(-\frac{e\varphi_{S0}}{3k_BT}\right)\right]\exp\left(-\frac{e\left(y/L_{eff}\right)V_D}{k_BT}\right) - N_Dt_n \tag{7.42}$$

7.3.2.1.2 Derivation of the Expression of Tunnel Current

Here, we calculate tunnel probability and tunnel current. As shown in Figure 7.10, we assume a triangular tunnel barrier whose height is equivalent to the energy gap (E_G). With the Wentzel, Kramers, Brillouin (WKB) approximation, the tunnel probability of electrons is given as

$$T_t[E] = T_{t0}\exp\left[-2\int_0^{\left(1 - \frac{E}{E_G}\right)\Delta} k(z)dz\right] \tag{7.43}$$

where Δ is the maximal tunnel path length; we assume $\Delta \sim t_n$ for simplicity in the following calculation.

Thus the tunnel current can be expressed as

$$I_T \cong eWL_{eff}\int_0^{t_n} dz\left\{n(z) < T_t[E]\cdot\left(\frac{1}{\hbar}\frac{\partial E}{\partial k}\right)/\Delta >_E\right\} \tag{7.44}$$

where $<g(E)>_E$ means the energy integral of $g(E)$. When we use Eq. (7.43) as the tunnel probability, we have

$$
<T_t[E]\cdot\left(\frac{1}{\hbar}\frac{\partial E}{\partial k}\right)/\Delta>_E \cong T_t[k_BT]\cdot\frac{1}{\Delta}\sqrt{\frac{2(e\varphi_{S0}-\delta)}{m^*}}
$$

$$
=T_{t0}\cdot\frac{1}{\Delta}\sqrt{\frac{2(e\varphi_{S0}-\delta)}{m^*}}\exp\left[-\frac{4(E_G-e\varphi_{S0}+\delta)^{3/2}\Delta\sqrt{2m^*}}{3E_G\hbar}\right]
$$

(7.45)

where m* is the effective mass of tunnel electrons, and parameter δ is given by

$$
\delta=\frac{E_G}{2}-k_BT\ln\left(\frac{N_D}{n_i}\right)
$$

(7.46)

δ is the offset energy between the Fermi level of the n-type layer and the conduction band bottom of the n-type layer. This yields the following expression for tunnel current:

$$
I_T(y)dy\cong dy\cdot eWT_{t0}\cdot\frac{1}{\Delta}\sqrt{\frac{2\left(e\varphi_{S0}+e(y/L_{eff})V_D-\delta\right)}{m^*}}
$$

$$
\times\exp\left[-\frac{4\left(E_G-e\varphi_{S0}-e(y/L_{eff})V_D+\delta\right)^{3/2}\Delta\sqrt{2m^*}}{3E_G\hbar}\right]\int_0^{t_n}n(z)dz
$$

(7.47)

As the electron density in the n-type layer is given by Eq. (7.40), the final expression of the tunnel current is given by

$$
I_T(y)\cong eWT_{t0}\cdot\left(\frac{3k_BTN_Dt_n}{4e\varphi_{S0}\Delta}\right)\sqrt{\frac{2\left(e\varphi_{S0}+e(y/L_{eff})V_D-\delta\right)}{m^*}}
$$

$$
\times\exp\left[-\frac{4\left(E_G-e\varphi_{S0}-e(y/L_{eff})V_D+\delta\right)^{3/2}\Delta\sqrt{2m^*}}{3E_G\hbar}+\frac{e\varphi_{S0}}{k_BT}\right]
$$

$$
\times\left\{1-\exp\left(-\frac{4e\varphi_{S0}}{3k_BT}\right)\right\}\exp\left(-\frac{e(y/L_{eff})V_D}{k_BT}\right)
$$

(7.48)

7.3.3   Comparing Model Results with TCAD Results

7.3.3.1.1   Calculation Results by the Model

In this section, we introduce some current versus voltage characteristics of a VTFET with Si substrate, and discuss important aspects of VTFET operation. In the calculations, we assume

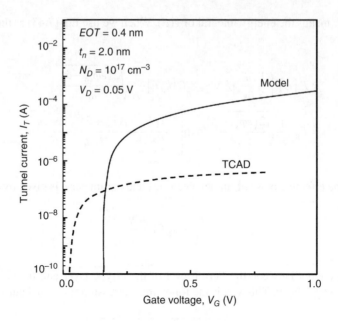

Figure 7.11 Calculated current versus voltage characteristics of vertical tunnel FET. $W_{eff}=1\,\mu m$, $L_{eff}=30\,nm$. Reproduced by permission of the Institute of Electronics, Information, and Communication Engineers (2014). Y. Omura, S. Sato, and A. Mallik, "Physics-Based Analytical Model for Gate-on-Germanium Source (GoGeS) TFET," IEICE Technical Report, SDM2014-97, pp. 7–12. Copyright © 2014 IEICE.

$T_{t0}=1$ and $\Delta=t_{n}$. The drain current of the VTFET is obtained by numerically integrating Eq. (7.48) from the source to the drain.

In Figure 7.11, the solid line plots the results of the model, and the dotted line plots the TCAD results. The model assumes that the work function of the gate electrode is 4.1 eV, whereas TCAD assumes it is 3.7 eV. The model overestimates the tunnel current by three orders. One reason for the discrepancy stems from the rough assumptions of $T_{t0}=1$ and $\Delta=t_{n}$. According to recent studies [28, 29], the tunnel path length of electrons is 10–15 nm, and is dependent on the substrate doping concentration. Therefore, we must carefully estimate the value of Δ in the model.

We examine how the value of Δ modulates the tunnel current level using the model. Calculation results are shown in Figure 7.12, where the work function of the gate material is 4.1 eV. We can see that the effective tunnel path length is about 3 nm; this value is larger than the t_{n} value assumed initially.

Figure 7.12 suggests that the tunnel path length assumed in Figure 7.10 is too short. The tunnel path length of electrons is considered to be larger than t_{n} as shown in Figure 7.13; it is anticipated that many electrons present at positions deep within the substrate tunnel to the top surface of the n-type layer.

In addition, it is seen that TCAD yields a smaller increase rate in tunnel current than the model. This point is discussed in the next section.

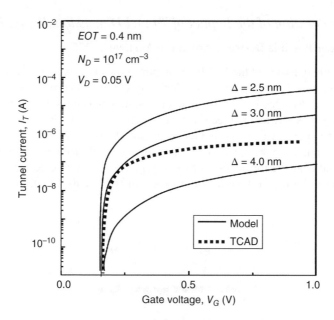

Figure 7.12 Calculated tunnel current characteristics for various Δ values. The work function of the gate material is 4.1 eV. Reproduced by permission of the Institute of Electronics, Information, and Communication Engineers (2014). Y. Omura, S. Sato, and A. Mallik, "Physics-Based Analytical Model for Gate-on-Germanium Source (GoGeS) TFET," IEICE Technical Report, SDM2014-97, pp. 7–12. Copyright©2014 IEICE.

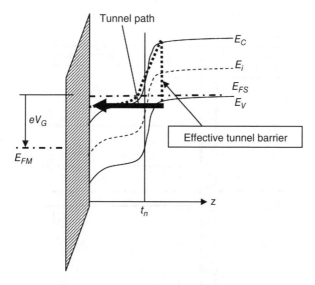

Figure 7.13 Possible tunnel path length image. It is expected that electrons existing at deep positions tunnel to the surface of the top n-type layer. Reproduced by permission of the Institute of Electronics, Information, and Communication Engineers (2014). Y. Omura, S. Sato, and A. Mallik, "Physics-Based Analytical Model for Gate-on-Germanium Source (GoGeS) TFET," IEICE Technical Report, SDM2014-97, pp. 7–12. Copyright © 2014 IEICE.

7.3.4   Consideration of the Impact of Tunnel Dimensionality on Drivability

7.3.4.1   Another Possible Device Geometry of Vertical TFET

This chapter initially assumed the VTFET structure shown in Figure 7.9a without qualification. Here we re-examine this point. For calculating the tunnel current, the model assumed Eqs. (7.42) and (7.48). This is valid because the top thin n-type layer acts with the electron accumulation layer as a virtually three-dimensional carrier system [30].

Some recent papers assume the device structure shown in Figure 7.14a [31]. In that case, electron tunneling takes place from the valence band of the p-type substrate to the conduction band near the top surface of the p-type substrate; an electron inversion layer is generated near the top surface of the p-type substrate. This situation is illustrated in Figure 7.15.

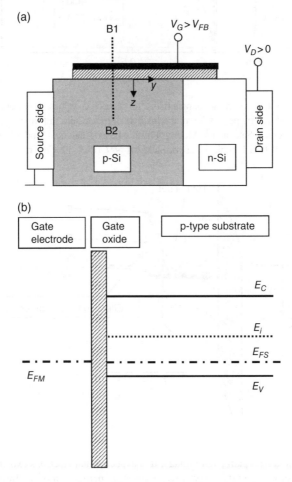

Figure 7.14   Schematic of vertical tunnel FET without a surface n-type layer and its energy band diagram. (a) Schematic of device structure and (b) energy band diagram along with the B1-B2 cut line at the flat-band condition

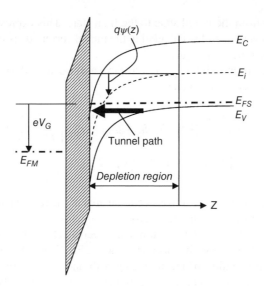

Figure 7.15   Schematic energy band diagram of vertical tunnel FET without a surface n-type layer, and the B1-B2 cut line, in the ON state.

As the surface inversion layer has quasi-two-dimensional (Quasi-2D) attributes, the tunnel property of this system differs from that shown in Figure 7.13. In the following, we consider the fundamental aspects of the tunnel current of the devices shown in Figures 7.9 and 7.14.

7.3.4.1.1   Theoretical Review of Tunnel Current

The probability (P) governing the transition between the two states of the two three-dimensional (3D) carrier systems on each side of the tunneling barrier is given by Fermi's golden rule [32]:

$$P = \left(\frac{2\pi}{\hbar}\right) \sum_{k_i, k_f} \left| M_{i \to f}^{3D-3D} \right|^2 \left\{ f(E_i) - f(E_f) \right\} \delta(E_f - E_i - eV_A) \tag{7.49}$$

where subscripts i and f mean *initial* and *final,* and the δ function represents conservation of the total energy assuming that the applied bias voltage V_A partially drops across the barrier; k_i and k_f denote wave numbers of the initial state and the final state, respectively.

Taking momentum conservation on the barrier surface along the x-axis and the y-axis based on the coordinate system shown in Figure 7.14a, we have the following transition matrix element [33]:

$$M_{i \to f}^{3D-3D} = \left(-\frac{\hbar^2}{2m^*}\right) \left\{ \psi_f^* \frac{\partial \psi_i}{\partial z} - \psi_i \frac{\partial \psi_f^*}{\partial z} \right\} \delta(k_{f,\parallel} - k_{i,\parallel}) \tag{7.50}$$

where $k_{i,\parallel}$ and $k_{f,\parallel}$ are the wave vectors on the energy barrier surface for the initial state and for the final state, respectively; ψ_i and ψ_f are the wave functions of electrons in the initial state (valence band) and the final state (conduction band), respectively, and m^* is the effective mass of tunnel electrons. This is the transition matrix element for unconfined electrons that alter their electronic states from the initial state to the final state. This expression of the transition matrix element is also very useful in calculating the tunnel current density in low-dimensional electron systems, as will be shown later.

When the material has approximately an isotropic band structure, the tunnel current ($I_{T(3D-3D)}$) between two 3D carrier systems is given as

$$
\begin{aligned}
I_{T(3D-3D)} &= eS\left(\frac{2\pi}{\hbar}\right)\int dk_{ix}\,dk_{iy}\,dk_{iz}\int dk_{fx}\,dk_{fy}\,dk_{fz}\left(\frac{\partial E_z}{\hbar\partial k_{fz}}\right)\left|M_{i\to f}^{3D-3D}\right|^2 \\
&\times\left\{f(E_i)-f(E_f)\right\}\delta(E_f-E_i-eV_A)
\end{aligned}
\tag{7.51}
$$

where q is electronic charge, S is the cross-sectional area of the tunnel path, and E_z is the component of electron energy along with the z-axis.

Next, we turn to calculating the tunnel current between a 3D carrier system and a 2D carrier system. To assess tunneling in low-dimensional systems, Eq. (7.50) must be carefully reconsidered. A schematic configuration of the tunneling path and the electron system is shown in Figure 7.16, where the carriers around the silicon surface are assumed to be physically confined on the z axis and carrier transport is parallel to the y axis as shown in Figures 7.14 and Figure 7.15. As suggested in Figure 7.16, we can expect many transition paths between the sub-band of the surface inversion layer (conduction band) and the valence band of the p-type substrate, depending on the Fermi level. Since the Fermi level modulates the tunnel spectra, this configuration is useful when characterizing the band structure of semiconductors.

The probability (P) of a transition between the two states of the two systems on each side of the tunneling barrier is given by Fermi's golden rule [32]:

$$
P=\left(\frac{2\pi}{\hbar}\right)\sum_n\sum_{k_i,k_f}\left|M_{i\to f}^{3D-2D}\right|^2\left\{f(E_i)-f(E_f)\right\}\delta(E_f-E_i-eV_A)
\tag{7.52}
$$

where n denotes the sub-band index attributed to the final states.

In the present case, shown in Figure 7.16, when the channel current along the y-axis is not so large, the matrix element $M_{i\to f}^{3D-2D}$ can be approximately written in terms of the current density operator given by Eq. (7.50) [33] such that

$$
M_{i\to f}^{3D-2D}=\left(-\frac{\hbar^2}{2m^*}\right)\left\{\psi_f^*\frac{\partial\psi_i}{\partial z}-\psi_i\frac{\partial\psi_f^*}{\partial z}\right\}\delta\left(k_{f,x}-k_{i,x}\right)\delta\left(k_{f,y}-k_{i,y}\right)
\tag{7.53}
$$

This is the same form as Eq. (7.50). When the channel current is large, we can't assume momentum conservation ($\delta(k_{fy}-k_{iy})$) on the y-axis.

When the material has approximately an isotropic band structure, the tunnel current ($I_{T(3D-2D)}$) between the two carrier systems is given as

$$
\begin{aligned}
I_{T(3D-2D)} &= eS\left(\frac{2\pi}{\hbar}\right)\sum_n\int dk_{ix}\,dk_{iy}\,dk_{iz}\int dk_{fx}\,dk_{fy}\left(\frac{\partial E_{z,n}}{\hbar\partial k_{z,n}}\right)\left|M_{i\to f}^{3D-2D}\right|^2 \\
&\times\left\{f(E_i)-f(E_f)\right\}\delta(E_f-E_i-eV_A)
\end{aligned}
\tag{7.54}
$$

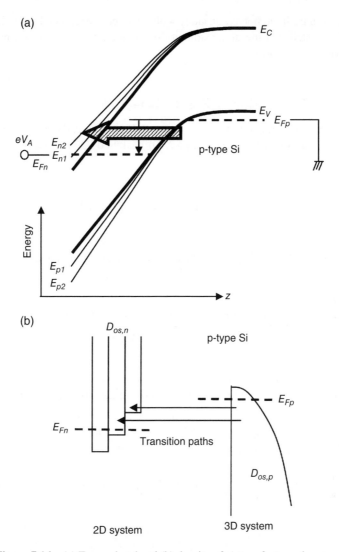

Figure 7.16   (a) Energy band and (b) density of states of a tunnel system.

Although Eqs. (7.51) and (7.54) are not finalized, we can predict that $I_{T(3D\text{-}2D)}$ would be smaller than $I_{T(3D\text{-}3D)}$ due to the lower density of states. This suggests that the performance of the VTFET shown in Figure 7.9 is superior to that of Figure 7.14.

7.3.5   Summary

In this chapter, we proposed a possible device model for the VTFET based on the WKB-based tunnel probability and the tunnel current theory. The model reproduced the fundamental characteristics of the VTFET, but didn't quantitatively reproduce the current vs. voltage characteristics because it did not use an analytically derived tunnel path length. In addition, we theoretically suggested that the dimensionality of tunnel carriers plays an important role in predicting drivability performance.

7.4   Appendix Integration of Eqs. (7.14)–(7.16)

Here we explain how the integrations are performed. First, we need to prepare for the calculation. As is suggested in section 7.2.2.2, Eq. (7.10) is altered. On assuming $E_x \ll E_G$, we have approximately

$$T_t\left(y, E_x, V_G, V_D\right) \cong \frac{8E_x}{E_G}\exp\left[-2k_t a\left(y, V_G\right)\right] \tag{7.55}$$

where we assume $E_x > E_y$ and E_z because the drain bias is applied to the device. In addition, as we assume $E_x \ll E_G$, the wave number k_t is also approximated as

$$k_t = \left[\frac{2m_x^*\left(E_G/2 - E_x\right)}{\hbar^2}\right]^{1/2} \approx \sqrt{\frac{m_x^* E_G}{\hbar^2}}\left(1 - \frac{E_x}{E_G}\right) \tag{7.56}$$

1. *Calculation of Eq. (7.14)*

   When we use Eqs. (7.55) and (7.56), we can perform the first integration in Eq. (7.14) as

$$\int_{-\frac{t_S}{2}}^{\frac{t_S}{2}} dy I_1\left(y\right) = 2\left(\frac{8E_{FS}}{E_G}\right)\int_{E_{FS}-eV_D+\delta}^{E_{FS}} E_x dE_x \int_{2}^{\frac{t_S}{2}} dy \exp\left(-2k_t\frac{K\left(V_G\right)}{y}\right)$$

$$= \left(\frac{16E_{FS}}{E_G}\right)\int_{E_{FS}-eV_D+\delta}^{E_{FS}} E_x dE_x \left\{\left[-\frac{1}{z}\exp\left(-2k_t K\left(V_G\right)z\right)\right]_{2/t_S}^{\infty} - 2k_t K\left(V_G\right)I\left(-1\right)\right\} \tag{7.57}$$

where $I(-1)$ denotes the integral-exponential function. Function $K(V_G)$ is explained later. When we limit the maximum body thickness to 20 nm, we can approximately estimate the magnitude of $I(-1)$ as

$$I\left(-1\right) = \int_{2/t_S}^{\infty} dz \frac{1}{z}\exp\left(-2k_t K\left(V_G\right)z\right) \approx \left[\frac{1}{-2k_t K\left(V_G\right)z}\exp\left(-2k_t K\left(V_G\right)z\right)\right]_{2/t_S}^{\infty}$$

$$= \frac{t_S}{4k_t K\left(V_G\right)}\exp\left(-4k_t K\left(V_G\right)/t_S\right) \tag{7.58}$$

Since the aspect of the integral exponential function suggests that it is smaller than the value of Eq. (7.58), we discard the term in Eq. (7.57) and introduce a fitting parameter ($\eta < 1$) to allow comparison with the device simulation results. Thus we have

$$\int_{-\frac{t_S}{2}}^{\frac{t_S}{2}} dy I_1(y) = \eta \left(\frac{16E_{FS}}{E_G} \right) \int_{E_{FS}-eV_D+\delta}^{E_{FS}} E_x dE_x \left[\frac{t_S}{2} \exp\left(-4k_t K(V_G)/t_S\right) \right]$$

$$= \left(\frac{\eta t_S}{2} \right) \left(\frac{16E_{FS}}{E_G} \right) \int_{E_{FS}-qV_D+\delta}^{E_{FS}} E_x \exp\left[-4k_t \frac{K(V_G)}{t_S} \right] dE_x \tag{7.59}$$

When we use Eq. (7.56), we can finalize the integration of Eq. (7.59).
Now we go back to explain function $K(V_G)$. The function is a component of function $a(V_G)$ as follows

- *For partial body depletion:*

$$a(y,V_G) = \frac{K(V_G)}{\left\{ y - \left(\frac{t_S}{2} - t_d \right) \right\}} \tag{7.60}$$

$$K(V_G) = \frac{E_G}{4e\varphi_S} \left(\frac{t_S}{2} - t_d \right)^2 \tag{7.61}$$

- *For full body depletion and the subthreshold condition:*

$$a(y,V_G) = \frac{t_S^2 E_G}{8ey(\varphi_S - \varphi_0)} = \frac{K(V_G)}{y} \tag{7.62}$$

$$K(V_G) = \frac{t_S^2 E_G}{8e\left\{ \frac{eN_A t_S^2}{8\varepsilon_S} + \frac{eN_s(V_G)}{C_{ox}} \right\}} \tag{7.63}$$

$$N_S(V_G) = \frac{n_i^2}{N_A} \frac{k_B T}{eF_S(V_G)} \exp\left\{ \frac{e\varphi_S}{k_B T} \right\} \tag{7.64}$$

$$F_S(V_G) = \frac{2(\varphi_S - \varphi_o)}{t_S} + \frac{k_B T}{eL_{DE}} \tag{7.65}$$

$$\varphi_S - \varphi_0 = \frac{eN_A t_S^2}{8\varepsilon_S} \tag{7.66}$$

- *For full body depletion and the inversion condition:*

$$K(V_G) = \frac{t_S^2 E_G}{8e\left\{ \frac{eN_A t_S^2}{8\varepsilon_S} + \frac{eN_s(V_G)}{C_{ox}} \right\}} \tag{7.67}$$

$$N_S\left(V_G\right)=\frac{C_{ox}}{e}\left(V_G-V_{TH}\right)+N_AL_{DE} \tag{7.68}$$

for $t_s>L_D$.

$$N_S\left(V_G\right)=\frac{C_{ox}}{e}\left(V_G-V_{TH}\right)+N_At_s \tag{7.69}$$

for $t_s<L_D$. L_D denotes the Debye length.

2. *Calculation of I_2*
In a similar way, we have

$$\int_{-\frac{t_S}{2}}^{\frac{t_S}{2}}dyI_2\left(y\right)=2\int_0^{\frac{t_S}{2}}dy\int_{E_{FS}-eV_D+\delta}^{E_{FS}}E_x\frac{8E_x}{E_G}\exp\left(-2k_t\frac{K\left(V_G\right)}{y}\right)dE_x$$

$$\approx\left(\frac{16\eta}{E_G}\right)\int_{E_{FS}-eV_D+\delta}^{E_{FS}}E_x^2dE_x\times\left(\frac{t_S}{2}\right)\exp\left(-4k_t\frac{K\left(V_G\right)}{t_S}\right) \tag{7.70}$$

When we use Eq. (7.56), we can finalize the integration of Eq. (7.70).

3. *Calculation of I_3*
In a similar way, we have

$$\int_{-\frac{t_S}{2}}^{\frac{t_S}{2}}dyI_3\left(y\right)=2\int_0^{\frac{t_S}{2}}\left(eV_D-\delta\right)dy\int_0^{E_{FS}-eV_D+\delta}\frac{8E_x}{E_G}\exp\left(-2k_t\frac{K\left(V_G\right)}{y}\right)dE_x$$

$$\approx\frac{8\left(eV_D-\delta\right)\eta t_S}{E_G}\int_0^{E_{FS}-eV_D+\delta}E_x\exp\left(-4k_t\frac{K\left(V_G\right)}{t_S}\right)dE_x \tag{7.71}$$

By using Eq. (7.56) we can finalize the integration of Eq. (7.71).

References

[1] T. Masuhara, "Quest for low-voltage and low-power integrated circuits: towards a sustainable future," *IEEE Solid-State Circuits Mag.*, vol. 5, pp. 8–26, 2013.
[2] C. Aydin, A. Zaslavsky, S. Luryi, S. Cristoloveanu, D. Mariolle, D. Fraboulet, and S. Deleonibus, "Lateral inter-band tunneling transistor in silicon-on-insulator," *Appl. Phys. Lett.*, vol. 84, pp. 1780–1782, 2004.
[3] P.-F. Wang, K. Hilsenbeck, T. Nirschl, M. Oswald, C. Tepper, M. Weiss, D. Schmitt-Landsiedel, and W. Hansch, "Complementary tunneling transistor for low power applications," *Solid State Electron.*, vol. 48, pp. 2281–2286, 2004.
[4] K. K. Bhuwalka, J. Schulze, and I. Eisele, "Performance enhancement of vertical tunnel field-effect transistor with SiGe in the δp+ layer," *Jpn. J. Appl. Phys.*, vol. 43, pp. 4073–4078, 2004.
[5] J. Appenzeller, Y.-M. Lin, J. Knoch, and P. Avouris, "Band-to-band tunneling in carbon nanotube field-effect transistors," *Phys. Rev. Lett.* vol 93, pp 196805 1 196805 4, 2004.

[6] T. Nirschl, P.-F. Wang, C. Weber, J. Sedlmeir, R. Heinrich, R. Kakoschke, K. Schrufer, J. Holz, C. Pacha, T. Schulz, M. Ostermayr, A. Olbrich, G. Georgakos, E. Ruderer, W. Hansch, and D. Schmitt-Landsiedel, "The Tunneling Field-Effect Transistor (TFET) as an Add-on for Ultra-Low Voltage Analog and Digital Processes," IEEE IEDM (Dec., 2004) pp. 195–198, 2004.

[7] K. R. Kim, H. H. Kim, K.-W. Song, J. I. Huh, J. D. Lee, and B.-G. Park, "Field-induced interband tunneling effect transistor (FITET) with negative-differential trans- conductance and negative-differential conductance," IEEE Trans. Nanotechnol., vol. 4, pp. 317–321, 2005.

[8] Q. Zhang, W. Zhao, and A. Seabaugh, "Low- subthreshold-swing tunnel transistors," IEEE Electron Device Lett., vol. 27, pp. 297–300, 2006.

[9] F. Mayer, C. Le Royer, J.-F. Damlencourt, K. Romanjek, F. Andrieu, C. Tabone, B. Previtali, and S. Deleonibus, "Impact of SOI, Si1-xGexOI and GeOI Substrates on CMOS Compatible Tunnel FET Performance," Tech. Dig. IEEE IEDM(San Francisco, 2008) pp. 163–166, 2008.

[10] R. Jhaveri, V. Nagavarapu, and J. C. Woo, "Effect of pocket doping and annealing schemes on the source-pocket tunnel field-effect transistor," IEEE Trans. Electron Devices, vol. 1, pp. 80–86, 2011.

[11] C. Shen, S.-L. Ong, C.-H. Heng, G. Samudra, and Y.-C. Yeo, "A variational approach to the two-dimensional nonlinear Poisson's equation for the modeling of tunneling transistors," IEEE Electron Device Lett., vol. 29, pp. 1252–1255, 2008.

[12] M. G. Bardon, H. P. Neves, R. Puers, and C. Van Hoof, "Pseudo-two-dimensional model for double-gate tunnel FETs considering the junctions depletion regions," IEEE Trans. Electron Devices, vol. 57, pp. 827–834, 2010.

[13] J. L. Padilla, F. Gámiz, and A. Godoy, "Impact of quantum confinement on gate threshold voltage and sub-threshold swings in double-gate tunnel FETs," IEEE Trans. Electron Devices, vol. 59, pp. 3205–3211, 2012.

[14] B. Bhushan, K. Nayak, and V. R. Rao, "DC compact model for SOI tunnel field-effect transistors," IEEE Trans. Electron Devices, vol. 59, pp. 2635–2642, 2012.

[15] L. Zhang, X. Lin, J. He, and M. Chan, "An analytical charge model for double-gate tunnel FETs," IEEE Trans. Electron Devices, vol. 59, pp. 3217–3223, 2012.

[16] Synopsys Inc., "Sentaurus Device User Guide," Version A-2007, 2007, ftp://147.46.117.90/synopsys/manuals/PDFManual/data/sdevice_ug.pdf (accessed June 10, 2016.)

[17] E. Merzbacher, "Quantum Mechanics," 3rd ed. (John Wiley & Sons, Inc., 1998), p. 97.

[18] J. G. Simmons, "Generalized formula for the electric tunnel effect between similar electrodes separated by a thin insulating film," J. Appl. Phys., vol. 34, pp. 1793–1803, 1963.

[19] Y. Omura, S. Horiguchi, M. Tabe, and K. Kishi, "Quantum-mechanical effects on the threshold voltage of ultra-thin-SOI nMOSFET's," IEEE Electron Device Lett., vol. 14, no. 12, pp. 569–571, 1993.

[20] K. Boucart and A. M. Ionescu, "A new definition of threshold voltage in tunnel FETs," Solid-State Electron., vol. 52, pp. 1318–1323, 2008.

[21] K. Boucart and A. M. Ionescu, "Double-gate tunnel FET with high-κ gate dielectric," IEEE Trans. Electron Devices, vol. 54, pp. 1725–1733, 2007.

[22] A. Mallik and A. Chattopadhyay, "Drain-dependence of tunnel field-effect transistor characteristics: the role of the channel," IEEE Trans. Electron Devices, vol. 58, pp. 4250–4257, 2011.

[23] D. Leonelli, A. Vandooren, R. Rooyackers, A. S. Verhulst, C. Huyghebaert, S. De Gendt, M. M. Heyns, and G. Groeseneken, "Novel Architecture to Boost the Vertical Tunneling in Tunnel Field Effect Transistors," Proc. IEEE Int. SOI Conf., pp. 1–2, 2011.

[24] R. Asra, M. Shrivastava, K. V. R. M. Murali, R. K. Pandey, H. Gossner, and V. R. Rao, "A tunnel FET for VDD scaling below 0.6 V with a CMOS-comparable performance," IEEE Trans. Electron Devices, vol. 58, no. 7, pp. 1855–1863, 2011.

[25] A. Mallik, A. Chattopadhyay, S. Guin, and A. Karmakar, "Impact of a spacer-drain overlap on the characteristics of a silicon tunnel field-effect transistor based on vertical tunneling," IEEE Trans. Electron Devices, vol. 60, no. 3, pp. 935–943, 2013.

[26] Y. Morita, T. Mori, S. Migita, W. Mizubayashi, A. Tanabe, K. Fukuda, T. Matsukawa, K. Endo, S. O'uchi, Y. X. Liu, M. Masahara, and H. Ota, "Synthetic Electric Field Tunnel FETs: Drain Current Multiplication Demonstrated by Wrapped Gate Electrode Around Ultrathin Epitaxial Channel," Tech. Dig. 2013 Int. Symp. VLSI Technol. (Kyoto) pp. 11–13, 2013.

[27] Silvaco International, Atlas User's Manual, 2011, http://ridl.cfd.rit.edu/products/manuals/Silvaco/atlas_users.pdf (accessed June 10, 2016).

[28] D. Verreck, A. S. Verhulst, K.-H. Kao, W. G. Vandenberghe, K. D. Meyer, and G. Groeseneken, "Quantum mechanical performance predictions of p-n-i-n versus pocketed line tunnel field-effect transistors," IEEE Trans. Electron Devices, vol. 60, pp. 2128–2134, 2013.

[29] K.-H. Kao, A. S. Verhulst, W. G. Vandenberghe, B. Soree, G. Groeseneken, and K. D. Meyer, "Direct and indirect band-to-band tunneling in germanium-based TFETs," *IEEE Trans. Electron Devices*, vol. 59, pp. 292–301, 2012.

[30] J. Sune, P. Olivo, and B. Rico, "Quantum-mechanical modeling of accumulation layers in MOS structure," *IEEE Trans. Electron Devices*, vol. 39, pp. 1732–1739, 1992.

[31] A. Mallik, A. Chattopadhyay, and Y. Omura, "A Gate-on-Germanium Source (GoGeS) tunnel field-effect transistor enabling sub-0.5-V operation," *Jpn. J. Appl. Phys.*, vol. 53, pp. 104201–104208, 2014.

[32] E. Merzbacher, "Quantum Mechanics," 3rd ed. (John Wiley & Sons, Ltd, 1997), Chapters 5 to 7.

[33] J. Bardeen, "Tunneling from a many-particle point of view," *Phys. Rev. Lett.*, vol. 6, pp. 57–59, 1961.

Part III

POTENTIAL OF CONVENTIONAL BULK MOSFETs

8

Performance Evaluation of Analog Circuits with Deep Submicrometer MOSFETs in the Subthreshold Regime of Operation

8.1 Introduction

Complementary metal oxide semiconductor (CMOS) technology is continuously scaled to achieve improved performance for digital and memory applications. The advancement in CMOS technology has enhanced its unity-gain frequency f_T in the gigahertz range, which has made this technology very attractive for system-on-chip (SoC) applications, where analog circuits are used with digital circuits and memories in the same integrated circuit in order to reduce cost and improve performance.

Use of the metal oxide semiconductor field-effect transistor (MOSFET) in the subthreshold regime of operation is a potential solution to meet the tremendous market demand for extremely low power (ultralow power) integrated circuits. For analog/SoC applications, such use of MOSFETs has the additional advantage of having a significantly higher gain due to the exponential behavior of the drain current in this region.

Investigation of both the analog performance and modeling based on the subthreshold operation of conventional (CON) long-channel devices have been reported in the literature [1]. This chapter reports the analog performance of the deep submicrometer MOSFETs biased in the subthreshold regime. Prediction of such performance using an improved drift-diffusion theory based current model is also presented.

MOS Devices for Low-Voltage and Low-Energy Applications, First Edition.
Yasuhisa Omura, Abhijit Mallik, and Naoto Matsuo.
© 2017 John Wiley & Sons Singapore Pte. Ltd. Published 2017 by John Wiley & Sons Singapore Pte. Ltd.

8.2   Subthreshold Operation and Device Simulation

As the MOSFET approaches the subthreshold region, $I_D - V_{GS}$ characteristics, where V_{GS} is the gate-to-source voltage, change from square to exponential. In the subthreshold region, g_m and I_D can be related as:

$$g_m = \frac{I_D}{(mk_BT)\big/q} \tag{8.1}$$

where m is a subthreshold slope factor $(1 < m < 3)$ that has a dependence on the channel doping concentration, k_B is Boltzmann's constant, T is the absolute temperature, and q is the charge of an electron. Equation (8.1) shows that g_m depends linearly on I_D in the subthreshold region. It may be recalled that g_m is proportional to the square root of I_D in the superthreshold region. It is also clear from Eq. (8.1) that the device performance parameter g_m/I_D is independent of the device geometry in the subthreshold regime.

The technology parameters and the supply voltage used for conventional n-MOS (metal oxide semiconductor) devices are in accordance with the International Technology Roadmap for Semiconductors (ITRS) as it relates to analog devices [2] and the 180 nm technology node. The mDRAW device structure simulator and grid generator, and the Device Simulation for Smart Integrated Systems (DESSIS) integrated systems engineering-technology computer-aided design (ISE-TCAD) device simulator are used for this study. Figure 8.1 shows I_D and g_m/I_D as a

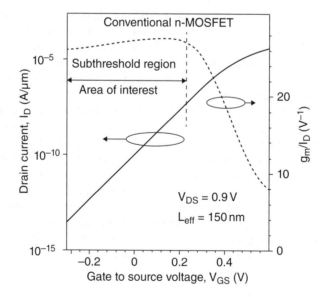

Figure 8.1   Variations of I_D and g_m/I_D with V_{GS} for a typical 150 nm effective channel length n-channel device at $V_{DS} = V_{DD}/2 = 0.9$ V. Copyright © 2006 IEEE. Reprinted, with permission, from S. Chakraborty, S. Baishya, A. Mallik, and C. K. Sarkar, "Performance evaluation of analog circuits with deep submicrometer MOSFETs in the subthreshold regime of operations," in *IEEE International Conference on Industrial and Information Systems (ICIIS)*, Aug. 8–11, 2006, Peradeniya, Sri Lanka, pp. 99–102 (2006).

function of V_{GS} for the 150 nm CON n-channel MOSFET. It shows that I_D indeed varies exponentially in the subthreshold regime – that is, when $V_{GS} < V_T$, where V_T is the threshold voltage. It is also evident from this figure that g_m/I_D is higher in the subthreshold region than the superthreshold region. As the intrinsic device gain is directly proportional to g_m/I_D, circuits based on the subthreshold operation of devices are expected to provide higher gain compared with their superthreshold counterparts.

8.3   Model Description

Based on the drift-diffusion theory, the subthreshold drain current density J_n and the effective conduction layer thickness δ for an nMOS can be written as [3]:

$$J_n = -q\mu_n\varphi_t N_A \frac{\exp\left(-\dfrac{\psi_s\left(L_{eff}\right)}{\varphi_t}\right) - \exp\left(-\dfrac{\psi_s(0)}{\varphi_t}\right)}{\displaystyle\int_0^{L_{eff}} \exp\left(-\dfrac{\psi_s(x)}{\varphi_t}\right)dx} \tag{8.2}$$

$$\delta = \varphi_t \sqrt{\varepsilon_{Si}/\left\{2qN_A\left(2\varphi_F + V_{GT}/\eta\right)\right\}} \tag{8.3}$$

where n is the inversion layer electron density, μ_n is the electron mobility in the inversion layer, φ_t is the thermal voltage, N_A is the channel doping density, ψ_s is the surface potential with respect to the interior of the bulk substrate, x is the position along the channel, L_{eff} is the effective channel length, ε_{Si} is the dielectric permittivity of Si, φ_F is the Fermi potential of the channel, $V_{GT} = V_{GS} - V_T$, and $\eta = 1.5$ is a fitting parameter [4].

For this analysis, the surface potential model proposed in [5] is used and V_T is evaluated by defining it to be the gate voltage at which the minimum value of ψ_s is equal to $2\varphi_F + V_{SB}$, where V_{SB} is the source-to substrate bias. It may be noted that the expression in Eq. (8.3) is only valid in the subthreshold regime of operation. The corresponding drain current for a channel width of W is obtained as:

$$I_D = J_n W\delta \tag{8.4}$$

It is clear that the accuracy of the computed value of the current using Eqs. (8.2)–(8.4) largely depends upon the accuracy of ψ_s and δ. For deep submicrometer devices, the value of δ computed using Eq. (8.3) needs to be modified. This is because the source and drain junction depletion layers occupy a large portion of the channel where ψ_s is higher than the middle portion of the channel, which gives rise to an overall higher value of ψ_s, whereas δ is computed using a lower value of ψ_s in the middle portion of the channel. Furthermore, the computed value of δ, applying Gauss's law, ignores the horizontal electric field component along the negative x direction, which degrades the estimated value. Therefore, an enhanced value of δ is required for deep submicrometer MOSFETs. It is true that higher substrate concentration – that is, φ_F – reduces the short-channel effects. An empirical correction factor for δ is, therefore, proposed to produce higher subthreshold current in term of φ_F as:

$$\alpha = \exp\left\{0.235\left(0.56 - \varphi_F\right)^2 / \varphi_t^2 - 2V_{GS} / \left(\varphi_F + 2V_T\right)\right\} \qquad (8.5)$$

Thus, an expression for the corrected subthreshold drain current is obtained as $I_D = J_n W \delta \alpha$. As regards the mobility model, the transverse field-dependent effective carrier mobility proposed in [6] is used for the model's predictions. Computation of the transverse field involves the surface potential, which is nonuniform. Therefore, the minimum value of ψ_s, which has a major role to play in producing the drain current, is used for the model calculations.

8.4   Results

The value of the g_m of the device is calculated by numerical differentiation of the model equation for the subthreshold drain current with respect to V_{GS} at a fixed drain-to-source voltage V_{DS} and this is compared with the simulated result in Figure 8.2. The value of g_m/I_D is compared between the simulated result and the model calculation in the Figure 8.3. An excellent agreement between the model predictions and the numerical simulation results is observed in both the figures. This justifies the introduction of the empirical correction factor α in Eq. (8.5).

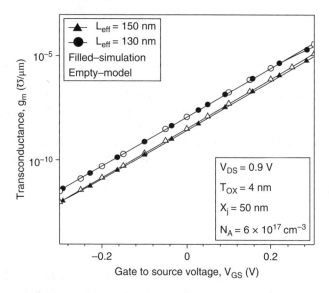

Figure 8.2 Variation of g_m with V_{GS} at $V_{DS} = 0.9$ V for 130 and 150 nm effective channel length devices. Both the simulated and the model curves are shown. Copyright © 2006 IEEE. Reprinted, with permission, from S. Chakraborty, S. Baishya, A. Mallik, and C. K. Sarkar, "Performance evaluation of analog circuits with deep submicrometer MOSFETs in the subthreshold regime of operations," in *IEEE International Conference on Industrial and Information Systems (ICIIS)*, Aug. 8–11, 2006, Peradeniya, Sri Lanka, pp. 99–102 (2006).

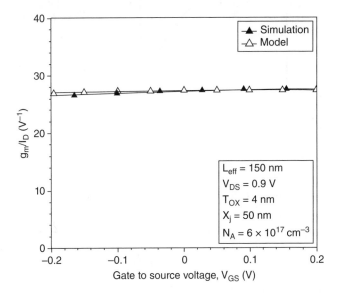

Figure 8.3 Variation of g_m/I_D with V_{GS} for a typical 150 nm effective channel length device at $V_{DS} = 0.9$ V. Both the simulated results and the model predictions are shown. Copyright © 2006 IEEE. Reprinted, with permission, from S. Chakraborty, S. Baishya, A. Mallik, and C. K. Sarkar, "Performance evaluation of analog circuits with deep submicrometer MOSFETs in the subthreshold regime of operations," in *IEEE International Conference on Industrial and Information Systems (ICIIS)*, Aug. 8–11, 2006, Peradeniya, Sri Lanka, pp. 99–102 (2006).

To observe the analog circuit performances, an inverter amplifier circuit is made as shown in the inset of Figure 8.4. The load resistance is chosen to be 30 MΩ so that the amplifier operates in the subthreshold region. The voltage transfer characteristic (VTC) of the circuit is shown in Figure 8.4. Load currents as predicted by the model and as obtained by the numerical simulations, to obtain the same VTC, are plotted in Figure 8.5 as a function of V_{in}. It is evident in Figure 8.5 that simulated results agree well with the model predictions and, hence, the model can be used effectively for analog circuit design using deep submicrometer devices.

Figure 8.6 shows the VTC of the circuit at different load resistances. It is clear that the lower the resistance value is the lower is the slope of the VTC as the circuit bias gradually moves toward the superthreshold region. By measuring the slope of the VTC, the gain of the amplifier is found. The plot of gain for different load resistances is shown in Figure 8.7. For load resistance above approximately 3 MΩ, the circuit operates in the subthreshold region and provides higher gain as compared to that in the superthreshold region. Both Figure 8.6 and Figure 8.7 show a good agreement between simulated curves and model predictions in the subthreshold region. The model predictions, however, deviate from the simulation results in the superthreshold regions as the model is valid only for the subthreshold regime of operation.

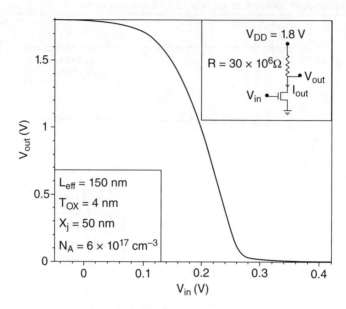

Figure 8.4 Voltage transfer characteristic (VTC) of an inverter amplifier in the subthreshold region with a 150 nm effective channel length n-MOS device. Copyright © 2006 IEEE. Reprinted, with permission, from S. Chakraborty, S. Baishya, A. Mallik, and C. K. Sarkar, "Performance evaluation of analog circuits with deep submicrometer MOSFETs in the subthreshold regime of operations," in *IEEE International Conference on Industrial and Information Systems (ICIIS)*, Aug. 8–11, 2006, Peradeniya, Sri Lanka, pp. 99–102 (2006).

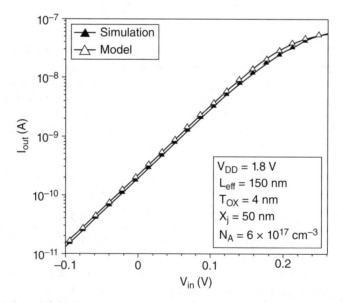

Figure 8.5 The variation of load current (I_{out}) of an inverter amplifier with input voltage (V_{in}). Both the simulated and the model results are used to get the same VTC. Copyright © 2006 IEEE. Reprinted, with permission, from S. Chakraborty, S. Baishya, A. Mallik, and C. K. Sarkar, "Performance evaluation of analog circuits with deep submicrometer MOSFETs in the subthreshold regime of operations," in *IEEE International Conference on Industrial and Information Systems (ICIIS)*, Aug. 8–11, 2006, Peradeniya, Sri Lanka, pp. 99–102 (2006).

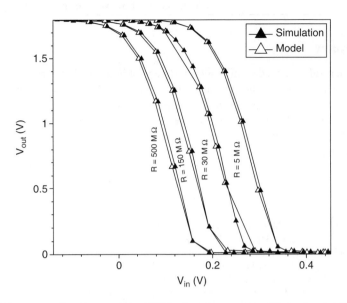

Figure 8.6 Voltage transfer characteristic (VTC) of an inverter amplifier using a 150 nm effective channel length n-MOS device for different load resistances. Copyright © 2006 IEEE. Reprinted, with permission, from S. Chakraborty, S. Baishya, A. Mallik, and C. K. Sarkar, "Performance evaluation of analog circuits with deep submicrometer MOSFETs in the subthreshold regime of operations," in *IEEE International Conference on Industrial and Information Systems (ICIIS)*, Aug. 8–11, 2006, Peradeniya, Sri Lanka, pp. 99–102 (2006).

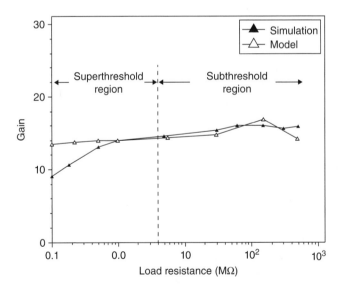

Figure 8.7 Variation of gain with the load resistance of an inverter amplifier with a 150 nm effective channel length device. Both the simulated and the model results are shown. Copyright © 2006 IEEE. Reprinted, with permission, from S. Chakraborty, S. Baishya, A. Mallik, and C. K. Sarkar, "Performance evaluation of analog circuits with deep submicrometer MOSFETs in the subthreshold regime of operations," in *IEEE International Conference on Industrial and Information Systems (ICIIS)*, Aug. 8–11, 2006, Peradeniya, Sri Lanka, pp. 99–102 (2006).

8.5   Summary

The simulation results showed significant improvement in the analog performance parameters like g_m, g_m/I_D, and so on, in the subthreshold regime of operation of deep submicrometer nMOS devices. Such performance had also been modeled accurately using an improved drift-diffusion based theory. The model may be very useful to perform circuit analysis in the subthreshold regime because the model predictions showed good agreement with the numerical simulation results.

References

[1] E. Vittoz and J. Fellrath, "CMOS analog integrated circuits based on weak inversion operation," *IEEE J. Solid-State Circuits*, vol. SC-12, no. 3, pp. 224–231, 1977.

[2] "International Technology Roadmap for Semiconductors," 2001, https://www.dropbox.com/sh/vxigcu48nfe4t81/AACuMvZEh1peQ6G8miYFCSEJa?dl=0 (accessed June 10, 2016).

[3] T. A. Fjeldly and M. Shur, "Threshold voltage modeling and the subthreshold regime of operation of short-channel MOSFET's," *IEEE Trans. Electron Devices*, vol. 40, no. 1, pp. 137–145, 1993.

[4] C. S. Ho, J. J. Liou, K.-Y. Huang, and C.-C. Cheng, "An analytical subthreshold current model for pocket-implanted NMOSFETs," *IEEE Trans. Electron Devices*, vol. 50, no. 6, pp. 1475–1479, 2003.

[5] S. Baishya, A. Mallik, and C. K. Sarkar, "A subthreshold surface potential model for short-channel MOSFET taking into account the varying depth of channel depletion layer due to source and drain junctions," *IEEE Trans. Electron Devices*, vol. 53, no. 3, pp. 507–514, 2006.

[6] H. Shin, G. M. Yeric, A. F. Tasch, and C. M. Maziar, "Physically-based models for effective mobility and local-field mobility of electrons in MOS inversion layers," *Solid-State Electron.*, vol. 34, no. 6, pp. 545–552, 1991.

9

Impact of Halo Doping on the Subthreshold Performance of Deep-Submicrometer CMOS Devices and Circuits for Ultralow Power Analog/Mixed-Signal Applications

9.1 Introduction

Continuous scaling of, and advances in, complementary metal oxide semiconductor (CMOS) technology, which has been the dominant technology for digital logic and semiconductor memories, have made it attractive for system-on-chip (SoC) applications [1]. In order to reduce the short-channel effects arising from aggressive scaling, various channel engineering techniques like double-halo (DH) and single-halo (SH) or lateral asymmetric channel (LAC) devices have been proposed [2–8] in the deep submicrometer regime. Quite a few reports are available on the performance of LAC and DH devices for both digital and analog or mixed-signal applications [2–4].

Complementary metal oxide semiconductor circuits biased on the subthreshold regime are very attractive for extremely low-power (ultralow power) applications. Analog circuits biased in the subthreshold regime of the devices have the additional advantage of having significantly higher gain due to the exponential behavior of the drain current I_D giving rise to higher g_m/I_D in this regime [9]. Various aspects of device design for subthreshold operation have been reported for digital applications [10]. Investigation of analog circuits based on the subthreshold operation of conventional (CON) long-channel devices have been reported in the literature [11–13].

MOS Devices for Low-Voltage and Low-Energy Applications, First Edition.
Yasuhisa Omura, Abhijit Mallik, and Naoto Matsuo.
© 2017 John Wiley & Sons Singapore Pte. Ltd. Published 2017 by John Wiley & Sons Singapore Pte. Ltd.

In this chapter, a systematic investigation of subthreshold performance is reported for deep-submicrometer CON, LAC, and DH MOSFETs for ultralow power analog applications. Performance parameters for 100 nm CON, LAC, and DH devices are investigated. The halo or the pocket implantation of LAC and DH devices is optimized. The performance of CMOS amplifiers is also studied, with n- and p-channel devices as driver and load, respectively, and with different combinations of CON, LAC, and DH devices.

9.2 Device Structures and Simulation

In this section, the processes and structures of CON, LAC, and DH MOSFETs are discussed. The device structures of CON, LAC, and DH devices used in this chapter are shown in Figure 9.1. The technology parameters and the supply voltage used for these devices are given in Table 9.1, which are in accordance with the analog roadmap of International Technology Roadmap for Semiconductors (ITRS) [14] for 100 nm technology node. The standard threshold adjustment implant is carried out before the gate oxidation by adjusting the dose of BF_2 and arsenic for the n- and p-channel CON devices, respectively, at a constant energy of 12 keV to set the threshold voltage $|V_T|$ at 300 mV. The process flows for LAC and DH devices are identical to that of CON device except the V_T adjustment implant. The V_T adjustment in LAC

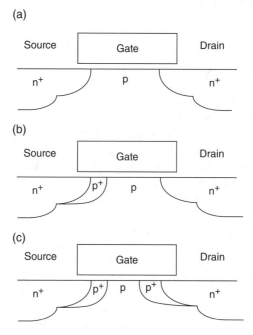

Figure 9.1 Structures of different n-channel MOSFETs. (a) CON structure, (b) LAC structure showing a p^+ pocket near the source end, and (c) DH structure showing a p^+ pocket near both the source and drain ends. Copyright © 2007 IEEE. Reprinted, with permission, from S. Chakraborty, A. Mallik, C. K. Sarkar, and V. R. Rao, "Impact of halo doping on the subthreshold performance of deep-submicrometer CMOS devices and circuits for ultralow power analog/mixed-signal applications," *IEEE Trans. Electron Devices,* vol. 54, no. 2, pp. 241–248, Feb. 2007.

Table 9.1 Technology parameters and the supply voltage used for the devices.

Parameters (units)	Values
L_g (μm)	0.1
V_{DD} (V)	1.2
t_{ox} (nm) (analog)	3
$|V_T|$ (V) (analog)	0.3
X_j (nm)	30

Copyright © 2007 IEEE. Reprinted, with permission, from S. Chakraborty, A. Mallik, C. K. Sarkar, and V. R. Rao, "Impact of halo doping on the subthreshold performance of deep-submicrometer CMOS devices and circuits for ultralow power analog/mixed-signal applications," *IEEE Trans. Electron Devices*, vol. 54, no. 2, pp. 241–248, Feb. 2007.

devices is done through an implantation at a low tilt angle (SH implant) with a constant energy (10 keV for n-channel MOSFETs and 100 keV for p-channel MOSFETs) after the lightly doped source and drain extensions are formed. The dose of this implantation (boron for n-MOSFETs and arsenic for p-MOSFETs) is adjusted to set $|V_T|$ at 300 mV. This results in an asymmetric channel with a pocket near the source end of the channel as shown in Figure 9.1b. For the DH MOSFETs, $|V_T|$ is adjusted at 300 mV by a similar process to that for the LAC but with two equal and opposite tilted implantations resulting in a symmetric channel with a pocket near both the source and the drain ends as shown in Figure 9.1c. No additional mask is used for the halo implantation of LAC or DH devices. The DIOS process simulator and two-dimensional device DESSIS (Device Simulation for Smart Integrated Systems) device simulator from Integrated System Engineering-Technology Computer Aided Design (ISE-TCAD) [15], are used for the realization and analysis of all the devices used in this chapter. An enhanced slicer is used to observe the doping profile along the channel. Energy balance models are used for device simulations in order to account for the nonlocal effects [15].

9.3   Subthreshold Operation

As the gate-to-source voltage (V_{GS}) approaches below the threshold voltage V_T from the supply voltage V_{DD}, a MOSFET goes into the subthreshold region from the superthreshold region. In the subthreshold region, the drain-current I_D is given by the following equation [10]:

$$I_D = \frac{W_{eff}}{L_{eff}} \mu \sqrt{\frac{q\varepsilon_{Si}N_{Ceff}}{2\varphi_s}} \left(\frac{k_BT}{q}\right)^2 \times exp\left(\frac{q(V_{GS}-V_T)}{mk_BT}\right)\left(1-exp\left(\frac{-qV_{DS}}{k_BT}\right)\right) \tag{9.1}$$

where μ is the effective carrier mobility, ε_{Si} is the permittivity of silicon, L_{eff} and W_{eff} are the effective channel length and width of the device, respectively, m is a subthreshold slope factor ($1 < m < 3$) that has a dependence on the channel doping concentration, k_B is Boltzmann's

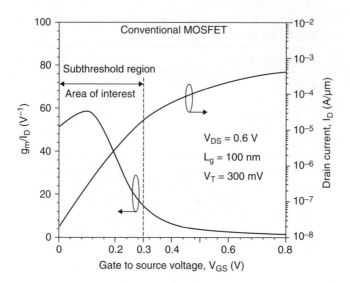

Figure 9.2 Variations of I_D and g_m/I_D with V_{GS} for a typical 100 nm n-channel CON device at $V_{DS} = V_{DD}/2 = 0.6\,\text{V}$. Copyright © 2007 IEEE. Reprinted, with permission, from S. Chakraborty, A. Mallik, C. K. Sarkar, and V. R. Rao, "Impact of halo doping on the subthreshold performance of deep-submicrometer CMOS devices and circuits for ultralow power analog/mixed-signal applications," *IEEE Trans. Electron Devices*, vol. 54, no. 2, pp. 241–248, Feb. 2007.

constant, T is the absolute temperature, q is the charge of an electron, N_{Ceff} is the effective channel doping, φ_s is the surface potential, and V_{DS} is the drain-to-source voltage.

From Eq. (9.1), the transconductance g_m can be derived as

$$g_m = \frac{I_D}{(mk_BT)/q} \tag{9.2}$$

From Eq. (9.2), it is clear that, in the subthreshold region, g_m depends linearly on I_D. It may be recalled that g_m is proportional to the square root of I_D in the superthreshold region. It is also clear from Eq. (9.2) that the device performance parameter g_m/I_D is independent of the device geometry in the subthreshold regime. Figure 9.2 shows I_D and g_m/I_D as a function of V_{GS} for a 100 nm CON n-channel MOSFET at $V_{DS} = V_{DD}/2 = 0.6\,\text{V}$. In conformity with Eq. (9.1), it is seen in Figure 9.2 that I_D indeed varies exponentially in the subthreshold regime. It is also evident in this figure that the value of g_m/I_D is high in the subthreshold region as compared to that in the superthreshold region. For this reason, analog circuits based on subthreshold operation of devices are expected to provide higher gain as compared to their superthreshold counterparts, as will be explained later.

The expression for the output resistance R_o, another device performance parameter of a MOSFET, is

$$R_o = \frac{V_A}{I_{Dsui}} \tag{9.3}$$

where V_A is the Early voltage and I_{Dsat} is the saturation drain current. As I_{Dsat} is very small in the subthreshold region, it is clear from Eq. (9.3) that R_o should be very high in this region as compared with the superthreshold region.

The intrinsic gain of a device can be written as

$$g_m R_o = \frac{g_m V_A}{I_{Dsat}} \tag{9.4}$$

From the above equation, it may be seen that the intrinsic gain of a device depends on g_m/I_D and V_A. It has already been seen from Figure 9.2 that g_m/I_D is high in the subthreshold region as compared to that in the superthreshold region. Circuits based on the subthreshold operation of devices are, therefore, expected to provide higher gain as compared to their superthreshold counterparts due to significantly higher g_m/I_D.

As unity current gain bandwidth (GBW) f_T depends linearly on g_m, and g_m depends linearly on I_D in the subthreshold region as can be seen from Eq. (9.2), it may be concluded that the f_T will depend linearly on I_D. As I_D is very small in this region, circuits implemented based on the subthreshold operation of devices therefore provide relatively low bandwidth as compared to their superthreshold counterparts. In summary, subthreshold analog circuits can be very attractive for ultralow-power, high-gain, and medium-frequency applications.

9.4   Device Optimization for Subthreshold Analog Operation

In a halo implant process, it is important to optimize the tilt angle of the implant. Figure 9.3a shows g_m/I_D and V_A as a function of V_{GS} at $V_{DS} = 0.6\,\text{V}$ for four different LAC devices in which the halos are implanted at tilt angles of 5°, 7°, 10°, and 15°. As can be observed in Figure 9.3a, both g_m/I_D and V_A become better with a lower tilt angle of the halo implant, which can be explained as follows. The channel doping is confined to a smaller region at the source side for a relatively low tilt angle, leaving an increasing portion of the channel lightly doped. The average doping concentration of the channel is, therefore, lower, which results in a corresponding low value of m in Eq. (9.2) thereby giving rise to higher g_m/I_D for such a tilt angle of the halo implant. A similar trend in g_m/I_D has been reported in [3] for superthreshold device operations. The higher value of $|V_A|$ at a relatively low tilt implant angle, as observed in Figure 9.3a, can be explained as follows. Due to higher doping concentration near the source region at lower tilt angles, electrons, on entering to the channel from the source, experience a higher electric field, which increases the average carrier velocity and produces the velocity overshot phenomenon.

Figure 9.3b shows $g_m R_o$ and R_o as a function of V_{GS} at $V_{DS} = 0.6\,\text{V}$ for the LAC devices with different tilt angles of the halo implant. As expected from the results of Figure 9.3a and Eqs. (9.3) and (9.4), both $g_m R_o$ and R_o were found to be better for the devices with a lower tilt angle of the halo implant.

Similarly, the effects of varying the tilt angle of the halo implant in DH devices on its analog performance parameters were also investigated. The different device parameters as a function of V_{GS} at $V_{DS} = 0.6\,\text{V}$ for four different tilt angles of 5°, 7°, 10°, and 15° are plotted in Figure 9.4a, and b. The effects of varying the tilt angle on different device parameters of DH devices are very similar to those observed in Figure 9.3 for the LAC devices; a lower tilt angle produces better performance.

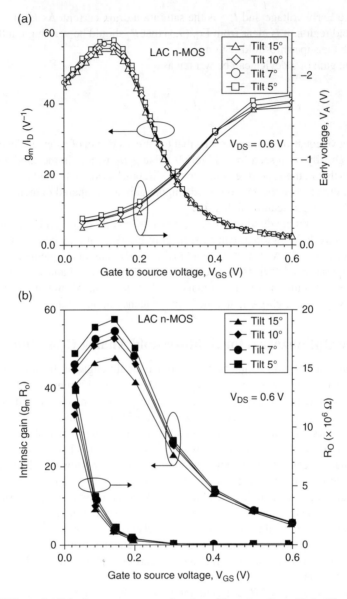

Figure 9.3   Different LAC device parameters as a function of the gate voltage (V_{GS}) at $V_{DS} = V_{DD}/2 = 0.6$ V for different tilt angles of the halo implant. (a) Transconductance generation factor (g_m/I_D) and Early voltage (V_A) and (b) intrinsic gain (g_mR_o) and output resistance (R_o). Copyright © 2007 IEEE. Reprinted, with permission, from S. Chakraborty, A. Mallik, C. K. Sarkar, and V. R. Rao, "Impact of halo doping on the subthreshold performance of deep-submicrometer CMOS devices and circuits for ultralow power analog/mixed-signal applications," *IEEE Trans. Electron Devices*, vol. 54, no. 2, pp. 241–248, Feb. 2007.

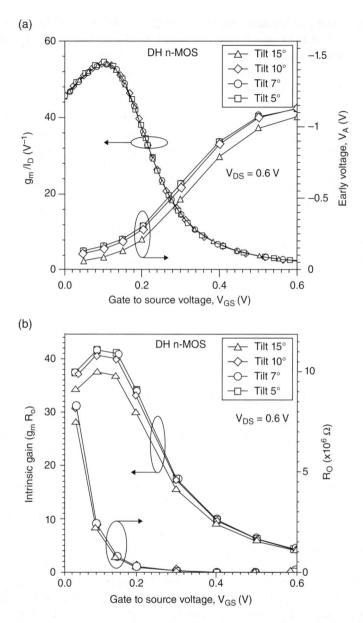

Figure 9.4 Different DH device parameters as a function of the gate voltage (V_{GS}) at $V_{DS} = V_{DD}/2 = 0.6$ V for different tilt angles of the halo implant. (a) Transconductance generation factor (g_m/I_D) and Early voltage (V_A) and (b) intrinsic gain ($g_m R_o$) and output resistance (R_o). Copyright © 2007 IEEE. Reprinted, with permission, from S. Chakraborty, A. Mallik, C. K. Sarkar, and V. R. Rao, "Impact of halo doping on the subthreshold performance of deep-submicrometer CMOS devices and circuits for ultralow power analog/mixed-signal applications," *IEEE Trans. Electron Devices,* vol. 54, no. 2, pp. 241–248, Feb. 2007.

The different analog performance parameters of CON, LAC, and DH devices as a function of V_{GS} at $V_{DS} = 0.6\,V$ are shown in Figure 9.5. Variations of g_m/I_D and V_A with V_{GS} are shown in Figure 9.5a while the same of R_o and $g_m R_o$ are shown in Figure 9.5b. In view of the results of the tilt angle optimization obtained earlier for the halo implanted devices, the tilt angle for both the LAC and DH devices in Figure 9.5 is kept at $5°$. It is evident in Figure 9.5 that there is a significant improvement in all the performance parameters for the entire subthreshold region for the halo-implanted devices in comparison with the CON devices. Amongst the halo-implanted devices, g_m/I_D is found to be higher for the LAC devices for most part of the subthreshold regime. The other parameters, namely V_A, R_o, and $g_m R_o$, are also significantly higher for the LAC devices as compared with those for the DH devices. In summary, subthreshold analog performance is better for the halo-implanted devices in comparison with the CON devices, and the performance of the LAC devices is the best.

The subthreshold performance in terms of R_o and V_A of a p-channel MOSFET is equally important and has to be investigated to study the subthreshold performance of a CMOS amplifier where the p-channel device is used as an active load. Figure 9.6 shows R_o and V_A as a function of V_{GS} for the p-channel LAC, DH, and CON devices at $V_{DS} = -0.6\,V$. The tilt angle for the LAC and DH devices in Figure 9.6 is kept at $5°$ following the results of tilt-angle optimization for the n-channel devices obtained earlier. It is clear in Figure 9.6 that the halo doping results in some improvement in both R_o and V_A for the p-channel device also, similar to that observed earlier for the n-channel device, with the LAC device showing the best performance.

9.5   Subthreshold Analog Circuit Performance

The current source CMOS inverter amplifier in Figure 9.7 with the n-channel device as a driver and the p-channel device, biased by a DC voltage V_{Bias}, as an active load is used to investigate the subthreshold analog performance of the LAC, DH, and CON CMOS devices. The mixed-mode capability of the ISE-TCAD device simulator DESSIS is used to investigate the performance of this circuit. The width of the p-MOS device is four times greater than that of the n-MOS device. Five different amplifiers with different combinations of LAC, DH, and CON devices in both the driver and load, out of the nine possible conditions, are used to compare the performance. The combinations taken are n-CON_p-CON, n-DH_ p-DH, n-LAC_ p-LAC, n-CON_p-LAC, and n-CON_p-DH.

Figure 9.8 shows the voltage transfer characteristics (VTC) of the n-CON_p-CON CMOS amplifier for different values of V_{Bias}. As the threshold voltages of n- and p-channel devices are set at $+0.3$ and $-0.3\,V$, respectively, the amplifier with a supply voltage of $1.2\,V$ works in the subthreshold regime for $V_{Bias} > 0.9\,V$. As can be seen in Figure 9.8, the transition of VTC becomes more and more sharp as V_{Bias} increases from $0.6\,V$ (superthreshold operation) to $1.2\,V$ (subthreshold operation). The exponential nature of the $I_D - V_{GS}$ characteristic in the subthreshold regime, as discussed earlier, makes the VTC near ideal in this regime. The V_{Bias} dependence of VTC for all five different combinations was studied and the results were very similar to that observed in Figure 9.8 for the n-CON_p-CON combination. The VTC corresponding to $V_{Bias} = 1.1855\,V$, which produces symmetrical transition about $V_{in} = 0\,V$ for the n-CON_p-CON combination, for all five combinations are shown in Figure 9.9. Figure 9.9 shows that the n-LAC_p-LAC combination produces the maximum slope in the transition region.

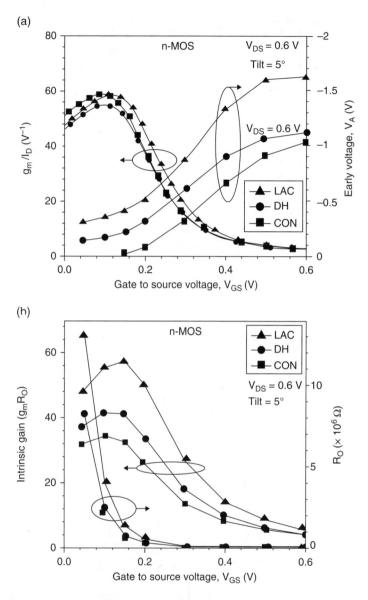

Figure 9.5 Comparison of the different device parameters between the CON, LAC, and DH devices. (a) Transconductance generation factor (g_m/I_D) and Early voltage (V_A) and (b) intrinsic gain ($g_m R_o$) and output resistance (R_o) as a function of gate voltage (V_{GS}) at $V_{DS} = V_{DD}/2 = 0.6$ V. Copyright © 2007 IEEE. Reprinted, with permission, from S. Chakraborty, A. Mallik, C. K. Sarkar, and V. R. Rao, "Impact of halo doping on the subthreshold performance of deep-submicrometer CMOS devices and circuits for ultralow power analog/mixed-signal applications," *IEEE Trans. Electron Devices,* vol. 54, no. 2, pp. 241–248, Feb. 2007.

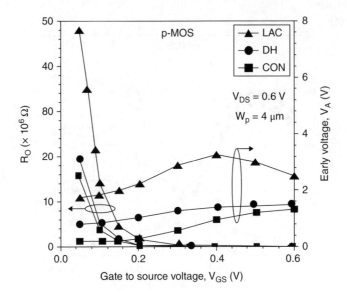

Figure 9.6   Comparisons of R_o and Early voltage (V_A) between 100 nm p-channel CON, LAC, and DH devices. Copyright © 2007 IEEE. Reprinted, with permission, from S. Chakraborty, A. Mallik, C. K. Sarkar, and V. R. Rao, "Impact of halo doping on the subthreshold performance of deep-submicrometer CMOS devices and circuits for ultralow power analog/mixed-signal applications," *IEEE Trans. Electron Devices*, vol. 54, no. 2, pp. 241–248, Feb. 2007.

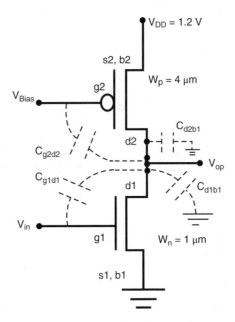

Figure 9.7   CMOS current source amplifier with a fixed V_{Bias} at the gate of the p-channel device. The different parasitic capacitances at the output node are also shown. Copyright © 2007 IEEE. Reprinted, with permission, from S. Chakraborty, A. Mallik, C. K. Sarkar, and V. R. Rao, "Impact of halo doping on the subthreshold performance of deep-submicrometer CMOS devices and circuits for ultralow power analog/mixed-signal applications," *IEEE Trans. Electron Devices*, vol. 54, no. 2, pp. 241–248, Feb. 2007.

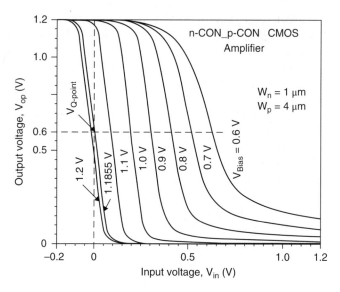

Figure 9.8 VTCs for different values of V_{Bias} for a CMOS amplifier with CON p- and n-channel devices. Copyright © 2007 IEEE. Reprinted, with permission, from S. Chakraborty, A. Mallik, C. K. Sarkar, and V. R. Rao, "Impact of halo doping on the subthreshold performance of deep-submicrometer CMOS devices and circuits for ultralow power analog/mixed-signal applications," *IEEE Trans. Electron Devices*, vol. 54, no. 2, pp. 241–248, Feb. 2007.

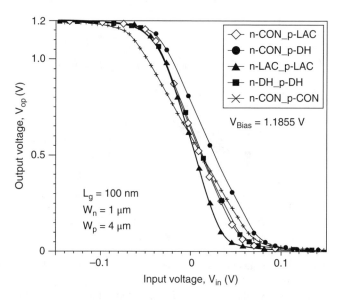

Figure 9.9 Comparison of the VTCs at $V_{Bias} = 1.1855$ V between CMOS amplifiers with different combinations of p- and n-channel CON, LAC, and DH devices. Copyright © 2007 IEEE. Reprinted, with permission, from S. Chakraborty, A. Mallik, C. K. Sarkar, and V. R. Rao, "Impact of halo doping on the subthreshold performance of deep-submicrometer CMOS devices and circuits for ultralow power analog/mixed-signal applications," *IEEE Trans. Electron Devices*, vol. 54, no. 2, pp. 241–248, Feb. 2007.

The voltage gain, which is the slope of the VTC in the transition region and may be calculated as $\Delta V_{op}/\Delta V_{in}$ in this region, is estimated for all five combinations for different values of V_{Bias}. Figure 9.10 shows the variation of the voltage gain with V_{Bias} for all the combinations. The following observations can be made from this figure. First, the gain increases with increasing V_{Bias} in the superthreshold regime (for $V_{Bias} < 0.9\,V$) and becomes more or less constant in the subthreshold regime ($V_{Bias} > 0.9\,V$). For increasing V_{Bias} in the superthreshold region, the current in the p-channel device that biases the n-channel driver decreases and, hence, g_m/I_D increases, resulting in an increase in the gain. As seen earlier, g_m/I_D is relatively constant in the subthreshold region, making the gain constant too in this region. Second, a significant improvement in the gain (more than 100%) is observed for the n-LAC_ p-LAC amplifier in the subthreshold regime as compared to that for the n-CON_p-CON combination, with the gain of all other combinations lying in between. This can be explained as follows. The voltage gain of the CMOS amplifier can be written as $g_m R_{out}$, with $g_m = g_{m1}$ as the transconductance of the n-channel driver and R_{out} is the parallel combination of the output resistances of the p- and n-channel devices. Hence,

$$R_{out} = \frac{|V_{A1} V_{A2}|}{I_{Dsat}\left(|V_{A1}|+|V_{A2}|\right)} \tag{9.5}$$

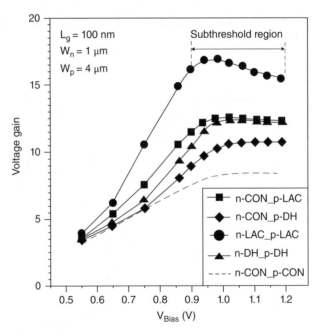

Figure 9.10   Voltage gain as a function of V_{Bias} for CMOS amplifiers with different combinations of p- and n-channel CON, LAC, and DH devices. Subthreshold region ($V_{Bias} > 0.9\,V$) shows higher gain than the superthreshold region. Copyright © 2007 IEEE. Reprinted, with permission, from S. Chakraborty, A. Mallik, C. K. Sarkar, and V. R. Rao, "Impact of halo doping on the subthreshold performance of deep-submicrometer CMOS devices and circuits for ultralow power analog/mixed-signal applications," *IEEE Trans. Electron Devices*, vol. 54, no. 2, pp. 241–248, Feb. 2007.

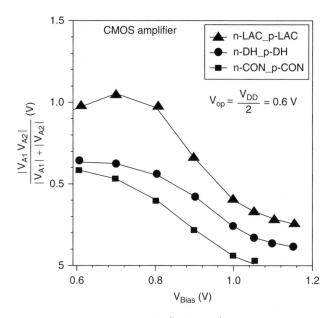

Figure 9.11   Effect of halo implant on $|V_{A1} V_{A2}| / (|V_{A1}| + |V_{A2}|)$ of the CMOS amplifier. The n-LAC_ p-LAC combination shows maximum improvement at $V_{op} = V_{DD}/2 = 0.6$ V. Copyright © 2007 IEEE. Reprinted, with permission, from S. Chakraborty, A. Mallik, C. K. Sarkar, and V. R. Rao, "Impact of halo doping on the subthreshold performance of deep-submicrometer CMOS devices and circuits for ultralow power analog/mixed-signal applications," *IEEE Trans. Electron Devices*, vol. 54, no. 2, pp. 241–248, Feb. 2007.

where V_{A1} and V_{A2} are the Early voltages of the n- and p-channel devices, respectively. Therefore,

$$g_m R_{out} = \left(\frac{g_{m1}}{I_{Dsat}} \right) \times \left(\frac{|V_{A1} V_{A2}|}{(|V_{A1}| + |V_{A2}|)} \right) \tag{9.6}$$

The variations of $|V_{A1} V_{A2}| / (|V_{A1}| + |V_{A2}|)$ with V_{Bias} for the n-LAC_p-LAC, n-DH_p-DH, and n-CON_p-CON amplifiers are shown in Figure 9.11. This figure shows the improvement in $|V_{A1} V_{A2}| / (|V_{A1}| + |V_{A2}|)$ for the halo-implanted devices, and the improvement is significant for the n-LAC_p-LAC combination over the n-CON_p-CON amplifier. This huge improvement in $|V_{A1} V_{A2}| / (|V_{A1}| + |V_{A2}|)$, together with some improvement in g_m/I_D for the n-channel LAC devices over its CON counterpart, as observed in Figure 9.5a, results in significant improvement in the gain for the n-LAC_p-LAC CMOS amplifier.

Different parasitic capacitances of the CMOS amplifier at the output node, as shown in Figure 9.7, are extracted from the AC simulation at $V_{op} = V_{DD}/2 = 0.6$ V to find $C_{Total} = C_{g1d1} + C_{g2d2} + C_{d1b1} + C_{d2b1}$. The BW and the GBW products of the amplifier are estimated as BW $= 1/(R_{out} C_{Total})$ and GBW $= g_{m1}/C_{Total}$ [13, 16]. The C_{Total} and GBW for the

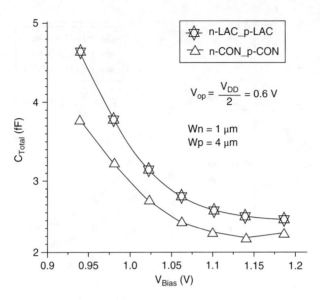

Figure 9.12 Total capacitance C_{Total} at the output node as a function of V_{Bias} for the n-CON_p-CON and n-LAC_p-LAC amplifiers at $V_{op} = V_{DD}/2 = 0.6\,V$. Copyright © 2007 IEEE. Reprinted, with permission, from S. Chakraborty, A. Mallik, C. K. Sarkar, and V. R. Rao, "Impact of halo doping on the subthreshold performance of deep-submicrometer CMOS devices and circuits for ultralow power analog/mixed-signal applications," *IEEE Trans. Electron Devices*, vol. 54, no. 2, pp. 241–248, Feb. 2007.

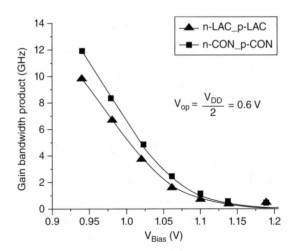

Figure 9.13 GBW product as a function of V_{Bias} for n-CON_p-CON and n-LAC_p-LAC amplifiers at $V_{op} = V_{DD}/2 = 0.6\,V$. Copyright © 2007 IEEE. Reprinted, with permission, from S. Chakraborty, A. Mallik, C. K. Sarkar, and V. R. Rao, "Impact of halo doping on the subthreshold performance of deep-submicrometer CMOS devices and circuits for ultralow power analog/mixed-signal applications," *IEEE Trans. Electron Devices*, vol. 54, no. 2, pp. 241–248, Feb. 2007.

n-LAC_ p-LAC and n-CON_p-CON amplifiers are plotted as a function of V_{Bias} in Figure 9.12 and Figure 9.13, respectively. It is clear in these figures that LAC doping increases the parasitic capacitance of the device, which results in a slight decrease in the GBW of the amplifier with LAC-doped devices.

9.6   CMOS Amplifiers with Large Geometry Devices

Although devices are being aggressively scaled for improved performance of digital circuits and memories, the undesirable short-channel effects degrade the analog performance of such devices. For mixed-signal SoC applications, analog circuits are implemented without many variations in the fabrication of their digital counterparts to reduce the complexity and cost. To reduce the short-channel effects in the analog circuitry, one may, however, use relatively large geometry devices without varying the technology parameters of the given fabrication node. The analog circuitry is generally a small part of the whole chip, so this does not significantly affect the packing density. Keeping this in view, the effect of LAC doping on the subthreshold performance of CMOS amplifiers with longer channel length devices implemented using the same technology parameters of 100 nm node is now investigated. Figure 9.14 shows the voltage gain as a function of the drawn gate length of the device in the CMOS amplifier for both the n-LAC_p-LAC and n-CON_p-CON combinations. Figure 9.14 shows that, although the gain is improved for relatively long channel-length devices in both the amplifiers,

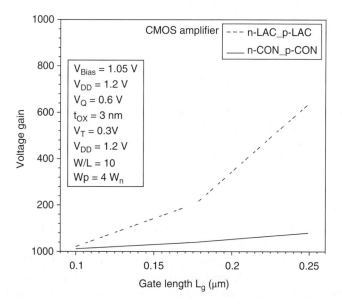

Figure 9.14   Voltage gain as a function of the channel length for the n-CON_p-CON and n-LAC_ p-LAC amplifiers at $V_{op} = V_{DD} / 2 = 0.6$ V. Copyright © 2007 IEEE. Reprinted, with permission, from S. Chakraborty, A. Mallik, C. K. Sarkar, and V. R. Rao, "Impact of halo doping on the subthreshold performance of deep-submicrometer CMOS devices and circuits for ultralow power analog/mixed-signal applications," *IEEE Trans. Electron Devices*, vol. 54, no. 2, pp. 241–248, Feb. 2007.

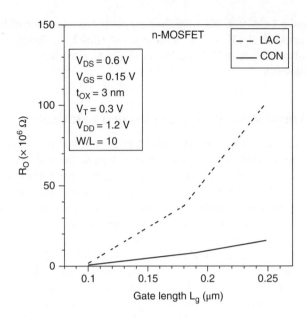

Figure 9.15 Output resistance as a function of the channel length for n-channel CON and LAC devices at $V_{DS} = 0.6$ V. Copyright © 2007 IEEE. Reprinted, with permission, from S. Chakraborty, A. Mallik, C. K. Sarkar, and V. R. Rao, "Impact of halo doping on the subthreshold performance of deep-submicrometer CMOS devices and circuits for ultralow power analog/mixed-signal applications," *IEEE Trans. Electron Devices*, vol. 54, no. 2, pp. 241–248, Feb. 2007.

the improvement is huge when LAC doping is used. It is interesting to note that one can get remarkably high gain (more than 600 for 0.25-μm LAC devices) using amplifiers in the subthreshold regime of operation. Figure 9.15 shows the plot of R_o as a function of gate length for both the CON and LAC n-MOSFETs. The similarity in the nature of curves in Figure 9.14 and Figure 9.15 confirms that the improvement in gain of the amplifier is due to the improved output resistance of the devices.

9.7 Summary

The effects of a halo implant on the subthreshold performance of 100 nm CMOS devices and circuits for ultralow power analog/mixed-signal applications were investigated systematically. Improvement in device performance parameters like g_m/I_D, V_A (and hence R_o), intrinsic gain, and so on, for such applications was observed in the halo implanted (both SH or LAC and DH) devices while the improvement was found to be significant for the LAC devices. The tilt angle of the halo implant had also been optimized and a low tilt angle had been found to yield best results for both the LAC and DH devices. A more than 100% improvement in the voltage gain was observed for a current source CMOS amplifier in the subthreshold regime of operation when the LAC doping was used for both the p- and n-channel devices.

References

[1] M. Saito, M. Ono, R. Fujimoto, H. Tanimoto, N. Ito, T. Yoshitomi, T. Ohguro, H. S. Momose, and H. Iwai, "0.15-μm RF CMOS technology compatible with logic CMOS for low-voltage operation," *IEEE Trans. Electron Devices*, vol. 45, no. 3, pp. 737–742, 1998.

[2] H. V. Deshpande, B. Cheng, and J. C. S. Woo, "Channel engineering for analog device design in deep submicron CMOS technology for system on chip applications," *IEEE Trans. Electron Devices*, vol. 49, no. 9, pp. 1558–1565, 2002.

[3] K. Narasimhulu, D. K. Sharma, and V. R. Rao, "Impact of lateral asymmetric channel doping on deep submicrometer mixed-signal device and circuit performance," *IEEE Trans. Electron Devices*, vol. 50, no. 12, pp. 2481–2489, 2003.

[4] H. V. Despande, B. Cheng, and J. C. S. Woo, "Analog device design for low power mixed mode application in deep submicron CMOS technology," *IEEE Electron Device Lett.*, vol. 22, no. 12, pp. 588–590, 2001.

[5] H. Shin and S. Lee, "An 0.1-μm asymmetric halo by large-angle-tilt implant (AHLATI) MOSFET for high performance and reliability," *IEEE Trans. Electron Devices*, vol. 46, no. 4, pp. 820–822, 1999.

[6] B. Cheng, V. R. Rao, and J. C. S. Woo, "Exploration of velocity overshoot in a high-performance deep sub-0.1-μm SOI MOSFET with asymmetric channel profile," *IEEE Electron Device Lett.*, vol. 20, no. 10, pp. 538–540, 1999.

[7] D. G. Borse, M. K. N. Rani, N. K. Jha, A. N. Chandorkar, J. Vasi, V. R. Rao, B. Cheng, and J. C. S. Woo, "Optimization and realization of sub-100-nm channel length single halo p-MOSFETs," *IEEE Trans. Electron Devices*, vol. 49, no. 6, pp. 1077–1079, 2002.

[8] K. Narasimhulu, M. P. Desai, S. G. Narendra, and V. R. Rao, "The effect of LAC doping on deep submicrometer transistor capacitances and its influence on device RF performance," *IEEE Trans. Electron Devices*, vol. 51, no. 9, pp. 1416–1423, 2004.

[9] Y. Tsividis, "Operation and Modeling of the MOS Transistor," 2nd ed. (McGraw-Hill, 1999).

[10] B. C. Paul, A. Raychowdhury, and K. Roy, "Device optimization for digital subthreshold logic operation," *IEEE Trans. Electron Devices*, vol. 52, no. 2, pp. 237–247, 2005.

[11] E. Vittoz and J. Fellrath, "CMOS analog integrated circuits based on weak inversion operation," *IEEE J. Solid-State Circuits*, vol. SC-12, no. 3, pp. 224–231, 1977.

[12] W. Steinhagen and W.L. Engl, "Design of integrated analog CMOS circuits — a multichannel telemetry transmitter [single-stage differential amplifier]," *IEEE J. Solid-State Circuits*, vol. 13, no. 6, pp. 799–805, 1978.

[13] M.G. Degrauwe, J. Rijmenants, E. A. Vittoz, and H. J. de Man, "Adaptive biasing CMOS amplifiers," *IEEE J. Solid-State Circuits*, vol. 17, no. 3, pp. 522–528, 1982.

[14] "International Technology Roadmap for Semiconductors," 2001, https://www.dropbox.com/sh/vxigcu48nfe4t81/AACuMvZEh1peQ6G8miYFCSEJa?dl=0 (accessed June 10, 2016).

[15] Synopsis, "Sentaurus Device User Guide," Version A-2007, 2007, ftp://147.46.117.90/synopsys/manuals/PDFManual/data/sdevice_ug.pdf (accessed June 10, 2016)

[16] P. E. Allen and D. R. Holberg, "CMOS Analog Circuit Design," 2nd ed. (Oxford Univ. Press, 2002).

10

Study of the Subthreshold Performance and the Effect of Channel Engineering on Deep Submicron Single-Stage CMOS Amplifiers

10.1 Introduction

The subthreshold performance of a current source amplifier built with 100 nm devices, as reported in [1], has been discussed in Chapter 9. The aggressive scaling of complementary metal oxide semiconductor (CMOS) technology gives rise to short-channel effects (SCEs). In the deep submicrometer regime, various channel engineering techniques like double halo (DH) and single halo (SH) or lateral-asymmetric channel (LAC) have been proposed, which can reduce SCEs [2]. The impact of both the SH and DH doping on the performance of the current source amplifier has also been discussed in Chapter 9. In this chapter, a comparative study between the subthreshold analog performance of a CMOS current source amplifier and a CMOS cascode amplifier, as shown in Figure 10.1a,b, is presented. As LAC doping has been found to produce the best performance for a CMOS current source amplifier in Chapter 9, the effects of such doping on the performance of a CMOS cascode amplifier are also presented in this chapter.

10.2 Circuit Description

The single-stage CMOS cascode amplifier in Figure 10.1a, consisting of the transistors MN1, MN2, and MP1, is similar to the CMOS current source amplifier in Figure 10.1b except for MN2. The cascode amplifier provides high output impedance and reduces the effect of

MOS Devices for Low-Voltage and Low-Energy Applications, First Edition.
Yasuhisa Omura, Abhijit Mallik, and Naoto Matsuo.
© 2017 John Wiley & Sons Singapore Pte. Ltd. Published 2017 by John Wiley & Sons Singapore Pte. Ltd.

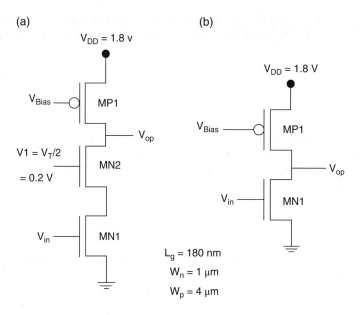

Figure 10.1 CMOS (a) cascode amplifier and (b) current source amplifier. Copyright © 2007 IEEE. Reprinted, with permission, from A. Debnath, S. Chakraborty, C. K. Sarkar, and A. Mallik, "Study of the subthreshold performance and the effect of channel engineering on deep submicron single stage CMOS amplifiers," in *IEEE TENCON 2007,* Oct. 30 to Nov. 2, pp. 140–143, 2007, Taipei.

Miller capacitance. The small-signal performance of the simple cascode amplifier in Figure 10.1a can be analyzed using the small-signal model [3]. Using nodal analysis, one may write:

$$\left(g_{ds1} + g_{ds2} + g_{m2}\right)v_1 - g_{ds2}v_o = -g_{m1}v_{in} \tag{10.1}$$

$$-\left(g_{ds2} + g_{m2}\right)v_1 + \left(g_{ds2} + g_{ds3}\right)v_o = 0 \tag{10.2}$$

where g_{ds1}, g_{ds2}, and g_{ds3} are the drain-to-source conductances of MN1, MN2, and MP1, respectively, and g_{m1} and g_{m2} are the transconductances of MN1 and MN2, respectively. Solving for v_o/v_{in} yields:

$$\frac{v_o}{v_{in}} = \frac{-\left(g_{ds2} + g_{m2}\right)g_{m1}}{g_{ds1}g_{ds2} + g_{ds1}g_{ds3} + g_{ds2}g_{m2}} \cong -\frac{g_{m1}}{g_{ds3}} \tag{10.3}$$

The total capacitance C_T at the output node is given by Allen and Holberg [3]:

$$C_T = C_{g2d2} + C_{b2d2} + C_{g3d3} + C_{b3d3} \tag{10.4}$$

where C_{g2d2}, C_{b2d2}, C_{g3d3}, and C_{b3d3} are the gate-to-drain and body-to-drain capacitance of MN2 and MP1, respectively. The gain bandwidth product (GBW) is given by Allen and Holberg [3]:

$$GBW = \left| -\frac{g_{m1}}{g_{ds3}} \right| \times \frac{1}{C_T \times r_{ds3}} = \frac{g_{m1}}{C_T} \qquad (10.5)$$

The output resistance of the amplifier in Figure 10.1a can be found by combining in parallel the output resistance of the cascode current sink, consisting of MN1 and MN2 in series, with the output resistance of MP1. Therefore, the small-signal output resistance of the cascode amplifier can be written as:

$$r_{out} = \left(r_{ds1} + r_{ds2} + g_{m2} r_{ds1} r_{ds2} \right) \| r_{ds3} \cong r_{ds3} \qquad (10.6)$$

From Eq. (10.6), it is observed that the output resistance of the cascode amplifier is nearly equal to the channel resistance of the p-load (MN1) in Figure 10.1a. On the other hand, the output resistance of the current source amplifier in Figure 10.1b is the parallel combination of the output resistances of the p-and n-channel devices [1]. It is, therefore, clear that the output resistance of the cascode amplifier is always greater (nearly double) than the output resistance of the current source amplifier.

10.3   Device Structure and Simulation

The device structures of the conventional (CON) and LAC devices used in this chapter have been described in Chapter 9 and are in accordance with the analog road map of International Technology Roadmap for Semiconductors (ITRS) [4] for 180 nm technology nodes. The standard threshold adjust implant is done before the gate oxidation by adjusting the dose of BF_2 and arsenic for the n- and p-channel CON devices, respectively, at a constant energy of 12 keV to set the threshold voltage $|V_T|$ at 400 mV.

The V_T adjustment for the LAC devices is done through a tilted implantation with a low tilt angle and a constant energy (10 keV for n-channel metal oxide semiconductor field-effect transistors (MOSFETs) and 100 keV for p-channel MOSFETs) after the lightly doped source and drain extensions are formed. The n- and p-channel device parameters such as g_m/I_D, output resistance (r_o), intrinsic gain ($g_m r_o$), and so on, are optimized for the SH (LAC) implant at a 5° tilt angle. The DIOS process simulator and the two-dimensional DESSIS (Device Simulation for Smart Integrated Systems) device simulator from Integrated Systems Engineering (ISE) Technology Computer-Aided Design (TCAD) [5] are used for the analysis of all the devices and circuits.

10.4   Results and Discussion

The voltage gain is estimated by measuring the ratio of the differential change in the output voltage to a differential change in the input voltage around the Q-point. In order to measure the voltage gain in the subthreshold regime, the Q point is set at $V_o = V_{DD}/2 = 0.9\,V$, the

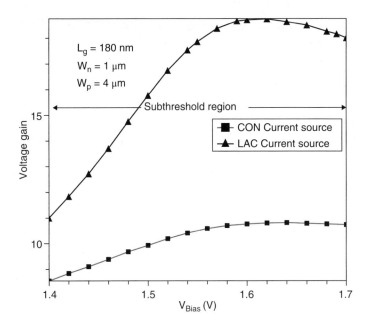

Figure 10.2   Voltage gain as a function of V_{Bias} for the CMOS current source amplifier with CON and LAC devices. Copyright © 2007 IEEE. Reprinted, with permission, from A. Debnath, S. Chakraborty, C. K. Sarkar, and A. Mallik, "Study of the subthreshold performance and the effect of channel engineering on deep submicron single stage CMOS amplifiers," in *IEEE TENCON 2007,* Oct. 30 to Nov. 2, pp. 140–143, 2007, Taipei.

input V_{MN2} of MN2 of Figure 10.1a is set at V1 = $V_T/2 = 0.2$ V, and the bias voltage of MP1 V_{Bias} is varied from 1.55 to 1.72 V, so that all of MN1, MN2, and MP3 operate in the sub-threshold region. For the current source amplifier in Figure 10.1b, V_{Bias} is varied from 1.40 to 1.72 V for the operation of the amplifier in the subthreshold regime. The voltage gain is plotted as a function of V_{Bias} for the CMOS current source amplifier and the CMOS cascode amplifier in Figure 10.2 and Figure 10.3, respectively. These figures show that the voltage gain for the cascode amplifier is almost twice of that for the current source amplifier. This is because the output resistance of the cascode amplifier is nearly twice of that of the current source amplifier [1], as discussed earlier. In both Figure 10.2 and Figure 10.3, the gain for the LAC devices is also found to be nearly twice of that for the CON devices, which is consistent with the results in Chapter 9. This is because SH or LAC implant reduces SCEs such as drain-induced barrier lowering, which results in higher output resistance and, hence, voltage gain for such devices.

The total output capacitance of the cascode amplifier is evaluated using Eq. (10.4) and ac DESSIS simulation at $V_0 = V_{DD}/2 = 0.9$ V. Figure 10.4 and Figure 10.5 show the plot of total output capacitance for the CMOS current source amplifier and CMOS cascode amplifier, respectively. It is evident from these figures that the total output capacitance for the cascode amplifier is less than that for the current source amplifier. This is due to the fact that a cascode amplifier reduces the Miller capacitance [3]. It is also evident in both Figure 10.4 and Figure 10.5 that LAC doping increases the total capacitance for both types of amplifiers.

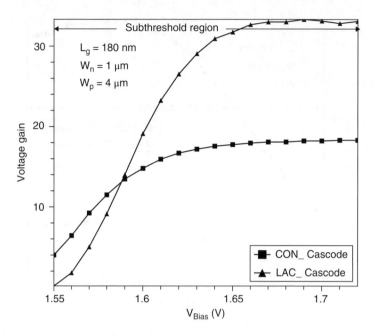

Figure 10.3 Voltage gain as a function of V_{Bias} for the CMOS cascode amplifier with CON and LAC devices. Copyright © 2007 IEEE. Reprinted, with permission, from A. Debnath, S. Chakraborty, C. K.Sarkar, and A. Mallik, "Study of the subthreshold performance and the effect of channel engineering on deep submicron single stage CMOS amplifiers," in *IEEE TENCON 2007*, Oct. 30 to Nov. 2, pp. 140–143, 2007, Taipei.

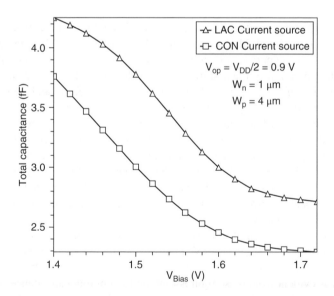

Figure 10.4 Variation of the total output capacitance as a function of V_{Bias} at $V_O = V_{DD}/2 = 0.9$ V for the current source amplifier. Copyright © 2007 IEEE. Reprinted, with permission, from A. Debnath, S. Chakraborty, C. K. Sarkar, and A. Mallik, "Study of the subthreshold performance and the effect of channel engineering on deep submicron single stage CMOS amplifiers," in *IEEE TENCON 2007*, Oct. 30 to Nov. 2, pp. 140–143, 2007, Taipei.

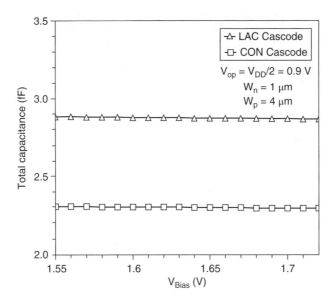

Figure 10.5   Variation of the total output capacitance as a function of V_{Bias} at $V_o = VDD/2 = 0.9$ V for the cascode amplifier. Copyright © 2007 IEEE. Reprinted, with permission, from A. Debnath, S. Chakraborty, C. K. Sarkar, and A. Mallik, "Study of the subthreshold performance and the effect of channel engineering on deep submicron single stage CMOS amplifiers," in *IEEE TENCON 2007*, Oct. 30 to Nov. 2, pp. 140–143, 2007, Taipei.

Due to an increase in the parasitic capacitance of the LAC device, output capacitance for such a device is higher than that for the CON device [1], which is also consistent with the results in Chapter 9.

The GBW product for the CMOS current source amplifier and CMOS cascode amplifier are shown in Figure 10.6 and Figure 10.7. The following observations can be made from these figures. First, GBW decreases as $V_{Bias}V_{Bias}$ increases for both types of amplifiers. As V_{Bias} increases, V_{GS} of the load MP1 in Figure 10.1a decreases causing a decrease in I_D flowing in the circuit. Since g_m is proportional to I_D in the subthreshold regime (Eq. (9.2)), g_m also decreases as V_{Bias} increases, resulting in GBW decreasing with increasing V_{Bias}. Second, GBW is either higher (for the current source amplifier in Figure 10.6) or comparable (for the cascode amplifier in Figure 10.7) for the amplifiers with LAC devices than CON devices. This is despite the fact that the total output capacitance for the amplifiers with LAC devices is higher than that with CON devices, as observed in Figure 10.4 and Figure 10.5. The improvement in GBW for the amplifiers with LAC devices may be attributed to the improvement in transconductance with LAC doping for such devices. Finally, the GBW of the current source amplifier in Figure 10.6 is found to be higher than that of the cascode amplifier in Figure 10.7. This is simply due to the difference in the ranges of V_{Bias} that ensure the subthreshold operation of the two different circuits. The range of V_{Bias} is higher for the cascode amplifier than the current source amplifier, as mentioned earlier. A lower V_{Bias} results in higher V_{GS} for the load of the current source amplifier than the cascode amplifier, which gives rise to higher I_D and hence higher g_m, resulting in a higher GBW for the current source amplifier than the cascode amplifier.

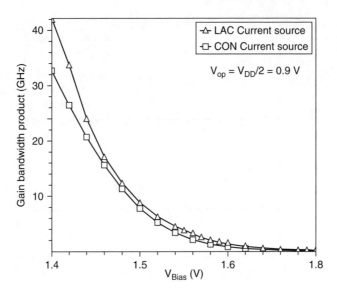

Figure 10.6 GBW as a function of V_{Bias} for the current source amplifier with CON and LAC devices at $V_o = VDD/2 = 0.9$ V. Copyright © 2007 IEEE. Reprinted, with permission, from A. Debnath, S. Chakraborty, C. K. Sarkar, and A. Mallik, "Study of the subthreshold performance and the effect of channel engineering on deep submicron single stage CMOS amplifiers," in *IEEE TENCON 2007,* Oct. 30 to Nov. 2, pp. 140–143, 2007, Taipei.

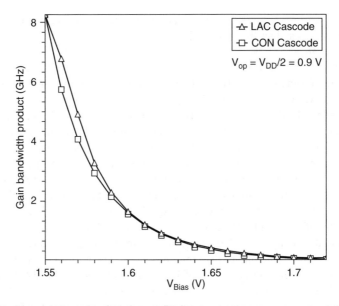

Figure 10.7 GBW as a function of V_{Bias} for the cascode amplifier with CON and LAC devices at $V_o = VDD/2 = 0.9$ V. Copyright © 2007 IEEE. Reprinted, with permission, from A. Debnath, S. Chakraborty, C. K. Sarkar, and A. Mallik, "Study of the subthreshold performance and the effect of channel engineering on deep submicron single stage CMOS amplifiers," in *IEEE TENCON 2007,* Oct. 30 to Nov. 2, pp. 140–143, 2007, Taipei.

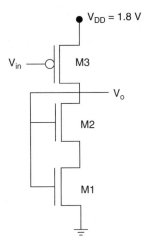

Figure 10.8 Proportional to absolute temperature sensor CMOS circuit. Copyright © 2007 IEEE. Reprinted, with permission, from A. Debnath, S. Chakraborty, C. K. Sarkar, and A. Mallik, "Study of the subthreshold performance and the effect of channel engineering on deep submicron single stage CMOS amplifiers," in *IEEE TENCON 2007*, Oct. 30 to Nov. 2, pp. 140–143, 2007, Taipei.

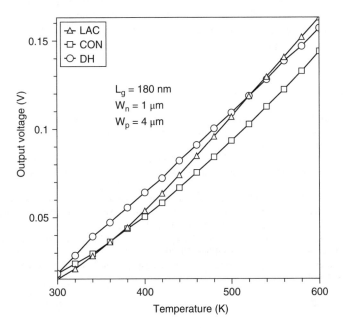

Figure 10.9 Output voltage as a function of temperature for the PTAT circuit implemented with the CON, LAC, and DH devices in the subthreshold regime of operation. Copyright © 2007 IEEE. Reprinted, with permission, from A. Debnath, S. Chakraborty, C. K. Sarkar, and A. Mallik, "Study of the subthreshold performance and the effect of channel engineering on deep submicron single stage CMOS amplifiers," in *IEEE TENCON 2007*, Oct. 30 to Nov. 2, pp. 140–143, 2007, Taipei.

10.5 PTAT as a Temperature Sensor

Circuits, such as a PTAT (proportional to absolute temperature) [6], can be designed to sense temperature using CMOS devices operating in the superthreshold region. Such a circuit, however, dissipates a large amount of power. For ultralow power applications, a PTAT circuit is proposed as a temperature sensor in Figure 10.8, which is implemented with CMOS devices operating in the subthreshold regime; 180 nm CMOS devices are used for the realization of the PTAT circuit. The impact of halo doping (both single halo or LAC and double halo) on the sensitivity of the circuit was also studied. The output voltage as a function of temperature for the PTAT circuit implemented with the CON, LAC, and DH devices in the subthreshold regime of operation is shown in Figure 10.9. This shows that both types of halo device result in higher output voltage than the CON device for the PTAT circuit. Temperature sensitivity, which is defined as the change in the output voltage per degree Kelvin change in the temperature, is the highest for circuit with LAC devices that also shows good linearity for temperature variation in the range of 350–600 K.

10.6 Summary

A comparative study between a cascode amplifier and a current source amplifier based on subthreshold operation of the devices revealed that the voltage gain is higher and the output capacitance is lower for the cascode stage. The use of LAC devices in such amplifiers results in improvement in the voltage gain for both types of amplifier. Although LAC doping increases the total output capacitance, GBW is either higher or comparable for the amplifiers with LAC devices. LAC doping also results in higher temperature sensitivity for the PTAT circuit based on the subthreshold operation of devices.

References

[1] S. Chakraborty, A. Mallik, C. K. Sarkar, and V. R. Rao, "Impact of halo doping on the subthreshold performance of deep-submicrometer CMOS devices and circuits for ultralow power analog/mixed-signal applications," *IEEE Trans. Electron Devices*, vol. 54, no. 2, pp. 241–248, 2007.

[2] H. V. Deshpande, B. Cheng, and J. C. S. Woo, "Channel engineering for analog device design in deep submicron CMOS technology for system on chip applications," *IEEE Trans. Electron Devices*, vol. 49, no. 9, pp. 1558–1565, 2002.

[3] P. E. Allen and D. R. Holberg, "CMOS Analog Circuit Design," 2nd ed., London, U.K.: Oxford Univ. Press, 2002.

[4] "International Technology Roadmap for Semiconductors," 2001, https://www.dropbox.com/sh/vxigcu48nfe4t81/AACuMvZEh1peQ6G8miYFCSEJa?dl=0 (accessed June 10, 2016).

[5] Synopsys Inc., "Sentaurus Device User Guide," Version A-2007, 2007, ftp://147.46.117.90/synopsys/manuals/PDFManual/data/sdevice_ug.pdf (accessed June 10, 2016.)

[6] E. A. Vittoz and O. Neyroud, "A low-voltage CMOS bandgap reference," *IEEE J. Solid-State Circuits*, vol. 14, no. 3, pp. 573–579, 1979.

11

Subthreshold Performance of Dual-Material Gate CMOS Devices and Circuits for Ultralow Power Analog/Mixed-Signal Applications

11.1 Introduction

As mentioned in previous chapters, the advancement in complementary metal oxide semiconductor (CMOS) technology has enhanced its f_T in the gigahertz range, which has made this technology very attractive for system-on-chip (SoC) applications. Apart from the halo doping discussed in the previous chapters, the dual-material gate (DMG) metal oxide-semiconductor field-effect transistor (MOSFET), in which two different materials with different work functions are used to form the gate, also shows great promise for alleviating the short-channel effects arising from aggressive technology scaling [1–9]. The novel attributes for such DMG devices are improved carrier transport efficiency and transconductance, and reduced drain-induced barrier lowering (DIBL) and drain conductance [1, 2]. Models for surface potential and electric field distribution in dual-/tri-material gate devices [6] and fully depleted (FD) DMG silicon-on-insulator (SOI) devices [7] have been proposed, which can predict such attributes accurately. An accurate model for the subthreshold surface potential has also been proposed [9].

Analog circuits based on the subthreshold operation of devices are not only attractive for ultralow power applications but are capable of producing higher gain [10]. The performance of subthreshold analog circuits has been reported in the literature for conventional long-channel devices [11–13] and also for conventional as well as halo-implanted short-channel devices [14].

MOS Devices for Low-Voltage and Low-Energy Applications, First Edition.
Yasuhisa Omura, Abhijit Mallik, and Naoto Matsuo.
© 2017 John Wiley & Sons Singapore Pte. Ltd. Published 2017 by John Wiley & Sons Singapore Pte. Ltd.

In this chapter, a systematic investigation of the subthreshold analog performance for deep submicrometer DMG CMOS devices for ultralow power analog/mixed-signal applications is reported. The performance of CMOS amplifiers is also studied, with n- and p-channel devices as the driver and load, respectively, and with different combinations of single-material gate (SMG) and DMG devices.

11.2   Device Structure and Simulation

In 1999, Long *et al.* [1] proposed a DMG field-effect transistor (FET) in which two different materials having different work functions are used to form the gate. The device structure of a DMG device is shown in the inset of Figure 11.1a, where the work function of metal-gate1 $W1$ is chosen to be higher than that of metal-gate2 $W2$ for an n-channel device. The technology parameters and the supply voltage used for device simulations are in accordance with the analog roadmap of International Technology Roadmap for Semiconductors (ITRS) [15] for 100 nm gate-length devices. The values of the threshold voltage V_T used for the n- and p-channel devices are +0.3 and −0.3 V, respectively, with a power supply voltage V_{DD} of 1.2 V. The devices have 3 nm gate oxide thickness T_{ox} and 50 nm source/drain junction depth X_j. The standard threshold adjust implant is done before the gate oxidation by adjusting the doses of BF_2 and arsenic, respectively, for the n- and the p-channel SMG devices at a constant energy of 12 keV. For SMG devices, the work function of the gate metal is set at 4.25 and 5.0 eV, respectively, for the n- and p-channel devices. No such V_T adjust implant is done for the DMG device, for which the lengths ($L1$ and $L2$) and the work functions ($W1$ and $W2$) of metal-gate1 and metal-gate2 are adjusted to achieve the desired V_T value. The values of $L1$ and $L2$ used for simulations are 0.035 and 0.065 μm, respectively. The value of $W1$ is set at 4.63 and 4.72 eV, respectively, for the n- and p-channel devices with $W2$ remaining the same as it was for the respective SMG device. As $L1 < L2$ and $W1 > W2$ for the n-channel device, the electrons experience a higher electric field on entering the channel from the source side, and therefore, accelerate more rapidly. This, in turn, improves their average carrier velocity [1–3]. The values of $L1$ and $L2$ can be optimized further to obtain better device performance. Accordingly, the values of $W1$ and $W2$ should be adjusted to achieve the desired V_T value. Several metal gate electrodes such as W/TiN, Mo, Ta, TaN, TiN, and $TaSi_xN_y$ have been studied by the researchers throughout the world in order to obtain the desired work functions with the required thermal and chemical stability. The work function value of (1 1 0)-Mo makes it attractive for the p-channel devices. Furthermore, its work function can also be reduced by implanting nitrogen to meet the requirements for the n-channel devices [16]. The change in the value of the work function of Mo can be controlled by the nitrogen implant parameters, making it a potential candidate for DMG technology. Gold, silver, and chromium, with work function values of 5.1, 4.26, and 4.6 eV, respectively, can also be used for the DMG technology. This so-called *gate work-function engineering* allows DMG devices to have a much lower doping concentration in the channel region as compare to their SMG counterpart to get the same V_T value. This, in turn, provides higher transconductance for the DMG devices [1, 6].

The DIOS process simulator and two-dimensional DESSIS (Device Simulation for Smart Integrated Systems) device simulator from Integrated Systems Engineering Technology Computer Aided Design (ISE-TCAD) [17] are used for the realization and analysis of all the devices used in this study. An enhanced slicer is used to observe the doping profile, electric field,

(a)

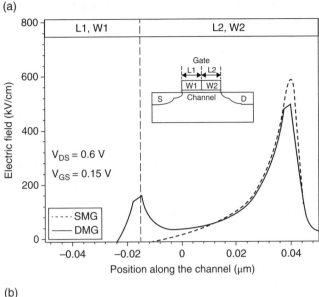

(b)

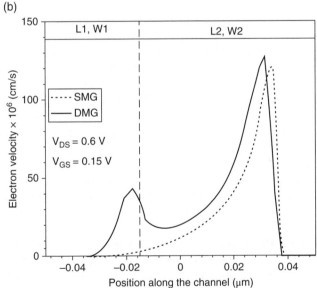

Figure 11.1 Comparison of different parameters for the DMG and SMG n-channel devices. (a) Lateral electric-field along the channel at $V_{DS} = V_{DD}/2 = 0.6$ V and $V_{GS} = V_T/2 = 0.15$ V. (b) Electron velocity along the channel at $V_{DS} = V_{DD}/2 = 0.6$ V and $V_{GS} = V_T/2 = 0.15$ V. Copyright © 2008 IEEE. Reprinted, with permission, from S. Chakraborty, A. Mallik, and C. K. Sarkar, "Subthreshold performance of dual-material gate CMOS devices and circuits for ultralow power analog/mixed-signal applications," *IEEE Trans. Electron Devices,* vol. 55, no. 3, pp. 827–832, Mar. 2008.

and carrier velocity along the channel. Energy balance models are used for device simulations in order to account for the nonlocal effects.

The conventional approach for the fabrication of such a DMG MOSFET requires the etching of the first deposited metal. This exposes the delicate thin gate oxide causing an undesirable reliability problem. An alternative method of forming dual materials over the gate dielectric in CMOS devices by metal interdiffusion has been proposed in [18]. In this, the two materials with different work functions are deposited, one after another; the second one is then selectively removed from the n- or the p-channel device as required. The two remaining metals on the p- or the n-channel device are subsequently allowed to interdiffuse so that, in some cases, they become mixed and produce a material with an intermediate work function. This eliminates the requirement of etching of the metal immediately over the gate oxide. Similarly, various fabrication techniques to produce CMOS devices with dual work function gate materials have been proposed in [19, 20]. We believe that a similar technique can also be used to form dual materials in a single metal oxide semiconductor (MOS) device.

11.3   Results and Discussion

The difference in the value of the work function of the two gate materials in a DMG device results in an additional peak in the lateral electric field near the source side in addition to the peak that is normally found near the drain side of the channel of a SMG device. Simultaneously, the peak value of the electric field near the drain side is somewhat reduced for the DMG device as compared to that for its SMG counterpart. This causes the DMG device to achieve simultaneous suppression of the short-channel effects and hot-carrier effects [1, 2]. A similar electric field distribution is also found for the DMG device under study in the subthreshold regime of operation, as shown in Figure 11.1a for drain-to-source voltage $V_{DS} = 0.6$ V and gate-to-source voltage $V_{GS} = V_T/2 = 0.15$ V. In this figure, the position along the channel is plotted in the X-axis direction where "0" indicates the center of the channel. The nature of the electron velocity distribution is found to be very similar to that of the electric field distribution in the subthreshold regime, as shown in Figure 11.1b. This kind of electron velocity near the source plays an important role in overall carrier transport efficiency of DMG devices [1, 6].

The different analog performance parameters for both the DMG and SMG n-channel devices were studied. The variations of I_D and g_m with V_{GS} for $V_{DS} = 0.1$ V are shown in Figure 11.2 for both the DMG and SMG devices. It is evident in Figure 11.2 that both I_D and g_m were higher in most part of the subthreshold regime for the DMG device as compared to that for the SMG device. The values of g_m found at $V_{GS} = 0.5$ V are 3.1×10^{-9} and 1.15×10^{-9} mho for the DMG and SMG devices, respectively. The lower doping concentration in the channel region and the peak in the electron velocity near the source region for the DMG device, as mentioned earlier, are responsible for the improvement in both I_D and g_m for such devices. The variations of g_m/I_D and the Early voltage V_A with V_{GS} for $V_{DS} = 0.6$ V are shown in Figure 11.3 for both types of devices. A slight reduction (<10%) in g_m/I_D and a huge improvement in V_A are observed in Figure 11.3 for the DMG device in the subthreshold regime. As g_m/I_D is viewed as the available gain per unit value of the power dissipation, a slight reduction of this, due to the higher I_D, should not affect performance much in the subthreshold regime because the power dissipation is much less in this regime. The improvement in V_A, and, hence, in R_o, as seen in Figure 11.4, is due to the reduction of short channel effects in the DMG

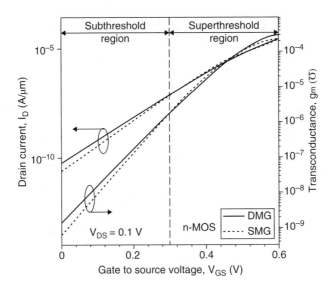

Figure 11.2 Comparison of drain–current (I_D) and transconductance (g_m) at $V_{DS} = 0.1$ V for the DMG and SMG n-channel devices as a function of V_{GS}. Copyright © 2008 IEEE. Reprinted, with permission, from S. Chakraborty, A. Mallik, and C. K. Sarkar, "Subthreshold performance of dual-material gate CMOS devices and circuits for ultralow power analog/mixed-signal applications," *IEEE Trans. Electron Devices*, vol. 55, no. 3, pp. 827–832, Mar. 2008.

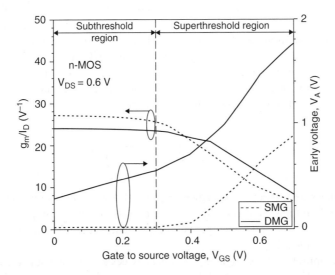

Figure 11.3 Comparison of transconductance generation factor (g_m/I_D) and Early voltage (V_A) at $V_{DS} = 0.6$ V for the DMG and SMG n-channel devices as a function of V_{GS}. Copyright © 2008 IEEE. Reprinted, with permission, from S. Chakraborty, A. Mallik, and C. K. Sarkar, "Subthreshold performance of dual-material gate CMOS devices and circuits for ultralow power analog/mixed-signal applications," *IEEE Trans. Electron Devices*, vol. 55, no. 3, pp. 827–832, Mar. 2008.

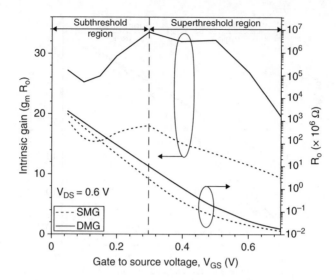

Figure 11.4 Comparison of intrinsic gain ($g_m R_o$) and output resistance (R_o) at $V_{DS} = V_{DD}/2 = 0.6$ V for the DMG and SMG n-channel devices as a function of V_{GS}. Copyright © 2008 IEEE. Reprinted, with permission, from S. Chakraborty, A. Mallik, and C. K. Sarkar, "Subthreshold performance of dual-material gate CMOS devices and circuits for ultralow power analog/mixed-signal applications," *IEEE Trans. Electron Devices*, vol. 55, no. 3, pp. 827–832, Mar. 2008.

device. Due to all of this, the intrinsic gain $g_m R_o$ of the DMG device is improved significantly, as can be seen in Figure 11.4. Figure 11.5 shows gate-to-source (C_{gs}) and gate-to-drain (C_{gd}) capacitance as a function of V_{GS} for $V_{DS} = 0.6$ V. All the capacitances are extracted from small-signal AC device simulations at a frequency of 1 MHz. Figure 11.5 shows that C_{gs} is considerably lower for the DMG device. This is because the DMG device has a lower electron concentration at the source side, as is evident in Figure 11.6. Similarly, due to the higher electron concentration at the drain side, the DMG device shows higher C_{gd}. The unity-gain cutoff frequency was also investigated; this is given by

$$f_T = \frac{g_m}{2\pi \left(C_{gs} + C_{gd} + C_{gb} \right)} \tag{11.1}$$

where C_{gb} is the gate-to-body capacitance. Figure 11.5 also shows that the DMG device provides higher f_T in the superthreshold region due to its higher g_m. But, in the subthreshold region, due to very low g_m and higher parasitic capacitances, the DMG device does not show significant improvement in f_T.

In a CMOS amplifier, with the n- and p-channel devices as the driver and load, respectively, the R_o (and, hence, V_A) of the p-channel device also plays an important role in determining the circuit performance. The plots of R_o and V_A as a function V_{GS} for the p-channel DMG and SMG devices are shown in Figure 11.7. It is evident in Figure 11.7 that the p-channel DMG device also shows significant improvement in these parameters similar to its n-channel counterpart, as expected.

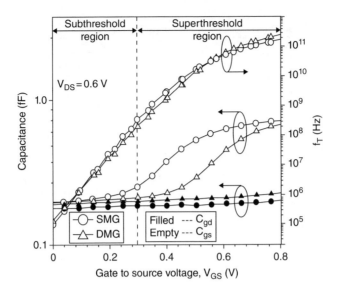

Figure 11.5 Comparison of gate-to-source capacitance (C_{gs}), gate-to-drain capacitance (C_{gd}), and unity-gain cutoff frequency (f_T) at $V_{DS} = 0.6$ V for the DMG and SMG n-channel devices as a function of V_{GS}. Copyright © 2008 IEEE. Reprinted, with permission, from S. Chakraborty, A. Mallik, and C. K. Sarkar, "Subthreshold performance of dual-material gate CMOS devices and circuits for ultralow power analog/mixed-signal applications," *IEEE Trans. Electron Devices,* vol. 55, no. 3, pp. 827–832, Mar. 2008.

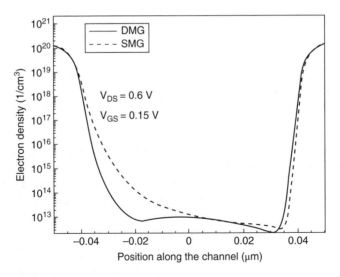

Figure 11.6 Electron density along the channel at $V_{DS} = 0.6$ V and $V_{GS} = 0.15$ V for the DMG and SMG n-channel devices. Copyright © 2008 IEEE. Reprinted, with permission, from S. Chakraborty, A. Mallik, and C. K. Sarkar, "Subthreshold performance of dual-material gate CMOS devices and circuits for ultralow power analog/mixed-signal applications," *IEEE Trans. Electron Devices*, vol. 55, no. 3, pp. 827–832, Mar. 2008.

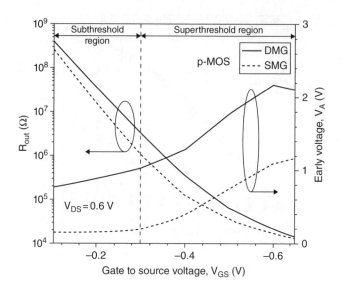

Figure 11.7 Comparison of output resistance (R_o) and Early voltage (V_A) at $V_{DS} = 0.6$ V for the DMG and SMG p-channel devices. Copyright © 2008 IEEE. Reprinted, with permission, from S. Chakraborty, A. Mallik, and C. K. Sarkar, "Subthreshold performance of dual-material gate CMOS devices and circuits for ultralow power analog/mixed-signal applications," *IEEE Trans. Electron Devices,* vol. 55, no. 3, pp. 827–832, Mar. 2008.

The performance of the DMG device in a CMOS circuit was studied using a CMOS amplifier, as shown in the inset of Figure 11.8a with the n-channel device as a driver and the p-channel device as a load. The width of the p-channel device W_p was four times the width of the n-channel device W_n in order to obtain almost the same subthreshold current in both the devices. The amplifier is operated in the subthreshold region when $V_{Bias} > 0.9$ V. The voltage gain as a function of V_{Bias} for such an amplifier, with different combinations of p- and n-channel SMG and DMG devices, is plotted in Figure 11.8a. The voltage gain is estimated by calculating the slope of the voltage transfer characteristics at $V_{out} = 0.6$ V. The circuit in which the DMGs are used for both the n- and p-channel devices shows significantly higher gain in comparison with the other combinations, as evident in Figure 11.8a. It may also be noted that the voltage gain is higher in the subthreshold region ($V_{Bias} > 0.9$ V) as compared to that in the superthreshold region for the reason explained earlier. The voltage gain of a CMOS amplifier can be written as $g_m \times R_{out}$, with $g_m = g_{m1}$ as the transconductance of the n-channel driver and R_{out} being the parallel combination of the output resistance of the p- and n-channel devices [21]. Hence, it can be shown that $g_m R_{out} = (g_{m1}/I_{Dsat}) \times |V_{A1}V_{A2}| / (|V_{A1}| + |V_{A2}|)$, where V_{A1} and V_{A2} are the Early voltages for the n- and p-channel devices, respectively. The variations of $|V_{A1}V_{A2}| / (|V_{A1}| + |V_{A2}|)$ with V_{Bias} for the CMOS amplifier with different combinations of DMG and SMG devices are also shown in Figure 11.8a. It is clear in this figure that the significant improvement in the voltage gain of the CMOS amplifier with DMG n- and p-channel devices is mainly due to the improvement in $|V_{A1}V_{A2}| / (|V_{A1}| + |V_{A2}|)$ for this combination. The devices may be optimized to further improve the voltage gain.

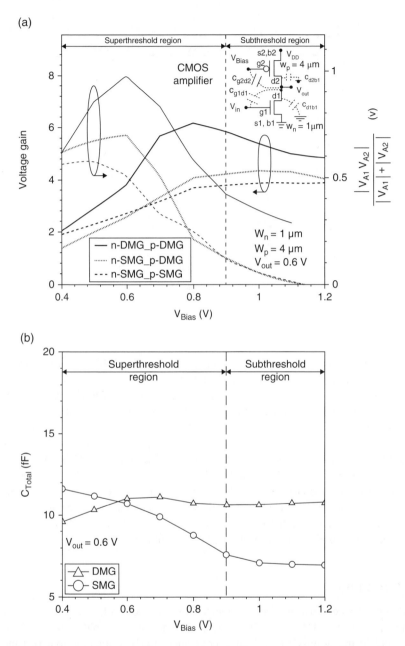

Figure 11.8 Comparison of different parameters of the CMOS amplifier with different combinations of p- and n-channel DMG and SMG devices as a function of V_{Bias} at $V_{out} = 0.6$ V. (a) Voltage gain and $|V_{A1}V_{A2}|/(|V_{A1}|+|V_{A2}|)$ (inset shows the CMOS amplifier circuit), (b) C_{Total}, and (c) gain–bandwidth product. Copyright © 2008 IEEE. Reprinted, with permission, from S. Chakraborty, A. Mallik, and C. K. Sarkar, "Subthreshold performance of dual-material gate CMOS devices and circuits for ultralow power analog/mixed-signal applications," *IEEE Trans. Electron Devices*, vol. 55, no. 3, pp. 827–832, Mar. 2008.

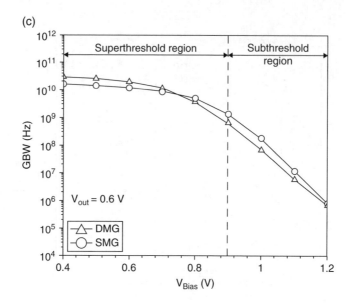

Figure 11.8   (Continued)

Different parasitic capacitances at the output node of the CMOS amplifier are extracted from the AC simulation at $V_{op} = V_{DD}/2 = 0.6\,V$ to find $C_{Total} = C_{g1d1} + C_{g2d2} + C_{d1b1} + C_{d2b1}$, where C_{g1d1} is the capacitance between the input gate of the n-channel driver and the output node, C_{g2d2} is the capacitance between the gate of the p-channel load and the output node, C_{d1b1} is the capacitance between the output node and the body of n-channel driver, and C_{d2b1} is the capacitance between the output node and the body of the p-channel load, as shown in the inset of Figure 11.8a. The bandwidth (BW) and the gain–bandwidth product (GBW) of the amplifier are estimated as $BW = 1/(R_{out} C_{Total})$ and $GBW = g_{m1}/C_{Total}$ [13, 14, 21]. C_{Total} and GBW for the n-DMG_p-DMG and n-SMG_p-SMG amplifiers are plotted as a function of V_{Bias} in Figure 11.8b,c, respectively. It is clear from these figures that a slight increase in the value of parasitic capacitance in the subthreshold regime for the DMG device results in a corresponding slight reduction in GBW for the respective CMOS amplifier.

11.4   Summary

The different device performance parameters for analog applications were investigated systematically for the DMG CMOS devices in the subthreshold regime of operation. Improvement in the drain–current, transconductance, Early voltage, output resistance, and intrinsic gain had been observed for the n-channel DMG devices. Similarly, the DMG p-channel device was found to outperform its SMG counterpart in terms of the Early voltage and output resistance. A CMOS amplifier built with DMG devices showed huge improvement (~ 70%) in the voltage gain in comparison with that built with SMG devices when biased in the subthreshold regime. Therefore, DMG CMOS devices can be very attractive for ultralow power moderate-frequency analog/mixed-signal applications.

References

[1] W. Long, H. Ou, J.-M. Kuo, and K. K. Chin, "Dual material gate (DMG) field-effect transistor," *IEEE Trans. Electron Devices*, vol. 46, no. 5, pp. 865–870, 1999.

[2] M. Saxena, S. Haldar, M. Gupta, and R. S. Gupta, "Physics-based analytical modeling of potential and electrical field distribution in dual material gate (DMG)-MOSFET for improved hot electron effect and carrier transport efficiency," *IEEE Trans. Electron Devices*, vol. 49, no. 11, pp. 1928–1938, 2002.

[3] A. Chaudhry and M. J. Kumar, "Investigation of the novel attributes of a fully depleted dual-material gate SOI MOSFET," *IEEE Trans. Electron Devices*, vol. 51, no. 9, pp. 1463–1467, 2004.

[4] R. Kaur, R. Chaujar, M. Saxena, and R. S. Gupta, "Performance investigation of 50-nm insulated-shallow-extension gate-stack (ISEGaS) MOSFET for mixed mode applications," *IEEE Trans. Electron Devices*, vol. 54, no. 2, pp. 365–368, 2007.

[5] R. Kaur, R. Chaujar, M. Saxena, and R. S. Gupta, "Hot-carrier reliability and analog performance investigation of DMG-ISEGaS MOSFET," *IEEE Trans. Electron Devices*, vol. 54, no. 9, pp. 2556–2561, 2007.

[6] K. Goel, M. Saxena, M. Gupta, and R. S. Gupta, "Modeling and simulation of a nanoscale three-region tri-material gate stack (TRIMGAS) MOSFET for improved carrier transport efficiency and reduced hot-electron effects," *IEEE Trans. Electron Devices*, vol. 53, no. 7, pp. 1623–1633, 2006.

[7] M. J. Kumar and A. Chaudhry, "Two-dimensional analytical modeling of fully depleted DMG SOI MOSFET and evidence for diminished SCEs," *IEEE Trans. Electron Devices*, vol. 51, no. 4, pp. 569–574, 2004.

[8] K. Goel, M. Saxena, M. Gupta, and R. S. Gupta, "Two-dimensional analytical threshold voltage model for DMG Epi-MOSFET," *IEEE Trans. Electron Devices*, vol. 52, no. 1, pp. 23–29, 2005.

[9] S. Baishya, A. Mallik, and C. K. Sarkar, "A pseudo two-dimensional subthreshold surface potential model for dual-material gate MOSFETs," *IEEE Trans. Electron Devices*, vol. 54, no. 9, pp. 2520–2525, 2007.

[10] Y. Tsividis, "Operation and Modeling of the MOS Transistor," 2nd ed., : (McGraw-Hill, 1999).

[11] E. Vittoz and J. Fellrath, "CMOS analog integrated circuits based on weak inversion operation," *IEEE J. Solid-State Circuits*, vol. SC-12, no. 3, pp. 224–231, 1977.

[12] W. Steinhagen and W. L. Engl, "Design of integrated analog CMOS circuits — a multichannel telemetry transmitter [single-stage differential amplifier]," *IEEE J. Solid-State Circuits*, vol. 13, no. 6, pp. 799–805, 1978.

[13] M. G. Degrauwe, J. Rijmenants, E. A. Vittoz, and H. J. de Man, "Adaptive biasing CMOS amplifiers," *IEEE J. Solid-State Circuits*, vol. 17, no. 3, pp. 522–528, 1982.

[14] S. Chakraborty, A. Mallik, C. K. Sarkar, and V. R. Rao, "Impact of halo doping on the subthreshold performance of deep-submicrometer CMOS devices and circuits for ultralow power analog/mixed-signal applications," *IEEE Trans. Electron Devices*, vol. 54, no. 2, pp. 241–248, 2007.

[15] "International Technology Roadmap for Semiconductors," 2001, https://www.dropbox.com/sh/vxigcu48nfe4t81/AACuMvZEh1peQ6G8miYFCSEJa?dl=0 (accessed June 10, 2016).

[16] Q. Lu, R. Lin, P. Ranade, T.-J. King, and C. Hu, "Metal Gate Work Function Adjustment for Future CMOS Technology," Proc. Int. Symp. VLSI Technol., Jun. 2001, pp. 45–46, 2001.

[17] Synopsys Inc., "Sentaurus Device User Guide," Version A-2007, 2007, ftp://147.46.117.90/synopsys/manuals/PDFManual/data/sdevice_ug.pdf (accessed June 10, 2016.)

[18] I. Polishchuk, P. Ranade, T.-J. King, and C. Hu, "Dual work function metal gate CMOS technology using metal interdiffusion," *IEEE Electron Device Lett.*, vol. 22, no. 9, pp. 444–446, 2001.

[19] Y.-C. Yeo, Q. Lu, P. Ranade, H. Takeuchi, K. J. Yang, I. Polishchuk, T.-J. King, C. Hu, S. C. Song, H. F. Luan, and D.-L. Kwong, "Dual-metal gate CMOS technology with ultrathin silicon nitride gate dielectric," *IEEE Electron Device Lett.*, vol. 22, no. 5, pp. 227–229, 2001.

[20] C. Ren, H. Y. Yu, J. F. Keng, X. P. Wang, H. H. H. Ma, Y.-C. Yeo, D. S. H. Chan, M.-F. Li, and D.-L. Kwong, "A dual-metal gate integration process for CMOS with sub-1-nm EOT HfO2 by using HfN replacement gate," *IEEE Electron Device Lett.*, vol. 25, no. 8, pp. 580–582, 2004.

[21] P. E. Allen and D. R. Holberg, "CMOS Analog Circuit Design," 2nd ed. (Oxford University Press, 2002.).

12

Performance Prospect of Low-Power Bulk MOSFETs

It follows from the discussions in Part I and Part II that dynamic power dissipation can be reduced by reducing the supply voltage. For the medium performance medium power region of the design spectrum, reduction in supply voltage affects performance unless the threshold voltage is suitably scaled. Reduction in the threshold voltage, however, increases the OFF-state leakage. A tradeoff between power and performance is therefore necessary for this region of the design spectrum.

Ultralow power circuits can be designed by using a supply voltage, which is below the threshold voltage of the devices, resulting in subthreshold circuits. High performance cannot be expected from such circuits as the drive current is extremely low in the subthreshold regime. This therefore leads us to one of the two extreme ends of the design spectrum that can deliver integrated circuits (ICs) with ultralow power but with only acceptable performance. Applications of such subthreshold circuits that do not require high performance include medical equipment such as pacemakers and hearing aids, self-power devices, amplification, and processing of the signal from different sensors, and so on.

Apart from ultralow power consumption, a subthreshold circuit can also have a few inherent advantages over its above threshold counterpart. First, the exponential relationship between I_D and V_{GS} (Eq. 9.1) results in extremely high transconductance g_m, which gives rise to near-ideal voltage transfer characteristics of an inverter gate. This in turn results in better static noise margins for the logic circuits and higher voltage gain for the analog circuits. Second, in the subthreshold regime, I_D becomes independent of V_{DS} for $V_{DS} > 3k_BT/q$, which results in an excellent current source. Third, the power-delay product of a subthreshold circuit has been reported [1] to be more than one order of magnitude lower than an above threshold circuit.

The simulation results in Chapter 8 also showed a significant improvement in the analog performance parameters, such as g_m, g_m/I_D, in the subthreshold regime of operation of deep

MOS Devices for Low-Voltage and Low-Energy Applications, First Edition.
Yasuhisa Omura, Abhijit Mallik, and Naoto Matsuo.
© 2017 John Wiley & Sons Singapore Pte. Ltd. Published 2017 by John Wiley & Sons Singapore Pte. Ltd.

submicrometer n-channel metal oxide-semiconductor (nMOS) devices. Such performance had also been modeled accurately using an improved drift-diffusion based theory.

In Chapter 9, halo doping (single-halo (SH) or lateral asymmetric channel (LAC) and double-halo (DH)) was found in to improve device performance parameters for analog applications, such as g_m/I_D, V_A (and hence R_o), intrinsic gain, and so on. The improvement was found to be significant for LAC devices. A low tilt angle for the halo implant was found to yield the best results for both the LAC and DH devices. A more than 100% improvement in the voltage gain was observed for a current source complementary metal oxide semiconductor (CMOS) amplifier in the subthreshold regime of operation when LAC doping was used for both the p- and n-channel devices of the amplifier.

A comparative study between a cascode amplifier and a current source amplifier in Chapter 10 revealed that the voltage gain is higher and the output capacitance is lower for the cascode stage for the subthreshold regime of operation of the devices. The use of LAC devices in such amplifiers results in improvement in the voltage gain for both types of amplifiers. Although LAC doping increases the total output capacitance, gain band width (GBW) is either higher or comparable for the amplifiers with LAC devices. Lateral asymmetric channel doping also results in higher temperature sensitivity for the proportional to absolute temperature (PTAT) circuit based on the subthreshold operation of devices.

Improvement in the drain–current, transconductance, Early voltage, output resistance, and intrinsic gain were also observed in Chapter 11 for the n-channel dual-material gate devices. Similarly, the dual-material gate (DMG) p-channel device was found to outperform its single-material gate (SMG) counterpart in terms of the Early voltage and output resistance. When biased in the subthreshold regime, a CMOS amplifier built with DMG devices showed a huge improvement (~70%) in the voltage gain in comparison with that built with SMG devices. Therefore, DMG CMOS devices can also be attractive for ultralow power moderate-frequency analog/mixed-signal applications.

Reference

[1] H. Soeleman and K. Roy, "Ultra-Low Power Digital Subthreshold Logic Circuits," Int. Symp. Low Power Electron. Design, 1999, pp. 94–96, 1999.

Part IV

POTENTIAL OF FULLY-DEPLETED SOI MOSFETs

13

Demand for High-Performance SOI Devices

As is well known, we have long anticipated that silicon-on-insulator (SOI) devices would continue to contribute to high-speed device technology because most demands from industry were expressed simply as "high-speed digital signal processing." Recently, however, the issue of energy consumption has become an international concern. Global warming, energy harvesting, and sensor networks for various social risks are pushing demands for lower energy consumption in every country. As a large part of this demand can be satisfied by electronics-based technology, several innovations are being targeted.

It is clear that SOI device technology can contribute to greater power savings. Actually, SOI devices are promising low-power devices when the device architecture is optimized for each application. We lack guidelines for matching devices to applications, so this part discusses "What are the technology issues?" and "What should be reconsidered?" We also address how the fully depleted SOI metal oxide semiconductor field effect transistor (MOSFET), and multigate SOI devices including Fin field effect transistors (FinFETs) and gate-all-around (GAA) MOSFETs, can contribute to low-energy device technologies.

MOS Devices for Low-Voltage and Low-Energy Applications, First Edition.
Yasuhisa Omura, Abhijit Mallik, and Naoto Matsuo.
© 2017 John Wiley & Sons Singapore Pte. Ltd. Published 2017 by John Wiley & Sons Singapore Pte. Ltd.

14

Demonstration of 100 nm Gate SOI CMOS with a Thin Buried Oxide Layer and its Impact on Device Technology

14.1 Introduction

In the twentieth century, separation by implanted oxygen (SIMOX) technology [1–3] attracted considerable attention for its applicability to ultrathin-film complementary metal oxide semiconductor (CMOS) integrated circuits (ICs), especially from the viewpoint of uniformity in top silicon film thickness and good reproducibility. The NTT laboratory has already successfully produced a 0.25 µm gate CMOS/SIMOX and has demonstrated the high performance potential of the ultrathin-film CMOS/SIMOX [4].

This chapter describes the device design concept for a subquarter micron gate SOI CMOS using a simplified analytical model that captures the short-channel effect in SOI devices; it elucidates the performance of these devices. A guideline for improving the performance of 100 nm gate SOI CMOS devices is also discussed.

14.2 Device Design Concept for 100 nm Gate SOI CMOS

Using two-dimensional simulations, Sano *et al.* [5] and Sekigawa and Hayashi [6] showed that reducing the thicknesses of the silicon active layer (t_s) and the buried oxide layer (t_{BOX}) is effective in suppressing short-channel effects. Their discussions, however, did not adequately address the physics of short-channel effects in subquarter micron gate metal oxide-semiconductor field-effect transistors (MOSFETs). Recently, a simplified analytical model was proposed in an effort

MOS Devices for Low-Voltage and Low-Energy Applications, First Edition.
Yasuhisa Omura, Abhijit Mallik, and Naoto Matsuo.
© 2017 John Wiley & Sons Singapore Pte. Ltd. Published 2017 by John Wiley & Sons Singapore Pte. Ltd.

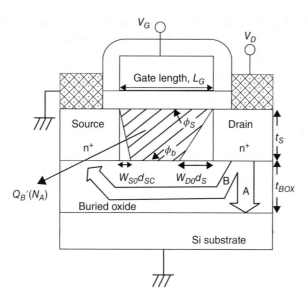

Figure 14.1 A schematic diagram to estimate the short-channel effect on threshold voltage $(V_{TH}) \cdot Q_B{}'(N_A) = f_{SC} \, Q_B(N_A)$. Copyright © 1991 IEEE. Reprinted, with permission, from Y. Omura, S. Nakashima, K. Izumi, and T. Ishii, "0.1-µm-Gate, Ultrathin-Film CMOS Devices Using SIMOX Substrate with 80-nm-Thick Buried Oxide Layer," 1991 IEEE Int. Electron Devices Meeting, Tech. Dig., pp. 675–678, 1991.

to devise advanced subquarter micron MOSFETs [7]. A schematic diagram depicting how to estimate the short-channel effects in threshold voltage (V_{TH}) is shown in Figure 14.1.

The impact on short-channel effects of the drain-induced electric field through the buried oxide layer can be considered by categorizing the drain-induced electric field into components A and B as shown in the figure. It is anticipated that the drain-to-source electric field through the buried oxide layer yields an effective substrate potential effect called an "buried-insulator-induced barrier lowering (BIIBL)" effect [8], which acts as an additional short-channel effect. An empirical consideration allows it to be incorporated into a comprehensive V_{TH} expression. The threshold voltage (V_{TH}) is given as:

$$V_{TH} = V_{FB} + \phi_s + f_{SC}\left[\frac{Q_B}{2C_{ox}} - (\phi_b - \phi_s)\frac{C_s}{C_{ox}}\right] \tag{14.1}$$

$$f_{sc} = 1 - \frac{(W_{s0} - W_{d0})}{2L_G}d_{sc} \tag{14.2}$$

$$d_{sc} = \frac{3t_{BOX}^2}{t_{BOX}^2 + L_G^2} \tag{14.3}$$

$$\phi_b = \frac{Q_B}{C_s + C_{BOX}} + \frac{C_{BOX}V_{sub}}{C_s + C_{BOX}} + \frac{C_s\phi_s}{C_s + C_{BOX}} \tag{14.4}$$

where f_{sc} is the usual fraction factor of the short-channel effect based on the charge-sharing effect, d_{sc} is the additional fraction factor of the short-channel effect ("BIIBL" effect) modulated by the buried oxide, φ_b is the back-surface potential [9], C_s is the body capacitance, C_{BOX} is the buried oxide capacitance, and C_{ox} is the gate oxide capacitance. The other symbols have their conventional meanings.

Suppressing the short-channel effect is equivalent to making the fraction f_{sc} approach unity; the width of the depletion layer stretching from source and drain junctions (W_{s0} and W_{d0}) must be reduced. Another requirement is reducing the new fraction d_{sc}; geometrically, the buried-oxide layer thickness (t_{BOX}) should be smaller than the gate length (L_G) in the model. Reducing the t_{BOX}-to-L_G ratio results in the suppression of drain-induced electric field component B (yielding the BIIBL effect) in Figure 14.1 because the geometric effect suppresses the drain-to-source electric field component B.

The above argument yields an advanced guideline for device design: the t_{BOX}-to-L_G ratio should be reduced as much as possible to suppress the short-channel effect.

14.3 Device Fabrication

A cross-sectional transmission electron microscope (XTEM) photomicrograph of a MOSFET/SIMOX with 30 nm thick active silicon film is shown in Figure 14.2. SIMOX substrates formed with a dose of 0.4×10^{18} O$^+$/cm^2 at 180 keV (substrate A) were used as the starting material [10]. These have very low dislocation densities of the order of 10^2/cm^2 [10]. A few SIMOX substrates formed with a dose of 2.0×10^{18} O$^+$/cm^2 at 180 keV (substrate B) were used for comparison [4]. All substrates were annealed at 1350 °C for 4 hours. Substrate A had an 80 nm thick buried oxide, while the buried oxide in substrate B was 480 nm thick.

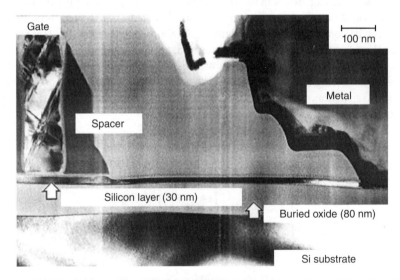

Figure 14.2 XTEM photograph of MOSFET/SIMOX with 30-nm-thick silicon film. Copyright © 1991 IEEE. Reprinted, with permission, from Y. Omura, S. Nakashima, K. Izumi, and T. Ishii, "0.1-μm-Gate, Ultrathin-Film CMOS Devices Using SIMOX Substrate with 80-nm-Thick Buried Oxide Layer," 1991 IEEE Int. Electron Devices Meeting, Tech. Dig., pp. 675–678, 1991.

Starting with the post-annealed superficial silicon layer thickness of 270 nm for substrate A, a silicon layer tens of nanometers thick was removed by oxidation followed by etching. Device-active areas were defined by mesa etching of the superficial silicon layer [4]. Boron ions for nMOSFET and phosphorus ions for pMOSFET were implanted into the superficial silicon layer to adjust the threshold voltage. A 7 nm thick gate oxide was formed at 850 °C, followed by deposition of 370 nm thick, phosphorus-doped amorphous silicon film to form the gate electrodes. After silicon gate formation, 80 nm thick silicon nitride spacers were introduced to adjust the location of the source/drain junction edges to form a graded low-doping drain (LDD) structure around the gate edges.

Source and drain regions were formed by ion implantation using boron ions for pMOSFET and arsenic ions for nMOSFET. Dopants in the source and drain regions were activated by rapid thermal annealing. Finally, the devices were connected through two-level metallization. Silicon island, gate electrode, and contact hole patterns were drawn by electron-beam lithography [11].

14.4   Performance of 100-nm- and 85-nm Gate Devices

14.4.1   Threshold and Subthreshold Characteristics

nMOS threshold-voltage shift (ΔV_{TH}) and subthreshold swing (SS) versus gate length (L_G) are shown in Figure 14.3 for various values of silicon-layer thickness (t_s) and in Figure 14.4 for the 80 and 480 nm thick buried oxide structures. The dashed lines were calculated by Eq. (14.1). Suppression of the short-channel effect in the threshold voltage (V_{TH}) and the SS is achieved by reducing both the silicon-layer thickness (t_s) and the buried-oxide layer thickness (t_{BOX}).

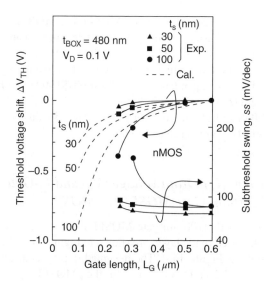

Figure 14.3 ΔV_{TH} and SS dependences on gate length for various values of t_s for substrate B. Copyright © 1991 IEEE. Reprinted, with permission, from Y. Omura, S. Nakashima, K. Izumi, and T. Ishii, "0.1-μm-Gate, Ultrathin-Film CMOS Devices Using SIMOX Substrate with 80-nm-Thick Buried Oxide Layer," 1991 IEEE Int. Electron Devices Meeting, Tech. Dig., pp. 675–678, 1991.

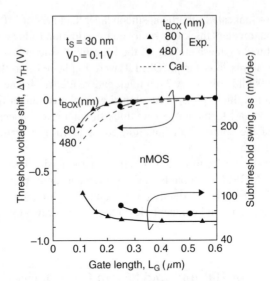

Figure 14.4 ΔV_{TH} and SS dependences on gate length for 80- and 480-nm thick buried oxide layers. Copyright © 1991 IEEE. Reprinted, with permission, from Y. Omura, S. Nakashima, K. Izumi, and T. Ishii, "0.1-µm-Gate, Ultrathin-Film CMOS Devices Using SIMOX Substrate with 80-nm-Thick Buried Oxide Layer," 1991 IEEE Int. Electron Devices Meeting, Tech. Dig., pp. 675–678, 1991.

All calculated curves roughly fit the measured results. Similar results were obtained for pMOS. This demonstrates that the present model is practical for subquarter micron gate SOI CMOS designs.

The threshold voltage (V_{TH}) dependences of MOSFET/SIMOX with 30 nm thick active silicon layer and 80 nm thick buried oxide layer on gate length (L_G) are shown in Figure 14.5 for two different substrate biases. In the case of nMOS, reverse short-channel behavior is observed due to the substrate bias, which might be due to a slight depletion in the built-in graded LDD region. The SS dependences of MOSFET/SIMOX with appropriate device parameters on gate length (L_G) are shown in Figure 14.6. SS values of the 100 nm gate device are 108 mV/dec for nMOS and 150 mV/dec for pMOS. The SS value is less than 90 mV/dec in devices with gate lengths (L_G) greater than 150 nm.

14.4.2   Drain Current (I_D)-Drain Voltage (V_D) and I_D-Gate Voltage (V_G) Characteristics of 100-nm-Gate MOSFET/SIMOX

The I_D-V_D characteristics of the 100 nm gate MOSFET, 16 µm gate width on SIMOX substrates with 80 nm thick buried oxide layer, are shown in Figure 14.7; the corresponding I_D-V_G characteristics are shown in Figure 14.8. The designed doping concentration of the body is 3×10^{17} cm^{-3} for the nMOSFET and 5×10^{16} cm^{-3} for the pMOSFET.

The threshold voltage of the nMOSFET/SIMOX takes a slightly negative value, even though the short-channel effect (SCE) was moderately suppressed, because of the thin buried oxide layer and excessively high body-doping concentration [7]. The threshold voltages were −0.15 V for the nMOSFET/SIMOX at $V_{sub} = 0$ and −0.8 V for the pMOSFET/SIMOX at

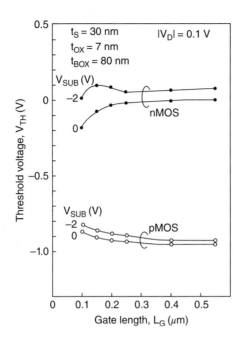

Figure 14.5 V_{TH} dependence of nMOS and pMOS on gate length for two different substrate biases (V_{SUB}). Copyright © 1991 IEEE. Reprinted, with permission, from Y. Omura, S. Nakashima, K. Izumi, and T. Ishii, "0.1-μm-Gate, Ultrathin-Film CMOS Devices Using SIMOX Substrate with 80-nm-Thick Buried Oxide Layer," 1991 IEEE Int. Electron Devices Meeting, Tech. Dig., pp. 675–678, 1991.

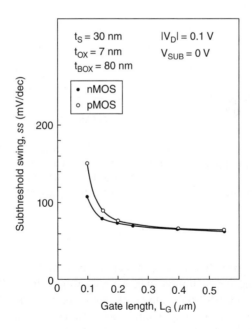

Figure 14.6 SS dependence of nMOS and pMOS on gate length. Copyright © 1991 IEEE. Reprinted, with permission, from Y. Omura, S. Nakashima, K. Izumi, and T. Ishii, "0.1-μm-Gate, Ultrathin-Film CMOS Devices Using SIMOX Substrate with 80-nm-Thick Buried Oxide Layer," 1991 IEEE Int. Electron Devices Meeting, Tech. Dig., pp. 675–678, 1991.

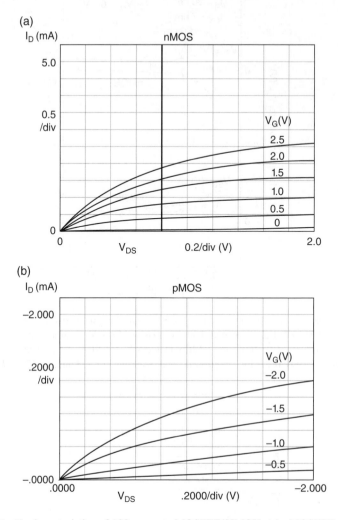

Figure 14.7 I_D–V_D characteristics of 100-nm-gate MOSFET/SIMOX. (a) nMOSFET; (b) pMOSFET. Copyright © 1991 IEEE. Reprinted, with permission, from Y. Omura, S. Nakashima, K. Izumi, and T. Ishii, "0.1-μm-Gate, Ultrathin-Film CMOS Devices Using SIMOX Substrate with 80-nm-Thick Buried Oxide Layer," 1991 IEEE Int. Electron Devices Meeting, Tech. Dig., pp. 675–678, 1991.

$V_{SUB}=0$V. The rather strong short-channel effect observed in the pMOSFET/SIMOX can be attributed to the doping concentration, which was lower in the pMOSFET/SIMOX body region than in the nMOSFET/SIMOX. The short-channel effect of the pMOSFET/SIMOX can be suppressed by increasing the body doping concentration, but this increases the threshold voltage imbalance in SOI CMOS devices. This imbalance can be resolved by employing a p-type gate surface-channel SOI pMOSFET or an n-type gate buried-channel SOI pMOSFET as discussed later

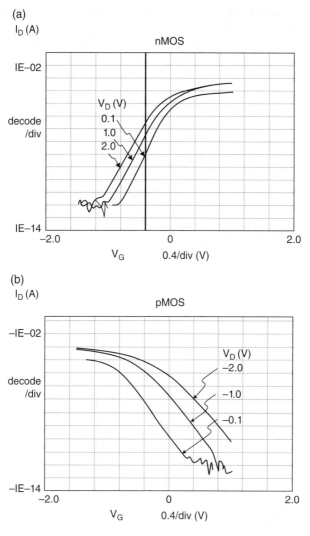

Figure 14.8 Subthreshold current characteristics of 100-nm-gate MOSFET/SIMOX with 30-nm-thick silicon layer. $W_G = 16\,\mu m$. (a) nMOSFET and (b) pMOSFET. Copyright © 1991 IEEE. Reprinted, with permission, from Y. Omura, S. Nakashima, K. Izumi, and T. Ishii, "0.1-μm-Gate, Ultrathin-Film CMOS Devices Using SIMOX Substrate with 80-nm-Thick Buried Oxide Layer," 1991 IEEE Int. Electron Devices Meeting, Tech. Dig., pp. 675–678, 1991.

The maximum transconductance values ($g_{m(max)}$) were 73 mS/mm for the nMOSFET/SIMOX at $V_D = 2\,V$ and 67 mS/mm for the pMOSFET/SIMOX at $V_D = -2\,V$. These values are much smaller than anticipated because of the high parasitic resistance in the source and drain regions, although the field-effect mobility was 440 cm²/Vs for the nMOSFET/SIMOX and 140 cm²/Vs for the pMOSFET/SIMOX. These mobility values are normal. The mobility might be affected slightly by surface-roughness scattering because the mobility decreased as the gate oxide became thinner. The sheet resistances of the 30 nm thick source and drain regions were

690 Ω/sq. for the nMOSFET/SIMOX and 1900 Ω/sq. for the pMOSFET/SIMOX. Intrinsic g_m values were estimated by subtracting the parasitic resistance effect; the intrinsic g_m was 550 mS/mm for the nMOSFET/SIMOX and 500 mS/mm for the pMOSFET/SIMOX. These parasitic resistances should be reduced by using silicide materials, or some other technique [12]. Mobility can be recovered by employing a higher temperature in the gate oxidation process, for example, rapid thermal oxidation processing.

14.4.3 I_D–V_D and I_D–V_G Characteristics of 85-nm-Gate MOSFET/SIMOX

We fabricated a 85 nm gate nMOSFET/SIMOX and a pMOSFET/SIMOX with an 8 nm thick silicon active layer on a trial basis following the guideline derived from the proposed device model [7]. I_D–V_D characteristics are shown in Figure 14.9. The drain current is very small because the source and drain regions have extremely high parasitic resistance, of the order of $10^5 \, \Omega$/sq. Subthreshold characteristics of the devices were also measured. Very small SS values were obtained – 95 mV/dec for the nMOSFET/SIMOX and 124 mV/dec for the pMOSFET/SIMOX – although the parasitic resistance is very large.

Even the thinnest SOI device was successfully fabricated and all devices with a 100 nm long gate throughout the wafer operated well, demonstrating that the new high-quality SIMOX substrates do offer sufficient uniformity in silicon-layer thickness and crystalline quality for future sub-100 nm scale devices [10].

14.4.4   Switching Performance

Propagation delay time dependences of CMOS ring oscillators on gate length (L_G) for various values of the silicon-layer thickness (t_s) are shown in Figure 14.10. For a 30 nm thick silicon layer (t_s), the propagation delay time does not decrease linearly with the reduction in gate length (L_G), and the delay time of a 100 nm gate CMOS/SIMOX is not so noticeable. The 150 nm gate CMOS/SIMOX with a 50 nm thick silicon layer, however, yielded a delay time of 35 ps/ stage and a power-delay product of 0.67 fJ/stage at a supply voltage of 1.5 V. The difference indicates that devices with a 30 nm thick silicon layer have relatively higher parasitic source and drain resistances than devices with a 50 nm thick silicon layer.

For 100 nm thick silicon-layer devices, which have a small parasitic resistance of 100 Ω/sq., the extrapolated dashed line in Figure 14.10 indicates the possibility of a 10 ps delay for a 100 nm gate SOI CMOS. It is anticipated that we will be able to develop high performance 100 nm gate SOI CMOS devices by reducing the parasitic resistance [13].

14.5   Discussion

14.5.1   Threshold Voltage Balance in Ultrathin CMOS/SOI Devices

In this work, we mostly consider CMOS/SIMOX devices consisting of an n-type gate surface-channel nMOSFET/SIMOX or an n-type gate surface-channel pMOSFET/SIMOX. Threshold voltage dependences on gate length are already shown in Figure 14.5 for two substrate voltages, V_{SUB}. The nMOSFET/SIMOX exhibits reverse-short-channel-effect-like behavior for the negative substrate bias case. This might be attributed to one or both of the following

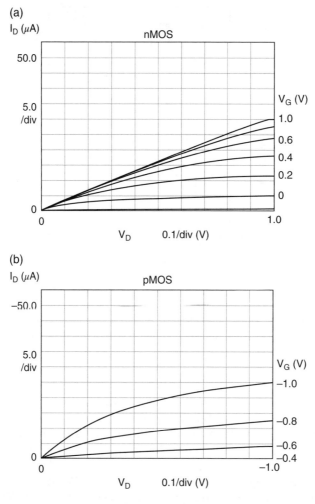

Figure 14.9 I_D–V_D characteristics of 85-nm-gate MOSFET/SIMOX with 8-nm-thick silicon layer. $W_G = 16\,\mu m$. (a) nMOSFET and (b) pMOSFET. Copyright © 1991 IEEE. Reprinted, with permission, from Y. Omura, S. Nakashima, K. Izumi, and T. Ishii, "0.1-μm-Gate, Ultrathin-Film CMOS Devices Using SIMOX Substrate with 80-nm-Thick Buried Oxide Layer," 1991 IEEE Int. Electron Devices Meeting, Tech. Dig., pp. 675–678, 1991.

mechanisms: one is the slight depletion of the LDD-like regions near source and drain junctions, the other is the offset of the drain-induced depletion layer in the thin silicon layer because of hole accumulation at the bottom of the silicon layer. Generally speaking, it is not easy to balance the threshold voltage of SOI nMOSFET and SOI pMOSFET devices even if the substrate bias effect in the SOI pMOSFET is taken into account.

If we employ a phosphorus-doped n-type silicon gate to stabilize device characteristics, there are two kinds of pMOSFETs that are applicable to SOI CMOS circuits: n-type gate surface-channel SOI pMOSFETs, and n-type gate buried-channel SOI pMOSFETs. Both are assumed to have a uniform doping profile in the body region. For comparison, n-type gate

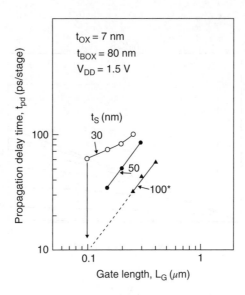

Figure 14.10 Propagation delay time dependence of CMOS/SIMOX ring oscillators on gate length (L_G) for various t_s values. * t_{BOX}=480 nm. Copyright © 1991 IEEE. Reprinted, with permission, from Y. Omura, S. Nakashima, K. Izumi, and T. Ishii, "0.1-μm-Gate, Ultrathin-Film CMOS Devices Using SIMOX Substrate with 80-nm-Thick Buried Oxide Layer," 1991 IEEE Int. Electron Devices Meeting, Tech. Dig., pp. 675–678, 1991.

buried-channel SOI pMOSFETs were also fabricated by almost the same process as that described in the previous section. Threshold voltage dependences on gate length for both devices are shown in Figure 14.11 for two substrate voltages. The buried-channel pMOSFET/ SIMOX is a good choice for CMOS/SIMOX circuits, considering its excellent threshold balance.

However, it has been reported that the buried-channel bulk MOSFET is inferior to the surface-channel bulk MOSFET in terms of the short-channel effect [14]. We found the same short-channel effect tendency in our measured results, see Figure 14.11. Regarding this problem, one of the authors (Omura) has suggested that the short-channel effect can be over-come in the buried-channel SOI MOSFET by thinning the silicon layer [15]. This suggests that device structures having excellent performance can be designed if all of the device parameters are rigorously considered.

14.6 Summary

A 100 nm gate CMOS/SIMOX was successfully fabricated using high-quality SIMOX substrates with an 80 nm thick buried oxide layer and an advanced design concept based on a simple device model. In addition, both 85 nm gate n- and pMOSFETs/SIMOX samples with 8 nm thick silicon layer were successfully fabricated for the first time.

The propagation delay time of a 100 nm gate CMOS/SIMOX was not so remarkable due to the high parasitic resistance in source and drain regions. However, it is expected that

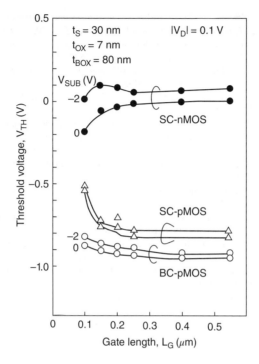

Figure 14.11 V_{TH} dependence of nMOS and pMOS on gate length for two different substrate biases (V_{SUB}). Threshold voltage dependence of surface-channel SOI pMOSFET and buried-channel SOI pMOSFET on gate length are compared for various substrate biases (V_{SUB}). Copyright © 1991 IEEE. Reprinted, with permission, from Y. Omura, S. Nakashima, K. Izumi, and T. Ishii, "0.1-μm-Gate, Ultrathin-Film CMOS Devices Using SIMOX Substrate with 80-nm-Thick Buried Oxide Layer," 1991 IEEE Int. Electron Devices Meeting, Tech. Dig., pp. 675–678, 1991.

10 ps delay 100 nm gate CMOS/SIMOX devices can be achieved by reducing the parasitic resistance.

A performance enhancement technique and threshold balancing for a 100 nm gate SOI CMOS were discussed in detail. The n-type gate buried-channel SOI pMOSFET is promising in terms of SOI CMOS device fabrication from the standpoint of threshold voltage balancing, but its short-channel effect problem must be addressed in future.

References

[1] K. Izumi, M. Doken, and H. Ariyoshi, "C.M.O.S. devices fabricated on buried Si02 layers formed by oxygen implantation into silicon," *Electron. Lett.*, vol. 14, pp. 593–594, 1978.

[2] K. Izumi, Y. Omura, and S. Nakashima, "Promotion of practical SIMOX technology by the development of a 100-mA-class high-current oxygen implanter," *Electron. Lett.*, vol. 22, pp. 775–776, 1986.

[3] K. Izumi, Y. Omura, and S. Nakashima, "Modifications of a 100-mA-class high-current oxygen implanter and its application to ultrathin film MOSFETISIMOX," *Nucl. Instrum. Methods Phys. Res.*, vol. B3738, pp. 299–303, 1989.

[4] Y. Omura and K. Izumi, "A 21.5 ps/Stage Ultra-Thin-Film CMOS/SIMOX at 2.5 V Using a 0.25-μm Polysilicon Gate," Proc. 4th Symp. on SOI Technol. Dev. (ECS, 1990, Montreal), vol. 90-6, pp. 509–517, 1990.

[5] E. Sano, R. Kasai, K. Ohwada, and H. Ariyoshi, "A two-dimensional analysis for MOSFETs fabricated on buried SiO$_2$ layer," *IEEE Trans. Electron Devices*, vol. ED-27, pp. 2043–2050, 1980.

[6] T. Sekigawa and Y. Hayashi, "Calculated threshold-voltage characteristics of an XMOS transistor having an additional bottom gate," *Solid-State Electron.*, vol. 27, pp. 827–828, 1984.

[7] Y. Omura and K. Izumi, "A Simple Model on Short-Channel Effect in Ultrathin Film MOSISIMOX and a Prospect for Down-Scaling," 1990 1EICE Spring Meeting (Japan), Abstract C-612, 1990.

[8] Y. Omura, H. Konishi, and S. Sato, "Quantum-mechanical suppression and enhancement of short-channel effects in ultra-thin SOI MOSFET's," *IEEE Trans. Electron Devices*, vol. 53, no. 4, pp. 677–684, 2006.

[9] Y. Omura, S. Nakashima, and K. Izumi, "Theoretical analysis on threshold characteristics of surface-channel MOSFET's fabricated on a buried oxide," *IEEE Trans. Electron Devices*, vol. ED-30, no. 12, pp. 1656–1662, 1983.

[10] S. Nakashima and K. Izumi, "Practical reduction of dislocation density in SIMOX wafers," *Electron. Lett*, vol. 26, pp. 1647–1648, 1990.

[11] T. Ishii and T. Matsuda, "Contrast enhancement of SAL resist by reducing residual solvent at prebake," *Jpn. J. Appl. Phys.*, vol. 30, pp. L1215–L1217, 1991.

[12] Y. Yamaguchi, Y. Inoue, T. Ipposhi, T. Nishimura, and Y. Akasaka, "Improved Characteristics of MOSFETs on Ultra Thin SIMOX," IEDM Tech. Dig., 1989, pp. 825–828, 1989.

[13] Y. Omura, S. Nakashima, and K. Izumi, "Investigation on high-speed performance of 0.1-μm-gate, ultrathin-film CMOS/SIMOX," *IEICE Trans. Electron.*, vol. E75-C, no. 12, pp. 1491–1497, 1992.

[14] G. J. Hu and R. H. Bruce, "Design tradeoffs between surface and buried-channel FET's," *IEEE Trans. Electron Devices*, vol. ED-32, pp. 584–588, 1985.

[15] Y. Omura, "A simple model for short-channel effects of a buried-channel MOSFET on the buried insulator," *IEEE Trans. Electron Devices*, vol. ED-29, pp. 1749–1755, 1982.

15

Discussion on Design Feasibility and Prospect of High-Performance Sub-50 nm Channel Single-Gate SOI MOSFET Based on the ITRS Roadmap

15.1 Introduction

The aggressive downscaling of metal oxide semiconductor field-effect transistors (MOSFETs) continues unabated, leading to advanced applications that will support various social demands. However, modern scaling approaches raise various issues that must be overcome, such as short-channel effects [1], significant gate leakage [2], and various parasitic difficulties including large gate fringing capacitance values [3]. Attention is being focused on the silicon-on-insulator metal oxide semiconductor field-effect transistor (SOI MOSFET) as it promises to overcome most of these difficulties [4]. The SOI MOSFET can reduce source and drain parasitic capacitance because it replaces the semiconductor-depletion region with a low-k insulator [5]. The SOI MOSFET has other significant benefits such as low power consumption, low threshold voltage, steep subthreshold swing (SS), and radiation hardness [4].

The International Technology Roadmap for Semiconductors (ITRS) roadmap of 2003 [6] noted that the conventional planar single-gate (SG) SOI MOSFET technology would not be applied to device generations beyond the 50 nm regime, but we note that the conclusion has been reconsidered several times in recent years. One of the authors (Omura) has already studied whether the sub-50 nm channel single-gate silicon-on insulator metal oxide semiconductor field-effect transistor (SG SOI MOSFET) is promising with regard to future applications [7, 8]; it was predicted that 20 nm channel SG SOI MOSFETs would support high-speed applications [7]. His papers introduced an SG SOI MOSFET design guideline that was based

MOS Devices for Low-Voltage and Low-Energy Applications, First Edition.
Yasuhisa Omura, Abhijit Mallik, and Naoto Matsuo.
© 2017 John Wiley & Sons Singapore Pte. Ltd. Published 2017 by John Wiley & Sons Singapore Pte. Ltd.

on the model of the minimum channel length (L_{min}) [7], but support for various applications was not addressed comprehensively. The previous model for L_{min} was constructed on the basis of the results of many simulations conducted from the viewpoint of high-end applications. However, we must reconsider the latest guidelines because low-standby power designs are still needed for many portable applications.

In this chapter, we propose an advanced design guideline for sub-50 nm channel planar SG SOI MOSFETs. We take into account the lateral diffusion length of source and drain diffusion regions (L_{ld}) because they play a significant role in suppressing the short-channel effect [9]. We use the 2-D device simulator, Synopsys-DESSIS (Device Simulation for Smart Integrated Systems) [10] with the hydrodynamic transport model. We propose new models to identify minimum channel length from the viewpoint of SS control – in other words, low-standby power designs. We also show how to design high-performance sub-50 nm channel SG SOI MOSFETs that have low-standby power consumption. Intrinsic delay time and power-delay product are examined on the basis of many simulation results. It is demonstrated that sub-50 nm channel SG SOI MOSFETs remain suitable for many applications, which contradicts previous predictions.

15.2 Device Structure and Simulations

The device structure assumed here is the fully depleted (FD) SG n-channel SOI MOSFET shown in Figure 15.1. Device parameter values are given in Table 15.1. In the simulations, the channel length (L_{eff}) was changed from 10 to 200 nm, the gate oxide thickness (t_{ox}) from 2 [11] to 5 nm, the buried oxide thickness (t_{BOX}) from 10 to 100 nm [12], the SOI doping (N_A) from 3×10^{15} to 1×10^{18} cm^{-3}, and the lateral diffusion length (L_{ld}) to suppress the short-channel effect from 5 to 30 nm. We assumed that the minimum value of SOI layer thickness is 5 nm, which obviated the necessity to consider distinct quantum-mechanical effects; the simulations used the hydrodynamic transport model. We will not consider the sub-2 nm thick gate oxide (or effective oxide thickness, EOT) in this chapter because the SOI MOSFET has a design

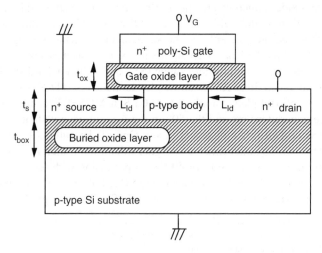

Figure 15.1 Schematic device structure assumed.

Table 15.1 Device parameters assumed in the simulations.

Device parameters	Values (unit)
Gate oxide thickness, t_{ox}	2–5 (nm)
SOI layer thickness, t_s	5–30 (nm)
Buried oxide layer thickness, t_{box}	10–100 (nm)
SOI doping concentration, N_A	3×10^{15}–1×10^{18} (cm^{-3})
Lateral diffusion length, L_{ld}	5–30 (nm)
Substrate doping concentration, N_{sub}	3×10^{17} (cm^{-3})
Source/drain doping concentration, $N_{S/D}$	4×10^{20} (cm^{-3})
Gate poly-Si doping concentration, $N_{S/D}$	4×10^{20} (cm^{-3})

margin in terms of suppressing short-channel effects [11]; this is very important not only from the viewpoint of production cost, but also for suppressing the band-to-band tunnel current near the drain junction. It is assumed for simplicity that the lateral out-diffusion of impurities from the source and drain region is Gaussian. All device characteristics are calculated at the drain voltage (V_D) of 1 V. We extract threshold voltage (V_{TH}) and SS from the simulation results; the threshold voltage is defined as the gate voltage at which the drain current is (W_{eff}/I_{eff}) 10^{-7} A, where W_{eff} is the channel width. The minimum channel length (L_{min}) means effective channel length in this chapter.

The minimum channel length (L_{min}) is extracted in the following two ways:

- *Extract A.* L_{min} is defined as the channel length at which the threshold voltage of the device is lower by 0.1 V than that of a long-channel device (L_{eff}=200 nm). This approach was proposed by Kawamoto *et al.* [7].
- *Extract B.* L_{min} is defined as the channel length at which the SS is larger than a certain value.

15.3   Proposed Model for Minimum Channel Length

15.3.1   Minimum Channel Length Model Constructed using Extract A

We investigated the dependence of L_{min} on individual device parameters. Since the previous model [7] tied the device parameters into a complex function, we tried to simplify the model. We obtained the following new expression.

$$\gamma_{TH} = \beta_{TH} \cdot In\left(\frac{N_A}{\alpha}\right) + 3.8 \cdot t_{ox} + 1.3 \cdot t_s + 0.13 \cdot t_{box} - 0.46 \cdot L_{ld} + \sigma_{TH} \tag{15.1}$$

$$L_{min} = 2 \cdot \gamma_{TH} \tag{15.2}$$

where γ_{TH} and L_{min} use nanometers as units and we assume that $\alpha = 1$ cm^{-3}, $\beta_{TH}=1.3$ nm, $\delta_{TH}=-46$ nm. N_A has units of cm^{-3} and the other device parameters have nanometers as units. The curve given by Eq. (15.2) is shown in Figure 15.2 where simulation results are also plotted for comparison. We can see that there is a good agreement between the simulation results and

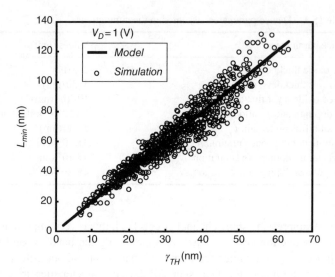

Figure 15.2 L_{min} dependence on γ_{TH} at $V_D=1\,\text{V}$ using *Extract-A* ($\Delta V_{TH}=0.1\,\text{V}$). Solid line/symbols denote the model/simulation results.

the model curve (Eq. (15.2)). Thus model Eq. (15.2) can yield an effective design guideline for device construction. Contrary to a previous prediction [7], it should be noted that we can realize a 10 nm channel SG SOI MOSFET by optimizing the device parameters. The performance that can be realized using *Extract A* is discussed later in detail.

15.3.2   Minimum Channel Length Model Constructed using Extract B

Here we consider how low standby power integrated circuits ICs can be designed. Since the low-standby power operation of these devices is primarily ruled by SS, we must determine the maximal swing value (SS_{max}) so that the low standby power performance can be reproduced. For convenience, we derive the following useful expression for L_{min}:

$$\gamma_S = C_{N_A} \cdot In\left(\frac{N_A}{\alpha}\right) + C_{t_{ox}} \cdot t_{ox} + C_{t_s} \cdot t_s + C_{t_{box}} \cdot t_{box} + C_{L_{ld}} \cdot L_{ld} + C_\delta \tag{15.3}$$

$$L_{min} = 2\gamma_S \tag{15.4}$$

where $\alpha=1\,[\text{cm}^{-3}]$ and C_{NA}, C_{tox}, C_{ts}, C_{tBOX}, C_{Lld} and C_δ are parameters depending on the maximal SS value (SS_{max}) allowed in the device design. These parameters are given by

$$C_{N_A} = P_1 \exp\left(P_2 SS_{max}\right) \tag{15.5}$$

$$C_{t_{ox}} = P_3 SS_{max} + 18.5 \tag{15.6}$$

$$C_{t_s} = P_6 SS_{max} + 4.89 \tag{15.7}$$

$$C_{t_{box}} = P_4 SS_{max}^{\ 2} + P_5 SS_{max} - 9.19 \times 10^{-1} \tag{15.8}$$

$$C_{L_{ld}} = P_7 SS_{max}^{\ 2} + P_8 SS_{max} + 1.02 \tag{15.9}$$

$$C_\delta = P_9 SS_{max} + 46.1 \tag{15.10}$$

where P_1 to P_9 are constants (see Table 15.2).

The dependency of L_{min} on γ_S for various SS_{max} values is shown in Figure 15.3 and Figure 15.4. It can be seen that model equations (15.3)–(15.10) successfully reproduce the simulation results. It should be noted that we can realize a 20 nm channel SG SOI MOSFET with $SS_{max} = 80\,\text{mV/dec}$ for low standby power applications by optimizing the device parameters.

Table 15.2   Constant values used in the model.

Constants	Values
P_1	$2.91 \times 10^{-2}\,(\text{nm})$
P_2	$3.59 \times 19^{-2}\,(\text{dec/mV})$
P_3	$-1.65 \times 10^{-1}\,(\text{dec/mV})$
P_4	$-1.00 \times 10^{-4}\,((\text{dec/mV})^2)$
P_5	$2.14 \times 10^{-2}\,(\text{dec/mV})$
P_6	$-3.76 \times 10^{-2}\,(\text{dec/mV})$
P_7	$1.00 \times 10^{-4}\,((\text{dec/mV})^2)$
P_8	$-3.01 \times 10^{-2}\,(\text{dec/mV})$
P_9	$-8.08 \times 10^{-1}\,(\text{dec/mV})$

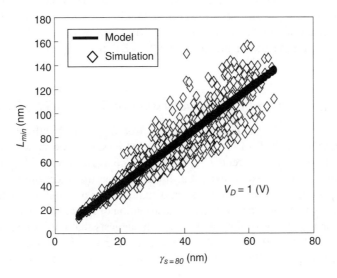

Figure 15.3 L_{min} dependence on γ_{TH} at $V_D = 1\,\text{V}$ using *Extract-B* ($SS_{max} = 80\,\text{mV/dec}$). Solid line/symbols denote the model/simulation results.

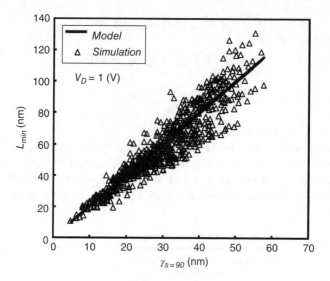

Figure 15.4 L_{min} dependence on γ_{TH} at $V_D = 1$ V using *Extract-B* ($SS_{max} = 90$ mV/dec). Solid line/symbols denote the model/simulation results.

A 10 nm channel SG SOI MOSFET with $SS_{max} = 90$ mV/dec can be realized in a similar manner. These results are the first confirmation of this level of design optimization.

15.4   Performance Prospects of Scaled SOI MOSFETs

We statistically analyze L_{min} dependency of drive current (I_{ON}), intrinsic delay time (τ), standby leakage current (I_{OFF}), and delay-time (τ)-power dissipation (P_D) product (τP_D) in order to examine the performance prospects of SG SOI MOSFETs with minimum channel length (L_{min}). A comparison of simulation results to the 2013 ITRS roadmap [13] yields a couple of new findings on the potential of the SG SOI MOSFET. In all figures, we assume $V_D = 1$ V, $V_G = 1$ V, and $V_{TH} = 0.1$ or 0.3 V.

15.4.1   Dynamic Operation Characteristics of Scaled SG SOI MOSFETs

In this section, we consider the impact of *Extract A* on L_{min} control with regard to a comprehensive performance analysis of scaled SG SOI MOSFETs. First we show on-current (I_{ON}) in Figure 15.5 and off-leakage current (I_{OFF}) in Figure 15.6 as functions of L_{min}. In Figure 15.5, I_{ON} increases as L_{min} decreases, as expected. It can be seen that there are many possibilities (many device parameter choices) to realize maximal I_D when a certain L_{min} value is assumed. On the other hand, it should be noted that the distribution of I_{ON} values for $V_{TH} = 0.1$ V and that of I_{ON} values for $V_{TH} = 0.3$ V overlap widely, which suggests that the intrinsic delay times of devices with $V_{TH} = 0.1$ or 0.3 V are almost identical. Figure 15.6 shows that I_{OFF} depends strongly on V_{TH} as expected. We can extract the following empirical equation from Figure 15.6:

$$I_{OFF} = I_t \, 10^{\frac{-V_{TH}}{SS}} \, L_{min}^{\sigma_{TH}} \qquad (15.11)$$

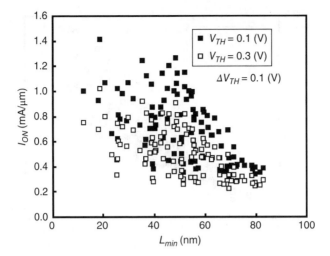

Figure 15.5 Drive current (I_{ON}) dependence on L_{min} at $V_D = 1$ V using *Extract A* ($\Delta V_{TH} = 0.1$ V) for various threshold voltages.

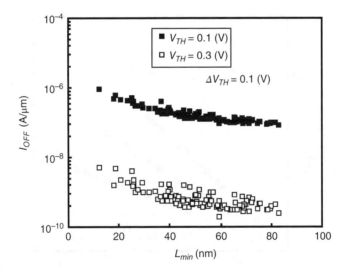

Figure 15.6 Standby leakage current (I_{OFF}) dependence on L_{min} at $V_D = 1$ V using *Extract A* ($\Delta V_{TH} = 0.1$ V) for various threshold voltages.

where I_L is the constant value of leakage current, SS is the subthreshold swing, and parameter $\sigma_{TH} = -6/5$ and $-7/5$ for $V_{TH} = 0.1$ and 0.3 [V], respectively. From the viewpoint of estimating the standby power dissipation, it is worthwhile noting that the L_{min} dependence of I_{ON} is steeper than first expected.

Figure 15.7 shows the intrinsic delay time (τ) dependence on L_{min} at $V_D = 1$ V, where $\tau = C_G V_d / I_{ON}$ and C_G is the physical gate oxide capacitance. From Figure 15.7, we can see that τ decreases as L_{min} decreases because of the increase in I_{ON}, and that the requirement for a small τ is not

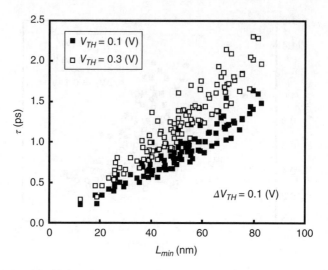

Figure 15.7 Intrinsic delay time (τ) dependence on L_{min} at $V_D = 1$ V using *Extract A* ($\Delta V_{TH} = 0.1$ V) for various threshold voltages.

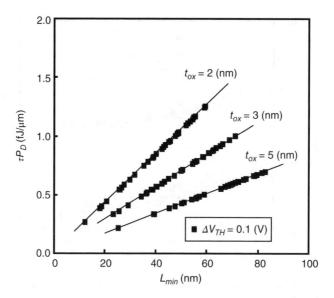

Figure 15.8 Dissipation-power-delay time product (τP_D) dependence on L_{min} at $V_D = 1$ V using *Extract A* ($\Delta V_{TH} = 0.1$ V) for the threshold voltage of 0.3 V.

always satisfied if we simply tune the threshold voltage as mentioned previously. As, in practice, scaled SOI MOSFETs have a relatively large fringe capacitance and capacitance overlap, the practical usefulness of calculating τ simply from simulation results should be reconsidered.

The dissipation power-delay product (τP_D) is often used in estimating the power efficiency of a device. We show the dependence of τP_D on L_{min} at $V_D = 1$ V in Figure 15.8 using data extracted from simulation results achieved with *Extract A*; τP_D is normalized by channel width.

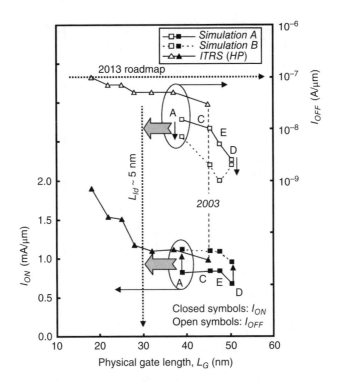

Figure 15.9 Comparison of simulated I_{ON} and I_{OFF} values with 2003–2013 ITRS roadmaps using *Extract A*. This viewgraph shows the potential of the single-gate SOI MOSFET. In addition, simulated I_{ON} and I_{OFF} values assuming the use of a high-κ gate insulator are also shown. *Simulation A:* $t_{ox} = 2$ nm and *simulation B*: EOT = 1 nm.

τP_D is proportional to L_{min}, and inversely proportional to t_{ox}. This is because τP_D is expressed by $C_G V_G^2$ in the present consideration. Curves show that τP_D decreases as t_{ox} increases because the total charge required in switching decreases as t_{ox} increases while τ itself increases because I_{ON} decreases.

In the following, we discuss the performance prospects of scaled SG SOI MOSFETs on the basis of the simulation results shown above. Simulated parameters are compared to the ITRS roadmaps published from 2003 to 2013 [13] (see Figure 15.9 and Figure 15.10); physical L_G equals $L_{min} + 2L_{ld} + 2$ nm and threshold voltage matches the roadmap value by adjusting the work function of the gate electrode (see Table 15.3); here, we follow the high-performance (HP) requirements of the roadmaps. Although shorter L_G values are assumed in the roadmaps because L_{ld} is basically not considered, we use this method in order to suppress the short-channel effect sufficiently. V_G and V_D are assumed to be 1.0 V in the simulations, whereas they range from 1.0 to 1.1 V in the 2003 ITRS roadmap and they assume sub-1 V in the 2013 ITRS roadmap. Simulated values and roadmap requirements of I_{ON} and I_{OFF}, are shown in Figure 15.9 for comparison; *Simulation A* shows values for the case of $t_{ox} = 2$ nm (SiO$_2$) and *Simulation B* shows values for the case of EOT = 1 nm. All simulated I_{ON} values for EOT = 2 nm are lower than the roadmap requirement, while simulated I_{OFF} satisfies the roadmap requirement over 2015 ($<10^{-7}$ A/μm) [13]. The lower values of simulated I_{ON} result from a thicker gate SiO$_2$

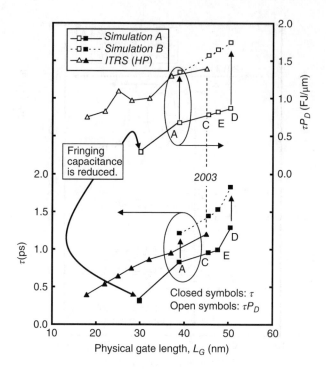

Figure 15.10 Comparison of simulated τ and τP_D values with 2003 to 2013 ITRS roadmaps using *Extract A*. This viewgraph shows the potential of the single-gate SOI MOSFET. In addition, simulated τ and τP_D value assuming the use of a high-κ gate insulator are also shown.

Table 15.3 Possible device parameters for each technology node (Figure 15.9 and Figure 15.10).

Technology node	L_{min} (nm)	t_s (nm)	N_A (cm^{-3})	t_{ox} (nm)	t_{box} (nm)	$2L_{ld}$ (nm)
A	30	5	3×10^{17}	2	10	10
B	26	5	3×10^{15}	2	100	30
C	24	5	3×10^{15}	3	50	30
D	21	10	3×10^{15}	2	10	30
E	18	5	3×10^{17}	2	10	30

film (2 nm) [11]; of course, sub-2 nm thick EOT is possible when a high-κ material is employed [14]. We investigated the impact of EOT on device performance; we compared the performance (I_{ON}) of a device with an EOT of 1 nm (high-κ insulator with a permittivity of $15.6\varepsilon_0$ [15]) with that with an EOT of 2 nm (2 nm thick SiO$_2$ film). Simulation results of I_{ON} for EOT of 1 nm are denoted by the broken line with closed squares in Figure 15.9; they almost satisfy the ITRS roadmap requirement for $L_G > 40$ nm ($L_{min} > 18$ nm). As a result, it is suggested that the condition of EOT < 2 nm is basically appropriate without any change in any other device parameter for $L_G > 40$ nm ($L_{min} > 18$ nm). In contrast to I_{ON}, I_{OFF} is drastically improved by the use of the high-κ gate insulator because the SS is sharpened; it is expected to yield I_{OFF} values that satisfy the roadmap requirement. When we assume $L_{ld} \sim 5$ nm and $V_D < 1$ V, data point "*D*" moves to the left side in Figure 15.9 ($L_G \sim 30$ nm). I_{ON} and I_{OFF} with EOT of 1 nm still satisfy the roadmap requirement for high performance (HP) devices.

Simulated values and roadmap requirements of τ and τP_D are shown in Figure 15.10 for comparison; all simulated values of τ basically meet the roadmap requirements, and all simulated values of τP_D are much lower than the roadmap requirements. This means that possible device parameters for HP applications have been identified. Table 15.3 shows the extracted device parameters for each technology node shown in Figure 15.9 and Figure 15.10. It should be noted that large L_{ld} values, ranging from 10 to 30 nm, are required in contrast to the initial expectation because the short-channel effect must be sufficiently suppressed. As shown in Table 15.3, a very thin SOI layer of 5 nm is required for realizing HP applications because short-channel effects must be sufficiently suppressed. In addition, a quite thin buried oxide layer ranging from 10 nm to sub-100 nm is assumed in the simulations. This suggests that a thinner gate EOT is required if we assume an approximately 100 nm thick buried oxide layer [13]. As shown in Figure 15.10, the high-κ insulator used for the gate insulator does not contribute to the high-speed shift because it enhances the contribution of the inversion layer capacitance, resulting in less improvement in I_{ON} and a degradation in τ. In order to improve the switching speed, we have to use the underlapped source and drain structure so as to maximize I_{ON} and minimize the fringing capacitance [16]. If the fringing capacitance is reduced with a small L_{ld} value, τ, and τP_D will be reduced as shown in Figure 15.10.

We can see that the SG SOI MOSFET still has the potential to support HP and low operation power (LOP) applications as shown in Figure 15.10. However, strained Si-on-insulator (sSOI) [17] or germanium-on-insulator (GOI) [18] will be required when the limits of the silicon-based SG SOI MOSFET are reached [13].

15.4.2 Tradeoff and Optimization of Standby Power Consumption and Dynamic Operation

In the previous section, we discussed the design feasibility of scaled SG SOI MOSFETs using *Extract A;* it was demonstrated that *Extract A* makes it possible to design devices for HP applications. In this section, we discuss the simulation results obtained by *Extract B;* we show on-current (I_{ON}) in Figure 15.11 and off-leakage current (I_{OFF}) in Figure 15.12 as functions of L_{min}. In Figure 15.11, I_{ON} increases as L_{min} decreases regardless of the SS_{max} value as expected. It can be seen that there are many configurations (many device parameter choices) that realize the maximal I_{ON} when a certain L_{min} value is assumed; the threshold voltage is also a parameter to be assumed. It should also be noted that the distribution of I_{ON} values for $SS_{max} = 70$ mV/dec and that of I_{ON} values for $SS_{max} = 90$ mV/dec do not significantly overlap unlike in *Extract A* (see Figure 15.5), which suggests that the intrinsic delay times of devices with $SS_{max} = 70$ mV/dec differ from those with 90 mV/dec. Figure 15.12 shows that I_{OFF} depends strongly on not only SS_{max} but also V_{TH} as expected. We can extract the following empirical equation from Figure 15.12:

$$I_{OFF} = I_L 10^{\frac{-V_{TH}}{SS}} L_{min}^{\sigma_s} \tag{15.12}$$

where we assume parameter $\sigma_s = -1.0$ and $-6/5$ for $V_{TH} = 0.1$ and 0.3 V, respectively; $SS = 70$ mV/dec. On the other hand, $\sigma_s = -6/5$ and $-7/5$ for $V_{TH} = 0.1$ and 0.3 V, respectively; $SS = 90$ mV/dec. It is worthwhile noting that a small swing value makes the standby leakage current relatively insensitive to L_{min}.

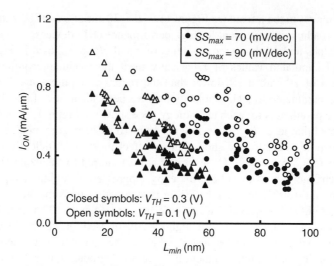

Figure 15.11   Drive current (I_{ON}) dependence on L_{min} at $V_D = 1$ V using *Extract B* ($SS_{max} = 70$ and 90 mV/dec) for various threshold voltages.

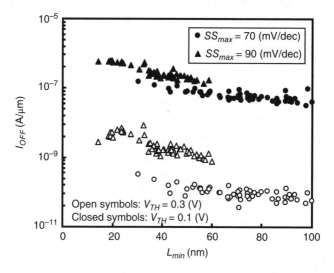

Figure 15.12   Standby leakage current (I_{OFF}) dependence on L_{min} at $V_D = 1$ V in the *Extract B* ($SS_{max} = 70$ and 90 mV/dec) for various threshold voltages.

Figure 15.13 shows the intrinsic delay time (τ) dependence on L_{min} at $V_D = 1$ V with parameters of V_{TH} and SS_{max}. From Figure 15.13, we can see that τ decreases as L_{min} decreases because of the increase in I_{ON} and that the requirement of small τ is not always satisfied if we restrict ourselves to merely tuning the threshold voltage as mentioned previously.

The dissipation power-delay product (τP_D) is often used in estimating the power efficiency of a device. We show the τP_D dependence on L_{min} at $V_D = 1$ V in Figure 15.14 using data

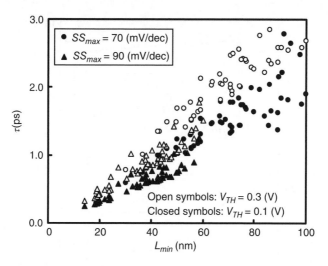

Figure 15.13 Intrinsic delay time (τ) dependence on L_{min} at $V_D = 1\,V$ using *Extract B* ($SS_{max} = 70$ and 90 mV/dec) for various threshold voltages.

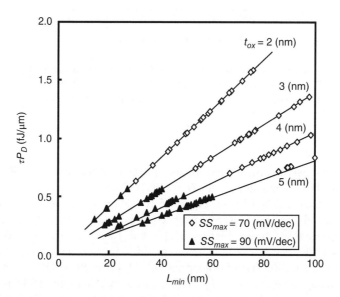

Figure 15.14 Delay-time-dissipation-power product (τP_D) dependence on L_{min} at $V_D = 1\,V$ using *Extract B* ($SS_{max} = 70$ and 90 mV/dec) for the threshold voltage of 0.3 V.

extracted from simulation results gained using *Extract-B*; τP_D is normalized by channel width. τP_D is proportional to L_{min}, and inversely proportional to t_{ox}. This is because τP_D is expressed by $C_G V_G^{\,2}$ in the present analysis. The curves show that τP_D decreases as t_{ox} increases because the total charge required in switching decreases as t_{ox} increases, while τ itself increases because I_{ON} decreases. The important point is that a large SS_{max} value yields a small τP_d value.

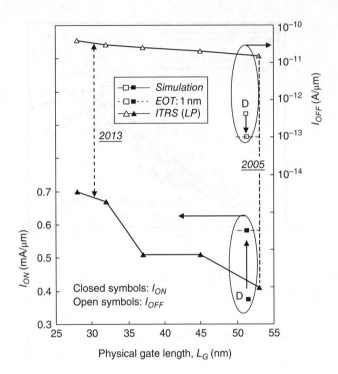

Figure 15.15 Comparison of simulated I_{ON} and I_{OFF} values with 2005–2013 ITRS roadmaps using *Extract B* (SS_{max} = 70 mV/dec). This viewgraph shows the potential of the single-gate SOI MOSFET. In addition, simulated I_{ON} and I_{OFF} values assuming the use of a high-κ gate insulator are shown (D).

In the following, we discuss the performance prospect of the scaled SG SOI MOSFET on the basis of the above simulation results (*Extract B*). Simulated parameters are compared to the ITRS roadmaps of 2003–2013 (see Figure 15.15 and Figure 15.16); physical L_G is equal to $L_{min} + 2L_{ld} + 2$ nm and threshold voltage is matched to the roadmap value by adjusting the work function of the gate electrode. Figures 15.15 and 15.16 were generated by assuming SS_{max} = 70 mV/dec. V_G and V_D are assumed to be 1.0 V in the simulations, while they range from 1.0 to 1.2 V in the 2003 ITRS roadmap and sub-1 V in the 2013 ITRS roadmap. Simulated values and roadmap requirements of I_{ON} and I_{OFF} are shown in Figure 15.15 for comparison; all simulated I_{ON} values are lower than the roadmap requirement after 2005, while simulated I_{OFF} satisfies the roadmap requirement even after 2015. Lower values of simulated I_{ON} result from a thicker gate SiO_2 film (2 nm) [11]; of course, sub-2 nm thick EOT is possible when a high-κ material is employed [14]. The potential for using a high-κ gate insulator with a permittivity of $15.6\varepsilon_0$ [15] is examined in Figure 15.15. The high-κ insulator contributes to the increase in I_{ON}, and I_{ON} values for devices with the high-κ insulator match the roadmap requirement, which is different from the result seen in Figure 15.9. In addition, I_{OFF} is quite improved by use of the high-κ gate insulator because the SS is sharpened; even after 2015, I_{OFF} values satisfy the roadmap requirement.

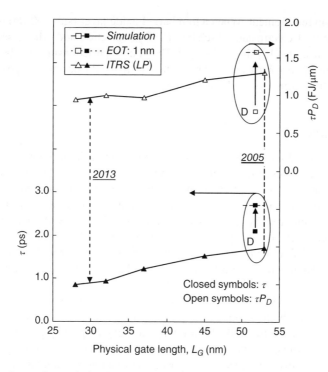

Figure 15.16 Comparison of simulated τ and τP_D values with 2005–2013 ITRS roadmaps using *Extract B* ($SS_{max} = 70\,mV/dec$). This viewgraph shows the potential of the single-gate SOI MOSFET. In addition, simulated τ and τP_D values assuming the use of a high-κ gate insulator are shown (D).

Table 15.4 Possible device parameters for each technology node (Figure 15.15 and Figure 15.16).

Technology node	L_{min} (nm)	t_s (nm)	N_A (cm^{-3})	t_{ox} (nm)	t_{box} (nm)	$2L_{ld}$ (nm)	WF (eV)
A	52	5	3×10^{17}	2	10	20	5.2
B	45	5	3×10^{17}	2	20	20	5.2
C	39	10	3×10^{15}	2	50	20	5.3
D	32	5	3×10^{15}	2	20	20	5.2

Simulated values and roadmap requirements of τ and τP_D are shown in Figure 15.16 for comparison; all simulated values of τ are greater than the roadmap requirements, while all simulated values of τP_D are much lower than the roadmap requirements. Accordingly, we extracted possible device parameters for low-standby power (LSTP) applications. Table 15.4 lists the extracted device parameters for each technology node shown in Figure 15.15 and Figure 15.16. It should be noted that a large L_{ld} value of 20 nm is required (confounding the simple expectation) in order to suppress the band-to-band tunnel current; we must reconsider the source-drain structure to minimize the switching time [16]. A very thin SOI layer of 5 nm is required to realize LSTP applications because the short-channel effects must be sufficiently suppressed. A moderately thin buried oxide layer ranging from 10 to 50 nm is assumed in the simulations. Use of the high-κ gate insulator allows a thicker buried oxide layer, which would match the roadmap requirement. When a thin buried oxide layer (sub-100 nm) is not allowed from the viewpoint of electrostatic-discharge (ESD) protection and other issues, a double-buried-insulator substrate [19] should be used. The double-buried insulator substrate has high potential with regard to robust circuit applications but would be more expensive than the present SOI substrate with a single buried insulator.

15.5 Summary

This chapter considered an advanced methodology to extract the minimum channel length. A new approach to determine the minimum channel length was proposed, based on the specific restriction of the value of SS. A comprehensive design guideline for scaled SG SOI MOSFETs was derived from the viewpoints of high-speed operation and low power consumption or low standby power consumption. Simulations suggest that the SG SOI MOSFET can achieve the minimum channel length of about 10 nm if we can assume the threshold voltage roll-off restriction ($\Delta V_{TH}=0.1$ V) (*Extract A*). We also find that the SG SOI MOSFET can achieve the minimum channel lengths of about 30, 20, or 10 nm when the maximal SS value is restricted to 70, 80, or 90 mV/dec, respectively (*Extract B*).

When *Extract A* is used in SG SOI MOSFET scaling, it has been demonstrated that it gives us a design guideline for high-performance devices. The simulation results shown here counter the negative opinion of the SG SOI MOSFET by strongly suggesting that it can offer high-speed performance sufficient to meet the 2014 roadmap requirements if device parameters are set appropriately. However, in practice, the post-2104 roadmaps require a sub-100 nm thick buried oxide layer. This limits the design window; a sub-1 nm thick EOT and/or a sub-5 nm thick SOI layer must be applied.

Extract-B was shown to yield a design guideline for low standby power devices. Simulation results strongly suggest that the corresponding SG SOI MOSFET can offer adequate low-standby power performance beyond 2015 if the appropriate device parameters are set. The ITRS roadmap requiring sub-100 nm thick buried oxide layer may limit the design window and require nonideal optimization of device parameters. In this case, the use of a high-κ gate insulator is one possible solution.

We have shown that the thickness of the buried oxide layer significantly restricts SG SOI MOSFET scaling, and so must be considered as a key issue to be resolved for advancing SOI devices and substrate technology.

References

[1] Y. Taur and T. H. Ning, "Fundamentals of Modern VLSI Devices," Cambridge University Press, 1998.
[2] H. S. Momose, M. Ono, T. Yoshitomi, T. Ohguro, S. Nakamura, M. Saito, and H. Iwai, "Tunneling Gate Oxide Approach to Ultra-High Current Drive in Small-Geometry MOSFETs," 1994 IEEE Int. Electron Devices Meeting (San Francisco, 1994) pp. 593–596, 1994.
[3] M. Wiatr, P. Seegebrecht, and H. Peters, "Charge based modeling of the inner fringing capacitance of SOI-MOSFETs," *Solid State Electron.*, vol. 45, pp. 585–592, 2002.
[4] J.-P. Colinge, "Silicon-on-Insulator Technology: Materials to VLSI," 3rd ed. (Kluwer Academic Publishers, 2004).
[5] M. Jurczak, T. Skotnicki, M. Paoli, B. Tormen, J. Martins, J. L. Regolini, D. Dutartre, P. Ribot, D. Lenoble, R. Pantel, and S. Monfray, "Silicon-on-Nothing (SON)–an innovative process for advanced CMOS," *IEEE Trans. Electron Devices, vol.* 47, pp. 2179–2187, 2000.

[6] P. M. Zeitzoff and J. E. Chung, "A perspective from the 2003 ITRS," *IEEE Circuits Devices Mag.*, vol. 21, pp. 4–15, 2005.

[7] A. Kawamoto, H. Mitsuda, and Y. Omura, "Design guideline for silicon-on-insulator MOSFET," *IEEE Trans. Electron Devices*, vol. 50, pp. 2303–2305, 2003.

[8] A. Kawamoto, S. Sato, and Y. Omura, "Engineering source and drain diffusion for sub-100-nm channel silicon-on-insulator MOSFETs," *IEEE Trans. Electron Devices*, vol. 51, pp. 907–913, 2004.

[9] A. Kawamoto and Y. Omura, "Optimization of Lateral Diffusion of Source and Drain for Sub-100-nm Channel Silicon-on-Insulator MOSFETs," Proc. 2nd ECS Int. Semicond. Technol. Conf. (ISTC-2002, Tokyo) pp. 128–134, 2003.

[10] Synopsys Inc., "Sentaurus Device User Guide," Version A-2007, 2007, ftp://147.46.117.90/synopsys/manuals/PDFManual/data/sdevice_ug.pdf (accessed June 10, 2016.)

[11] Y. Omura, "Silicon-on-Insulator (SOI) MOSFET Structure for Sub-50-nm Channel Regime," Proc. of 10th Int. Symp. on Silicon-on-Insulator Technology and Devices (The Electrochem. Soc., Washington DC, 2001) PV2001-3, pp. 205–210, 2001.

[12] Y. Omura, "Silicon-on-insulator MOSFET structure for sub-100 nm channel regime and performance perspective," *J. Electrochem. Soc.*, vol. 148, no. 12, pp. G476–G479, 2001.

[13] "International Technology Roadmap for Semiconductors," 2001 and 2013, http://www.itrs2.net/itrs-reports.html (accessed June 10, 2016); see "2003 ITRS" and "2013 ITRS"

[14] Y. Nakamori, K. Komiya, and Y. Omura, "Physics-based model of quantum mechanical wave function penetration into thin dielectric films for evaluating modern MOS capacitors," *Solid State Electron.*, vol. 49, pp. 1118–1126, 2005.

[15] D. J. Frank and H.-S. P. Wong, "Analysis of the Design Space Available for High-κ Gate Dielectrics in Nanoscale MOSFETs," Abstr., IEEE 2000 Si Nanoelectronics Workshop (Hawaii, June, 2000) pp. 47–48, 2000.

[16] V. P. Trivedi and J. G. Fossum, "Source/Drain Engineering for Optimal Nanoscale FinFET Design," Proc. 2004 IEEE Int. SOI Conf. (California, 2004) pp. 192–194, 2004.

[17] T. A. Langdo, M. T. Currie, A. Lochtefeld, R. Hammond, J. A. Carlin, M. Erdtmann, G. Braithwaite, V. K. Yang, C. J. Vineis, H. Badawi, and T. Bulsara, "SiGe-free strained Si on insulator by wafer bonding and layer transfer," *Appl. Phys. Lett.*, vol. 82, pp. 4256–4258, 2003.

[18] C. H. Huang, M. Y. Yang, A. Chin, W. J. Chen, C. X. Zhu, B. J. Cho, M.-F. Li, and D.-L. Kwong, "Very Low Defects and High Performance Ge-On-Insulator p-MOSFETs with Al_2O_3 Gate Dielectrics," Proc. Int. Symp. VLSI Technol., (Kyoto, 2003) pp. 119–120, 2003.

[19] T. Ohno, S. Matsumoto, and K. Izumi, "An intelligent power IC with double buried-oxide layers formed by SIMOX technology," *IEEE Trans. Electron Devices*, vol. 40, pp. 2074–2080, 1993.

16

Performance Prospects of Fully Depleted SOI MOSFET-Based Diodes Applied to Schenkel Circuits for RF-ID Chips

16.1 Introduction

As RF-ID chips have no internal power supply, they need a way to use the received signal as an energy source; a common approach is the Schenkel circuit [1, 2]. The basic Schenkel circuit is shown in Figure 16.1. It usually consists of capacitors and pn diodes (PNDs) or Schottky-barrier diodes (SBDs). Modern RF applications such as RF-ID chips often use SBDs in this circuit [2–4]. Unfortunately, generally speaking, the reverse-biased current (I_R) of an SBD is not significantly lower than the forward-biased current (I_F) because the requirement for high drive currents results in a low barrier height. The reverse-biased current should be extremely low because the AC signal voltage received is very small in RF-ID systems. Overall, generally speaking, Schenkel circuits that use SBDs fail to offer high boost-up efficiency, resulting in boost-up circuit blockage.

Recently, RF-ID chips were applied to various systems in which the emphasis was on public security and safety rather than limiting production costs [5]. In these cases, performance and reliability are primarily important. So, a new market for RF-ID chips is growing.

On the other hand, the silicon-on insulator metal oxide semiconductor field-effect transistor (SOI MOSFET) is a promising device that can be applied to RF circuit applications [6] because a high-resistivity substrate can be easily introduced [7]. As the high-resistivity substrate provides not only a low loss transmission of the RF signal [7] but also a low cross-talk in digital circuits [8], various applications have already been reported [9, 10].

In this chapter, we discuss using the SOI-MOSFET-based quasidiode (SOI-QD) to replace the SBD in Schenkel circuits. First we propose the expression of boost-up efficiency for a low-frequency range using the experimental DC characteristics of SOI-QD made from various

MOS Devices for Low-Voltage and Low-Energy Applications, First Edition.
Yasuhisa Omura, Abhijit Mallik, and Naoto Matsuo.
© 2017 John Wiley & Sons Singapore Pte. Ltd. Published 2017 by John Wiley & Sons Singapore Pte. Ltd.

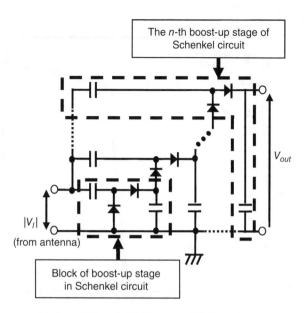

Figure 16.1 Schematic of Schenkel circuit. Y. Omura and Y. Iida, "Performance Prospects of Fully-Depleted SOI MOSFET-Based Diodes applied to Schenkel Circuit for RF-ID Chips," Circuits and Systems, Vol. 4 No. 2, 2013, pp. 173–180. doi: 10.4236/cs.2013.42024.

SOI MOSFETs, and AC analyses of SOI-QD are conducted using a two-dimensional (2D) device-simulator (Integrated Systems Engineering (ISE) *DESISS* [11]) to investigate operation stability in the RF band. We also define another expression of boost-up efficiency in the RF band, and examine its availability on the basis of AC simulation results. The RF-band potential of a SOI-MOSFET-based quasidiode is addressed from the viewpoint of future RF applications.

16.2 Remaining Issues with Conventional Schenkel Circuits and an Advanced Proposal

We used the PSpice [12, 13] circuit simulator to examine the performance of a Schenkel circuit that used the SBD, PND, or conventional bulk MOSFET-based quasidiode (CB-QD); in the CB-QD variant, the gate terminal, and the drain terminal are connected and the source terminal and the substrate terminal are connected. We assumed that the SBD and PND had a junction area of $46.1\,\mu m^2$, and that the gate width (W_G) and the gate length (L_G) of the bulk MOSFET were 20.6 and $0.32\,\mu m$. All devices had identical active areas; the junction area of bulk MOSFET is $39.5\,\mu m^2$. The circuit simulations employed the empirical model (level = 3) for simplicity [12]. To obtain a realistic device performance from the PND and SBD variants, we introduced the minority carrier lifetime model shown in section 16.5.

Figure 16.2 shows simulated rectifier characteristics of the various diodes in a low voltage input anode range (V_A). We can see that the SBD with a barrier height (φ_b) of $0.15\,eV$ has the largest driving current among the three diodes, but it has the highest reverse-biased current ($4.43\,\mu A$). Figure 16.3 shows the performance of five-stage Schenkel circuits that use

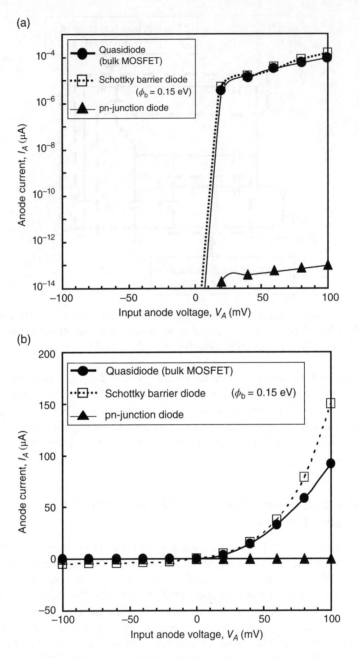

Figure 16.2 Forward and reverse characteristics of various diodes (PSpice simulation results). (a) Log scale plot and (b) linear scale plot. Y. Omura and Y. Iida, "Performance Prospects of Fully-Depleted SOI MOSFET-Based Diodes applied to Schenkel Circuit for RF-ID Chips," *Circuits and Systems*, Vol. 4 No. 2, 2013, pp. 173–180. doi: 10.4236/cs.2013.42024.

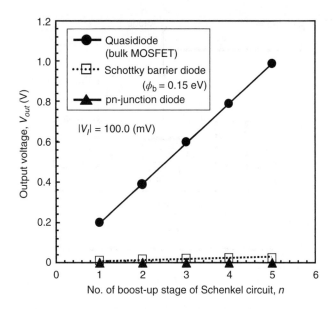

Figure 16.3 Boost-up performance of Schenkel circuit with various diodes (PSpice simulation results). Y. Omura and Y. Iida, "Performance Prospects of Fully-Depleted SOI MOSFET-Based Diodes applied to Schenkel Circuit for RF-ID Chips," Circuits and Systems, Vol. 4 No. 2, 2013, pp. 173–180. doi: 10.4236/cs.2013.42024.

the three different diodes for an input voltage $|V_I|$ of 100 mV. It shows that the CB-QD successfully boosts the input signal from a very low level to an acceptable level, while the SBD and pn-junction diode fail to do so. The SBD failed to match this despite its large driving current; as current SBD designs have a high I_R value, almost identical to I_F, in the low voltage range of 100 mV, the high leakage current (I_R) degrades the signal boost process. On the other hand, the I_F and I_R values of the CB-QD are much smaller than those of the SBD. However, CB-QD offers an acceptable level of boost. The main reason is that the I_F of the CB-QD is larger than its own I_R, which means that the effective boost efficiency (η) of a Schenkel circuit should not be determined by the direct value of driving current, but by the ratio of I_F to I_R defined as

$$\eta = I_F / \left(I_F + |I_R| \right) \tag{16.1}$$

In the voltage boost-up mechanism, the leakage current of the diode prevents the capacitor from carrying out its charge-up process. A large I_R value therefore loses a part of the charge stored in a capacitor connected in series. So, the above definition of boost-up efficiency is available independently of the number of boost stages in low-frequency conditions where nonlocal effects are negligible. Moreover, the voltage across the capacitor rises when the capacitance of the rectifier is larger that of the capacitor connected in series; however, this results in a low-level current source in contrast to the purpose. Therefore, we have to optimize the capacitance of the capacitor so as to fit the performance objectives.

Unfortunately, we cannot apply the above CB-QD to a practical Schenkel circuit as is because it has a crucial drawback: a CB-QD made on an n-channel bulk MOSFET has a

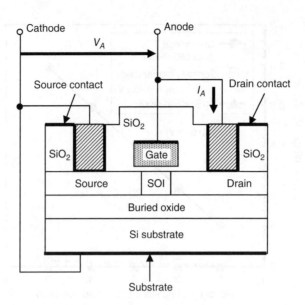

Figure 16.4 SOI-MOSFET and terminal nodes for quasidiode operation. Device structure is also assumed for a. c. analyses. Device parameters are carefully designed to take account of practical evaluate gate overlap capacitance, S/D parasitic resistance, fringing capacitance, and other parasitic effects in the device. Y. Omura and Y. Iida, "Performance Prospects of Fully-Depleted SOI MOSFET-Based Diodes applied to Schenkel Circuit for RF-ID Chips," Circuits and Systems, Vol. 4 No. 2, 2013, pp. 173–180. doi: 10.4236/cs.2013.42024.

parasitic pn-junction diode between the drain and the substrate that can work when the drain is negatively biased. This effective reverse current I_R of CB-QD that passes through the parasitic pn-junction diode degrades the η value.

Our solution is to base the quasidiode on an SOI MOSFET instead of a bulk MOSFET to raise the η value. Figure 16.4 shows the device structure assumed here and the terminal nodes of an SOI-MOSFET-based quasidiode (SOI-QD). The n-channel fully depleted (FD) SOI MOSFETs used here for evaluation of device performance had gate lengths (L_G) of 0.32 and 1.0 µm (see Table 16.1); these devices are used only for a feasibility test and they are not well tempered for the present purpose. Later we perform AC simulations for SOI-QD. In this case it is anticipated that the gate-to-source capacitance and the gate-to-drain capacitance play important roles in the AC analyses because they yield various parasitic capacitances including fringing capacitances [14]. So, we consider a realistic device structure to obtain reliable AC simulation results. All physical parameters to draw the cross-section of device are determined on the basis of the 0.4 µm complementary metal oxide semiconductor (CMOS) design rule [15]; they are identical to those of the device used in the present experiments. In simulations described later, the doping level of the SOI layer, $N_{a,ch}$, was changed from 6.0×10^{16} to 3.0×10^{17} cm^{-3} to adjust the threshold voltage.

Figure 16.5 shows the I_D-V_G characteristics of the FD-SOI MOSFETs (L_G=0.32 and 1.0 µm) measured at V_D=50 mV; the substrate bias was 0V. The subthreshold swing (SS) values of the two devices are quite different. It can be seen that, because of short channel effects, the SS value is larger at L_G=0.32 µm than at L_G=1.0 µm.

Table 16.1   Device parameters used in experiments.

Parameters	Values (units)
L_G	0.32, 1.0 (μm)
W_G	20.6 (μm)
t_S	50.0 (nm)
t_{BOX}	80.0 (nm)
t_{ox}	7.0 (nm)
$N_{a,ch}$	3.0×10^{17} (cm^{-3})

Y. Omura and Y. Iida, "Performance Prospects of Fully-Depleted SOI MOSFET-Based Diodes applied to Schenkel Circuit for RF-ID Chips," Circuits and Systems, Vol. 4 No. 2, 2013, pp. 173–180. doi: 10.4236/cs.2013.42024.

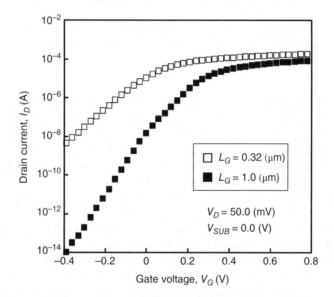

Figure 16.5 I_D-V_G characteristics of fully depleted SOI MOSFET (experimental results). Y. Omura and Y. Iida, "Performance Prospects of Fully-Depleted SOI MOSFET-Based Diodes applied to Schenkel Circuit for RF-ID Chips," Circuits and Systems, Vol. 4 No. 2, 2013, pp. 173-180. doi: 10.4236/cs.2013.42024.

It is anticipated that the device characteristics are sensitive to substrate bias (V_{SUB}) because of the thin buried oxide layer, so we can produce SOI-QDs with various rectification characteristics by modifying the substrate bias (V_{SUB}) applied to the FD-SOI MOSFET. It should be noted that this technique is introduced to examine how the threshold voltage of FD-SOI MOSFET modulates the η value through the change of I_F and I_R. With the substrate bias, we can easily vary the SS value and the current level as well as the threshold voltage. As a result, we can find the best solution for the SOI-QDs rectification characteristics to be tuned. In practical applications, we cannot assume the substrate bias to the device because the assumed

Table 16.2 I_F, I_R, R_{ch}, and η values of SOI-QD at various operation conditions.

Group	L_G=0.32 (µm)	SS (mV/dec)/V_{TH} (V)	Forward bias		Reverse bias		η (%)[a]
			I_F (µA)	R_{ch} (kΩ)	I_R (µA)	R_{ch} (kΩ)	
A	V_{SUB}=0.25 (V)	211/0.01	90.0	1.1	−47.0	2.2	64.4
B	V_{SUB}=0.0 (V)	142/0.10	2.1	47.0	−0.89	110	70.7
C	V_{SUB}=−1.5 (V)	132/0.27	1.2	81.0	−0.55	180	69.3
Group	L_G=1.0 (µm)	SS (mV/dec)/V_{TH} (V)	Forward bias		Reverse bias		η (%)[a]
			I_F (µA)	R_{ch} (kΩ)	I_R (µA)	R_{ch} (kΩ)	
A	V_{SUB}=3.0 (V)	169/0.01	29.0	3.4	−11.0	8.8×10^{-3}	72.0
B	V_{SUB}=0.0 (V)	71.8/0.20	0.25	401.0	−0.016	6.2	94.0
C	V_{SUB}=−3.0 (V)	71.5/0.92	7.2×10^{-3}	1.4×10^4	-2.6×10^{-4}	3.9×10^3	96.6

[a]Amplitude of input signal is 100 mV.
Y. Omura and Y. Iida, "Performance Prospects of Fully-Depleted SOI MOSFET-Based Diodes applied to Schenkel Circuit for RF-ID Chips," Circuits and Systems, Vol. 4 No. 2, 2013, pp. 173–180. doi: 10.4236/cs.2013.42024.

RF-ID chip does not have a voltage supplier; at the stage of device design, the threshold voltage must be tuned by positive voltage parameter if possible. In this chapter we apply a negative, zero, or a positive V_{SUB} value to the device; these conditions are labeled "A," "B," or "C" (see Table 16.2), respectively. It should be noted that the V_{SUB} value at different L_{eff} values is not identical when the threshold voltage (V_{TH}) is adjusted to be the same. Figure 16.6 shows the I_A-V_A characteristics of SOI-QD "A," "B," and "C."

First, we consider the impact of the L_{eff} value as shown in Figure 16.6. In condition "A," the SOI-QD works near the threshold voltage (V_{TH}) because V_{SUB} is positive. S value is larger in condition "A" than in condition "B," resulting in a smaller ratio of I_F/I_R and thus a smaller η value (see Table 16.2 and Figure 16.6a,b). In condition "A," however, since the channel resistance (R_{ch}) is reduced due to the lowering of the threshold voltage (V_{TH}), the driving current of the device increases. In contrast, in condition "C," the device works in the sub-threshold region because of the negative substrate bias (V_{SUB}). The SS value is smaller in condition "C" than in condition "B"; the channel resistance (R_{ch}) in condition "C" increases due to the raising of the threshold voltage (V_{TH}), resulting in a lower drive current, but identical η value to that seen in condition "B" (see Figure 16.6c).

Next we compare the performance of the two devices (L_G=1.0 and 0.32 µm) shown in Figure 16.6. As mentioned above, the SS value with L_G=1.0 µm is smaller than that with L_G=0.32 µm. Accordingly, I_F/I_R is larger with L_G=1.0 µm than that with L_G=0.32 µm, and the η value with L_G=1.0 µm is larger than that with L_G=0.32 µm as shown in Table 16.2. It should be noted that the difference in forward current level (I_F) at the same bias comes from not only the different W_G/L_G value but also the different SS value. This means that we must take account of the tradeoff between η and I_F – that is, it is necessary to select the most suitable operation bias condition and to reduce the SS value. Table 16.2 suggests that the SS value should be less than 75 mV/dec to reduce I_R value and threshold voltage should be ranging from 0.1 to 0.2 V to gain a high η value simultaneously

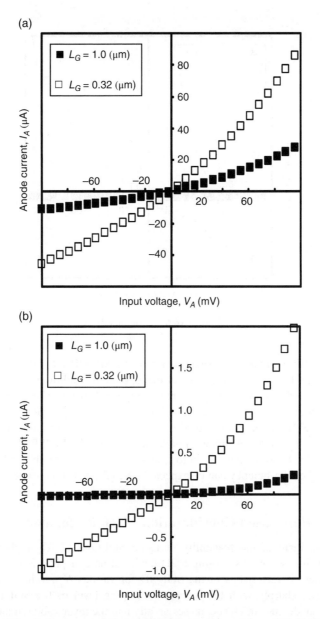

Figure 16.6 Rectifier I_A-V_A characteristics of SOI-QD (experimental results). Y. Omura and Y. Iida, "Performance Prospects of Fully-Depleted SOI MOSFET-Based Diodes applied to Schenkel Circuit for RF-ID Chips," Circuits and Systems, Vol. 4 No. 2, 2013, pp. 173-180. doi: 10.4236/cs.2013.42024. (a) Operation condition "A" ($V_{SUB}>0$), (b) operation condition "B" ($V_{SUB}=0$V), and (c) operation condition "C" ($V_{SUB}<0$).

(c)

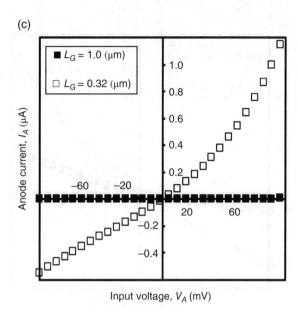

Figure 16.6　(Continued)

Finally, we briefly compare the performance of the SBD, PND and SOI-QD; Table 16.2 shows the I_F, I_R, and η of the SOI-QD at various operation conditions when the input signal amplitude ($|V_A|$) is 100 mV. According to a recent report [3], when $V_A = 100$ mV, the SBD has an I_F value of about 1.0 nA and the pn-junction diode has an I_F value of about 0.1 pA, this indicates that the SOI-QD has an identical I_F to the other devices or has a larger I_F at all conditions for the two L_G values examined. In addition, as the SOI MOSFET has lower leakage current than the SBD and the subthreshold characteristic of well-tempered SOI MOSFET is excellent, a high boost efficiency can be expected when the SOI-QD is used.

16.3　Simulation-Based Consideration of RF Performance of SOI-QD

In order to investigate of the feasibility of using SOI-QDs in RF applications [16], we conducted extensive AC analyses using a 2D device simulator [11]. Figure 16.7 shows the simulated rectifier characteristics of the SOI-QD for various $N_{a,ch}$ values. As the threshold voltage (V_{TH}) rises sharply with the increase in doping level of the SOI layer ($N_{a,ch}$), the forward-biased anode current (I_F) decreases greatly and the reverse-biased anode current (I_R) also decreases. However, η increases on the basis of Eq. (16.1) as $N_{a,ch}$ increases because the reduction of I_R overwhelms that of I_F, which is as expected. A simple estimation method of η value of SOI MOSFET is shown in section 16.6.

When a high-frequency operation is considered, the AC response to the applied signal amplitude should be evaluated; this is very important in SOI MOSFET because it is anticipated that the floating-body effect delays the current response to the applied signal [17]. We think that anode conductance g_A ($= dI_A/dV_A$) successfully traces response capability of the SOI MOSFET because g_A is extracted from a small signal analysis; we think that a large signal

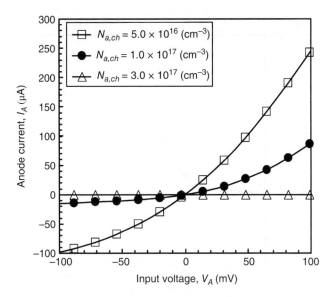

Figure 16.7   Simulated *I-V* characteristics of quasidiodes (device simulations). Y. Omura and Y. Iida, "Performance Prospects of Fully-Depleted SOI MOSFET-Based Diodes applied to Schenkel Circuit for RF-ID Chips," Circuits and Systems, Vol. 4 No. 2, 2013, pp. 173–180. doi: 10.4236/cs.2013.42024.

analysis is not always required because the Fourier transformation results of signals are effectively considered. Then we define the boost-up efficiency to AC signals as

$$\eta = \left\{ \Delta I_{AF}\left(f\right) / \Delta V_A\left(f\right) \right\} / \left\{ \Delta I_{AF}\left(f\right) / \Delta V_A\left(f\right) + \Delta I_{AR}\left(f\right) / \Delta V_A\left(f\right) \right\}$$
$$= g_{AF}\left(f\right) / \left\{ g_{AF}\left(f\right) + \left| g_{AR}\left(f\right) \right| \right\}$$
(16.2)

where g_{AF} and g_{AR} mean the anode conductance at the forward and the backward bias conditions, respectively; I_{AF} and I_{AR} are the anode current at the forward and the backward bias conditions, respectively.

Figure 16.8 shows the simulated g_A-V_A curves for various $N_{a,ch}$ values, where $g_A = dI_A/dV_A$. Figure 16.8a is for $N_{a,ch} = 5 \times 10^{16}\,\mathrm{cm}^{-3}$, Figure 16.8b for $N_{a,ch} = 1 \times 10^{17}\,\mathrm{cm}^{-3}$, and Figure 16.8c for $N_{a,ch} = 3 \times 10^{17}\,\mathrm{cm}^{-3}$. Frequency f was changed from 1 Hz to 10 GHz. In the case of $N_{a,ch} = 5.0 \times 10^{16}\,\mathrm{cm}^{-3}$ (see Figure 16.8a), the g_A-V_A characteristic is not sensitive to frequency (1 Hz to 10 GHz). In the cases of $N_{a,ch} = 1.0 \times 10^{17}\,\mathrm{cm}^{-3}$ (see Figure 16.8b) and $3.0 \times 10^{17}\,\mathrm{cm}^{-3}$ (see Figure 16.8c), however, the g_A-V_A characteristic is sensitive to frequency. In particular, for $N_{a,ch} = 3.0 \times 10^{17}\,\mathrm{cm}^{-3}$, the g_A-V_A characteristic reacts strongly to frequency.

As the SOI layer is 50 nm thick, the fully depleted condition is satisfied in two cases (Figure 16.8a,b) [18]. Majority carriers (holes) are basically not responsible for device operation, and so the parasitic bipolar action is not expected at the bias condition used ($|V_A| < 0.3\,\mathrm{V}$). As the threshold voltage is very low (~0 V), electrons in the inversion layer rule device operation; in this case, the frequency dependence of the dielectric response of electrons ("majority carriers" near the surface) limits the g_A-V_A characteristic. The limit of the frequency

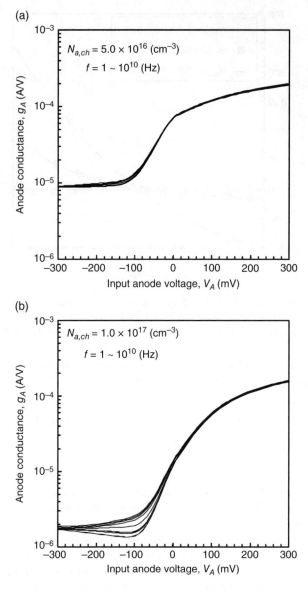

Figure 16.8 Simulated g_A-V_A characteristics (device simulations). (a) $N_{a,ch}=5.0\times10^{16}$ (cm⁻³), (b) $N_{a,ch}=1.0\times10^{17}$ (cm⁻³), and (c) $N_{a,ch}=3.0\times10^{17}$ (cm⁻³). Y. Omura and Y. Iida, "Performance Prospects of Fully-Depleted SOI MOSFET-Based Diodes applied to Schenkel Circuit for RF-ID Chips," Circuits and Systems, Vol. 4 No. 2, 2013, pp. 173–180. doi: 10.4236/cs.2013.42024.

response of electrons in Si is higher than 100 GHz [16], so the simulation results shown in Figure 16.8a and b are acceptable. The smaller variation in g_A-V_A characteristics seen in Figure 16.8b is related to the remaining hole density near the bottom of SOI layer, which should be higher than that in Figure 16.8a.

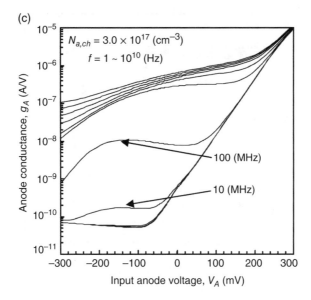

Figure 16.8 (Continued)

We note that the fully depleted condition is not satisfied in the case of $N_{a,ch}=3.0\times10^{17}$ cm^{-3} [18]. That is, the majority carriers, remaining near the SOI/buried oxide interface, play an important role in determining device operation, which corresponds to the typical floating body effect [17]. The g_A-V_A characteristics at frequencies above 10 MHz differ from those below 10 MHz as seen in Figure 16.8c; in other words, the g_A-V_A characteristics at frequencies above 10 MHz are not normal. In the present case, it is easily anticipated that the generation process of majority carriers (holes) (around the junction and inside the depleted body) rules the dynamic operation of SOI-QD. As the generation-recombination time constant is about 0.1 μs in the present simulations (see section 16.5), the hole-generation process does not respond at frequencies above 10 MHz, resulting in the body floating effect [17].

Figure 16.9 shows the simulated η-f characteristics of the SOI-QD from $f=1$ Hz to 1 THz with the parameter of $N_{a,ch}$, although the simulated value of η is not reliable for $f>100$ GHz because physical models for devices in the device simulator [11] are not proposed for such a high frequency; we simply focus on the behavior of η. The η values shown in Figure 16.9 are calculated by Eq. (16.2). In Figure 16.9, it should be noted that η values calculated by g_A (Eq. (16.2)) at low frequency region are almost identical to η values calculated from DC current (experimental results) for $L_G=1.0$ μm by Eq. (16.1). So this suggests that the η value can be estimated using Eq. (16.2) at the RF region. In Figure 16.9, we can see that the η value remains higher than 90% independently of $N_{a,ch}$ up to 10 MHz; this is supported by the fact that the SS value in the active range of I_A of SOI-QD is sufficiently small. However, when f is higher than 100 MHz, especially when $N_{a,ch}=3.0\times10^{17}$ cm^{-3}, η falls to 60%. This comes from the body-floating effect as mentioned previously. On the other hand, at $N_{a,ch}=5.0\times10^{16}$ and 1.0×10^{17} cm^{-3}, η remains high up to 10 GHz. As a result, SOI-QDs with $N_{a,ch}=5.0\times10^{16}$ and 1.0×10^{17} cm^{-3} can be used in RF applications; when $N_{a,ch}=3.0\times10^{17}$ cm^{-3}, the SOI-QD is no longer suitable because of the significant body floating effect.

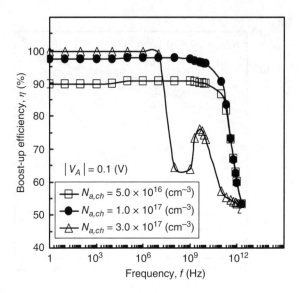

Figure 16.9 η-f characteristics for various body doping levels (device simulations). Y. Omura and Y. Iida, "Performance Prospects of Fully-Depleted SOI MOSFET-Based Diodes applied to Schenkel Circuit for RF-ID Chips," Circuits and Systems, Vol. 4 No. 2, 2013, pp. 173–180. doi: 10.4236/cs.2013.42024.

Figure 16.10 shows the simulated η-$N_{a,ch}$ characteristics of the SOI-QD at 1 MHz and 4 GHz. In the low frequency range (~1 MHz) η slightly increases with $N_{a,ch}$ because of the reduction of SS at the driving point, and finally reaches its upper limit when the reduction in SS value ceases. As $N_{a,ch}$ increases, driving current (I_A) decreases because the threshold voltage of FD-SOI MOSFET rises. Thus, the doping level of the SOI layer of SOI-QD must be optimized to realize a high-performance Schenkel circuit. In the radio frequency range (~4 GHz), η peaks because of the floating body effect mentioned above. This strongly suggests that the value of $N_{a,ch}$ must be selected so as to hold the SOI layer in the fully depleted condition of the SOI layer. As for the present simulation, $N_{a,ch} = 1.0 \times 10^{17}$ cm^{-3} yields the best SOI-QD performance in RF applications up to 100 GHz; when SOI layer is thinned to a range of sub-50 nm, a $N_{a,ch}$ value higher than 1×10^{17} cm^{-3} can be applied to devices [19].

In the above simulations, we assumed that $V_A = 0.1$ V because we considered the case of short-distance communications. Since the above simulations show that the proposed SOI-QD produces almost identical performance at $V_A < 0.1$ V, we think that Schenkel circuits with SOI-QDs are also applicable to long-distance communications.

16.4   Summary

The feasibility of replacing SBDs in the Schenkel circuit with SOI-MOSFETs as quasidiodes was examined by experiments and simulations. The reverse-biased current (I_R) of the SOI-QD is much lower than its forward-biased current (I_F) and the driving current (I_F) is high because of the excellent SS value provided by the SOI-MOSFET arrangement; we noted that the tradeoff between boost efficiency (η) and I_F should be taken into account. In addition,

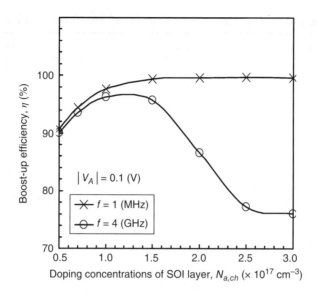

Figure 16.10 η-$N_{a,ch}$ characteristics for various frequencies (device simulations). Y. Omura and Y. Iida, "Performance Prospects of Fully-Depleted SOI MOSFET-Based Diodes applied to Schenkel Circuit for RF-ID Chips," Circuits and Systems, Vol. 4 No. 2, 2013, pp. 173–180. doi: 10.4236/cs.2013.42024.

AC analyses using a two-dimensional device simulator showed that the body doping concentration ($N_{a,ch}$) of the SOI layer should be optimized so as to hold the fully-depleted condition for RF applications up to 100 GHz.

16.5   Appendix: A Simulation Model for Minority Carrier Lifetime

Here we introduce the model for minority carrier lifetimes [11] used in PSpice simulations to ensure consistency with Device Simulation for Smart Integrated Systems (DESSIS) simulation results:

$$\tau = \tau_{max} + \frac{\tau_{max} - \tau_{min}}{1 + \left(N / N_{ref} \right)^{\gamma}} \qquad (16.3)$$

where N is the doping density, N_{ref} is the doping parameter, γ is the fitting parameter, τ_{max} and τ_{min} are lifetime parameters. Parameter values used here are summarized in Table 16.3.

16.6   Appendix: Design Guideline for SOI-QDs

We rewrite the expression of boost-up efficiency given by Eq. (16.1):

$$\eta = I_F / \left(I_F + |I_R| \right) \qquad (16.4)$$

Table 16.3   Physical parameters assumed in device simulations (DESSIS).

Parameters	Values (units)	Comments
τ_{max}	1.0×10^{-5} (s)	Electrons
	3.0×10^{-5} (s)	Holes
τ_{min}	0.0 (s)	Electrons
	0.0 (s)	Holes
N_{ref}	1.0×10^{16} (cm^{-3})	—
γ	1.0	—

Y. Omura and Y. Iida, "Performance Prospects of Fully-Depleted SOI MOSFET-Based Diodes applied to Schenkel Circuit for RF-ID Chips," Circuits and Systems, Vol. 4 No. 2, 2013, pp. 173–180. doi: 10.4236/cs.2013.42024.

When no short-channel effect is assumed, I_F and I_R can be expressed approximately as

$$I_F = \left(\frac{W_G}{2(1+\alpha)L_G} \right) \mu_n C_{ox} \left(V_A - V_{TH} \right)^2 + I_{TH} \quad \left(V_A > V_{TH} \right) \tag{16.5}$$

$$I_R = I_{TH} 10^{\frac{V_A - V_{TH}}{SS}} \quad \left(V_A < V_{TH} \right) \tag{16.6}$$

where most notations are conventional, and SS stands for the subthreshold swing for the fully depleted SOI MOSFET [17]. Parameters α and I_{TH} (threshold current) are given by

$$\alpha = \frac{C_{SOI} C_{BOX}}{C_{OX} \left(C_{SOI} + C_{BOX} \right)} \tag{16.7}$$

$$I_{TH} = \mu_n \left(\frac{W_G}{L_G} \right) q N_A \left(\frac{k_B T}{q} \right)^2 \frac{1}{E_S} \tag{16.8}$$

where C_{SOI} is the SOI layer capacitance [17], C_{BOX} is the buried oxide layer capacitance [17], and E_s is the surface electric field. The derivation of α is given in [17], and that of I_{TH} is given in [20, 21].

References

[1] U. Karthaus and M. Fischer, "Fully integrated passive UHF RFID transponder IC with 16.7-μW minimum RF input power," *IEEE J. Solid-State Circuits*, vol. 38, no. 10, pp. 1602–1608, 2003.

[2] M. Usami and M. Ohki, "The μ-chip: ultra-small 2.45 GHz RFID chip for ubiquitous recognition applications," *IEICE Trans. Electron.*, vol. E86-C, no. 4, pp. 521–528, 2003.

[3] W. Jeon, T. M. Firestone, J. C. Rodgers, and J. Melngailis "Design and fabrication of Schottky diode, on-chip RF power detector," *Solid-State Electron.*, vol. 48, no. 10–11, pp. 2089–2093, 2004.

[4] B. Strassner and K. Chang, "Passive 5.8-GHz radio-frequency identification tag for monitoring oil drill pipe," *IEEE Trans. Microwave Theory Tech.* vol. 51, no. 2, pp. 356–363, 2003.

[5] K. Ahsan, H. Shah, and P. Kingston, "RFID applications: an introductory and exploratory study," *Int. J. Comput. Sci. Issues*, vol. 7, Issue 1, no. 3, pp. 1–7, 2010.

[6] Y. Kado, M. Suzuki, K. Koike, Y. Omura, and K. Izumi, "A 1 GHz/0.9mW CMOS/SIMOX divide-by-128/129 dual-modulus prescaler using a divide-by-2/3 synchronous counter," *IEEE J. Solid-State Circuits*, vol. 28, no. 4, pp. 513–517, 1993.

[7] O. Rozeau, J. Jomaah, J. Boussey, and Y. Omura, "Comparison between high- and low-dose separation by implanted oxygen MOS transistors for low-power radio-frequency applications," *Jpn. J. Appl. Phys.*, vol. 39, no. 4B, pp. 2264–2267, 2000.

[8] J. P. Raskin, A. Viviani, D. Flandre, and J.-P. Colinge, "Substrate crosstalk reduction using SOI technology," *IEEE Trans. Electron Devices*, vol. 44, no. 12, pp. 2252–2261, 1997.

[9] Y. Omura, "Negative conductance properties in extremely thin silicon-on-insulator insulated-gate pn-junction devices (silicon-on-insulator surface tunnel transistor)," *Jpn. J. Appl. Phys.*, vol. 35, no. 11A, pp. L1401–L1403, 1996.

[10] Y. Omura and T. Tochio, "Significant aspects of minority carrier injection in dynamic-threshold SOI MOSFET at low temperature," *Cryogenics*, vol. 49, no. 11, pp. 611–614, 2009.

[11] Synopsys Inc., "Sentaurus Device User Guide," Version A-2007, 2007, ftp://147.46.117.90/synopsys/manuals/PDFManual/data/sdevice_ug.pdf (accessed June 10, 2016.)

[12] Microsim Corp., "Microsim PSpice 8 Free Download," http://getintopc.com/softwares/circuit-designing/microsim-pspice-8-free-download/#_blank (accessed June 10, 2016).

[13] K. Takahashi, S. Y. Wang, and M. Mizunuma, "Complementary charge pump booster," *Electron. Commun. Jpn. Part II-Electron.*, vol. 82, 1999, pp. 73–81.

[14] Y. Taur and T. H. Ning, "Fundamentals of Modern VLSI Devices," (Cambridge University Press, 1998).

[15] Y. Kado, M. Suzuki, K. Koike, Y. Omura, and K. Izumi, "An experimental full-CMOS multi-gigahertz PLL LSI using 0.4-μm gate ultrathin-film SIMOX technology," *IEICE Trans. Electron.*, vol. E76-C, no. 4, pp. 562–571, 1993.

[16] C. Wann, F. Assaderaghi, L. Shi, K. Chan, S. Cohen, H. Hovel, K. Jenkins, Y. Lee, D. Sadana, R. Viswanathan, S. Wind, and Y. Taur, "High-performance 0.07-μm CMOS with 9.3-ps gate delay and 150 GHz f_T," *IEEE Electron Device Lett.*, vol. 18, no. 12, pp. 625–627, 1997.

[17] J.-P. Colinge, "Silicon-on-Insulator: Materials to VLSI," 3rd ed. (Kluwer Academic Publishers, 2004).

[18] Y. Omura, S. Nakashima, and K. Izumi, "Investigation on high-speed performance of 0.1-μm-gate, ultrathin-film CMOS/SIMOX," *IEICE Trans. Electron.*, vol. E75-C, no. 12, pp. 1491–1497, 1992.

[19] Y. Omura, S. Nakashima, K. Izumi, and T. Ishii, "0.1-μm-gate, ultrathin-film CMOS devices using SIMOX substrate with 80-nm-thick buried oxide layer," *IEEE Trans. Electron Devices*, vol. 40, no. 5, pp. 1019–1022, 1993.

[20] J. R. Brews, "A charge-sheet model for the MOSFET," *Solid State Electron.*, vol. 21, no. 2, pp. 345–355, 1978.

[21] "Fully-Depleted SOI CMOS Circuits and Technology for Ultralow-Power Applications," ed. By Y. Sakurai, A. Matsuzawa, and T. Dozeki (Springer, 2006), Chapter 2.

17

The Potential and the Drawbacks of Underlap Single-Gate Ultrathin SOI MOSFET

17.1 Introduction

The short-channel effects of the sub-30 nm silicon-on-insulator metal oxide semiconductor field-effect transistor (SOI MOSFET) appear to limit the performance improvements possible with simple downscaling. It is thought, however, that multiple-gate SOI MOSFETs, such as SOI FinFET [1] and triple-gate (TG) SOI MOSFET [2], will make a technical breakthrough possible. However, it is already understood that a simple fin structure does not give a desirable solution for SOI FinFETs or TG SOI MOSFETs [3], which suggests that optimization of device parameters or the proposal of an advanced device structure is needed for practical applications.

On the other hand, an SOI device with a source/drain (S/D) away from the gate edge (so-called "underlap") is interesting because the short-channel effects can be suppressed through the relaxation effect of the drain-induced electric field [4, 5]. It has been predicted recently [6] that underlapped single-gate ultrathin (USU) SOI MOSFET may offer the sharp subthreshold swing (SS) and short intrinsic delay time (τ) that we would expect from a sub-50 nm channel device, although USU device parameters will have to be well tempered [7]. High-κ gate material must also be introduced to advance underlapped single-gate ultrathin silicon-on insulator (USU SOI) device performance in future. However, the impact of the use of a high-κ gate dielectric has not yet been extensively discussed. In modern applications of SOI MOSFETs, the simultaneous RF performance of USU SOI devices is required because fast AD/DA transformation is necessary in circuitry. For this reason, analog performance should also be discussed.

In this chapter, we will discuss fundamental DC and AC characteristics of USU SOI MOSFETs and the impact of the use of high-κ gate dielectric on the DC and switching

MOS Devices for Low-Voltage and Low-Energy Applications, First Edition.
Yasuhisa Omura, Abhijit Mallik, and Naoto Matsuo.
© 2017 John Wiley & Sons Singapore Pte. Ltd. Published 2017 by John Wiley & Sons Singapore Pte. Ltd.

characteristics of the USU SOI MOSFET with a view to optimizing the device structure by tuning device parameters, such as the underlap length (L_{no}). We will address the tradeoff between DC and AC characteristics, and how to optimize the device performance.

17.2   Simulations

We used the DESSIS (Device Simulation for Smart Integrated Systems) device simulator [8] for the hydrodynamic transport model. The device structures assumed are shown in Figure 17.1: (a) a conventional USU SOI device (device A), (b) a USU SOI device with a high-κ gate dielectric (device B) [9], and (c) a USU SOI device with a stacked gate dielectric (device C). The equivalent oxide thickness (EOT) of the gate dielectric was fixed at 2 nm, and the buried oxide layer (BOX) thickness (t_{BOX}) is 100 nm. Four different values of L_{no} and their doping profiles were considered, as shown in Figure 17.2. When the peak position of the doping concentration in the source/drain region is away from the gate edge, L_{no} increases. The physical gate length (70 nm) and the metallurgical channel length ($L_{met} = 30$ nm) remain invariant when L_{no} increases; that is, the overlap length is 20 nm. Here, we don't use the term *effective channel length,* because the *effective channel length* is significantly modulated with the gate voltage of USU SOI devices. As the electron density of the low-doped *under-lapped region* beneath the gate electrode is a strong function of the gate voltage, the effective channel length increases with the gate voltage. This mechanism is well known in the

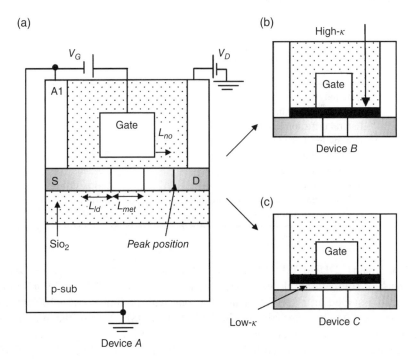

Figure 17.1   Schematic view of USU SOI MOSFET assumed in simulations. L_{ld} is the distance from the doping peak to the junction.

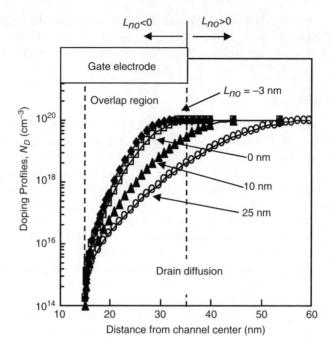

Figure 17.2   Drain diffusion profiles. Drain and source diffusions are symmetric.

Table 17.1   Device parameters assumed.

Parameters	Values
Gate length, L_G	70 nm
Metallurgical channel length, L_{met}	30 nm
Lateral diffusion length, L_{ld}	17–45 nm
Doping level of SOI layer, N_A	3×10^{15} cm^{-3}
Gate SiO$_2$ thickness, t_{ox}	2 nm
Buried oxide layer thickness, t_{BOX}	100 nm
SOI layer thickness, t_S	5–15 (nm)
EOT of high-κ gate insulator thickness	2.0 nm
Dielectric constant of high-κ film, $\varepsilon_{high\text{-}\kappa}$	3.9–50

lightly doped drain (LDD) MOSFET [10]. The body doping concentration (N_A) is 3.0×10^{15} cm^{-3}, and the peak of source/drain doping concentration (N_D) is 1.0×10^{20} cm^{-3}. For simplicity's sake we also assumed an n$^+$-poly-Si gate electrode with doping of 1.0×10^{20} cm^{-3}. As we must extract reliable characteristics of the USU SOI MOSFET from the simulations, we assumed device parameters that were applied to the 90 nm technology generation. Major device parameters are summarized in Table 17.1. The threshold voltage (V_{TH}) is defined as the gate voltage (V_G) at which the drain current I_D is equal to (W_{eff}/L_{met}) $\times 10^{-7}$ A. On-state drain current (I_{ON}) was calculated at $V_G = V_{TH} + 0.9$ V, with the drain voltage $V_D = 1.0$ V.

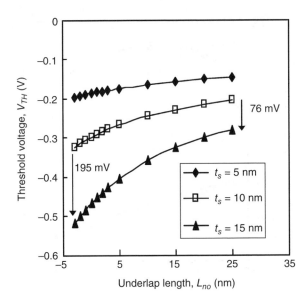

Figure 17.3   Threshold voltage dependence on underlap length (L_{no}).

17.3   Results and Discussion

17.3.1   DC Characteristics and Switching Performance: Device A

First, we will discuss the characteristics of device A, as described in Figure 17.1a. In Figure 17.3, threshold voltage (V_{TH}) dependencies on underlap length (L_{no}) are shown for various t_S values; it is assumed that $L_{met} = 30$ nm. It can be seen that the threshold voltage rises as the underlap length increases because the short-channel effect is suppressed. In addition, the incremental value of threshold voltage decreases as the underlap length increases. At $L_{no} = 20$ nm, a 5 nm deviation of L_{no} yields a threshold variation of, at most, 25 mV, even for $t_S = 15$ nm, where the short-channel effect is the most significant. The influence of a local variation in the SOI layer thickness on the threshold voltage is shown by the two arrows: a thinner SOI layer results in a smaller variation of the threshold voltage.

In the case of $L_{no} = -3$ nm, the threshold voltage decreases by 195 mV as t_S increases from 10 to 15 nm. In the case of $L_{no} = 25$ nm, on the other hand, the decrease of threshold voltage is as small as 76 mV when t_S increases from 10 to 15 nm. The impact of an increase in L_{no} results from the relaxation of the lateral electric field along the underlap region [10]. This means that the increase in L_{no} suppresses the threshold voltage variation. The underlap source/drain structure is very useful for fabrication of short-channel devices. In the fabrication of USU SOI MOSFET, the influence of doping fluctuation in the underlap region should be examined. S.-I. Chang *et al.* investigated the sensitivity of the doping level on GIDL current [9]; their simulation results suggest that a "worst-case-based" design is needed.

In Figure 17.4, drive current (I_{ON}) dependence of underlap length (L_{no}) is shown for various t_S values; device A is assumed in simulations. For $t_S = 5$ and 10 nm, the drive current decreases monotonously as L_{no} increases. This is due to the increase in parasitic resistance of source and drain regions near the gate electrode edge, while the short-channel effects are

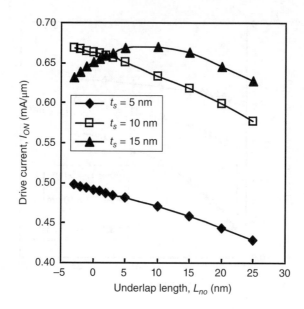

Figure 17.4   Drive current (I_{ON}) dependence on underlap length (L_{no}).

well suppressed [11]. On the other hand, for $t_S = 15$ nm, the drive current has a local peak as L_{no} increases. When L_{no} is still in a small value range, an increase in L_{no} leads to the suppression of short-channel effects. However, when L_{no} is in the large value range, the increase in L_{no} leads to an increase of parasitic resistance of source and drain diffusions near the gate electrode edge. This suggests that the setting of the t_S value directly impacts the maximization of drive current (i.e., minimization of intrinsic delay time [5]). In the present case, it is suggested that a better value of t_S is between 10 and 15 nm. A variation of the I_{ON} value stemming from a local variation of the t_S value is minimzed for $L_{no} = 3$ nm.

17.3.2   RF Analog Characteristics: Device A

In Figure 17.5, the cutoff frequency (f_c) and g_m/g_d ratio dependencies on underlap length (L_{no}) are shown for various t_S values. The g_m/g_d ratio means a small signal voltage gain (A_v) for device A – an intrinsic gain. If the cutoff frequency is extracted from the curve of the g_m/g_d ratio versus frequency, we get a 3-dB roll-off of the g_m/g_d ratio at the frequency. The g_m/g_d ratio increases as L_{no} increases, independently of t_S values. This is due primarily to the reduction of g_d as L_{no} increases, because the increase in L_{no} results in the suppression of short-channel effects and an increase in parasitic resistance.

The cutoff frequency rises constantly as L_{no} increases for $t_S = 5$ nm. However, the cutoff frequency reaches its maximum at around $L_{no} = 20$ nm for $t_S = 15$ nm. The mechanism is discussed below in detail. In order to analyze the analog behavior, in Figure 17.6 the g_m value and the effective gate capacitance (C_{geff}) are shown as a function of L_{no}. Figure 17.6a shows the simulation result for $t_S = 5$ nm and Figure 17.6b, for $t_S = 15$ nm. For $t_S = 5$ nm, the recession rate of C_{geff} is higher than that of g_m; the reduction of fringe capacitance (C_{fringe}) results in a

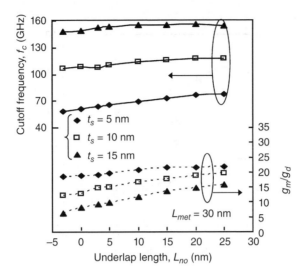

Figure 17.5   Cutoff frequency (f_c) and g_m/g_d dependencies on underlap length (L_{no}).

rapid decrease in C_{geff}. Subsequently, the cutoff frequency rises constantly as L_{no} increases. On the other hand, for $t_S = 15\,\text{nm}$, the recession rate of g_m is higher than that of C_{geff} for $L_{no} > 20\,\text{nm}$; the cutoff frequency curve shows a peak at around $L_{no} = 20\,\text{nm}$.

In simulations, the effective gate capacitance (C_{geff}) primarily consists of the gate oxide capacitance ($S_G \times C_{ox}$), where S_G is the gate electrode area covering the active region of the device, and the two fringe capacitances (C_{fringe} (source) $+ C_{fringe}$ (drain)). Briefly, we have $C_{geff} = S_g C_{ox} + C_{fringe}$ (source) $+ C_{fringe}$ (drain). As the doping level of the underlap region is very low, this underlap region is readily depleted by the built-in potential of the junction. In addition, the underlap region of the drain is considerably depleted by the effective gate-to-drain bias. As a result, C_{fringe} is composed of gate-side-wall-insulator capacitance (C_{swi}), SOI layer capacitance (δC_S), and buried oxide layer capacitance (δC_{BOX}). Briefly, we have $C_{fringe} = \{1/C_{swi} + 1/\delta C_S + 1/\delta C_{BOX}\}^{-1}$. As the dimensions of the device are quite small, the fringe capacitance shares C_{geff}, as anticipated from the above description. A large L_{no} lowers the doping level of the underlap region. So, C_{geff} decreases rapidly as L_{no} increases, although it seems to approach a specific minimal value.

17.3.3   Impact of High-κ Gate Dielectric on Performance of USU SOI MOSFET Devices: Devices B and C

In the following, three different devices A, B, and C are examined in order to consider the influence of a high-κ gate dielectric on device characteristics. A band-to-band tunneling model [12, 13] is introduced in simulations. Figure 17.7 shows the I_D-V_G characteristics of USU SOI MOSFET for $L_{met} = 30\,\text{nm}$; HfO$_2$ film ($\varepsilon_{high-\kappa} = 25$ [14]) is assumed as a high-κ gate dielectric. The horizontal axis is given by V_G-V_{TH} for the sake of comparing the gate-induced drain leakage (GIDL) current. When L_{no} increases, or a high-κ gate dielectric is introduced,

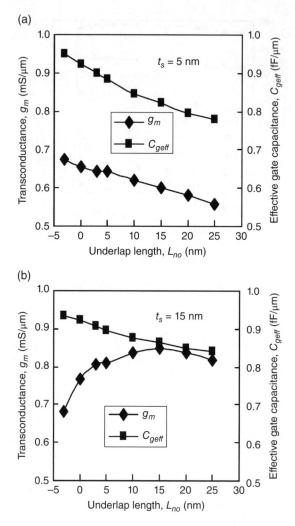

Figure 17.6 Transconductance (g_m) and effective gate capacitance (C_{geff}) dependencies on underlap length (L_{no}). (a) $t_s = 5$ nm and (b) $t_s = 15$ nm.

GIDL current is reduced. In the former case, it seems that, as L_{no} increases, the source/drain electric field relaxation and low-impurity-density region of the drain diffusion reduces the GIDL current [13]; and in the latter case, that reduction of the source/drain fringe electric-field results in a low GIDL current. Therefore, use of a high-κ gate dielectric can be seen to reduce *GIDL* current [9]. However, Figure 17.8 shows that GIDL current reduces as the relative dielectric constant of high-κ material ($\varepsilon_{high-\kappa}$) increases, and that it increases again with the $\varepsilon_{high-\kappa}$ value. This last behavior of the GIDL current looks strange because it can be anticipated that a high $\varepsilon_{high-\kappa}$ value simply results in a significant short-channel effect [15]. In order to investigate the mechanism of the above behavior of the GIDL current, we evaluated the source-to-drain profile of the surface electric-field of the device with a high-κ gate

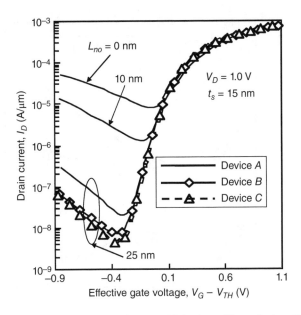

Figure 17.7 I_D-V_G characteristics of USU SOI MOSFET for L_{met}=30 nm; devices B and C have a high-κ gate dielectric with $\varepsilon_{high-\kappa}$=25.

dielectric. Simulation results for device B are shown in Figure 17.9 for $\varepsilon_{high-\kappa}$ values of 3.9, 10, and 50, respectively. In the case of $\varepsilon_{high-\kappa}$=50, we can see a very high electric field region far from the gate edge, and it is this region that causes a high GIDL current. As the peak in the electric-field profile is located near the maximal-doping position of drain diffusion, we can conclude that the background physics of the high GIDL current is the same as the original one [12]. Thus we must find the optimal $\varepsilon_{high-\kappa}$ value, so as to reduce the GIDL current [16]. Hafnium(IV) oxide is a good candidate, and a stacked-gate dielectric relaxes the short-channel effects.

The impact of L_{no} on SS and threshold voltage roll off (ΔV_{TH}) is shown in Figure 17.10a, b, where it is assumed that t_s=15 nm, L_{met}=30 nm, and ΔV_{TH}=V_{TH} (L_{met}=30 nm)−V_{TH} (L_{met}=50 nm). Hafnium(IV) oxide ($\varepsilon_{high-\kappa}$=25) is assumed in Figure 17.10. A high-κ dielectric degrades SS and ΔV_{TH}. The stacked gate dielectric (device C) effectively suppresses the undesired degradation of SS and ΔV_{TH}. As the physical thickness of the gate insulator of device C is thinner than that of device B, the fringe electric field of device C is higher than that of device B; the stacked-gate insulator has successfully suppressed both the drain-induced barrier lowering (DIBL) and the buried insulator-induced barrier lowering (BIIBL) [17].

Figure 17.11 shows dependencies of intrinsic delay time (τ) and I_{ON} on L_{no}, where τ=$C_g V_{ON}$/I_{ON}, C_g is the physical gate capacitance, and V_{ON}=1.0 V. When L_{no} increases, I_{ON} increases because of the slight suppression in short-channel effects. After reaching a peak it then decreases. The intrinsic delay time (τ) shows a certain minimal value. So, we must carefully consider the introduction of high-κ gate dielectric as the intrinsic delay time (τ) increases due to short-channel effects. When we require high-speed devices with a low GIDL current, the stacked gate dielectric should be applied to the devices.

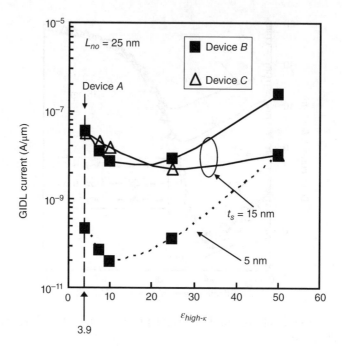

Figure 17.8 Relationship between GIDL current and $\varepsilon_{high\text{-}\kappa}$ at $V_G - V_{TH} = -0.4\,\text{V}$ and $V_D = 1.0\,\text{V}$. The solid line represents $t_S = 15\,\text{nm}$, and the dotted line, $t_S = 5\,\text{nm}$. The following device parameters are assumed: $L_{met} = 30\,\text{nm}$, $L_{no} = 25\,\text{nm}$, $EOT = 2\,\text{nm}$, $t_{BOX} = 100\,\text{nm}$.

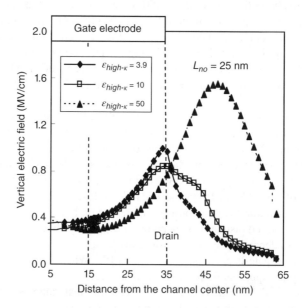

Figure 17.9 Surface vertical electric-field profiles of USU SOI MOSFET (device B) for three-different $\varepsilon_{high\text{-}\kappa}$ values at $V_G = V_{TH} - 0.4\,\text{V}$ and $V_D = 1.0\,\text{V}$. The following device parameters are assumed: $L_{met} = 30\,\text{nm}$, $L_{no} = 25\,\text{nm}$, $EOT = 2\,\text{nm}$, $t_{BOX} = 100\,\text{nm}$, and $t_S = 15\,\text{nm}$.

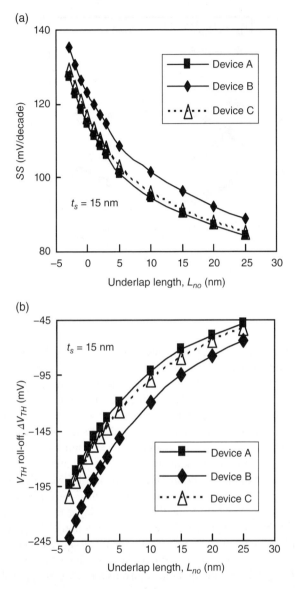

Figure 17.10   Impact of the value of L_{no} on short-channel effects of devices with $\varepsilon_{high\text{-}\kappa} = 25$ at $V_D = 1.0\,$V. (a) Subthreshold swing (SS) at $L_{met} = 30\,$nm. (b) Threshold voltage roll-off (ΔV_{TH}). $\Delta V_{TH} = V_{TH}\,(L_{met} = 30\,nm) - V_{TH}(L_{met} = 50\,$nm).

17.3.4   Impact of Simulation Model on Simulation Results

The simulation model often has a strong impact on simulation results, and, in the case of ultrathin SOI MOSFET, the simulation results should be carefully considered [17]. We can anticipate that the influence of underlap region on device characteristics will be significant when quantum effects are incorporated in simulations, because quantum-mechanical surface

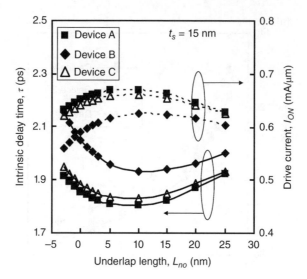

Figure 17.11 Behavior of $\tau\ (= C_g V_{ON}/I_{ON})$ and I_{ON} shown as a function of L_{no} for various device structures. It is assumed $L_{met} = 30\,\text{nm}$, $V_{ON} = 1.0\,\text{V}$, and $\varepsilon_{high-\kappa} = 25$.

depletion (surface "dark space" [18]) enhances the resistance value of the underlap region. In order to estimate the impact of quantum effects, as mentioned above, we utilized a density-gradient model (DGM) [19] as well as a hydrodynamic transport model in our *DESSIS* simulations. Recently, the impact of quantum effect on FinFET with an underlap-gate structure has been discussed [7], where primarily DC characteristics are evaluated.

In Figure 17.12, we show drive current (I_{ON}) and effective gate capacitance (C_{geff}) as a function of underlap length (L_{no}); it is assumed $L_{met} = 30\,\text{nm}$, $t_{ox} = 2\,\text{nm}$, $t_{BOX} = 100\,\text{nm}$, and $t_s = 15\,\text{nm}$. It can be seen that the fundamental behavior of both I_{ON} and C_{geff} are almost the same, regardless of the simulation model used, and that DGM reduces both I_{ON} and C_{geff}. Surface "dark space" reduces the effective gate capacitance (~6%) because C_{geff} can be estimated by $\varepsilon_{ox}/\{t_{ox} + (\varepsilon_{ox}/\varepsilon_s)t_{inv}\}$, where t_{inv} is the inversion layer thickness and ε_s is the silicon dielectric permittivity. So, we might conclude that the reduction of C_{geff} due to DGM primarily results from the surface "dark space." As the reduction rate of I_{ON} due to DGM (~8%) is almost the same as that of C_{geff}, it can be concluded that the reduction of I_{ON} is primarily responsible for the reduction of C_{geff}. It is also thought that a surplus reduction of I_{ON} of about 2% is responsible for the parasitic resistance of source and drain diffusions.

In Figure 17.13, we show the cut-off frequency (f_c) and small-signal gain (g_m/g_d) as a function of underlap length (L_{no}); it is assumed that $L_{met} = 30\,\text{nm}$, $t_{ox} = 2\,\text{nm}$, $t_{BOX} = 100\,\text{nm}$, and $t_s = 15\,\text{nm}$. It can be seen that the fundamental behavior of f_c and g_m/g_d is almost identical, regardless of simulation model, and that DGM promotes an increase in f_c and a slight decrease in small-signal gain (g_m/g_d). As C_{geff} is reduced, due to quantum effects, f_c naturally rises. The influence of quantum effects on intrinsic gain (g_m/g_d) appears somewhat complicated, and further investigation is needed. In the present simulation results, the reduction rate of g_m is shown to be slightly higher than that of g_d. In other words, the impact of parasitic resistance on g_m is stronger than that on g_d.

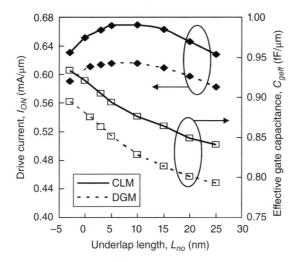

Figure 17.12   Drive current (I_{ON}) and effective gate capacitance (C_{geff}) as a function of underlap length (L_{no}). It is assumed $L_{met} = 30$ nm, $t_{ox} = 2$ nm, $t_{BOX} = 100$ nm, and $t_S = 15$ nm.

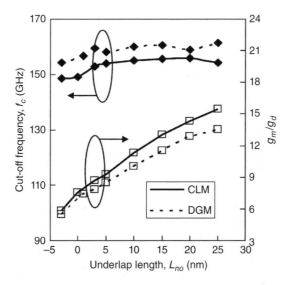

Figure 17.13   Cutoff frequency (f_c) and intrinsic gain (g_m/g_d) as a function of underlap length (L_{no}). It is assumed $L_{met} = 30$ nm, $t_{ox} = 2$ nm, $t_{BOX} = 100$ nm, and $t_S = 15$ nm.

From the above simulation results, we can conclude that electrical characteristics are strongly influenced by DGM, but that fundamental behaviors of those characteristics are almost independent of the transport model. Much of the discussion described above will be useful in creating a design guideline for underlap gate SOI MOSFET.

17.4   Summary

This chapter describes the performance prospect of USU SOI MOSFET with a low-κ or high-κ gate dielectric from the viewpoint of digital and analog applications. As is already known, the impact of an increase in underlap length on device characteristics results from the relaxation of the lateral electric field along the underlap region. This means that the increase in underlap length suppresses the threshold voltage variation as well as suppressing short-channel effects, suggesting that the underlap source/drain structure would be very useful for the fabrication of short-channel devices. In addition, the thickness of the SOI layer directly affects the maximization of drive current (i.e., minimization of intrinsic delay time). So, the SOI layer thickness and the underlap length must be determined carefully.

On the other hand, analog RF characteristics are also influenced in the use of the underlap region. As the fringe capacitance is reduced in the introduction of the underlap region, the effective gate capacitance is also reduced, while the voltage gain of the device rises. As the apparent rise in the cutoff frequency stems from the reduction of the voltage gain, the advancement of analog performance is inherently limited.

The use of a high-κ gate dielectric reduces the *GIDL* current, while the short-channel effects are degraded. In the case of a very high dielectric constant, however, a very high electric-field region appearing far from the gate edge becomes a new source of a high *GIDL* current. So, the optimization of the dielectric constant of the gate insulator is required.

References

[1] D. Hisamoto, W. Lee, J. Kedzierski, E. Anderson, H. Takeuchi, K. Asano, T. King, J. Bokor, and C. Hu, "A Folded-Channel MOSFET for Deep-Sub-Tenth Micron Era," Tech. Dig. IEEE IEDM (San Francisco, 1998) pp. 1032–1034, 1998.

[2] B. Doyle, B. Boyanov, S. Datta, M. Doczy, S. Hareland, B. Jin, J. Kavaieros, T. Linton, R. Rios, and R. Chau, "Tri-Gate Fully-Depleted CMOS Transistors: Fabrication, Design and Layout," Int. Symp. VLS. Tech. Dig. (Kyoto, 2003) pp. 133–134, 2003.

[3] H. Konishi and Y. Omura, "Robust engineering of S/D diffusion doping and metal contact layouts for multi-fin triple-gate FETs," *IEEE Electron Device Lett.*, vol. 27, pp. 472–475, 2006.

[4] C. Yin and P. C. H. Chan, "Investigation of the source/drain asymmetric effects due to gate misalignment in planar double-gate MOSFETs," *IEEE Trans. Electron Devices*, vol. 52, pp. 85–90, 2005.

[5] H. Lee, J. Lee, and H. Shin, "DC and AC characteristics of sub-50-nm MOSFETs with source/drain-to-gate nonoverlapped structure," *IEEE Trans. Nanotechnol.*, vol. 1, pp. 219–225, 2002.

[6] V. P. Trivedi and J. G. Fossum, "Source/Drain-Doping Engineering for Optimal Nanoscale FinFET Design," Proc. IEEE Int. SOI Conf., pp. 192–194, 2004.

[7] K. Tanaka, K. Takeuchi, and M. Hane, "Practical FinFET Design Considering GIDL for LSTP (Low Standby Power) Devices," Tech. Dig. IEEE IEDM (Washington DC, 2005) pp. 980–983, 2005.

[8] Synopsys Inc., "Sentaurus Device User Guide," Version A-2007, 2007, ftp://147.46.117.90/synopsys/manuals/PDFManual/data/sdevice_ug.pdf (accessed June 10, 2016.)

[9] S.-I. Chang, J. Lee, and H. Shin, "GIDL Currents in MOSFETs with High-κ Dielectric," Ext. Abstr. Int. Conf. Solid State Devices and Materials, pp. 234–235, 2001.

[10] S. Ogura, P. J. Tang, W. W. Walker, D. L. Critchlow, and J. F. Shepard, "Design and characteristics of the Lightly Doped Drain-Source (LDD) insulated gate field-effect transistor," *IEEE J. Solid-State Circuits*, vol. SC-15, pp. 424–432, 1980.

[11] A. Bansal, B. C. Paul, and K. Roy, "Modeling and optimization of fringe capacitance of nanoscale DGMOS devices," *IEEE Trans. Electron Devices*, vol. 52, pp. 256–262, 2005.

[12] J. Chen, T. Y. Chen, I. C. Chen, P. K. Ko, and C. Hu, "Subbreakdown drain leakage current in MOSFET," *IEEE Electron Device Lett.*, vol. EDL 8, pp. 515 517, 1987.

[13] T. Ishiyama and Y. Omura, "Influences of superficial Si layer thickness on band-to-band tunneling current characteristics in ultra-thin n-channel metal-oxide-semiconductor field-effect-transistor by separation by implanted oxygen (nMOSFET/SIMOX)," *Jpn. J. Appl. Phys.*, vol.36, pp. L264–L267, 1997.

[14] M. Balog, M. Schieber, M. Michman, and S. Patai, "Chemical vapor deposition and characterization of HfO$_2$ films from organo-hafnium compounds," *Thin Solid Films*, vol. 41, pp. 247–259, 1977.

[15] G. C. F. Yeap, S. Krishnan, and M. R. Lin, "Fringing-induced barrier lowering (FIBL) in sub-100 nm MOSFETs with high-κ gate dielectrics," *Electron. Lett.*, vol. 34, pp. 1150–1152, 1998.

[16] D. J. Frank and H.-S. P. Wong, "Analysis of the Design Space Available for High-κ Gate Dielectrics in Nanoscale MOSFETs," IEEE Silicon Nanoelectron. Workshop, 2000, pp. 47–48, 2000.

[17] Y. Omura, H. Konishi, and S. Sato, "Quantum-mechanical suppression and enhancement of short-channel effects in ultra-thin SOI MOSFET's," *IEEE Trans. Electron Devices*, vol. 53, no. 4, pp. 677–684, 2006.

[18] H. Nakatsuji and Y. Omura, "Semi-empirical and practical model for low-electric field direct tunneling current estimation in nanometer-thick SiO$_2$ films," *Superlattices Microstruct.*, vol. 28, no. 5–6, pp.425–428, 2000.

[19] M. G. Ancona and H. F. Tiersten, "Macroscopic physics of the silicon inversion layer," *Phys. Rev. B*, vol. 35, pp. 7959–7965, 1987.

18

Practical Source/Drain Diffusion and Body Doping Layouts for High-Performance and Low-Energy Triple-Gate SOI MOSFETs

18.1 Introduction

Silicon-on-insulator metal oxide semiconductor field-effect transistors (SOI MOSFETs) are considered as promising for effectively suppressing short-channel effects and achieving high drivability and low standby power consumption [1]. For aggressive scaling, it is anticipated that fin-type SOI MOSFETs are essential as they offer footprint reduction, high drivability [2], and short-channel effect suppression [3]. However, parasitic resistance, such as source and drain (S/D) diffusion resistance and contact resistance, has been discovered to be a serious problem in triple-gate (TG) SOI FinFETs [4–6].

One of the editors (Omura) previously introduced the following two guidelines for TG SOI FinFETs [7]. (i) In order to suppress short-channel effects (i.e., buried-insulator-induced barrier lowering (BIIBL) effect [8]), the S/D extension region should be shallow. (ii) In order to achieve high drivability, the cross-sectional area of the major S/D diffusion region, which carries the carrier path, should be large, but the diffusion region should not touch the buried oxide layer as otherwise the BIIBL effect would be significant. Thus a shallow junction (one that does not touch the buried oxide layer) is preferred to realize low standby power (LSTP); a remaining issue is how to simultaneously realize high drivability. In addition, how parasitic capacitance, attributed to the fin structure, can be reduced is also attracting attention; the underlap-gate device structure is being studied in order to reduce the fringing capacitance [9].

MOS Devices for Low-Voltage and Low-Energy Applications, First Edition.
Yasuhisa Omura, Abhijit Mallik, and Naoto Matsuo.
© 2017 John Wiley & Sons Singapore Pte. Ltd. Published 2017 by John Wiley & Sons Singapore Pte. Ltd.

This chapter examines how the S/D diffusion layout impacts the suppression of the BIIBL effect and the drivability of n-channel triple-gate silicon-on insulator metal oxide semiconductor field-effect transistor (TG SOI MOSFETs). Following a consideration of some physical aspects, many configurations of the p- or n-region beneath the S/D n⁺-region are examined with the goal of achieving both high drivability and the suppression of the BIIBL effect.

18.2   Device Structures and Simulation Model

Since the TG SOI MOSFET inherently has three-dimensional (3-D) structure-oriented aspects, as suggested in Figure 18.1a, 3-D device simulations were conducted using TCAD-*DESSIS* (Synopsys) [10]; this study assumes a single-fin n-channel TG SOI MOSFET (as a component of multiple-fin TG MOSFETs [7]) with a 2 nm thick gate-oxide layer uniformly covering the fin (see Figure 18.1b) because it has higher drivability than the alternatives [7]. The device is formed on a 100 nm thick buried oxide layer [7]. The gate length (L_G), 50 nm, is given by the sum of the 30 nm long channel (L_{eff}) and twice the 10 nm long extension length (L_{ov}). It is assumed that metal electrodes for S/D diffusion are placed on the top surface of the fin due to the difficulty of depositing the electrodes on its side surfaces [11]. L_{SC} and L_{DC} are the

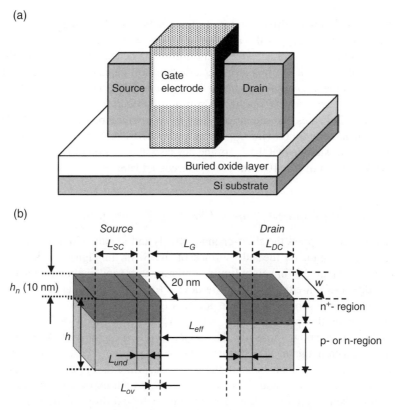

Figure 18.1   Schematic of n-channel TG SOI MOSFET with shallow n⁺-S/D diffusion region. (a) Bird's-eye view of TG SOI MOSFET. (b) Definition of device parameters for fin body.

electrode contact lengths on the 10 nm deep S/D diffusion region (h_n), where $L_{SC}=L_{DC}=30$ nm. L_{und} stands for the length of the region not covered by either the gate electrode or S/D contact. Si-fin height (h) and width (w) are basically 40 and 20 nm, respectively. The p-type body-doping concentration in the Si-fin beneath the gate is 1×10^{15} cm^{-3}; this value is selected so that the channel length is definitely determined. The simulations first examine a p-region (boron: 1×10^{15} cm^{-3}) beneath the S/D diffusion layer, and then an n-region beneath the layer (phosphorus: $1\times10^{16}\sim1\times10^{20}$ cm^{-3}).

Threshold voltage (V_{TH}) is fixed at 0.25 V regardless of device structure so as to eliminate the influence of device parameters on threshold voltage [7]. Drain-to-source voltage (V_D) is fixed at 1.0 V, and the source and the substrate are grounded for simplicity; the substrate bias effect is very limited in practice. The simulations used the hydrodynamic transport model with the band-to-band (BTB) tunnel model [10]. Note that the BTB model is not restrictive because the 20 nm wide fin yields only a moderate electric field around the drain junction [12, 13].

18.3 Results and Discussion

18.3.1 Impact of S/D-Underlying Layer on I_{ON}, I_{OFF}, and Subthreshold Swing

Figure 18.2 plots simulated drive current (I_{ON}) and subthreshold swing (SS) versus doping concentration of n-region; I_{ON} is the I_D value at $V_G=V_D=1$ V. The broken line indicates the simulated value of I_{ON} or SS for a device with the p-region beneath the n$^+$-S/D diffusion layer. The SS value is degraded as N_D increases because the short-channel effect (i.e., BIIBL effect [8]) becomes prominent. On the other hand, the I_{ON} value first decreases as N_D increases, and then rapidly increases because of the short-channel effect; I_{ON} and SS have a tradeoff relationship. This indicates that the introduction of an n-region beneath the n$^+$-S/D diffusion layer is not suitable for improving I_{ON} because the BIIBL effect is enhanced [7, 8, 11]. In addition, it seems strange that the I_{ON} of the device with n-region of $N_D=10^{18}$ cm^{-3} is lower than that with p-region of $N_A=10^{15}$ cm^{-3}; this is discussed later.

18.3.2 Tradeoff of Short-Channel Effects and Drivability

To consider how to suppress the short-channel effects and to gain a high I_{ON}, we show a 2D view of the current path on the side surface of the Si-fin in Figure 18.3a–c, where the vertical axis is along the Si-fin height (units of micrometers) and the horizontal axis is along the source-edge-to-drain-edge of the entire device (units of micrometers). The horizontal coordinate "0" indicates the center of the device. Figure 18.3a shows the simulation results of the device with the p-region of $N_A=10^{15}$ cm^{-3} beneath the n$^+$-S/D diffusion layer; Figure 18.3b and c show those of the devices with n-regions of $N_D=10^{18}$ cm^{-3} and $N_D=10^{19}$ cm^{-3} beneath the n$^+$-S/D diffusion layer, respectively. The current path of the devices with n-regions expands beneath the n$^+$-region, while that of devices with p-regions is basically limited to that inside the n$^+$-region; current path expansion strengthens as the doping level of the n-region increases. As the drain potential drops along the current path, the n-region beneath the n$^+$-region enhances the BIIBL effect [8]. As a result, the SS value is degraded due to the BIIBL effect.

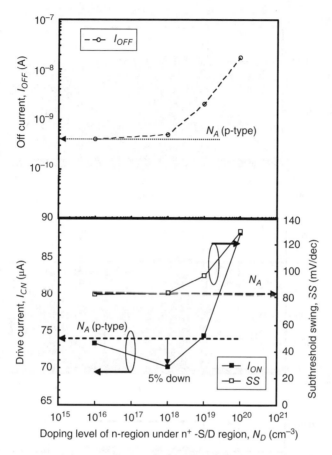

Figure 18.2   Drive current (I_{ON}), off-current (I_{OFF}), and subthreshold swing (SS) versus n-region doping level. Broken lines show values for the device with the p-region of $N_A = 1 \times 10^{15}\,\mathrm{cm}^{-3}$ beneath the n⁺-region. It is assumed that $h = 40\,\mathrm{nm}$ and $w = 20\,\mathrm{nm}$.

In addition, we calculated the resistivity, the electron current density, the electron density, and the electron mobility along cut line A–A′ shown in Figure 18.3a. Calculated results for various devices (shown in Figure 18.3) are shown in Figure 18.4. Figure 18.4a shows the resistivity profile along cut line A–A′, Figure 18.4b shows the electron current density profile along cut line A–A′, Figure 18.4c shows the electron density profile along cut line A–A′, and Figure 18.4d shows the carrier mobility profile along cut line A–A′. It shows that the resistivity of the device with p-region ($N_A = 10^{15}\,\mathrm{cm}^{-3}$) beneath the n⁺-region is lower than that of the device with n-region ($N_D = 10^{18}\,\mathrm{cm}^{-3}$); this interesting result stems from the fact that electron mobility in the p-region with $N_A = 10^{15}\,\mathrm{cm}^{-3}$ is three times as large as that in the n-region with $N_D = 10^{18}\,\mathrm{cm}^{-3}$ (see Figure 18.4d); in other words, electron mobility in the electron accumulation layer is smaller than bulk electron mobility. As shown in Figure 18.4b, it should be noted that the electron current density beneath the n⁺-S/D diffusion layer with $N_A = 10^{15}\,\mathrm{cm}^{-3}$ is smaller than that with $N_D = 10^{19}$ and $10^{20}\,\mathrm{cm}^{-3}$. This low current density beneath the n⁺-S/D diffusion

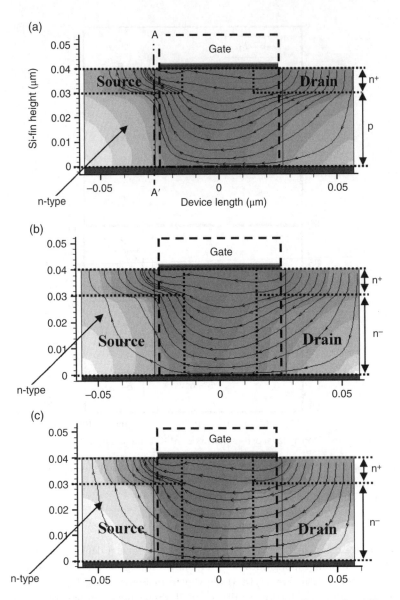

Figure 18.3   2-D schematic view of current path. It is assumed that $h = 40\,\text{nm}$ and $w = 20\,\text{nm}$. (a) A p-type region with $N_A = 1 \times 10^{15}\,(\text{cm}^{-3})$ beneath the n-type S/D diffusion. (b) A n-type region with $N_D = 1 \times 10^{18}\,(\text{cm}^{-3})$ beneath the n-type S/D diffusion. (c) A p-type region with $N_D = 1 \times 10^{19}\,(\text{cm}^{-3})$ beneath the n-type S/D diffusion.

layer stems from the low electron density beneath the n⁺-S/D diffusion layer. These results suggest that the p-type doping level must be optimized from the viewpoints of drivability and short-channel effect suppression, and that the n-region should not be put beneath the n⁺-S/D diffusion layer.

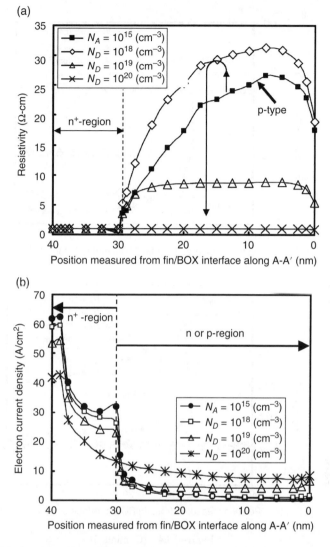

Figure 18.4 Calculated profiles of physical parameters along cut line A-A' for various doping conditions beneath n⁺-S/D diffusion layer. (a) Resistivity profiles, (b) electron current density profiles, (c) electron density profiles, and (d) electron mobility profiles.

Here we consider the impact of the h_n/h ratio to investigate how the depth of n⁺-S/D region affects drivability, off-current, and subthreshold swing. Calculated results are summarized in Table 18.1 for fin heights (h) ranging from 40 to 100 nm. Important points are (i) on current (I_{ON}), off current (I_{OFF}), and the SS take maximal values for higher fins when the depth (h_n) of n⁺-S/D region increases; (ii) the steepest SS values and the lowest off-current values are obtained for the most shallow n⁺-S/D diffusion; and (iii) the lowest off-current (I_{OFF}) value is realized with $h = 100$ nm and $h_n = 10$ nm, not with $h = 40$ nm and $h_n = 10$ nm, which means that the path length from the bottom surface of the source diffusion, via the bottom surface of

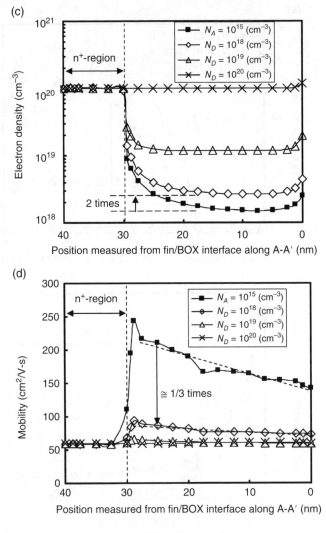

Figure 18.4   (Continued)

the fin, to the bottom surface of the drain diffusion is the longest among the present devices. In other words, the BIIBL effect is suppressed effectively. It should be noted that $I_{ON}/(2h+w)$ takes its maximal value with $h=40$ nm, not $h=100$ nm, which stems from the fact that the effective channel length at the side surface of fin with $h=40$ nm is shorter than that with $h=100$ nm. This suggests that layouts that use multiple low fins promise both a better I_{ON} and the lowest I_{OFF} with the steepest swing [14–16]; the resulting increase in TG SOI MOSFET footprint can be countered (as one example) by using high-κ dielectric materials in device fabrication.

Criteria for enhancing TG SOI MOSFET scaling and drivability have already been discussed [11], and it has been demonstrated that $w < L_{eff}/3$ is the guideline for TG SOI MOSFET scaling.

Table 18.1 Simulated device characteristics for various device parameters at $V_G = V_D = 1$ V.

h (nm)	h_n (nm)	N_A (cm^{-3})	I_{ON} (A)	I_{ON} (A/µm)	I_{OFF} (A)	I_{OFF} (A/µm)	SS (mV/dec)
40	10	1×10^{15}	7.40E−05	7.40E−04	4.21E−10	4.21E−09	83.6
		1×10^{18}	6.72E−05	6.72E−04	2.15E−10	2.15E−09	82.7
	20	1×10^{15}	8.47E−05	8.47E−04	2.40E−09	2.40E−08	101.3
		1×10^{18}	8.06E−05	8.06E−04	1.40E−09	1.40E−08	95.6
	30	1×10^{15}	8.93E−05	8.93E−04	7.32E−09	7.32E−08	116.7
		1×10^{18}	8.73E−05	8.73E−04	5.45E−09	5.45E−08	114.0
	40	N/A	8.80E−05	8.80E−04	1.00E−08	1.00E−07	120.0
100	10	1×10^{15}	7.70E−05	3.50E−04	1.92E−10	8.73E−10	78.8
		1×10^{18}	6.76E−05	3.07E−04	3.30E−10	1.50E−09	78.8
	25	1×10^{15}	1.00E−04	4.55E−04	5.51E−09	2.50E−08	98.4
		1×10^{18}	9.02E−05	4.10E−04	4.33E−09	1.97E−08	99.3
	50	1×10^{15}	1.29E−04	5.86E−04	1.46E−08	6.64E−08	112
		1×10^{18}	1.19E−04	5.41E−04	1.37E−08	6.23E−08	112
	75	1×10^{15}	1.45E−04	6.59E−04	1.71E−08	7.77E−08	114
		1×10^{18}	1.41E−04	6.41E−04	1.68E−08	7.61E−08	114
	100	N/A	1.47E−04	6.68E−04	1.51E−08	6.86E−08	115

The present case violates this guideline. Therefore, the swing value of the 40 nm-high TG SOI MOSFET is not so low. Well tuned device parameters for even more scaled TG SOI MOSFETs must be reviewed in light of the above guideline.

18.4 Summary

This chapter described the impact of taking a holistic approach to designing the S/D structures of TG SOI MOSFETs and the performance that can be expected. The fundamental tradeoff between drivability and short-channel effects was demonstrated for devices with various S/D structures. Simulation results yielded the following guidelines for designing shallow junctions for TG SOI MOSFETs:

- To suppress the BIIBL effect (steep swing and low-standby power), the n$^+$-S/D diffusion layer should be shallow. Placing an n-region beneath the n$^+$-S/D diffusion layer is not recommended for n-channel TG SOI MOSFETs.
- To improve drivability and suppress the BIIBL effect, a low-doped p-region should be put beneath the n$^+$-S/D diffusion layer.
- A low fin is recommended to optimize the performance tradeoffs of the TG SOI MOSFET.

References

[1] J.-P. Colinge, "Silicon-on-Insulator Technology: Materials to VLSI," 3rd ed. (Kluwer Academic Publishers, 2004).
[2] J. Kedzierski, D. M. Fried, E. J. Nowak, T. Kanarsky, J. H. Rankin, H. Hanafi, W. Natzle, D. Boyd, Y. Zhang, R. A. Roy, J. Newbury, C. Yu, Q. Yang, P. Saunders, C. P. Willets, A. Johnson, S. P. Cole, H. E. Young, N. Carpenter, D. Rakowski, B. A. Rainey, P. E. Cottrel, M. Ieong, and H.-S. Philip Wong, "High-Performance Symmetric-Gate and CMOS-Compatible Vt Asymmetric-Gate FinFET Devices," Ext. Abstr. IEEE IEDM (Washington DC, Dec., 2001) pp. 437–440, 2001.

[3] Y. Liu, K. Ishii, M. Masahara, T. Tsutsumi, H. Takashima, and E. Suzuki, "An Experimental Study of the Cross-Sectional Channel Shape Dependence of Short-Channel Effects in Fin-Type Double-Gate MOSFETs," Ext. Abstr. 2003 Int. Conf. Solid State Devices and Materials (Tokyo, Sept., 2004) pp. 284–285, 2004.

[4] H. Kam, L. Chang, and T.-J. King, "Impact of 3D Source-Drain Doping Profiles and Contact Schemes on FinFET Performance in the Nanoscale Regime," Abstr., IEEE 2004 Silicon Nanoelectron. Workshop (Hawaii, June, 2004) pp. 9–10, 2004.

[5] A. Dixit, K. G. Anil, R. Rooyackers, F. Leys, M. Kaiser, R. Weemaes, I. Ferain, A. De Keersgieter, N. Collaert, R. Surdeanu, M. Goodwin, P. Zimmerman, R. Loo, M. Caymax, M. Jurezak, S. Biesemans, and K. De Meyer, "Minimization of the MuGFET Contact Resistance by Integration of NiSi Contacts on Epitaxially Raised Source/Drain Regions," Proc. 2005 European Solid-State Devices Res. Conf. (ESSDERC) (Grenoble, Sept., 2005) pp. 443–446, 2005.

[6] A. Dixit, A. Kottantharayil, N. Collaert, M. Goodwin, M. Jurczak, and K. De Meyer, "Analysis of the parasitic S/D resistance in multiple-gate FETs," *IEEE Trans. Electron Devices*, vol. 52, pp. 1132–1140, 2005.

[7] H. Konishi and Y. Omura, "Robust engineering of S/D diffusion doping and metal contact layouts for multifin triple-gate FETs," *IEEE Trans. Electron Device lett.*, vol. 27, pp. 472–475, 2006.

[8] Y. Omura, H. Konishi, and S. Sato, "Quantum-mechanical suppression and enhancement of short-channel effects in ultra-thin SOI MOSFET's," *IEEE Trans. Electron Devices*, vol. 53, pp. 677–684, 2006.

[9] V. P. Trivedi and J. G. Fossum, "Source/Drain-Doping Engineering for Optimal Nanoscale FinFET Design," Proc. 2004 IEEE Int. SOI Conf. (Oct., 2004) pp. 192–194, 2004.

[10] Synopsys Inc., "Sentaurus Device User Guide," Version A-2007, 2007, ftp://147.46.117.90/synopsys/manuals/PDFManual/data/sdevice_ug.pdf (accessed June 10, 2016.)

[11] Y. Omura, H. Konishi, and K. Yoshimoto, "Impact of fin aspect ratio on short-channel control and drivability of multiple-gate SOI MOSFETs," *J. Semicond. Technol. Sci.*, vol. 8, no. 4, pp. 302–309, 2008.

[12] T. Ishiyama and Y. Omura, "Influence of superficial Si layer thickness on band-to-band tunneling current characteristics in ultra-thin n-channel metal-oxide-semiconductor field-effect-transistor by separation by implanted oxygen (nMOSFET/SIMOX)," *Jpn. J. Appl. Phys.*, vol. 36, pp. L264–L267, 1997.

[13] Y. Omura, T. Ishiyama, M. Shoji, and K. Izumi, "Quantum Mechanical Transport Characteristics in Ultimately Miniaturized MOSFETs/SIMOX," Proc. of the 7th Int. Symp. on Silicon-on-Insulator Technology and Devices (The Electrochem. Soc., Los Angeles), PV96-3, pp. 199–211, 1996.

[14] K. Tanaka, K. Takeuchi, and M. Hane, "Practical FinFET Design Considering GIDL for LSTP (Low Standby Power) Devices," Tech. Dig. IEEE Int. Electron Devices Meeting (Washington DC, 2005) pp. 980–983, 2005.

[15] M. J. Van Dal, G. Vellianitisb, R. Duffyc, B. Pawlakc, K. Laid, A. Y. Hikavyye, N. Collaerte, M. Jurczake, and R. Landerf, "Material aspects and challenges for SOI FinFET integration," *ECS Trans.*, vol. 13, no. 1, 2008; doi: 10.1149/1.2911503.

[16] K.-I. Seo, B. Haran, D. Gupta, D. Guo, T. Standaert, R. Xie, H. Shang, E. Alptekin, D.-I. Bae, G. Bae, C. Boye, H. Cai, D. Chanemougame, R. Chao, K. Cheng, J. Cho, K. Choi, B. Hamieh, J. G. Hong, T. Hook, L. Jang, J. Jung, R. Jung, D. Lee, B. Lherron, R. Kambhampati, B. Kim, H. Kim, K. Kim, T. S. Kim, S.-B. Ko, F. L. Lie, D. Liu, H. Mallela, E. Mclellan, S. Mehta, P. Montanini, M. Mottura, J. Nam, S. Nam, F. Nelson, I. Ok, C. Park, Y. Park, A. Paul, C. Prindle, R. Ramachandran, M. Sankarapandian, V. Sardesai, A. Scholze, S.-C. Seo, J. Shearer, R. Southwick, R. Sreenivasan, S. Stieg, J. Strane, X. Sun, M. G. Sung, C. Surisetty, G. Tsutsui, N. Tripathi, R. Vega, C. Waskiewicz, M. Weybright, C.-C. Yeh, H. Bu, S. Burns, D. Canaperi, M. Celik, M. Colburn, H. Jagannathan, S. Kanakasabaphthy, W. Kleemeier, L. Liebmann, D. McHerron, P. Oldiges, V. Paruchuri, T. Spooner, J. Stathis, R. Divakaruni, T. Gow, J. Iacononi, J. Jenq, R. Sampson, and M. Khare, "A 10 nm Platform Technology for Low Power and High Performance Application Featuring FINFET Devices with Multi Workfunction Gate Stack on Bulk and SOI," Tech. Dig. IEEE Int. Symp. VLSI Technology (Honolulu, 2014), paper No. 2.2, 2014.

19

Gate Field Engineering and Source/Drain Diffusion Engineering for High-Performance Si Wire Gate-All-Around MOSFET and Low-Power Strategy in a Sub-30 nm-Channel Regime

19.1 Introduction

Ultrathin silicon-on-insulator (SOI) MOSFETs and various similar advanced SOI devices are being extensively studied to realize high-performance large-scale integrated circuits because they offer high drivability and high immunity to short-channel effects [1]. The gate-all-around silicon-on-insulator metal oxide semiconductor field-effect transistor (GAA SOI MOSFET) [2], as well as the Fin-type MOSFET [3], is a promising device; its great potential derives from its strong electrostatic confinement of carriers in the Si body. Therefore, it is expected to offer higher performance than the alternatives and many studies have focused on its physics [4–11]; some people have examined ballistic conduction [12, 13]; others have looked at the parasitic capacitance issue [14, 15], and so on. However, no plausible design guideline has yet to be proposed because of the apparent drawback of parasitic resistance in the source and drain regions as well as mobility degradation due to the surface roughness effect [10] and strong confinement effects [7].

Device engineering on the GAA SOI MOSFET is also attracting attention from the viewpoint of 3-D electronics used in smaller footprint devices, such as retina applications [16] and tera-bit memory applications for robotics engineering [17]. As extremely low-power

MOS Devices for Low-Voltage and Low-Energy Applications, First Edition.
Yasuhisa Omura, Abhijit Mallik, and Naoto Matsuo.
© 2017 John Wiley & Sons Singapore Pte. Ltd. Published 2017 by John Wiley & Sons Singapore Pte. Ltd.

consumption is always a prerequisite in these applications for long-term usage, short-channel effects have to be strongly suppressed in order to reduce the standby power consumption; other demands are performance reproducibility and stability.

Omura has already investigated the drain current versus gate voltage characteristics of various 100 nm long channel GAA MOSFETs [18]. It was shown that a GAA device with relatively thick gate oxide covering the low-doped junction region has better drivability than a GAA device with a thin gate oxide covering the low-doped junction region. It was discovered that the lateral extension of the high-potential contour raises the electron density of the Si wire over the low-doped diffusion region because the lateral extension of contours creates an additional transversal field that impinges on the majority carrier density of source and drain diffusion regions. The lateral extension of gate potential yields suppression of the drain-induced barrier lowering (DIBL) effects and the buried-insulator-induced barrier lowering (BIIBL) of SOI MOSFETs [19, 20] because the electric field along the source diffusion is lowered. Such a phenomenon is quite obvious because the wire has a very small cross-sectional area, which means that the electric field inside the oxide layer can change the field of the whole wire.

Since the above consideration assumes a 100 nm long channel GAA MOSFET, it should be examined whether such physics are valid in sub-30 nm long channel GAA MOSFETs. This chapter reconsiders the design methodology of the sub-30 nm channel GAA SOI MOSFET and proposes an advanced concept to enhance its performance for future very large-scale integrated circuits. This chapter demonstrates the simulation results of various GAA MOSFETs with a 20 nm long or 15 nm long channel [21]. The new ideas are based on gate field engineering and source and drain diffusion engineering. Their validity, even in the sub-30 nm channel regime, is demonstrated by various device simulation results. We also address how to realize the low-power performance of the devices proposed here and show a possible device structure that should achieve low-power consumption and high drivability.

19.2   Device Structures Assumed and Physical Parameters

This study basically simulates three different device structures as shown in Figure 19.1. Figure 19.1 shows schematic bird's-eye views (Figure 19.1a) and cross-sections (Figure 19.1b) of various 20 nm long channel GAA SOI MOSFETs with a Si body with $10\,nm \times 10$ nm cross-section: device A has a 20 nm thick uniform gate SiO_2 layer, device B has a 1 nm thick uniform gate SiO_2 layer, and device C has a 1 nm thick gate SiO_2 layer that covers one-half of the channel length. The simulations use the hydrodynamic transport model and quantum-correction model of potential (density-gradient model) [22, 23]. Section 19.6 provides brief descriptions of the physical models used in the simulations.

All the devices have a 2 nm long gate overlap at every junction edge and source and drain (S/D) diffusion regions have 100 nm long graded profiles to suppress short-channel effects. Basically, it is assumed that the doping profile of the S/D diffusion regions is given by the error function; the slope of doping profile ranges from 16 to 37 nm/dec except for Figure 19.9; the doping profile $N_D(x)$ is assumed to be [22]:

$$N_D(x) = \frac{N_{Dmax}}{2}\left(1 - erf\left(\frac{x - x_{sym}}{ELength}\right)\right) \qquad (19.1)$$

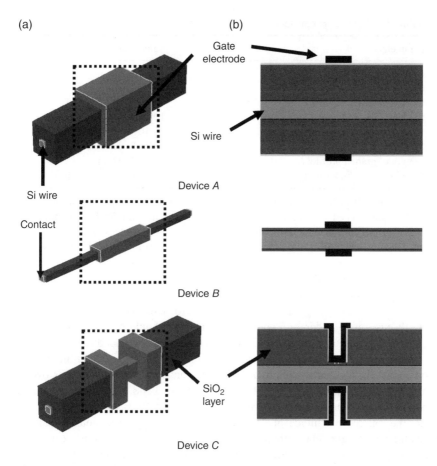

Figure 19.1 Schematic bird's views and cross sections of 20 nm long channel GAA SOI MOSFETs examined in the simulations. Copyright © 2011 IEEE. Reprinted, with permission, from Y. Omura, S. Nakano, and O. Hayashi, "Gate Field Engineering and Source/Drain Diffusion Engineering for High-Performance Si Wire GAA MOSFET and Low-Power Strategy in sub-30-nm-Channel Regime," IEEE Trans. Nanotechnol., vol. 10, pp. 715–726, 2011. (a) Bird's view of devices A, B, and C and (b) cross sections of devices A, B, and C.

where parameter *ELength* is defined with the following equation:

$$erf\left(\frac{x_{depth} - x_{sym}}{ELength}\right) = \frac{2 \times N_{D\min}}{N_{D\max}} - 1 \tag{19.2}$$

where, in the present case, it is assumed that $N_{Dmax} = 1 \times 10^{20}\,\text{cm}^{-3}$, $N_{Dmin} = 3 \times 10^{16}$ cm^{-3}, $x_{sym} = 2\,$nm, and $x_{depth} = 124$ nm. $N_D(0)\,(=N_{Dmax})$ means the doping concentration of n-type S/D regions at the metal contact, and N_{Dmin} means the doping level of the p-type active body. Other device parameters are shown in Table 19.1. Channel length (L_{eff}) is defined as the space between the two metallurgical junctions. It is assumed that the gate material is Al for simplicity. Such graded diffusion profiles usually yield undesired parasitic resistance in addition to the

Table 19.1 Device parameters.

Parameters	Values (units)
Gate length, L_G	24, 19 (nm)
Channel length, L_{eff}	20, 15 (nm)
Cross section of Si wire, $t_s \times t_s$	10×10, 20×20 (nm²)
Source diffusion length. L_s	100 (nm)
Body doping, N_A	3×10^{16} (cm⁻³)
S/D doping at contact, $N_{S/D}$	1×10^{20} (cm⁻³) (@metal contact)
Length of graded profile, L_{dif}	2–32, 100 (nm)
Gate insulator thickness, t_{in}	1.0, 20 (nm) (SiO_2 film)

resistance stemming from the narrow wire dimension. Therefore, this disadvantage must be suppressed or eliminated by some additional techniques. We propose a new technique that allows the advantages of the graded diffusion profile to be exploited.

Before discussion, we address a couple of issues that should be taken into account in the simulation results. The simulations provided by Sentaurus [22] do not directly take the ballistic effects in the transport regime into account; they only do this partially in the hydrodynamic transport model. Unfortunately, the hydrodynamic transport model sometimes overestimates the carrier velocity [24]. We must therefore establish a realistic assessment of velocity evolution along the 20 nm channel if precise drivability is required. However, as this chapter primarily considers the relative magnitude of drivability, the following discussion remains valid. We think that the density-gradient method used in the simulations of Sentaurus [22] automatically describe the volume inversion effect in such thin-body devices because the quantum confinement of carriers enhances the volume inversion. The mobility model of Sentaurus [22] is based on the physics of a 3-D bulk system. A recent article suggested that quantum-wire devices have lower carrier mobility than bulk devices [7] in contrast to a previous prediction [25]. In addition, it is well known that the quantum wire has much lower density of states than the 3-D system. Therefore, it is anticipated that the simulation results of Sentaurus basically overestimate carrier mobility and carrier density, which yields an overestimation of device drivability in terms of the present purpose. The gate leakage current density is of the order of ~10^2 A/cm² for the 1 nm thick oxide. However, its practical value is at most 1 nA at 1 V in device B because of the very small (1 nm thick) gate oxide area on the Si wire. Accordingly, for simplicity, we disregard the influence of the gate leakage current on drivability.

19.3 Simulation Results and Discussion

19.3.1 Performance of Sub-30 nm-Channel Devices and Aspects of Device Characteristics

In this section, we evaluate the feasibility of an extended design methodology for sub-30 nm channel GAA MOSFETs. Although we can expect that graded diffusion profiles yield high internal fields useful to the acceleration of carriers [18], the high parasitic resistance of the

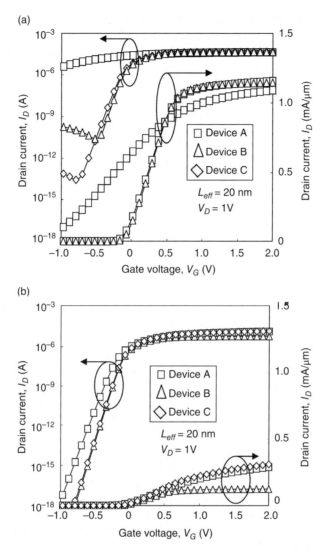

Figure 19.2 Current vs. voltage characteristics of 20 nm long-channel GAA MOSFETs (devices *A, B,* and *C*) at $V_D = 1$ V. $L_{dif} = 100$ nm and $t_s \times t_s = 10 \times 10$ nm². Copyright © 2011 IEEE. Reprinted, with permission, from Y. Omura, S. Nakano, and O. Hayashi, "Gate Field Engineering and Source/Drain Diffusion Engineering for High-Performance Si Wire GAA MOSFET and Low-Power Strategy in sub-30-nm-Channel Regime," IEEE Trans. Nanotechnol., vol. 10, pp. 715–726, 2011. (a) Devices with the abrupt junction and (b) Devices with the 100-nm-long graded junction.

graded diffusion regions hinders the realization of devices with high drivability. So, two device groups are assumed in this section; the first employ 100 nm long graded profiles to suppress short-channel effects and the others use an abrupt junction to reduce the parasitic resistance.

Figure 19.2 shows I_D-V_G characteristics of the three different GAA MOSFETs at $V_D = 1$ V; Figure 19.2a shows those of the devices with the abrupt junction and Figure 19.2b shows those devices with the 100 nm long graded diffusion regions. Here we assume that the effective surface channel is 40 nm wide. In Figure 19.2a, device A with 20 nm thick gate SiO_2 film shows a punch-through-like characteristic because of heavy short-channel effects, although device C does not despite the 20 nm thick SiO_2 film covering the source and drain junction regions. In Figure 19.2b, on the other hand, the drain current (I_D) of device C is higher than that of device B and short-channel effects of device C are suppressed as well as those of device B. This behavior stems from mechanisms described earlier [18]; even in the case of an abrupt junction, channel length (L_{eff}) is increased slightly by the lateral extension of the gate-induced field as discussed later. One more point that demands attention is the magnitude of the gate-induced drain leakage (GIDL) current [21] of devices B and C. In Figure 19.2a, it very important to note that the GIDL current of device C (~pA/μm) is more suppressed than that of device B (~nA/μm) because of the thick insulator covering the junction, it lowers the electric field normal to the Si wire surface [26, 27]. This is a very important point in realizing the goal of lowering the standby power consumption.

In Figure 19.2b, none of the devices shows a punch-through like characteristic stemming from heavy short-channel effects. Devices A and B have a sharp swing of ~60 mV/dec even at $V_D = 1$ V. Device C has a swing of at most 63 mV/dec. It follows that we must consider that the short-channel effects of device A, which has a thick gate oxide layer, do not raise its drain current. As the cross-sectional area of the Si wire is 10 nm × 10 nm, electrostatic confinement of carriers is still significant for such short-channel devices, and the thick gate SiO_2 layer does not allow short-channel effects, even in device A with its uniform 20 nm thick gate oxide film [28, 29]. Figure 19.2b shows that device C has almost the same drain current as device A. As the thin gate oxide layer of device C does not result in a low drain current, it is suggested that the graded diffusion region suppresses the short-channel effect, due to the potential penetration of the drain diffusion by the long-graded profile, and that the thick gate oxide film covering the junctions plays an important role in maintaining high drivability as discussed previously. In addition, it should be stated that the devices with 100 nm long graded diffusion region have drain current levels that are about one fourth those of devices with the abrupt junction. This issue is discussed again later.

The potential contours of devices A, B, and C around the source junction are shown in Figure 19.3 and Figure 19.5; Figure 19.3 is for the graded junction devices (devices A and C) and Figure 19.5 is for the abrupt junction devices (devices B and C). In Figure 19.3 the devices have a thick SiO_2 layer covering the low-doped source and drain regions. In the illustration of the potential contours of devices A and C, the potential contour of 1.3 V around the gate edge is shown as a bold black arch.

Figure 19.4a shows the potential profiles along the source diffusion region of devices A and C. Figure 19.4b shows the electric field profiles of device C, where the transversal field is normal to the wire surface and the longitudinal field is parallel to the wire surface; the observation point is illustrated in the upper side of Figure 19.4a. In devices A and C, the longitudinal field around the center of the diffusion region ($S1$) is higher than those near the edges and this accelerates the electron velocity. In addition, the lateral extension of the 1.3 V potential contour raises the transversal field of the Si wire over the width of $S2$; the additional transversal field impinges on the source diffusion. As a result, the transversal field raises the electron density. This feature is shown in Figure 19.4b,c; Figure 19.4c shows the electron density profile along the source diffusion region of devices A and C. This profile reduces the parasitic resistance of the "low-doped" diffusion region of $S2$ shown in Figure 19.3. In other words, the channel length of devices A and C appears to be longer than the nominal length [30].

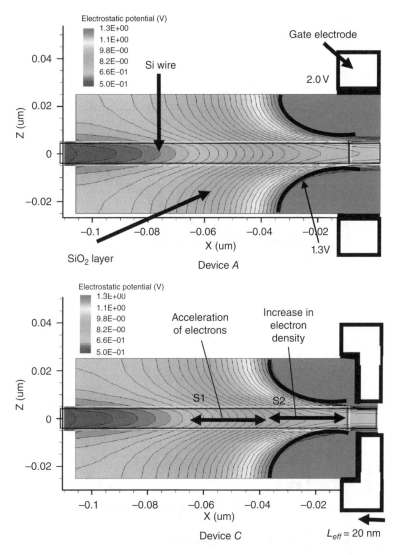

Figure 19.3 Potential contours of 20 nm long-channel devices *A* and *C* with the graded junction around the source junction at $V_D = 1$ V and $V_G = 2$ V. $L_{dif} = 100$ nm and $t_s \times t_s = 10 \times 10$ nm². (© 2011 IEEE. Reprinted, with permission, from Y. Omura, S. Nakano, and O. Hayashi, "Gate Field Engineering and Source/Drain Diffusion Engineering for High-Performance Si Wire GAA MOSFET and Low-Power Strategy in sub-30-nm-Channel Regime," IEEE Trans. Nanotechnol., vol. 10, pp. 715–726, 2011.

This yields the suppression of the short-channel effects. On the other hand, in device *B,* the lateral extension of the gate field is very limited because of the thinness of the oxide layer covering the junctions (not shown here).

Figure 19.5 shows the potential contours of devices *B* and *C* with abrupt junction for comparison to Figure 19.3. We can see somewhat different aspects from those in Figure 19.3. In the illustration of the potential contours of device *C,* the potential contour of 1.3 V around

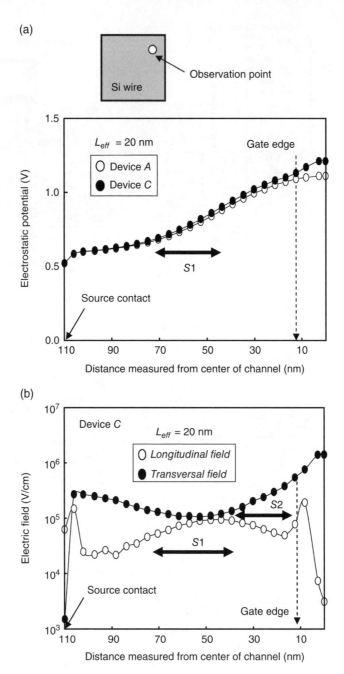

Figure 19.4 Potential profile, field profiles, and electron density profiles of Si wire along the source-to-drain direction at $V_D = 1$ V and $V_G = 2$ V. Calculated results of devices A and C with the graded junction are compared. $L_{eff} = 20$ nm, and $t_S \times t_S = 10 \times 10$ nm^2. Copyright © 2011 IEEE. Reprinted, with permission, from Y. Omura, S. Nakano, and O. Hayashi, "Gate Field Engineering and Source/Drain Diffusion Engineering for High-Performance Si Wire GAA MOSFET and Low-Power Strategy in sub-30-nm-Channel Regime," IEEE Trans. Nanotechnol., vol. 10, pp. 715–726, 2011. (a) Potential profiles (devices A and C), (b) electric field profiles (device C), and (c) electron density profiles (devices A and C).

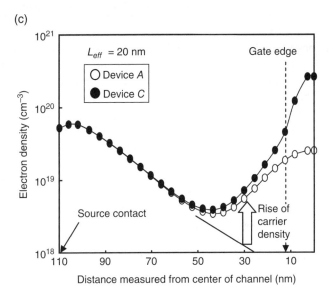

Figure 19.4 (Continued)

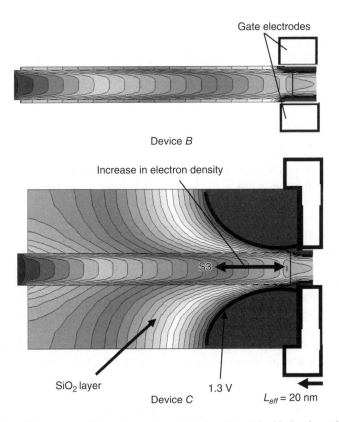

Figure 19.5 Potential contours of 20 nm long-channel devices B and C with the abrupt junction around the source junction at $V_D = 1$ V and $V_G = 2$ V. $L_{dif} = 100$ nm and $t_S \times t_S = 10 \times 10$ nm^2. Copyright © 2011 IEEE. Reprinted, with permission, from Y. Omura, S. Nakano, and O. Hayashi, "Gate Field Engineering and Source/Drain Diffusion Engineering for High-Performance Si Wire GAA MOSFET and Low-Power Strategy in sub-30-nm-Channel Regime," IEEE Trans. Nanotechnol., vol. 10, pp. 715–726, 2011.

the gate edge is shown as a bold black arch. Figure 19.6a shows the potential profiles along the source diffusion region of devices *B* and *C*. As the doping level of the diffusion region is very high, the potential profile is almost linear. The lateral extension of the 1.3 V potential contour shown in Figure 19.5 raises the electron density of the Si wire over the width of *S3* due to the same mechanism described in the previous paragraph; this behavior is shown in Figure 19.6b. The transversal electric field is enhanced over a distance of about 30 nm from the gate edge of device *C*. The electron density is raised by 30–50% in the source diffusion region near the junction of device *C*. This reduces the remaining parasitic resistance of the "high-doped," but narrow, diffusion region; this reduces the entire parasitic resistance by about 10% in the present case. The transversal field at the junction edge of device *C* is not so high because of the thick gate oxide layer covering the junction, resulting in the low GIDL current of device *C*. In the case of device *B*, however, the transversal field at the junction is much higher than that of device *C* because the 1 nm thick SiO_2 film covers an abrupt junction, resulting in the high GIDL current. Figure 19.6c illustrates the longitudinal field profiles of devices *B* and *C* along the source diffusion region. Specific local enhancement of the longitudinal field, which may be expected in device *C*, is not seen in Figure 19.6c; this is due to the uniformly high doping levels.

Finally, we address the fact that device *C*, which has an abrupt junction, does not suffer from short-channel effects. This suggests that the coverage provided by the 10 nm long thin gate insulator on the body is sufficient to trigger the suppression of short-channel effects. Therefore, it can be expected that sub-20 nm channel GAA devices with less short-channel effects and a lower GIDL current can be designed for a Si body with cross-section of 10 nm × 10 nm; the issue of the parasitic resistance remains.

19.3.2   Impact of Cross-Section of Si Wire on Short-Channel Effects and Drivability

Figure 19.2 suggests that devices with a graded junction have much larger parasitic resistance than devices with the abrupt junction. However, the parasitic resistance of the 100 nm long source/drain diffusion region of the GAA MOSFET is still large, even in the abrupt junction. Accordingly, we investigated the impact of the cross-sectional area of the Si wire on short-channel effects and the drivability of various 20-nm long channel GAA MOSFETs. Three devices, with different parameters, were considered as shown in Figure 19.7a; the first one is the control device (device *B*), the second one is device *BX* (cross-section is 20 nm × 20 nm), and the last one is device *BY* (10 nm × 10 nm in the channel region and 20 nm × 20 nm in the source/drain diffusion region). Most device parameters are identical to those shown in Table 19.1 except for the channel length and the local cross-sectional area of the Si wire. The graded junction consists of the 100 nm long graded profile.

In Figure 19.7b, simulated I_D-V_G characteristics of various devices with the abrupt junction and the graded junction are shown for $V_D = 1$ V. Device *BX* (abrupt junction) gives four times the drivability of device *B* (abrupt junction); this can be attributed to a reduction in the parasitic resistance of source/drain diffusion and the suppression of model-based mobility degradation. When the cross-section of the device body is reduced in such narrow semiconductors, the multiple gates raise the inversion carrier concentration because of the geometrical effect and eventual elimination of the doping effect [31]. This increases the effective field and the

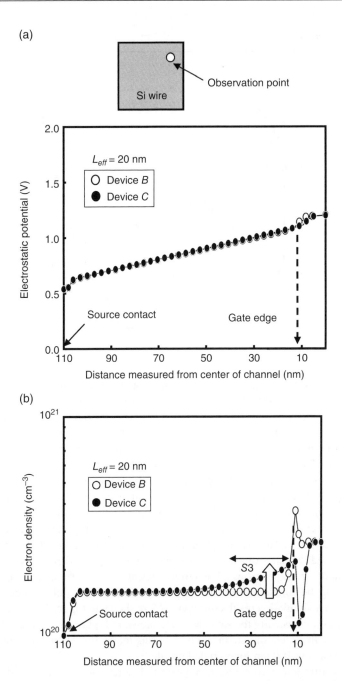

Figure 19.6 Potential profile, electron density profiles, and field profiles of Si wire along the source-to-drain direction at $V_D = 1$ V and $V_G = 2$ V. Calculated results of devices B and C with the abrupt junction are compared. $L_{eff} = 20$ nm, and $t_S \times t_S = 10 \times 10$ nm^2. Copyright © 2011 IEEE. Reprinted, with permission, from Y. Omura, S. Nakano, and O. Hayashi, "Gate Field Engineering and Source/Drain Diffusion Engineering for High-Performance Si Wire GAA MOSFET and Low-Power Strategy in sub-30-nm-Channel Regime," IEEE Trans. Nanotechnol., vol. 10, pp. 715–726, 2011. (a) Potential profiles (devices B and C), (b) electron density profiles (devices A and C), and (c) longitudinal electric field profiles (devices B and C).

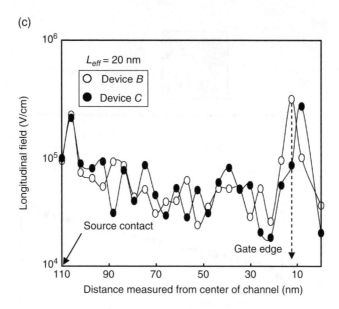

Figure 19.6   (Continued)

model-based mobility degradation. The reduction of parasitic resistance raises the junction-to-junction voltage drop, and the wider Si wire weakens the electrostatic carrier confinement. This results in the enhanced short-channel effects of device *BX* (abrupt junction). On the other hand, the cross-section of the channel region of device *BY* (abrupt junction) is the same as that of device *B* (abrupt junction); this suggests that the electrostatic confinement of carriers in the channel region is effective. Therefore, the short-channel effects of device *BY* (abrupt junction) are not degraded, and indeed this device has three times the drivability of device *B* (abrupt junction) because of the reduction in parasitic resistance. This means device *BY* (abrupt junction) is superior to devices *B* and *BX* (both abrupt junction). A similar point is discussed again later. However, the issue of high GIDL current (~nA/μm) cannot be resolved in the above family of device *B*. Therefore, if a device *B* variant is to be applied, the graded junction design must be used, which promises a drastic reduction (~1/7) in drivability.

We know that device *C* promises a low GIDL current in the off state because the thick gate SiO$_2$ film covering the junction lowers the gate-induced transversal field. For comparison with Figure 19.7, similar simulation results for device *C* are shown in Figure 19.8. The fundamental device parameters are basically the same as those shown in Figure 19.7a. The difference from Figure 19.7a is the thickness of gate insulator covering the body and the junctions. The graded junction consists of the 100 nm long graded profile.

Device *CX* (abrupt junction) shows heavy short-channel effects similar to those of device *BX* in Figure 19.7b. Major aspects of I_D-V_G characteristics of devices *C* and *CY* (both abrupt junction) shown in Figure 19.8 are similar to those shown in Figure 19.7b except for the GIDL current level. The GIDL current level is reduced to ~pA/μm for device *C* (abrupt junction) and to ~0.1 pA/μm for device *CY* (abrupt junction). Short-channel effects of devices *C* and *CY* are

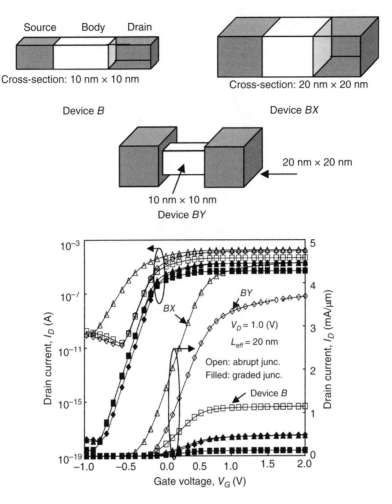

Figure 19.7 Current vs. voltage characteristics of 20 nm long-channel GAA MOSFETs with the abrupt junction or the graded junction (device B family) at $V_D = 1$ V. $L_{dif} = 100$ nm for the graded junction device and $L_{dif} = 0$ nm for the abrupt junction device. $t_s \times t_s = 10 \times 10$ nm² for device B and $t_s \times t_s = 20 \times 20$ nm² for device BX. Copyright © 2011 IEEE. Reprinted, with permission, from Y. Omura, S. Nakano, and O. Hayashi, "Gate Field Engineering and Source/Drain Diffusion Engineering for High-Performance Si Wire GAA MOSFET and Low-Power Strategy in sub-30-nm-Channel Regime," IEEE Trans. Nanotechnol., vol. 10, pp. 715–726, 2011. (a) Variations of device structures and (b) I_D vs. V_G characteristics of devices.

also well suppressed because the subthreshold swing is less than 80 mV/dec. On the other hand, devices C, CX, and CY (all graded junction) show the weakest short-channel effects and promise an extremely low GIDL current but at the cost of low drivability. This suggests that we can select device C or CY (both abrupt junction) from the viewpoint of high drivability, although the level of standby power consumption must be carefully reconsidered.

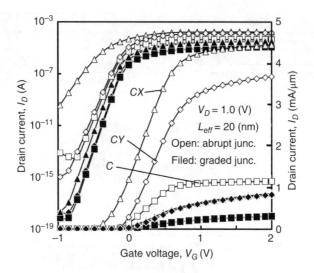

Figure 19.8 Current vs. voltage characteristics of 20 nm long-channel GAA MOSFETs with the abrupt junction or the graded junction (device C family) at $V_D = 1$ V. $L_{dif} = 100$ nm for the graded junction device and $L_{dif} = 0$ nm for the abrupt junction device. $t_s \times t_s = 10 \times 10$ nm^2 for device C and $t_s \times t_s = 20 \times 20$ nm^2 for device CX. Copyright © 2011 IEEE. Reprinted, with permission, from Y. Omura, S. Nakano, and O. Hayashi, "Gate Field Engineering and Source/Drain Diffusion Engineering for High-Performance Si Wire GAA MOSFET and Low-Power Strategy in sub-30-nm-Channel Regime," IEEE Trans. Nanotechnol., vol. 10, pp. 715–726, 2011.

19.3.3 Minimizing Standby Power Consumption of GAA SOI MOSFET

The previous section demonstrated that using source and drain diffusion regions with large cross-sections and channel regions with small cross-sections can eliminate the major issues raised by scaled GAA MOSFETs. However, such optimistic prospects for designing scaled devices fade when the high levels of power dissipation are considered.

The introduction of graded source and drain diffusion regions suppresses short-channel effects and reduces the off-leakage current as well as the GIDL current, as shown in Figures 19.7b and 19.8. The GAA devices with the graded source and drain diffusion regions shown in Figures 19.7b and 19.8 have 100 nm long graded profiles. As the source and drain diffusion regions have very large parasitic resistance, the graded region length must be as short as possible to raise drivability.

Figure 19.9 shows a simulated relationship between I_{OFF} level of devices B and device C (15 nm and 20 nm channels) and I_{ON} for various L_{dif} values, where L_{dif} represents the length of the graded profile region. I_{OFF} is defined at $V_G = V_{TH} - 0.2$ V and I_{on} is defined at $V_G = V_{TH} + 0.8$ V, where the threshold voltage (V_{TH}) is assumed to be 0.2 V. In the simulations, the threshold voltage (V_{TH}) is defined as the gate voltage that gives the maximal value of dg_m/dV_G. The cross-section of the entire Si wire for all devices is 10 nm × 10 nm. Subthreshold swing values extracted from I_D-V_G characteristics are also shown near each data point. As is expected, I_{ON} and I_{OFF} decrease as L_{dif} increases. The latest ITRS roadmap [32] requests I_{OFF} level of <10^{-9} A/μm for low-standby power applications. Simulation results suggest that a GAA MOSFET with the wire cross-section of 10 nm × 10 nm must use the L_{dif} value of ~50 nm, at least to

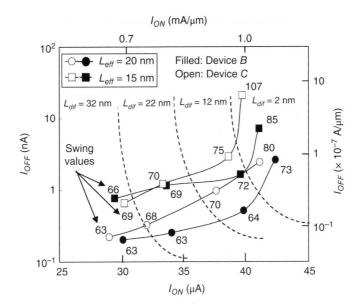

Figure 19.9 I_{ON}-I_{OFF} characteristics of devices B and C under various diffusion gradation conditions. Fifteen nm long and 20 nm long-channel devices with the abrupt junction or various graded junctions are examined; graded diffusion length (L_{dif}) is varied from 2 to 32 nm. I_{ON} is defined at $V_G = V_{TH} + 0.8$ V and I_{OFF} is at $V_{TH} - 0.2$ V, where it is assumed $V_{TH} = 0.2$ V. The cross-section of wire is 10 nm × 10 nm. Subthreshold swing values are also shown for each data in units of mV/dec. Copyright © 2011 IEEE. Reprinted, with permission, from Y. Omura, S. Nakano, and O. Hayashi, "Gate Field Engineering and Source/Drain Diffusion Engineering for High-Performance Si Wire GAA MOSFET and Low-Power Strategy in sub-30-nm-Channel Regime," IEEE Trans. Nanotechnol., vol. 10, pp. 715–726, 2011.

satisfy the roadmap request despite its small swing value. The curves reveal the superiority of 20 nm channel device B over the 20 nm channel device C regarding I_{ON} and I_{OFF} values for $L_{dif} > 2$ nm. On the other hand, the I_{ON} and I_{OFF} values of the 15 nm channel device C are superior to those of the 15 nm channel device B for L_{dif} of 32 nm. In addition, it should be noted that the I_{OFF} level shown in Figure 19.9 is not always limited by the GIDL current level. As a result, the choice of device structure strongly depends on the I_{OFF} level and/or I_{ON} level demanded for specific integrated circuits (ICs). For every device structure examined here, the use of the optimal graded diffusion profile ($L_{dif} > 50$ nm) reduces I_{ON} by about 60%, but I_{OFF} by two orders of magnitude.

19.3.4 Prospective Switching Speed Performance of GAA SOI MOSFET

We evaluated the intrinsic switching performance of 20 nm channel GAA MOSFETs. Figure 19.10 shows the intrinsic switching time ($C_g V_D / I_{ON}$) for various device structures, where C_g is the input gate capacitance, V_D is the power supply voltage, and I_{on} is the on-current. Here, the families of device B (see Figure 19.7a) and device C (see Figure 19.8) are compared with each other. It is assumed that the graded-junction devices have the 100 nm long graded diffusion region. The estimation of C_g of device B assumes a simple one-dimensional plate.

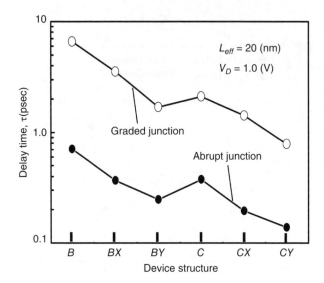

Figure 19.10   Intrinsic delay time for various device families with the 20 nm long channel. The delay time is calculated at $V_D = 1$ V. Copyright © 2011 IEEE. Reprinted, with permission, from Y. Omura, S. Nakano, and O. Hayashi, "Gate Field Engineering and Source/Drain Diffusion Engineering for High-Performance Si Wire GAA MOSFET and Low-Power Strategy in sub-30-nm-Channel Regime," IEEE Trans. Nanotechnol., vol. 10, pp. 715–726, 2011.

The estimation of C_g of device C uses a simple summation of one-dimensional piecewise plates for simplicity. Precise fringing capacitance was not calculated.

It is revealed that device BY and device CY offer the smallest delay time among their respective families; this is because they offer the lowest parasitic resistance regardless of the diffusion profile of source and drain regions. Device CY (abrupt junction) promises the highest speed. However, the I_{ON} vs. I_{OFF} relation shown in Figure 19.9 strongly suggests that its source and drain diffusion regions should use a graded profile to suppress the off-current level. Accordingly, device CY with the graded junction is the best device structure in terms of low power and low standby power consumption.

19.3.5   Parasitic Resistance Issues of GAA Wire MOSFETs

As the previous section suggested, device CY offers the best device structure to meet the design goal of the smallest delay time or low standby power consumption. However, the parasitic resistance of source and drain diffusion regions of the GAA MOSFET with the abrupt junction is large, as shown in Figure 19.6a because the large voltage drop across the diffusion region means that the device is still has a large parasitic resistance. We plot parasitic resistance values of source (R_S), drain (R_D), and channel regions (R_{ch}) of the devices in Figure 19.11 for the device B and device C families. R_S is calculated from the difference between the potential at the source metal electrode and the averaged potential at the source junction edge, R_D by the difference between the averaged potential at the drain junction edge and the potential at the drain metal electrode, and R_{ch} by the difference between the averaged potential at the source junction edge and the averaged potential at the drain junction edge. Figure 19.11a

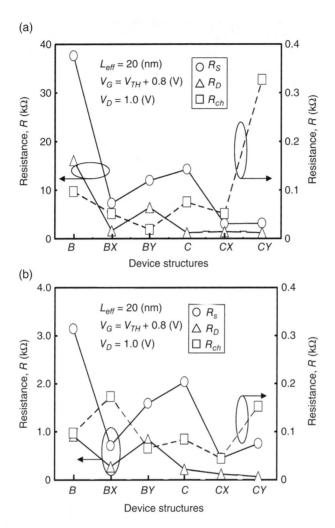

Figure 19.11 Parasitic resistance values of various device families with the 20 nm long channel at $V_D = 1$ V and $V_G = V_{TH} + 0.8$ V. Copyright © 2011 IEEE. Reprinted, with permission, from Y. Omura, S. Nakano, and O. Hayashi, "Gate Field Engineering and Source/Drain Diffusion Engineering for High-Performance Si Wire GAA MOSFET and Low-Power Strategy in sub-30-nm-Channel Regime," IEEE Trans. Nanotechnol., vol. 10, pp. 715–726, 2011. (a) Parasitic resistance values of devices with the 100-nm-long graded junction and (b) parasitic resistance values of devices with the abrupt junction.

shows the parasitic resistance values of the devices with the 100 nm long graded junction and Figure 19.11b shows the devices with the abrupt junction. The devices with the abrupt junction have much smaller R_s and R_D values than the equivalent devices with the graded junction. Devices CX and CY show relatively low resistance regardless of the diffusion profile of source and drain regions. Therefore, devices *CX and CY* are candidates for high-speed applications because they show roughly identical switching speed in Figure 19.10. When low standby power is needed, device *CY* is the best solution.

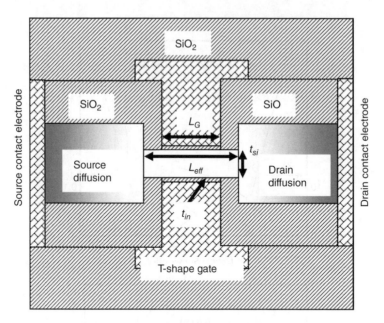

Figure 19.12 Proposal of desirable GAA wire MOSFET with 20 nm long channel. The device has a channel region with cross-section of 10 nm × 10 nm; its source and drain diffusion regions have the cross-section of 20 nm × 20 nm. In the illustration, t_{in} stands for the gate insulator thickness, and t_s stands for the thickness of the Si body. Copyright © 2011 IEEE. Reprinted, with permission, from Y. Omura, S. Nakano, and O. Hayashi, "Gate Field Engineering and Source/Drain Diffusion Engineering for High-Performance Si Wire GAA MOSFET and Low-Power Strategy in sub-30-nm-Channel Regime," IEEE Trans. Nanotechnol., vol. 10, pp. 715–726, 2011.

19.3.6   Proposal for Possible GAA Wire MOSFET Structure

Finally, we propose the sub-30 nm channel GAA device structure, which promises high performance in terms of switching speed and standby power consumption. Its cross-section for a 20 nm channel GAA MOSFET is shown in Figure 19.12. The device body is covered with a SiO$_2$ film, the cross-section of the channel region is 10 nm × 10 nm, and the source and drain diffusion regions have a cross-section of 20 nm × 20 nm (i.e. device CY); the wide diffusion regions used lower the parasitic resistance. For the assumption of the double-gate MOSFET, this device structure is similar to those examined theoretically by Laux *et al.* [33] and Likharev *et al.* [34], although they assumed infinitely wide source/drain diffusion regions and so a 3 nm thick [33] or a 1.5 nm thick [34] gate SiO$_2$ film was placed between the gate electrode and the semiconductor Si. Device CY has a T-shape gate electrode where the 10 nm long bottom of the T-shape gate is on the 1 nm thick gate SiO$_2$ film. The gate SiO$_2$ film can be replaced with a *high-κ* insulator [35]. The thick insulator around the channel edge reduces the gate field normal to the Si body, which suppresses the GIDL current. A comparison of Figure 19.7 and Figure 19.8 shows that the device C family has almost the same drivability as the device B family. Therefore, it can be concluded that the device C family is superior to the device B family for low standby power applications. The guideline for the silicide layout has been described earlier [36].

19.4 Summary

This chapter reconsidered the design methodology of the short-channel GAA SOI MOSFET and proposed an advanced technique to enhance its performance. The new ideas are based on gate field engineering and source and drain diffusion engineering. Simulations suggest that covering the source and drain diffusion paths with a thick insulator yields wide lateral extension of the gate field, which effectively increases carrier density in the low-doped source and drain diffusion regions; this raises the drivability of the short-channel GAA MOSFET with the low-doped source and drain diffusion regions. Simulations revealed that this effect can be expected even in devices that use an abrupt junction architecture.

Simulation results also suggested that the sub-30 nm channel GAA SOI MOSFET with body cross-section of 10 nm × 10 nm is promising, although its device structure must be optimized for the application to guarantee drivability, low GIDL currents, and low off-current. The key points are to introduce the graded junction in order to reduce the standby power, and cover the junctions with a thick gate insulator in order to reduce the parasitic resistance of source and drain diffusions and the GIDL current. It has been demonstrated that GAA devices with the graded junction promise one-order lower standby power consumption at the cost of only a 60% degradation in drivability. Finally, we proposed an attractive sub-30 nm channel GAA device structure.

19.5 Appendix: Brief Description of Physical Models in Simulations

• *Gate-induced drain leakage current.*
The simulations in this chapter are based on the Hurkx' trap-assisted tunneling model [37]. The following summarizes the model.

The assumption is that the band-to-band tunneling inducing the GIDL is supported by the trap-assisted tunneling. In this theory, the tunneling current density must satisfy the current continuity including the recombination process. The recombination rate they assumed has a field enhancement factor $g(F)$, where $g(F) = \Gamma_{tat}$. Γ_{tat} is the function of the electric field and it is given by

$$\Gamma_{tat} = \int_0^{\tilde{E}_n} \exp\left[u - \frac{2}{3} \frac{\sqrt{u^3}}{\tilde{E}} \right] du \qquad (19.3)$$

where $\tilde{E}$ and $\tilde{E}_n$ are defined as:

$$\tilde{E} = \frac{E}{E_0}, \qquad (19.4)$$

$$E_0 = \frac{\sqrt{8m_0 m_t k_B^3 T^3}}{qh}, \qquad (19.5)$$

$$\tilde{E}_n = \frac{E_n}{k_B T} = \begin{cases} 0, & k_B T \ln \dfrac{n}{n_i} > 0.5 E_g \\[2ex] \dfrac{0.5 E_g}{k_B T} - \ln\left(\dfrac{n}{n_i}\right) & E_{trap} \le k_B T \ln \dfrac{n}{n_i} \le 0.5 E_g \\[2ex] \dfrac{0.5 E_g}{k_B T} - E_{trap} & E_{trap} > k_B T \ln \dfrac{n}{n_i} \end{cases} \tag{19.6}$$

Notation used here takes its conventional meaning.

• *Hydrodynamic transport*
The energy-relaxation process is expressed by the following equations [22]:

$$\left.\frac{\partial W_n}{\partial t}\right|_{coll} = -H_n - \zeta_n \frac{W_n - W_{n0}}{\tau_{en}} \tag{19.7}$$

for electrons and

$$\left.\frac{\partial W_p}{\partial t}\right|_{coll} = -H_p - \zeta_p \frac{W_p - W_{n0}}{\tau_{ep}} \tag{19.8}$$

for holes. In the simulations, the energy relaxation time (τ_{en}) for electrons is 0.3 (ps) and that (τ_{ep}) for holes is 0.25 (ps). Parameters ζ_n and ζ_p are introduced to improve the convergence. ζ_n is expressed as

$$\zeta_n = 1 + \frac{n_{min}}{n} \left(\frac{n_0}{n_{min}}\right)^{\max\left[0,(T-T_n)/100\,K\right]} \tag{19.9}$$

where n_{min} and n_0 are the density parameters that are adjusted. They favor fast relaxation at low carrier densities and automatically approach unity at high carrier densities.

• *On the density gradient model*
Sentaurus simulations employ the local density approximation given by Paasch and Uebensee [23]. This model is easily applied to simulations, but it yields the quantum confinement effect without discrete energy levels. In this sense, the model is almost equivalent to the density gradient model (DGM) proposed by M. G. Ancona and his collaborators [38, 39]. The appropriateness of using the DGM was verified by Sato and Omura [40], who demonstrated that the DGM successfully reproduces fundamental quantum effects in devices. Therefore, we will not compare the present simulation results to the solutions of the Schrödinger equation.

• *Mobility models used in simulations*
  – *Masetti model*
We used the model proposed by Masetti *et al.* [41] with respect to doping dependent mobility:

Table 19.2 Simulation parameters used in the Masetti model.

Parameters	Electrons	Holes	(units)
μ_{min1}	52.2	44.9	(cm^2/Vs)
μ_{min2}	52.2	0	(cm^2/Vs)
μ_1	43.4	29.0	(cm^2/Vs)
μ_L	1417	470.5	(cm^2/Vs)
Pc	0	9.23×10^{16}	(cm^{-3})
Cr	9.68×10^{16}	2.23×10^{18}	(cm^{-3})
Cs	3.34×10^{20}	6.10×10^{20}	(cm^{-3})
α	0.680	0.719	–
β	2.0	2.0	–
ζ	2.5	2.2	–

$$\mu = \mu_{min1} \exp\left(-\frac{P_c}{N_i}\right) + \frac{\mu_{const} - \mu_{min2}}{1 + \left(\dfrac{N_i}{C_r}\right)^{\alpha}} - \frac{\mu_1}{1 + \left(\dfrac{C_s}{N_i}\right)^{\beta}} \tag{19.10}$$

$$\mu_{const} = \mu_L \left(\frac{T}{T_0}\right)^{-\zeta} \tag{19.11}$$

where $N_i \ (= N_D + N_A)$ is the total concentration of ionized impurities, $T_0 \ (= 300\,\text{K})$ and T is the lattice temperature. All other parameters are as shown in Table 19.2.

– *Lombardi model*
We used the model proposed by Lombardi *et al.* [42] with respect to mobility degradation at the Si/SiO$_2$ interface:

$$\frac{1}{\mu} = \frac{1}{\mu_b} + \frac{1}{\mu_{ac}} + \frac{1}{\mu_{sr}} \tag{19.12}$$

where μ_b is carrier mobility in bulk silicon, μ_{ac} is the surface-acoustic-phonon-limited carrier mobility, and μ_{sr} is the carrier mobility limited by surface roughness scattering.
The surface contribution due to acoustic phonon scattering is given by:

$$\mu_{ac} = \frac{B}{F_{\perp}} \frac{C \left(N_i / N_0\right)^{\lambda}}{F_{\perp}^{1/3} \left(T / T_0\right)^{k}} \tag{19.13}$$

where $F_\perp$ is the transverse electric field normal to the Si/SiO$_2$ interface. The contribution attributed to surface roughness scattering is given by:

$$\mu_{sr} = \left(\frac{\left(F_\perp / F_{ref} \right)^A}{\delta} + \frac{F_\perp^3}{\eta} \right)^{-1} \tag{19.14}$$

where the reference field F_{ref} is 1 V/cm. All other parameters are as shown in Table 19.3.

– *Hydrodynamic Canali model*

Our model for hydrodynamic simulations is based on the model proposed by Canali *et al.* [43] with respect to mobility degradation imposed by high-field velocity saturation:

$$\mu = \frac{\mu_{low}}{\left[\sqrt{1 + \alpha^2 \left(w_c - w_0 \right)^\beta} + \alpha \left(w_c - w_0 \right)^{\beta/2} \right]^{2/\beta}} \tag{19.15}$$

Table 19.3 Simulation parameters used in the Lombardi model.

Parameters	Electrons	Holes	(units)
B	4.75×10^6	9.925×10^6	(cm/s)
C	5.80×10^2	2.947×10^3	(cm$^{5/3}$/V$^{2/3}$s)
N$_0$	1	1	(cm^{-3})
λ	0.125	0.0317	—
k	1	1	—
δ	5.82×10^{14}	2.0546×10^{14}	(cm^2/Vs)
A	2	2	—
η	5.82×10^{30}	2.0546×10^{30}	(V^2/cms)

Copyright © 2011 IEEE. Reprinted, with permission, from Y. Omura, S. Nakano, and O. Hayashi, "Gate Field Engineering and Source/Drain Diffusion Engineering for High-Performance Si Wire GAA MOSFET and Low-Power Strategy in sub-30-nm-Channel Regime," IEEE Trans. Nanotechnol., vol. 10, pp. 715–726, 2011.

Table 19.4 Simulation parameters used in the hydrodynamic Canali model.

Parameters	Electrons	Holes	(units)
β_0	1.109	1.213	—
β_{exp}	0.66	0.17	—
$v_{sat,0}$	1.07×10^7	8.37×10^6	(cm/s)
$v_{sat,exp}$	0.87	0.52	—

Copyright © 2011 IEEE. Reprinted, with permission, from Y. Omura, S. Nakano, and O. Hayashi, "Gate Field Engineering and Source/Drain Diffusion Engineering for High-Performance Si Wire GAA MOSFET and Low-Power Strategy in sub-30-nm-Channel Regime," IEEE Trans. Nanotechnol., vol. 10, pp. 715–726, 2011.

$$\alpha = \frac{1}{2}\left(\frac{\mu_{low}}{q\tau_{e,c}v_{sat}^{2}}\right)^{\beta/2} \tag{19.16}$$

$$\beta = \beta_0\left(\frac{T}{T_0}\right)^{\beta_{exp}} \tag{19.17}$$

$$v_{sat} = v_{sat,0}\left(\frac{T_0}{T}\right)^{v_{sat,exp}} \tag{19.18}$$

where μ_{low} is low field mobility, $w_c = 3k_BT_c/2$ is average carrier thermal energy, $w_0 = 3k_BT/2$ is equilibrium thermal energy, and $\tau_{e,c}$ is energy relaxation time. T_c is carrier temperature. All other parameters are as shown in Table 19.4.

References

[1] J.-P. Colinge (ed.), "FinFETs and Other Multi-Gate Transistors" (Springer Science, 2008).
[2] J.-P. Colinge, H. M. Gao, A. Romano, H. Maes, and C. Claeys, "Silicon-on-Insulator 'Gate-All-Around Device'," Ext. Abstr. 1990 IEEE IEDM (San Francisco, 1990) pp. 595–598, 1990.
[3] B. Doyle, B. Boyanov, S. Datta, M. Doczy, S. Hareland, B. Jin, J. Kavaieros, T. Linton, R. Rios, and R. Chau, "Tri-Gate Fully-Depleted CMOS Transistors: Fabrication, Design and Layout," Int. Symp. VLS. Tech. Dig. (Kyoto, 2003) pp. 133–134, 2003.
[4] H. Majima, H. Ishikuro, and T. Hiramoto, "Threshold Voltage Increase by Quantum Mechanical Narrow Channel Effects in Ultra-Narrow MOSFETs," IEEE IEDM (Washington DC, 1999) pp. 379-382, 1999.
[5] F.-L. Yang, D.-H. Lee, H.-Y. Chen, C.-Y. Chang, S.-D. Liu, C.-C. Huang, T.-X. Chung, H.-W. Chen, C.-C. Huang, Y.-H. Liu, C.-C. Wu, C.-C. Chen, S.-C. Chen, Y.-T. Chen, Y.-H. Chen, C.-J. Chen, B.-W. Chan, P.-F. Hsu, J.-H. Shieh, H.-J. Tao, Y.-C. Yeo, Y. Li, J.-W. Lee, P. Chen, M.-S. Liang, and C. Hu, "5 nm-Gate Nanowire FinFET," 2004 Int. Symp. VLSI Tech. (Honolulu, June, 2004) pp. 196–197, 2004.
[6] J. Wang, A. Rahman, A. Ghosh, G. Klimeck, and M. Lundstrom, "On the validity of the parabolic effective-mass approximation for the I-V calculation of silicon nanowire transistors," IEEE Trans. Electron Devices, vol. 52, pp. 1589–1595, 2005.
[7] E. B. Ramayya, D. Vasiliska, S. M. Goodnick, and I. Knezevic, "Electron mobility in silicon nanowires," IEEE Trans. Nanotechnol., vol. 6, pp. 113–117, 2007.
[8] J.-P. Colinge, "Quantum-wire effects in trigate SOI MOSFETs," Solid-State Electron., vol. 51, pp. 1153–1160, 2007.
[9] R. Granzner, F. Schwierz, and V. M. Polyakov, "An analytical model for the threshold voltage shift caused by two-dimensional quantum confinement in undoped multiple-gate MOSFETs," IEEE Trans. Electron Devices, vol. 54, pp. 2562–2565, 2007.
[10] S. Jin, M. V. Fischetti, and T.-W. Tang, "Modeling of electron mobility in gated silicon nanowires at room temperature: surface roughness scattering, dielectric screening, and band nonparabolicity," J. Appl. Phys., vol. 102, pp. 083715–083728, 2007.
[11] M. Bescond, N. Cavassilas, and M. Lannoo, "Effective-mass approach for n-type semiconductor nanowire MOSFETs arbitrarily oriented," Nanotechnology, vol. 18, pp. 255201–255206, 2007.
[12] K. Natori, "Ballistic metal-oxide-semiconductor field effect transistor," J. Appl. Phys., vol. 76, pp. 4879–4890, 1994.
[13] V. Barral, T. Poiroux, J. Saint-Martin, D. Munteanu, J.-L. Autran, and S. Deleonibus, "Experimental investigation on quasi-ballistic transport: part I –Determination of a new backscattering coefficient extraction methodology," IEEE Trans. Electron Devices, vol. 56, pp. 408–419, 2009.

[14] M. Jagadesh Kumar, S. K. Gupta, and V. Venkataraman, "Compact modeling of the effects of parasitic internal fringe capacitance on the threshold voltage of high-κ, gate-dielectric nanoscale SOI MOSFETs," *IEEE Trans. Electron Devices*, vol. 53, pp. 706–711, 2006.

[15] A. Bansal, B. C. Paul, and K. Roy, "Modeling and optimization of fringe capacitance of nanoscale DGMOS devices," *IEEE Trans. Electron Devices*, vol. 52, pp. 256–262, 2005.

[16] M. Koyanagi, Y. Nakagawa, K. Lee, T. Nakamura, Y. Yamada, K. Inamura, K. Park, and K. Kurino, "Neuromorphic Vision Chip Fabricated Using Three-Dimensional Integration Technology," Proc. IEEE Int. Solid-State Circ. Conf. (ISSCC) (San Francisco, Feb., 2001) pp. 270–271, 2001.

[17] V. A. Gritsenko, "Design of SONOS Memory Transistor for Terabit Scale EEPROM," IEEE Conf. Electron Devices and Solid-State Circuits (Hong Kong, Dec., 2003), pp. 345–348.

[18] S. Nakano, O. Hayashi, Y. Omura, S. Yamakawa, and H. Wakabayashi, "Advanced design methodology of high-performance sub-100-nm-channel GAA MOSFET," *ECS Trans.*, vol. 19, no. 4, pp. 121–126, 2009.

[19] Y. Omura, S. Nakashima, K. Izumi, and T. Ishii, "0.1-μm-Gate, Ultrathin-Film CMOS Devices Using SIMOX Substrate with 80-nm-Thick Buried Oxide Layer," 1991 IEEE Int. Electron Devices Meeting, Tech. Dig., pp. 675–678, 1991.

[20] Y. Omura, H. Konishi, and S. Sato, "Quantum-mechanical suppression and enhancement of short-channel effects in ultra-thin SOI MOSFET's," *IEEE Trans. Electron Devices*, vol. 53, pp. 677–684, 2006.

[21] J. Chen, T. Y. Chan, I. C. Chen, P. K. Ko, and C. Hu, "Subbreakdown drain leakage current in OSFET," *IEEE Electron Device Lett.*, vol. 8, pp. 515–517, 1987.

[22] Synopsys Inc., "Sentaurus Device User Guide," Version A-2007, 2007, ftp://147.46.117.90/synopsys/manuals/PDFManual/data/sdevice_ug.pdf (accessed June 10, 2016.)

[23] G. Paasch and H. Uebensee, "A modified local density approximation," *Phys. Status Solidi B*, vol. 113, pp. 165–178, 1982.

[24] A. Kawamoto, S. Sato, and Y. Omura, "Engineering S/D diffusion for sub-100-nm channel SOI MOSFETs," *IEEE Trans. Electron Devices*, vol. 51, pp. 907–913, 2004.

[25] H. Sakaki, "Scattering suppression and high-mobility effect of size-quantized electrons in ultrafine semiconductor wire structures," *Jpn. J. Appl. Phys.*, vol. 19, pp. L735–L738, 1980.

[26] T. Ishiyama and Y. Omura, "Influences of superficial Si layer thickness on band-to-band tunneling current characteristics in ultra-thin n-channel metal-oxide-semiconductor field-effect-transistor by separation by implanted oxygen (nMOSFET/SIMOX)," *Jpn. J. Appl. Phys.*, vol. 36, pp. L264–L267, 1997.

[27] Y. Omura, T. Ishiyama, M. Shoji, and K. Izumi, "Quantum Mechanical Transport Characteristics in Ultimately Miniaturized MOSFETs/SIMOX," Proc. 7th Int. Symp. on Silicon-on-Insulator Technology and Devices (The Electrochem. Soc., Los Angeles), PV96-3, pp. 199–211, 1996.

[28] Y. Omura, "Silicon-on-Insulator (SOI) MOSFET Structure for Sub-50-nm Channel Regime," 10th Int. Symp. SOI Technol. Dev. (ECS, 2001) PV2001-3, pp. 205–210, 2001.

[29] Y. Omura, H. Konishi, and K. Yoshimoto, "Impact of fin aspect ratio on short-channel control and drivability of multiple-gate SOI MOSFETs," *J. Semicond. Technol. Sci.*, vol. 8, pp. 302–310, 2008.

[30] H. Noda, F. Murai, and S. Kimura, "Threshold Voltage Controlled 0.1-μm MOSFET Utilizing Inversion Layer as Extreme Shallow Source/Drain," Ext. Abstr. 1993 IEEE IEDM (Washington DC, 1993) pp. 123–126, 1993.

[31] Y. Tahara and Y. Omura, "Empirical quantitative modeling of threshold voltage of sub-50-nm double-gate SOI MOSFET's," *Jpn. J. Appl. Phys.*, vol. 45, pp. 3074–3078, 2006.

[32] 32."International Technology Roadmap for Semiconductors," 2001 and 2013, http://www.itrs2.net/itrs-reports.html (accessed June 10, 2016); see "2007 ITRS."

[33] D. J. Frank, S. E. Laux, and M.V. Fischetti, "Monte Carlo Simulation of a 20 nm Dual-Gate MOSFET: How Short Can Si Go?," Ext. Abstr. 1992 IEEE IEDM (San Francisco, Dec., 1992) pp. 553–556, 1992.

[34] V. A. Sverdlov, T. J. Walls, and K. K. Likharev, "Nanoscale silicon MOSFETs: a theoretical study," *IEEE Trans. Electron Devices*, vol. 50, pp. 1926–1933, 2003.

[35] D. J. Frank, Y. Taur, and H.-S. P. Wong, "Generalized scale length for two-dimensional effects in MOSFETs," *IEEE Electron Device Lett.*, vol. 19, pp. 385–387, 1998.

[36] Y. Omura, K. Yoshimoto, O. Hayashi, H. Wakabayashi, and S. Yamakawa, "Impact of metal silicide layout covering source/drain diffusion region on minimization of parasitic resistance of triple-gate SOI MOSFET and proposal of practical design guideline," *Solid State Electron.*, vol. 53, pp. 959–971, 2009.

[37] G. A. M. Hurkx, D. B. M. Klaassen, and M. P. G. Knuvers, "A new recombination model for device simulation including tunneling," *IEEE Trans. Electron Devices*, vol. 39, no. 2, pp. 331–338, 1992.

[38] M. G. Ancona and H. F. Tiersten, "Macroscopic physics of the silicon inversion layer," *Phys. Rev. B*, vol. 35, pp. 7959–7965, 1987.

[39] M. G. Ancona and G. J. Iafrate, "Quantum correction to the equation of state of an electron gas in a semiconductor," *Phys. Rev. B*, vol. 39, no. 13, pp. 9536–9540, 1989.

[40] S. Sato and Y. Omura, "Physics-based determination of the carrier effective mass assumed in density gradient model," *Jpn. J. Appl. Phys.*, vol. 45, pp. 689–693, 2006.

[41] G. Masetti, M. Severi, and S. Solmi, "Modeling of carrier mobility against carrier concentration in arsenic-, phosphorus-, and boron –doped silicon," *IEEE Trans. Electron Devices*, vol. ED-30, pp. 764–769, 1983.

[42] C. Lombardi, S. Manzini, A. Saporito, and M. Vanzi, "A physically based mobility model for numerical simulation of nonplanar devices," *IEEE Trans. Comput. Aided Des.*, vol. 7, pp. 1164–1171, 1988.

[43] C. Canali, G. Majni, R. Minder, and G. Ottaviani, "Electron and hole drift velocity measurements in silicon and their empirical relation to electric field and temperature," *IEEE Trans. Electron Devices*, vol. ED-22, pp. 1045–1047, 1975.

20

Impact of Local High-κ Insulator on Drivability and Standby Power of Gate-All-Around SOI MOSFET

20.1 Introduction

Ultrathin silicon-on-insulator (SOI) metal oxide-semiconductor field-effect-transistors (MOSFETs) and various similar advanced SOI devices are being studied extensively with the aim of realizing high-performance large-scale integrated circuits because they show high drivability and high immunity to short-channel effects [1]. The gate-all-around (GAA) SOI MOSFET [2] is one such promising device; its great potential is due to its electrostatic confinement of carriers in the Si body. It is therefore expected to offer higher performance than alternative devices, and many discussions have focused on its physical properties [3–10]. However, no plausible design guidelines have yet been proposed for GAA MOSFETs because of the apparent drawback of parasitic resistance in the source and drain regions as well as mobility degradation due to the surface roughness effect [9] and strong confinement effects [6]. Recently, Omura proposed an advanced design methodology for scaled Si-wire GAA MOSFETs, with better performance [11, 12]. In these studies, the lateral extension of the gate-induced electric field was utilized in order to induce the excess majority carriers in the source and drain diffusion regions to have a graded profile. It has already been demonstrated that a long gate overlap raises the drivability of a MOSFET [13]. In the authors' previous papers, however, this aspect was achieved using a locally thick SiO_2 film slightly covering the junction, while the main part of the gate SiO_2 film was very thin to suppress short-channel effects. This change in the device design guidelines is important in suppressing the undesirable overlap capacitance that degrades the switching performance. Nevertheless, it is anticipated that the fabrication of such a gate structure will not be easy in the sub-50 nm technology node. Thus, it is necessary to propose a more realistic gate structure.

MOS Devices for Low-Voltage and Low-Energy Applications, First Edition.
Yasuhisa Omura, Abhijit Mallik, and Naoto Matsuo.
© 2017 John Wiley & Sons Singapore Pte. Ltd. Published 2017 by John Wiley & Sons Singapore Pte. Ltd.

In this chapter, we consider the design methodology of a scaled GAA SOI MOSFET and propose a more advanced concept to enhance its performance as well as to realize its fabrication. The new idea of using a high-κ gate insulator is based on gate-field engineering and source and drain diffusion engineering [11, 12]. The validity of our design methodology is demonstrated by various results of three-dimensional device simulation.

20.2 Device Structure and Simulations

In this chapter we simulate two different device geometries as shown in Figure 20.1. Figure 20.1 shows schematic bird's-eye views of various 100 nm long channel GAA SOI MOSFETs with a Si body with $10 \times 10\,nm^2$ cross-section: device A has a 20 nm thick uniform gate SiO_2 layer, and device B has a 2.6 nm thick gate SiO_2 layer, which covers part of the channel length (=equivalent oxide thickness (EOT) of 20 nm thick HfO_2 layer); the other part of the SiO_2 film is 20 nm thick. Although the recent scaling of MOSFET devices has resulted in a sub-30 nm channel, we do not assume sub-30 nm channel GAA devices in this chapter because, first of all, a fundamental consideration with almost no short-channel effects is needed [11]. The scaling issues of sub-30 nm channel GAA devices are discussed in detail elsewhere [14].

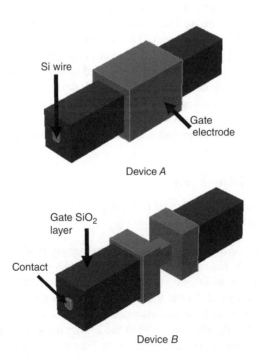

Figure 20.1 Device geometries assumed here. Reproduced by permission of the Japan Society of Applied Physics (2010). Yasuhisa Omura, Osanori Hayashi, and Shunsuke Nakano, "Impact of local high-κ insulator on drivability and standby power of gate-all-around silicon-on-insulator metal-oxide-semiconductor field-effect transistor," Jpn. J. Appl. Phys., Vol. 49, pp. 044303-1–044303-6, 2010.

Table 20.1   Device parameters.

Parameters	Values (units)
Gate length, L_G	104 (nm)
Channel length, L_{eff}	100 (nm)
Cross section of Si wire, $t_s \times t_s$	10×10 (nm^2)
Source diffusion length, L_S	100 (nm)
Body doping, N_A	3×10^{16} (cm^{-3})
S/D doping at contact, $N_{S/D}$	1×10^{20} (cm^{-3})
Insulator thickness, t_{in}	2.6, 20 (nm)
Permittivity of HfO$_2$	$30\varepsilon_0$
Length of HfO$_2$ region, L_{Hf}	0–300 (nm)
Length of thin SiO$_2$ region, L_{thin}	0–300 (nm)

Reproduced by permission of the Japan Society of Applied Physics (2010).
Yasuhisa Omura, Osanori Hayashi, and Shunsuke Nakano, "Impact of
local high-κ insulator on drivability and standby power of gate-all-around
silicon-on-insulator metal-oxide-semiconductor field-effect transistor,"
Jpn. J. Appl. Phys., Vol. 49, pp. 044303-1–044303-6, 2010.

Here, we performed three-dimensional simulations based on the hydrodynamic transport model and quantum-correction model of potential (density-gradient model) [15, 16]. In addition, a model of the band-to-band tunneling effect was taken into account in order to evaluate the gate-induced drain leakage (GIDL) current [17].

All the devices have a 2 nm long gate overlap at every junction edge, and the source and drain (S/D) diffusion regions have 100 nm long graded profiles to suppress the short-channel effects of scaled devices; the gate electrode length of all devices is fixed at 104 nm throughout this paper. The doping concentration of n-type S/D regions is 1×10^{20} cm^{-3} at the metal contact, and that in the p-type active body is 3×10^{16} cm^{-3}. It is assumed that the gate material is Al for simplicity. Details of the parameters and notation are summarized in Table 20.1.

20.3   Results and Discussion

20.3.1   Device Characteristics of GAA Devices with Graded-Profile Junctions

In this chapter, we consider the impact of a high-κ gate insulator on the drain current versus gate voltage (I_D-V_G) characteristics of a Si-wire GAA MOSFET. Thus, as shown in Figure 20.2, we assume an additional variation of device A for comparison; device AH has a 20 nm thick uniform gate insulator but the central part of the gate insulator is composed of a HfO$_2$ layer and the other part of the gate insulator is a SiO$_2$ layer. L_{Hf} denotes the width of the HfO$_2$ film region, and L_{thin} denotes the width of the 2.6 nm thick SiO$_2$ film region.

The simulated I_D-V_G characteristics are shown in Figure 20.3; the simple solid line shows the I_D-V_G curve of device A, and open and filled symbols show those of devices B and AH, respectively, for two different values of L_{Hf} and L_{thin}. All the devices exhibit very steep swing values of ~60 mV/dec; the devices do not suffer from the short-channel effects. It should be noted that the drain current (I_D) for device B is lower than that for device A when L_{thin} is larger

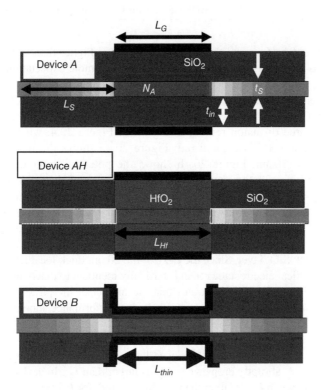

Figure 20.2 Cross-sectional views of devices for simulations. Reproduced by permission of the Japan Society of Applied Physics (2010). Yasuhisa Omura, Osanori Hayashi, and Shunsuke Nakano, "Impact of local high-κ insulator on drivability and standby power of gate-all-around silicon-on-insulator metal-oxide-semiconductor field-effect transistor," Jpn. J. Appl. Phys., Vol. 49, pp. 044303-1–044303-6, 2010.

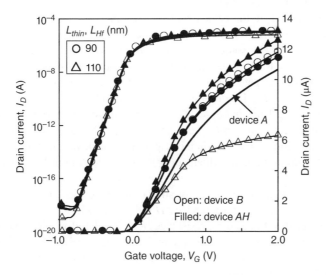

Figure 20.3 I_D-V_G characteristics at $V_D = 1$ V for devices A, B, and AH. All devices have graded profile junctions. Reproduced by permission of the Japan Society of Applied Physics (2010). Yasuhisa Omura, Osanori Hayashi, and Shunsuke Nakano, "Impact of local high-κ insulator on drivability and standby power of gate-all-around silicon-on-insulator metal-oxide-semiconductor field-effect transistor," Jpn. J. Appl. Phys., Vol. 49, pp. 044303-1–044303-6, 2010.

than the channel length (L_{eff}), and that I_D for device AH is higher than that for device B for $L_{Hf} > L_{eff}$. As this is an interesting result in contrast to the simple expectation, we discuss the mechanism of the drivability enhancement of device AH in the following.

Omura has already discussed the impact of gate-field lateral extension on the drivability of GAA MOSFET [11, 12]. Thus, here we examined the potential contours of devices AH and B near the source diffusion region, as shown in Figure 20.4, where the region with potential higher than 1.3 V is colored red; Figure 20.4a shows the potential contours of device AH for $L_{Hf} = 110$ nm, Figure 20.4b shows the potential contours of device B for $L_{thin} = 90$ nm, and Figure 20.4c shows the potential contours of device B for $L_{thin} = 110$ nm. In device AH ($L_{Hf} = 110$ nm), the lateral extension of the high gate potential covers the source diffusion region [18], while a very limited extension of high gate potential is observed in device B ($L_{thin} = 110$ nm). In device B with L_{thin} of 90 nm, however, the lateral extension of gate-induced potential covers the low-doped source diffusion region because the gate elec-trode covers the thick SiO$_2$ layer. Since the EOT of the 20 nm thick local HfO$_2$ layer of device AH is 2.6 nm, its vertical electrostatic effect is basically identical to that of device B regardless of the L_{thin} value. Figure 20.4 demonstrates that the impact of the physical thickness of the insulator around the gate electrode edge on the device performance should be focused on in analysis [18].

Figure 20.5 shows the drain current dependence on L_{Hf} and L_{thin} of devices AH and B, respectively. Here, we assume that all devices have a threshold voltage (V_{TH}) of 0.2 V at $V_D = 1$ V. In device B, I_D abruptly drops when L_{thin} is larger than L_{eff}. In device AH, on the other hand, I_D rises steeply when L_{HF} is larger than L_{eff}. In device B, for $L_{thin} > L_{eff}$, the lateral extension of gate potential almost disappears, as shown in Figure 20.4, resulting in no reduction of the parasitic resistance of low-doped source diffusion. On the other hand, in device AH, the lateral extension of gate potential still exists for $L_{Hf} > L_{eff}$ due to the geometrical effect mentioned above, as shown in Figure 20.4, resulting in a marked reduction of the parasitic resistance of low-doped source diffusion [11, 12].

In order to further analyze the behavior of I_D, the parasitic resistance dependences on L_{Hf} and L_{thin} are shown in Figure 20.6. Here, we assume that all devices have a threshold voltage (V_{TH}) of 0.2 V at $V_D = 1$ V. The source diffusion resistance (R_S) and drain diffusion resistance (R_D) of device B increase for $L_{thin} > L_{eff}$; the mechanism of the abrupt increase in R_S is based on a slight lateral extension of gate potential. This is the reason why the I_D value of device B is abruptly reduced for $L_{thin} > L_{eff}$. In addition, it can be seen in Figure 20.6 that the increase in R_D is smaller than that of R_S. Thus, in Figure 20.7, the potential contours of two different versions of device B with different values of L_{thin} are shown at $V_D = 1$ V and $V_G = 2$ V. It should be noted that the lateral extension of the gate-induced potential of device B with $L_{thin} = 110$ nm still survives around the drain junction, which suppresses the increase in R_D (see dotted circles in Figure 20.7).

In Figure 20.6, the nominal resistance of source diffusion for devices B and AH is about 110 kΩ. In contrast to device B, R_S for device AH falls to ~30 kΩ as L_{Hf} increases. However, R_S and R_D are still larger than the channel resistance (R_{ch}). We therefore have to enlarge the cross section of the S/D diffusion regions in order to maintain higher drivability; a simple estimation of R_S and R_D suggests that the cross-sectional area of the source and drain diffusion regions should be about 25 × 25 nm [2, 14].

Although the degraded junction has a high parasitic resistance, its use is necessary to sup-press the short channel effects of scaled MOSFETs. As already discussed for the resistance

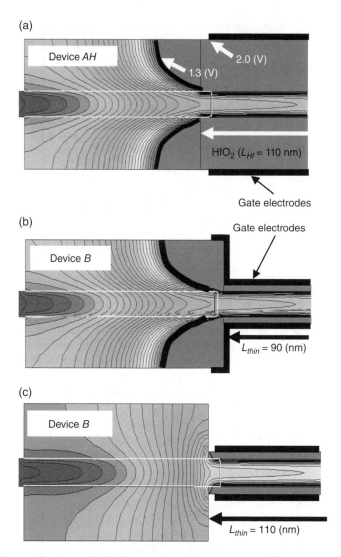

Figure 20.4 Potential contours around the source diffusion region. All devices have graded profile junctions at $V_G = V_{TH} + 2V$ and $V_D = 1V$. (a) *Device AH* ($L_{Hf} = 110$ nm). (b) *Device B* ($L_{thin} = 90$ nm). (c) *Device B* ($L_{thin} = 110$ nm). Reproduced by permission of the Japan Society of Applied Physics (2010). Yasuhisa Omura, Osanori Hayashi, and Shunsuke Nakano, "Impact of local high-κ insulator on drivability and standby power of gate-all-around silicon-on-insulator metal-oxide-semiconductor field-effect transistor," Jpn. J. Appl. Phys., Vol. 49, pp. 044303-1–044303-6, 2010.

characteristics shown in Figure 20.6, we have to minimize the parasitic resistance. As such scaled MOSFETs must have a graded diffusion profile for the source and drain regions in order to suppress short-channel effects, the method of gate-field engineering we previously proposed [11, 12] is very useful from the viewpoint of device design.

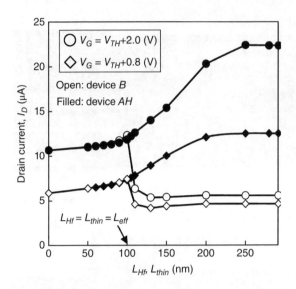

Figure 20.5 Drain current dependence on L_{Hf} and L_{thin} at $V_G = V_{TH} + 2$ V or $V_G = V_{TH} + 0.8$ V, and $V_D = 1$ V. All devices have graded profile junctions. Reproduced by permission of the Japan Society of Applied Physics (2010). Yasuhisa Omura, Osanori Hayashi, and Shunsuke Nakano, "Impact of local high-κ insulator on drivability and standby power of gate-all-around silicon-on-insulator metal-oxide-semiconductor field-effect transistor," Jpn. J. Appl. Phys., Vol. 49, pp. 044303-1–044303-6, 2010.

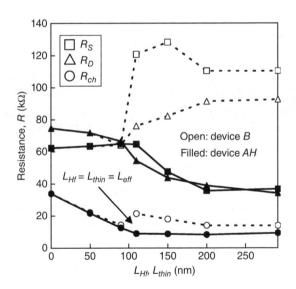

Figure 20.6 Resistance values of source diffusion (R_S), channel (R_{ch}), and drain diffusion (R_D) for devices B and AH at $V_G = V_{TH} + 2$ V and $V_D = 1$ V. All devices have graded profile junctions. Reproduced by permission of the Japan Society of Applied Physics (2010). Yasuhisa Omura, Osanori Hayashi, and Shunsuke Nakano, "Impact of local high-κ insulator on drivability and standby power of gate-all-around silicon-on-insulator metal-oxide-semiconductor field-effect transistor," Jpn. J. Appl. Phys., Vol. 49, pp. 044303-1–044303-6, 2010.

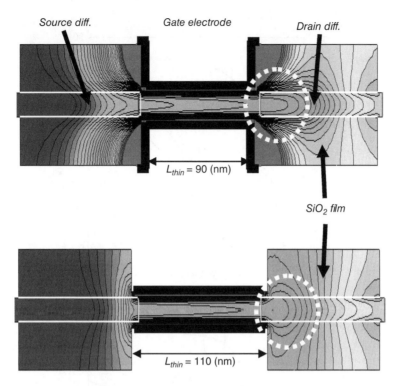

Figure 20.7 Potential contours of device B for $L_{thin}=90$ and 110 nm at $V_G=2$ V and $V_D=1$ V. All devices have graded profile junctions. Reproduced by permission of the Japan Society of Applied Physics (2010). Yasuhisa Omura, Osanori Hayashi, and Shunsuke Nakano, "Impact of local high-κ insulator on drivability and standby power of gate-all-around silicon-on-insulator metal-oxide-semiconductor field-effect transistor," Jpn. J. Appl. Phys., Vol. 49, pp. 044303-1–044303-6, 2010.

20.3.2 Device Characteristics of GAA Devices with Abrupt Junctions

Before discussing the dynamic performance of GAA devices with the graded-profile junctions, we show the I_D-V_G characteristics of devices having abrupt junctions in Figure 20.8; the I_D curves of device B are shown in Figure 20.8a and the I_D curves of device AH are shown in Figure 20.8b. The parameter is L_{thin} in Figure 20.8a and L_{Hf} in Figure 20.8b. The features of the I_D-V_G characteristics are summarized as follows: (i) I_D increases monotonically as L_{thin} for device B (or L_{Hf} of device AH) increases. (ii) The GIDL current increases rapidly as L_{thin} for device B (or L_{Hf} for device AH) increases. It is therefore anticipated that the standby power is higher in the device with the abrupt junction. The scaled devices should also be subjected to heavy short-channel effects [12].

On the other hand, Figure 20.9 shows the drain current dependence on L_{Hf} and L_{thin} for devices AH and B at $V_G=V_{TH}+2$ V and $V_{TH}+0.8$ V, respectively. Here, we assume that all devices have a threshold voltage (V_{TH}) of 0.2 V at $V_D=1$ V. In devices B and AH, I_D monotonically rises as L_{thin} or L_{Hf} increases. In device B, the lateral extension of gate potential almost disappears for $L_{thin}>L_{eff}$ as shown in Figure 20.4; however, this results in no reduction of the parasitic resistance of source diffusion because the parasitic resistance is lower than that of

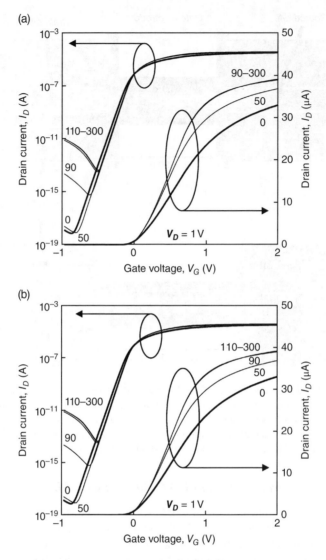

Figure 20.8 I_D-V_G characteristics at $V_D = 1\,$V for devices B and AH. All devices have abrupt junctions. (a) Device B for various L_{thin} values. Numbers mean L_{thin} values. (b) Device AH for various L_{Hf} values. Numbers mean L_{Hf} values. Reproduced by permission of the Japan Society of Applied Physics (2010). Yasuhisa Omura, Osanori Hayashi, and Shunsuke Nakano, "Impact of local high-κ insulator on driv-ability and standby power of gate-all-around silicon-on-insulator metal-oxide-semiconductor field-effect transistor," Jpn. J. Appl. Phys., Vol. 49, pp. 044303-1–044303-6, 2010.

devices having the graded-profile junction. In device $AH,$ the lateral extension of gate poten-tial still exists for $L_{Hf} > L_{eff}$ due to the aforementioned geometrical effect shown in Figure 20.4. However, this does not result in the reduction of the parasitic resistance of source diffusion for the same reason as described above [11, 12].

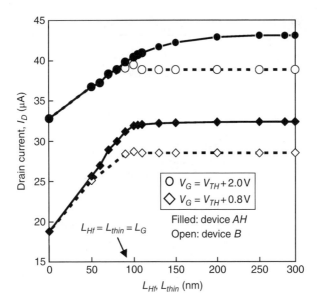

Figure 20.9 Resistance values of source diffusion (R_s), channel (R_{ch}), and drain diffusion (R_D) for devices B and AH at $V_G = V_{TH} + 2$ V and $V_G = V_{TH} + 0.8$ V, and $V_D = 1$ V. All devices have abrupt junctions. L_G denotes the gate length. Reproduced by permission of the Japan Society of Applied Physics (2010). Yasuhisa Omura, Osanori Hayashi, and Shunsuke Nakano, "Impact of local high-κ insulator on drivability and standby power of gate-all-around silicon-on-insulator metal-oxide-semiconductor field-effect transistor," Jpn. J. Appl. Phys., Vol. 49, pp. 044303-1–044303-6, 2010.

20.3.3 Behaviors of Drivability and Off-Current

We demonstrate the I_{ON}-I_{OFF} characteristics of devices B and AH with graded profile junctions in Figure 20.10a to consider how the parameter values of L_{Hf} and L_{thin} improve or degrade the drivability (I_{ON}) and the off-current (I_{OFF}). I_{ON} is defined as the on-current at $V_G = V_D = 1$ V with $V_{TH} = 0.2$ V, and I_{OFF} is defined as the off-current at $V_G = V_{TH} - 0.2$ V $= 0$ V and $V_D = 1$ V. Numbers in Figure 20.10 denote values of L_{Hf} for device AH and L_{thin} for device B; underlined numbers reveal the conditions under which the devices should show the best performance. It should be noted that the lowest I_{OFF} value of device AH is lower than that of device B. From Figure 20.10a, it is expected that device AH has a higher I_{ON} value and a lower I_{OFF} value than those of device B. The high I_{OFF} of device B at $L_{thin} = 0$ nm and that of device AH at $L_{Hf} = 0$ nm stem from a slight drain-induced barrier-lowering (DIBL) effect because the 20 nm thick gate insulator (SiO$_2$) covers the whole Si wire. On the other hand, the high I_{OFF} of device AH at $L_{Hf} = 300$ nm stems from the GIDL current because the EOT value of the HfO$_2$ layer covering the whole Si wire is very small (2.6 nm) [19]. However, the GIDL current is very low because the normal field at the interface is suppressed by the doping gradation [20]. On the other hand, we demonstrate the I_{ON}-I_{OFF} characteristics of devices B and AH with abrupt junctions in Figure 20.10b to consider how the parameter values of L_{Hf} and L_{thin} improve or degrade the drivability (I_{ON}) and the off-current (I_{OFF}); for comparison, the I_{ON}-I_{OFF} characteristics of the graded-profile junctions are also shown in Figure 20.10b.

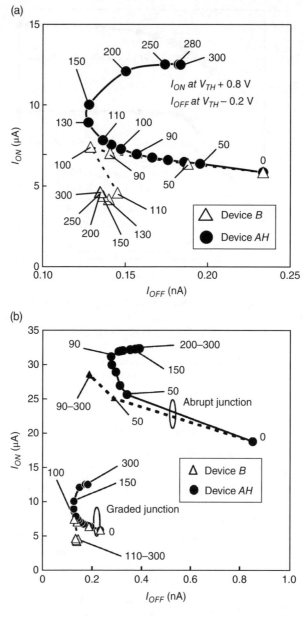

Figure 20.10 I_{ON}-I_{OFF} characteristics of devices B and AH at $V_D = 1\,\mathrm{V}$. I_{ON} is defined at $V_G = 1\,\mathrm{V}$ with $V_{TH} = 0.2\,\mathrm{V}$, and I_{OFF} at $V_G = V_{TH} - 0.2\,\mathrm{V} = 0\,\mathrm{V}$. Reproduced by permission of the Japan Society of Applied Physics (2010). Yasuhisa Omura, Osanori Hayashi, and Shunsuke Nakano, "Impact of local high-κ insulator on drivability and standby power of gate-all-around silicon-on-insulator metal-oxide-semiconductor field-effect transistor," Jpn. J. Appl. Phys., Vol. 49, pp. 044303-1–044303-6, 2010. (a) I_{ON}-I_{OFF} characteristics of devices having the graded-profile junctions. (b) Comparison of I_{ON}-I_{OFF} characteristics of devices.

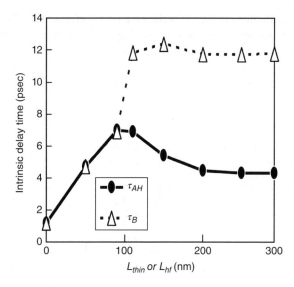

Figure 20.11 Intrinsic delay time (*CV/I*) dependence on L_{thin} and L_{Hf} at $V_G = V_{TH} + 0.8\,V = 1\,V$ and $V_D = 1.0\,V$. Reproduced by permission of the Japan Society of Applied Physics (2010). Yasuhisa Omura, Osanori Hayashi, and Shunsuke Nakano, "Impact of local high-κ insulator on drivability and standby power of gate-all-around silicon-on-insulator metal-oxide-semiconductor field-effect transistor," Jpn. J. Appl. Phys., Vol. 49, pp. 044303-1–044303-6, 2010.

20.3.4 Dynamic Performance of Devices with Graded-Profile Junctions

Finally, we briefly address the switching performance of devices *B* and *AH* with graded-profile junctions. As a measure of switching performance, the intrinsic delay time ($C_G V_D/I_{ON}$) is estimated and shown in Figure 20.11, where $C_G = L_{eff} W_{eff} C_{ox}$, $V_D = 1\,V$, and I_{ON} is equal to I_D at $V_G = 1$ and $0.8\,V$ with $V_{TH} = 0.2\,V$. W_{eff} denotes the channel width.

The intrinsic delay time of device *B* increases monotonically with L_{thin} for $L_{thin} < L_{eff}$ and saturates for $L_{thin} > L_{eff}$. This behavior is responsible for the behavior of the drain current shown in Figure 20.5. On the other hand, the intrinsic delay time of device *AH* increases monotonically with L_{Hf} for $L_{Hf} < L_{eff}$ and decreases for $L_{Hf} > L_{eff}$. This behavior is also responsible for the behavior of the drain current shown in Figure 20.5. Therefore, the advantage of device *AH* for $L_{Hf} > L_{eff}$ is clearly demonstrated. The results in Figure 20.11 suggest that the best performance is at $L_{Hf} = 150\,nm$, which is consistent with the behavior of the intrinsic delay time shown in Figure 20.11.

20.4 Summary

In this chapter, we studied the impact of a local high-κ gate insulator on the drivability and off-current of a GAA SOI MOSFET. The replacement of part of the gate insulator (SiO_2) with a high-κ insulator (HfO_2) results in the high drivability of the GAA MOSFET, which stems from the lateral extension of high gate-induced potential. However, it was also shown that the full replacement of the SiO_2 layer with a HfO_2 layer yields a high off-current, which is due to the increase in GIDL current. As a result, a simulation was performed to determine the optimal length of the high-κ insulator in order to realize the best performance. The simulation results revealed that the parasitic resistance of the low-doped source and drain diffusion regions is markedly reduced by the lateral extension of gate-induced potential; it has been demonstrated that this effect is expected when a thick high-κ insulator is used as a gate insulator. The reliability of GAA SOI MOSFETs is a future issue to be investigated.

References

[1] "FinFETs and Other Multi-Gate Transistors," ed. by J.-P. Colinge (Springer, 2008).

[2] J.-P. Colinge, H. M. Gao, A. Romano, H. Maes, and C. Claeys, "Silicon-on-Insulator gate-all-around device," IEDM Tech. Dig., 1990, pp. 595–598, 1990.

[3] H. Majima, H. Ishikuro, and T. Hiramoto, "Threshold Voltage Increase by Quantum Mechanical Narrow Channel Effect in Ultra-Narrow MOSFETs," IEDM Tech. Dig., 1999, pp. 379–382, 1999.

[4] F.-L. Yang, D.-H. Lee, H.-Y. Chen, C.-Y. Chang, S.-D. Liu, C.-C. Huang, T.-X. Chung, H.-W. Chen, C.-C. Huang, Y.-H. Liu, C.-C. Wu, C.-C. Chen, S.-C. Chen, Y.-T. Chen, Y.-H. Chen, C.-J. Chen, B.-W. Chan, P.-F. Hsu, J.-H. Shieh, H.-J. Tao, Y.-C. Yeo, Y. Li, J.-W. Lee, P. Chen, M.-S. Liang, and C. Hu, "5nm-Gate Nanowire FinFET," Tech. Dig. Int. Symp. VLSI Technology (Honolulu, Hawaii, 2004) pp. 196–197, 2004.

[5] J. Wang, A. Rahman, A. Ghosh, G. Klimeck, and M. Lundstrom, "On the validity of the parabolic effective-mass approximation for the I-V calculation of silicon nanowire transistors," *IEEE Trans. Electron Devices*, vol. 52, pp. 1589–1595, 2005.

[6] E. B. Ramayya, D. Vasiliska, S. M. Goodnick, and I. Knezevic, "Electron mobility in silicon nanowires," *IEEE Trans. Nanotechnol.*, vol. 6, pp. 113–117, 2007.

[7] J.-P. Colinge, "Quantum-wire effects in trigate SOI MOSFETs," *Solid-State Electron.*, vol. 51, pp. 1153–1160, 2007.

[8] R. Granzner, F. Schwierz, and V. M. Polyakov, "An analytical model for the threshold voltage shift caused by two-dimensional quantum confinement in undoped multiple-gate MOSFETs," *IEEE Trans. Electron Devices*, vol. 54, pp. 2562–2565, 2007.

[9] S. Jin, M. V. Fischetti, and T.-W. Tang, "Modeling of electron mobility in gated silicon nanowires at room temperature: surface roughness scattering, dielectric screening, and band nonparabolicity," *J. Appl. Phys.*, vol. 102, pp. 083715–083728, 2007.

[10] M. Bescond, N. Cavassilas, and M. Lannoo, "Effective-mass approach for n-type semiconductor nanowire MOSFETs arbitrarily oriented," *Nanotechnology*, vol. 18, pp. 255201–255206, 2007.

[11] S. Nakano, O. Hayashi, Y. Omura, S. Yamakawa, and H. Wakabayashi, "Advanced Design Methodology of High-Performance Sub-100-nm-Channel GAA MOSFET," 14th Int. Symp. Silicon-on-Insulator Technology and Devices, San Francisco, 2009 (The Electrochem. Soc. Trans.) Vol. 19, pp. 121–126, 2009.

[12] S. Nakano, Y. Omura, S. Yamakawa, and H. Wakabayashi, "Impact of Gate-Edge Field Extension on Performance of Deep Sub-50-nm-Channel Gate-All-Around MOSFET," Abstr. 2009 IEEE Silicon Nanoelectron. Workshop (Kyoto, June, 2009) pp. 31–32, 2009.

[13] T. Y. Chan, A. T. Wu, P. K. Ko, and C. Hu, "Effects of the gate-to-drain/source overlap on MOSFET characteristics," *IEEE Electron Device Lett.*, vol. 8, pp. 326–328, 1987.

[14] Y. Omura, S. Nakano, and O. Hayashi, "Gate field engineering and source/drain diffusion engineering for high-performance Si wire GAA MOSFET and low-power strategy in sub-30-nm-channel regime," *IEEE Trans. Nanotechnol.*, vol. 10, pp. 715–726, 2011.

[15] Synopsys Inc., "*Sentaurus Device User Guide*," Version A-2007, 2007, ftp://147.46.117.90/synopsys/manuals/PDFManual/data/sdevice_ug.pdf (accessed June 10, 2016.)

[16] G. Paasch and H. Uebensee, "A modified local density approximation. Electron density in inversion layers," *Phys. Status Solidi B*, vol. 113, pp. 165–178, 1982.

[17] A. Schenk, "Rigorous theory and simplified model of the band-to-band tunneling in silicon," *Solid-State Electron.*, vol. 36, pp. 19–34, 1993.

[18] H. Noda, F. Murai, and S. Kimura, "Threshold Voltage Controlled 0.1-/spl mu/m MOSFET Utilizing Inversion Layer as Extreme Shallow Source/Drain," Tech. Dig. IEEE IEDM (Washington DC, 1993) pp. 123–126, 1993.

[19] Y. Yoshioka, M. Hamada, and Y. Omura, "The potential and the drawbacks of underlap single-gate ultrathin SOI MOSFET," *Sci. Tech. Rep. Kansai Univ.*, vol. 50, pp. 17–27, 2008.

[20] T. Ishiyama and Y. Omura, "Influences of superficial Si layer thickness on band-to-band tunneling current characteristics in ultra-thin n-channel metal-oxide-semiconductor field-effect-transistor by separation by IMplanted OXygen (nMOSFET/SIMOX)," *Jpn. J. Appl. Phys.*, vol. 36, pp. L264-L267, 1997.

Part V

POTENTIAL OF PARTIALLY DEPLETED SOI MOSFETs

21

Proposal for Cross-Current Tetrode (XCT) SOI MOSFETs: A 60 dB Single-Stage CMOS Amplifier Using High-Gain Cross-Current Tetrode MOSFET/SIMOX

21.1 Introduction

It is well known that analog amplifier gain is expressed approximately by the transconductance-to-drain conductance ratio. Historically, a complementary metal oxide semiconductor (CMOS) has been used frequently in analog integrated circuits (ICs), because a high gain property is easily obtained. A way to obtain a higher gain is to make the gate length longer because the channel-length modulation is suppressed. However, this leads to a reduction in maximum operation frequency.

The gain of the conventional CMOS amplifier was around 30 dB/stage, for example, when a 2 or 3 μm design rule was employed. This chapter describes a new device that realizes a high-gain property. This device will be referred to as high-gain cross-current tetrode metal oxide semiconductor field-effect transistor (XCT MOSFET). This device requires a silicon-on-insulator structure to achieve its operation mode. A high-gain amplifier has been successfully fabricated using complimentary XCT MOSFETs and Separation by IMplanted OXygen) (SIMOX) technology [1].

MOS Devices for Low-Voltage and Low-Energy Applications, First Edition.
Yasuhisa Omura, Abhijit Mallik, and Naoto Matsuo.
© 2017 John Wiley & Sons Singapore Pte. Ltd. Published 2017 by John Wiley & Sons Singapore Pte. Ltd.

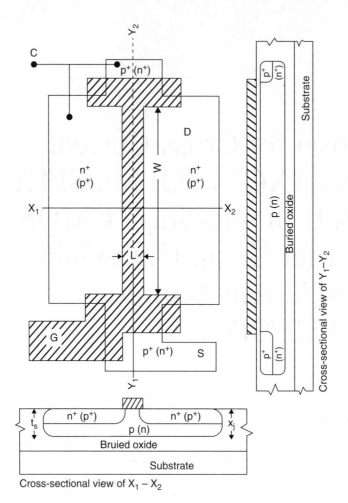

Figure 21.1 Schematic diagram of device structure. Reproduced by permission of the Japan Society of Applied Physics (1986). Yasuhisa Omura and Katsutoshi Izumi, "High-Gain Cross-Current Tetrode MOSFET/SIMOX and its Application," The 18th International Conf. on Solid State Devices and Material, pp. 715–716, 1986.

21.2 Device Fabrication

The device was fabricated using SIMOX technology [1]. Top view and cross-sectional view of the device are shown in Figure 21.1. In the figure, "D," "G," "S," and "C" represent drain, gate, source, and the fourth control terminals, respectively. The major processing steps are summarized in Table 21.1. Step- and-repeat photolithography was used for all the levels. P-type (100) Si wafers were used as the starting substrates.

A new self-aligned planar oxidation technology (SPOT) [2] was employed to obtain a flat surface after the LOCal Oxidation of Silicon (LOCOS) process. Field oxide thickness was 1.3 μm. The minimum gate length was 2 μm. After oxidizing the delineated gate polysilicon, boron, or phosphorus was implanted for source and drain diffusion formation. Device parameters are summarized in Table 21.2.

Table 21.1 Process sequence of XCT CMOS/SIMOX.

Process steps	Detail of conditions	
1. Oxygen implantation	80 keV, 1.0×10^{18} cm^{-2}	
2. Anneal	1150 C, 120 min, N$_2$	
3. Epitaxial silicon growth	1050 C, 1 μm	
4. Active area definition (LOCOS)	—	
5. Ion implant for active area	^{31}P$^+$, 195 keV (pMOS)	^{11}B$^+$, 80 keV (nMOS)
6. Grow gate oxide	20 nm, 900 C, dry-O$_2$	
7. Deposit of nondoped polysilicon	370 nm	
8. Etch polysilicon	Reactive ion etching technique	
9. Implant source-drain area	B, 50 keV, 3×10^{15} cm^{-2} (pMOS)	P, 25 keV, 3×10^{15} cm^{-2} (nMOS)
10. Deposit PSG film	600 nm	
11. PSG flow	1000 C, 10 min	
12. Contact hole etching	Reactive ion etching technique	
13. Metallization	Two levels (Al-Si and Al)	
14. Postmetallization anneal	H$_2$, 400 °C	

Reproduced by permission of The Japan Society of Applied Physics (1986). Yasuhisa Omura and Katsutoshi Izumi, "High-Gain Cross-Current Tetrode MOSFET/SIMOX and Its Application," The 18th International Conf. on Solid State Devices and Material, pp. 715–716, 1986.

Table 21.2 Device parameters.

Parameters	Values
Gate length (L_G)	2.0 (μm)
Gate width (W_G)	100 (μm)
Gate oxide thickness (t_{ox})	20 (nm)
Active region thickness (t_s)	1.0 (μm)
Junction depth (x_j)	0.7 (μm)
Doping concentration	
nMOS (N_A)	7.0×10^{16} (cm^{-3})
pMOS (N_D)	6.0×10^{16} (cm^{-3})
Buried oxide thickness (t_{BOX})	0.16 (μm)

Reproduced by permission of the Japan Society of Applied Physics (1986). Yasuhisa Omura and Katsutoshi Izumi, "High-Gain Cross-Current Tetrode MOSFET/SIMOX and Its Application," The 18th International Conf. on Solid State Devices and Material, pp. 715–716, 1986.

21.3 Device Characteristics

Typical drain current (I_D) versus drain voltage (V_D) characteristics are shown in Figure 21.2. Negative conductance appears at the I_D saturation region. This property is not due to the thermal effect [3]. The thermal effect results in the negative drain conductance only at a large current level. The thermal effect does not reduce the overall current because the effect is not due to the parasitic resistance. On the other hand, the drain current of XCT MOSFET is reduced across all the range. This is due to a parasitic device in the XCT MOSFET.

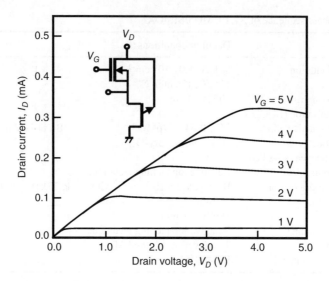

Figure 21.2 I_D vs. V_D characteristics of XCT MOSFET/SIMOX. Reproduced by permission of the Japan Society of Applied Physics (1986). Yasuhisa Omura and Katsutoshi Izumi, "High-Gain Cross-Current Tetrode MOSFET/SIMOX and its Application," The 18th International Conf. on Solid State Devices and Material, pp. 715–716, 1986.

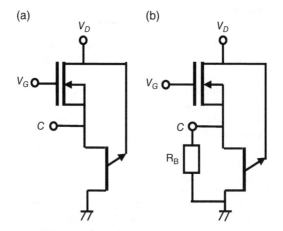

Figure 21.3 Simplified equivalent circuit of XCT MOSFET/SIMOX. (a) Original XCT SOI MOSFET and (b) XCT SOI MOSFET with a bypath resistor. Reproduced by permission of the Japan Society of Applied Physics (1986). Yasuhisa Omura and Katsutoshi Izumi, "High-Gain Cross-Current Tetrode MOSFET/SIMOX and its Application," The 18th International Conf. on Solid State Devices and Material, pp. 715–716, 1986.

An equivalent n-channel XCT MOSFET circuit is shown in Figure 21.3a. An n-channel XCT MOSFET consists of an n-channel MOSFET and a p-channel junction-gate field-effect transistor (JFET). The negative conductance property is caused because the gate bias of the parasitic JFET increases with the increasing drain bias of XCT MOSFET. A parasitic

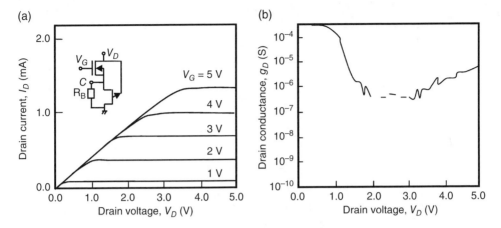

Figure 21.4 Characteristics of XCT MOSFET/SIMOX controlled by a bypass element ($R_B = 3.2\,\mathrm{k\Omega}$). Reproduced by permission of the Japan Society of Applied Physics (1986). Yasuhisa Omura and Katsutoshi Izumi, "High-Gain Cross-Current Tetrode MOSFET/SIMOX and its Application," The 18th International Conf. on Solid State Devices and Material, pp. 715–716, 1986. (a) I_D vs. V_D characteristics and (b) g_D vs. V_D characteristics.

resistance between terminals "C" and "S" can be controlled by putting a bypass circuit between those terminals.

The drain conductance changes from negative to positive by reduction of the parasitic resistance. Because of this feature, an extremely small positive drain conductance can be performed in the short-channel MOSFET/SIMOX when a proper bypass circuit between the terminals is employed (Figure 21.3b).

This is quite different from conventional hybrid negative conductance devices [4]. Examples of I_D vs. V_D and drain conductance (g_D) versus V_D characteristics controlled by the bypass element are shown in Figure 21.4. Drain current curves have better flatness at the saturation region. In the case shown in Figure 21.4b, drain conductance at 2 V gate bias and 2.5 V drain bias was less than 1 μS. This was less than 1/2000 of the conventional MOSFET/SIMOX operation mode value, although transconductance (g_m) was 1/3 that of conventional MOSFET/SIMOX operation. As a result, the g_m-to-g_D ratio was larger than 600.

21.4   Performance of CMOS Amplifier

It was expected that the above feature of the XCT MOSFET could be used in an analog amplifier. To explore this possibility, a single-stage CMOS amplifier was fabricated using complimentary XCT MOSFET/SIMOX. The diagram of the circuit block used is shown in Figure 21.5. The block diagram consists of the amplifier and the source follower. A CMOS inverter was used as the amplifier. Figure 21.6 shows gain phase versus frequency characteristics of the amplifier. An output signal amplitude of 1.0 V was obtained from an input signal amplitude of 1.0 mV at 500 Hz: that is, a gain of 60 dB.

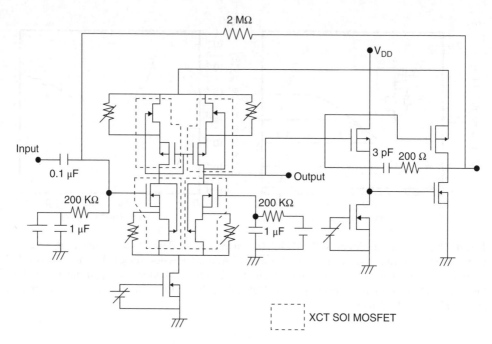

Figure 21.5 Schematic circuit block diagram of amplifier. Reproduced by permission of the Japan Society of Applied Physics (1986). Yasuhisa Omura and Katsutoshi Izumi, "High-Gain Cross-Current Tetrode MOSFET/SIMOX and its Application," The 18th International Conf. on Solid State Devices and Material, pp. 715–716, 1986.

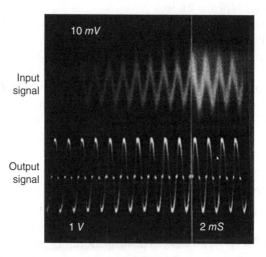

Figure 21.6 Gain/phase vs. frequency characteristics. Reproduced by permission of the Japan Society of Applied Physics (1986). Yasuhisa Omura and Katsutoshi Izumi, "High-Gain Cross-Current Tetrode MOSFET/SIMOX and its Application," The 18th International Conf. on Solid State Devices and Material, pp. 715–716, 1986.

The equivalent input noise voltage was almost the same level as the value on the conventional operation mode. This demonstrates that the XCT MOS/SIMOX has excellent potential for application to analog circuits.

21.5   Summary

Negative drain conductance was performed in the drain current saturation region in a MOSFET/SIMOX. It was controlled by a bypass circuit using the fourth control terminal. By means of the bypass element, an extremely small, positive drain conductance was obtained. A 60 dB single-stage CMOS on-chip amplifier was successfully realized using the features of this device.

References

[1] K. Izumi, M. Doken, and H. Ariyoshi, "CMOS devices fabricated on buried SiO2 layers formed by oxygen implantation into silicon," *Electron. Lett*, vol. 14, pp.593–594, 1978.

[2] K. Sakuma, Y. Arita, and M. Doken, "A new self-aligned planar oxidation technology," *J. Electrochem. Soc.*, vol. 134, pp. 1503–1507, 1987.

[3] D. K. Sharma and K. V. Ramanathan, "Modification of MOST I-V characteristics by self-heating," *Solid-State Electron.*, vol. 27, pp. 989–994, 1984.

[4] G. Kano and H. Iwasa, "A new Λ-type negative resistance device of integrated complementary FET structure," *IEEE Trans. Electron Devices*, vol. ED-21, pp. 448–449, 1974.

22

Device Model of the XCT-SOI MOSFET and Scaling Scheme

22.1 Introduction

It is widely accepted that the silicon-on-insulator (SOI) device is a promising approach to achieving higher integration levels, and it has recently been used to realize high-performance integrated circuits [1]. Shrinking the channel lengths of conventional metal oxide semiconductor field-effect transistors (MOSFETs) triggers a drastic increase in the off-leakage current because of the short-channel effects [1]. These effects, including degraded subthreshold swing, are crucial issues in device scaling [1]. Various SOI device structures are able to suppress such short-channel effects (SCEs) in comparison to bulk MOSFETs. Fortunately, the current production costs of these SOI devices are lower than they were two or three decades ago.

More recently, various families of SOI devices have been proposed and extensively studied [2]: triple-gate (TG) silicon-on-insulator metal oxide semiconductor field-effect transistor (SOI MOSFET), four-gate (G4), and other multiple-gate SOI MOSFETs [3–9] are attracting attention as well as the conventional single-gate silicon-on-insulator metal oxide semiconductor field-effect transistor (SG SOI MOSFET) and the double-gate (DG) SOI MOSFET. By using the partially depleted (PD) SG SOI MOSFET, Omura proposed the cross-current tetrode silicon-on-insulator metal oxide semiconductor field-effect transistor (XCT-SOI MOSFET) and examined its analog performance in 1986 as shown in the previous chapter [10]. The scaling feasibility of similar devices has been studied recently [7, 9, 11]. From the viewpoint of circuit applications, on the other hand, some may expect cross-current tetrode (XCT) devices to yield new applications such as high-voltage devices because the local voltage drop across the *pn* junction is lowered [12, 13] when a low specific on-resistance is not requested as a simple high-voltage switching device. Others may consider that XCT devices are

MOS Devices for Low-Voltage and Low-Energy Applications, First Edition.
Yasuhisa Omura, Abhijit Mallik, and Naoto Matsuo.
© 2017 John Wiley & Sons Singapore Pte. Ltd. Published 2017 by John Wiley & Sons Singapore Pte. Ltd.

applicable to static random access memory (SRAM) cells with high static noise margin because SCEs of XCT devices are automatically suppressed by the substrate-bias effect [14] resulting from the parasitic resistance effect of self-merged device [10, 12] and the negative differential conductance effect [10, 12]. In addition, discussion in the literature [15–20] has suggested that such apparent "low drivability" is not a crucial shortcoming for "low energy-dissipation" circuit applications, such as medical implant applications [15, 16, 21, 22] and sensor network device applications [23, 24], as mentioned below.

Extremely low energy-dissipation circuits are required in the fields of robotics [25] and medical implant applications [15, 16, 21, 22] because the batteries driving the circuits must operate within the human or animal body for long periods of time [21, 22] – of the order of tens of years. Although the drivability of the XCT-SOI MOSFET is about one hundredth of that of the conventional SOI MOSFET [12], the output voltage swing amplitude of the XCT-SOI MOSFET is identical to that of the conventional SOI MOSFET [12, 26] because the XCT-SOI MOSFET apparently works like the conventional MOSFET except for drivability and the differential negative conductance. Such a voltage-drive type device is very useful in extremely low energy-dissipation devices because we can expect a high signal-to-noise ratio in many digital circuits and the conventional circuit design methodology can be basically applied without any change [27]; it has been revealed that cross-current tetrode-silicon-on-insulator complementary metal oxide semiconductor (XCT-SOI CMOS) devices use only 1/10th of the energy needed by the corresponding conventional silicon-on-insulator complementary metal oxide semiconductor (SOI CMOS) devices. In order to discuss those applications in sufficient detail, a reliable analytical device model is needed to perform valid circuit simulations. However, modeling the XCT device is not so easy because of the three-dimensionality of its operations [12, 26].

This chapter proposes a quasi-three-dimensional (3-D) semianalytical device model for the XCT SOI MOSFET that allows detailed consideration of its various device aspects. As the purpose of this chapter is to propose a preliminary analytical device model of the XCT MOSFET, the chapter adopts the depletion approximation and simplifies the potential coupling effects between MOSFET and JFET. The device configuration and assumptions for modeling are described in section 22.2. Calculation results are shown in section 22.3 in a comparison with measured *I-V* characteristics. In addition, design guidelines for the XCT MOSFET, scaling issues, and preliminary circuit simulation results are discussed in section 22.4 with the aid of device simulations.

22.2   Device Structure and Assumptions for Modeling

22.2.1   Device Structure and Features of XCT Device

Figure 22.1 shows a bird's-eye view of the SOI MOSFET. It also shows the terminal configuration that realizes the XCT SOI MOSFET. The original SOI MOSFET is a transistor on a buried oxide (BOX) layer; that is, it has an in-depth structure of gate (poly-Si)/oxide (SiO_2)/body (Si)/oxide (SiO_2)/substrate (Si). On the other hand, the XCT device is composed of the conventional SOI MOSFET with two body contacts (*B1* and *B2*) as shown in Figure 22.1; the source diffusion region of SOI MOSFET is connected to the body contact *B1*. The source potential of the SOI MOSFET is transmitted to body contact *B1*. Body contact *B2* is grounded, resulting in the unique characteristics of the XCT device as described later.

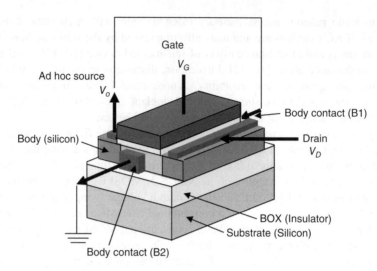

Figure 22.1 Schematic bird's eye view of *n*-channel XCT SOI MOSFET. Copyright (2011). Reprinted with permission from Elsevier from Yasuhisa Omura, Yu Azuma, Yoshimasa Yoshioka, Kyota Fukuchi, and Daishi Ino, "Proposal of preliminary device model and scaling scheme of cross-current tetrode SOI MOSFET aiming at low-energy circuit applications," Solid-State Electronics, Vol. 64, pp. 18–27, 2011.

We illustrate the structure of an *n*-channel XCT device in detail before discussing an analytical model of a MOSFET that consists of an *n*-channel SOI MOSFET and a *p*-channel JFET. The basic components of the *n*-channel SOI MOSFET are *p*-type SOI body, n^+ diffusion (source/drain of MOSFET) and two p^+ body contacts *B1* and *B2;* the p^+ body contacts (*B1* and *B2*) are also regarded as basic components (source and drain terminals) of the *p*-channel JFET. This means that the original *n*-channel SOI MOSFET and the parasitic *p*-channel JFET are self-merged with regard to the *n*-channel XCT device.

Figure 22.1 also identifies the terminal voltages. Gate voltage and drain voltage are denoted by V_G and V_D, respectively, and body contact *B2* is grounded. When V_G is higher than threshold voltage, V_{TH}, which is slightly higher than that of the original SOI *n*-channel MOSFET because the source potential of the original SOI *n*-channel MOSFET has a positive value due to the potential drop between *B1* and *B2*, the electron current of the *n*-channel MOSFET gradually increases with the drain bias of the XCT device (V_D). The potential of active body contact *B1* rises because of the potential drop between *B1* and *B2* (the drain-to-source voltage drop of the *p*-channel JFET). Thus the electron current of the *n*-channel MOSFET is transferred to the hole current of *p*-channel JFET in series via *B1*. This current flow traces the character "α," hence the device has the name of "*cross current.*" One important aspect of the XCT device is that it offers negative differential conductance (NDC) in the saturation region of the drain current [10, 12]. The fundamental mechanism of NDC is summarized below:

• When the MOSFET is in the "ON-state" with a high gate voltage (V_G), the parasitic JFET works as a parasitic source resistance. This reduces the drivability of the MOSFET.
• When the drain voltage (V_D) of the MOSFET is raised, MOSFET operation changes from nonsaturation mode to saturation mode. One of the two junction-gate terminals (n^+ diffusions) of the parasitic JFET is the drain terminal of the MOSFET. In saturation mode

operation, therefore, the channel width of the parasitic JFET (the distance between the front edges of the depletion layers as expanded from the junction gates) is reduced by the rise in the MOSFET drain voltage (V_D). This reduces the channel current of the parasitic JFET.

- As a result, the MOSFET channel current decreases as the drain voltage rises because the channel resistance of the parasitic JFET increases; the NDC appears in the saturation regime.

When the XCT MOSFET was proposed and its test circuit fabricated in 1986 [10], the major purpose of the proposal was to realize the ideal tetrode device with "zero drain conductance" in the saturation regime [10]. However, it has been experimentally demonstrated that the NDC is not a simple function of the channel length (L_{eff}) and the channel width (W_{eff}) of the MOSFET because L_{eff} represents the channel width of the parasitic JFET and W_{eff} represents the channel length of the parasitic JFET, see section 22.7. Therefore, the first study was not so successful.

In 2002, on the other hand, the SOI-based four-gate (G^4) FET was proposed [7] as a majority carrier device. The G^4-FET has two junction gates and two metal oxide semiconductor (MOS) gates [7]. The channel of the device is simply composed of a majority carrier like JFET, and so the device does not reveal NDC because the majority-carrier channel is simply opened and closed by the depletion layers controlled by the junction gates and the MOS gates. The interesting point underlying the purpose of the G^4-FET is to create a nano-scale cross-section of the majority-carrier channel. This feature will be useful in various circuit applications in the future. Therefore, the device operation of the G^4-FET is quite different from that of the XCT MOSFET.

22.2.2   Basic Assumptions for Device Modeling

This chapter models the drain current by assuming that the drift-diffusion process rules the channel current. Figure 22.2 defines three-different cut planes in order to build up an analytical model of the n-channel XCT device; the A-A cut plane is used for modeling the dimensions of depletion regions in MOSFET operation, the BB cut plane is used for modeling the p-type body potential, and the C-C cut plane is used for modeling the dimensions of depletion regions in JFET operation. In modeling the channel current of the XCT device, we make the following assumptions [28]:

- The long-channel approximation is introduced as the first step.
- Abrupt pn junctions are assumed for the source and drain junctions.
- Uniform SOI body doping (N_A) is assumed.
- The depletion region widths and the gradual-channel potential are insensitive to the substrate potential when the BOX layer thickness (t_{BOX}) is large.

In formulating mathematical expressions for channel-current components, Cartesian coordinates are used for simplicity, as shown in Figure 22.3. It is assumed that the channel current of the n-channel MOSFET flows along the x axis, the channel current of the p-channel JFET flows along the y axis, and the z axis is parallel to the depth of the SOI body. The parameter notations of the XCT device assumed here are also shown in Figure 22.3. In the illustrations, L_{eff} and W_{eff} are MOSFET channel length and channel width, and these are the full channel width and channel length for JFET, respectively. Parameter t_{ox} denotes the gate oxide layer

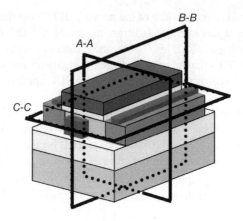

Figure 22.2 *N*-channel XCT SOI MOSFET and *A-A, B-B*, and *C-C* cut planes for device modeling. Copyright (2011). Reprinted with permission from Elsevier from Yasuhisa Omura, Yu Azuma, Yoshimasa Yoshioka, Kyota Fukuchi, and Daishi Ino, "Proposal of preliminary device model and scaling scheme of cross-current tetrode SOI MOSFET aiming at low-energy circuit applications," Solid-State Electronics, Vol. 64, pp. 18–27, 2011.

thickness, t_S denotes the SOI body thickness, and t_{BOX} denotes the BOX layer thickness. The ad-hoc source of the MOSFET (connected to *B1*) has a certain potential value (V_0) when an *n*-channel XCT device is active; here, $V_0 < V_D$.

Regarding the G⁴-FET, some analytical models of the internal potential profile in the SOI body and the threshold voltage have been proposed [29]. The models are very useful in calculating the MOSFETs threshold voltage but they are not helpful in deriving the channel current because the models can't be easily combined with the current continuity equation. The drain-current model for the depletion-all-around (DAA) mode of the G⁴-FET has also been proposed on the basis of the JFET model [30]. The theoretical form of the drain current is based on the conventional model [28]; in the following section, we also discuss the drain current model for the parasitic JFET of the XCT MOSFET based on the conventional model [28]. In Ref. [30], K Akarvardar *et al.* address the mobility model. In contrast, we simply assume constant mobility in the following section in order to simplify the theoretical expression of the drain current as the first step.

22.2.3   Derivation of Model Equations

22.2.3.1   Channel Current of SOI MOSFET

The channel current of SOI MOSFET is analytically calculated in this section based on the assumptions described in section 22.2.2. Basically, since the channel current of the XCT device is the channel current of the intrinsic MOSFET and the channel current of the parasitic JFET, it should be possible to derive the entire channel current of the XCT device from the assumptions in the manner of the well studied procedure of the MOSFET and JFET device models [10, 26]. In an early study [10], however, the proposed model was found to be limited to providing a qualitative, but not quantitative, description of XCT behavior; therefore, the model is not suitable for circuit simulations. We expect that an advanced model will yield

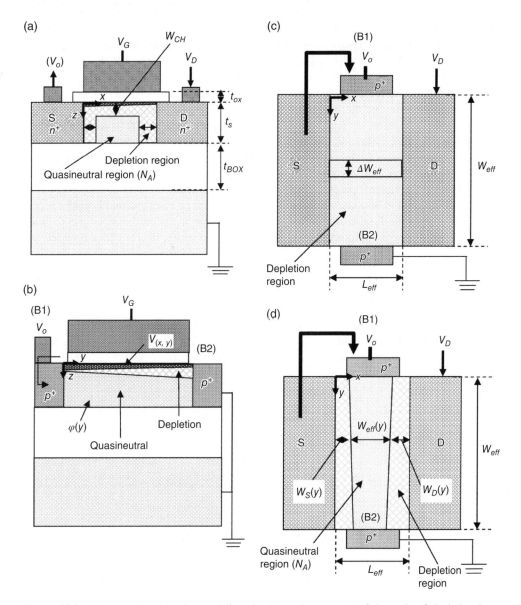

Figure 22.3 Device schematics for modeling the channel current. (a) Schematic of depletion layer layout (refer to the *A-A* cross-section in Figure 22.2). W_{CH} is the local depletion layer width spreading below the gate SiO$_2$/Si interface. It is assumed that the quasi-neutral region remains in the SOI body region so that the parasitic JFET works well. (b) Body potential effect in the SOI MOSFET (refer to the *B-B* cross-section in Figure 22.2). $V(x,y)$ and $\varphi(y)$ denote inversion-channel potential of MOSFET and JFET channel potential, respectively. (c) Definition of ΔW_{eff}. This is introduced to calculate the MOSFET current component through a small slice of channel width ΔW_{eff} (refer to the *C-C* cross-section in Figure 22.2). (d) Definitions of local channel width, $W_{eff}(y)$, and local depletion region widths, $W_S(y)$ and $W_D(y)$ (refer to the *C-C* cross-section in Figure 22.2). $W_S(y)$ and $W_D(y)$ are the depletion layer widths expanding from source and drain junctions of the SOI MOSFET. Copyright (2011). Reprinted with permission from Elsevier from Yasuhisa Omura, Yu Azuma, Yoshimasa Yoshioka, Kyota Fukuchi, and Daishi Ino, "Proposal of preliminary device model and scaling scheme of cross-current tetrode SOI MOSFET aiming at low-energy circuit applications," Solid-State Electronics, Vol. 64, pp. 18–27, 2011.

successful simulations by appropriately taking into account the behavior of the SOI body potential. Note that only the drift-based channel current of the n-channel XCT is considered here for simplicity. When calculation of the channel current of a p-channel XCT is needed, the polarity of doping and biases should be replaced with that of their p-channel counterparts.

The channel current of the n-channel MOSFET, I_{MOSFET} is formulated in the following manner. Figure 22.3a shows a schematic of the depletion region layout (refer to the A-A cross-section in Figure 22.2), where W_{CH} is the local depletion layer width spreading below the gate SiO$_2$/Si interface. Figure 22.3b visualizes the body potential effect in the SOI MOSFET (refer to the B-B cross-section in Figure 22.2), where $V(x, y)$ and $\varphi(y)$ denote MOSFET channel potential and JFET channel potential, respectively. As the inversion layer at the gate SiO$_2$/Si interface is assumed to be a charge sheet, $V(x, y)$ is taken to be independent of depth (z coordinate) for simplicity. It is assumed that the MOSFET has a uniformly doped p-type body region, so the MOS-gate-induced channel depletion region width, $W_{CH}(x, y)$, is expressed as

$$W_{CH}(x,y) = \sqrt{\frac{2\varepsilon_s \left(\phi_s + V(x,y) - \varphi(y) \right)}{qN_A}} \tag{22.1}$$

where ε_s denotes the permittivity of Si, q is the elementary charge of an electron, and ϕ_S is the surface potential near the ad hoc source junction of the MOSFET. From charge neutrality, the inversion layer charge per unit area (Q_n) is given by

$$\begin{aligned}
Q_n(x,y) &= -Q_s(x,y) + Q_B(x,y) \\
&= -C_{ox}\left[V_G - V_{FB} - \phi_s - V(x,y)\right] + \sqrt{2\varepsilon_s qN_A \left(\phi_s + V(x,y) - \varphi(y) \right)}.
\end{aligned} \tag{22.2}$$

When a constant electron mobility, μ_n, is assumed, which is reasonable in a long-channel device, the current component, ΔI_{MOSFET} through a small slice of channel width ($\Delta W(y)$) shown in Figure 22.3c (refer to the C-C cross-section in Figure 22.2), can be calculated as shown below [28]:

$$\begin{aligned}
\Delta I_{MOSFET} = \frac{\Delta W_{eff}}{L_{eff}} \mu_n & \left[C_{ox}\left\{ \left(V_G - V_{FB} - \phi_s\right) - \frac{1}{2}\left(V_D + V_o\right) \right\} \left(V_D - V_o\right) \right. \\
& \left. - \frac{2}{3}\sqrt{2\varepsilon_s qN_A} \left\{ \left(\phi_s + V_D - \varphi(y)\right)^{3/2} - \left(\phi_s + V_o - \varphi(y)\right)^{3/2} \right\} \right]
\end{aligned} \tag{22.3}$$

The total channel current of MOSFET, I_{MOSFET}, is obtained by integrating ΔI_{MOSFET} from $y=0$ to $y = W_{eff}$. This yields

$$\begin{aligned}
I_{MOSFET} &= \sum_{\Delta W_{eff}} \Delta I_{MOSFET} \\
&= \int_0^{W_{eff}} \frac{\mu_n}{L} \left[C_{ox}\left\{ \left(V_G - V_{FB} - \phi_s\right) - \frac{1}{2}\left(V_D + V_o\right) \right\} \left(V_D - V_o\right) \right. \\
& \quad \left. - \frac{2}{3}\sqrt{2\varepsilon_s qN_A} \left\{ \left(\phi_s + V_D - \varphi(y)\right)^{3/2} \left(\phi_s + V_o - \varphi(y)\right)^{3/2} \right\} \right] dy
\end{aligned} \tag{22.4}$$

where, for simplicity, we discard the $\varphi(y)$ dependence of surface potential, ϕ_s. The mathematical form of Eq. (22.4) must be simplified because this integration is not simple. The algebra in Eq. (22.4) is described in section 22.6.

22.2.3.2   Channel Current of JFET

The pJFET channel current of I_{JFET} is derived in the following manner. Figure 22.3d shows definitions of $W_S(y)$ and $W_D(y)$ (refer to the C-C cross-section in Figure 22.2), where they are the depletion-layer widths expanding from source and drain junctions of the SOI MOSFET. Following the same approach as for the MOSFET, the two depletion-region widths are given, assuming abrupt junctions, as

$$W_S(y) = \sqrt{\frac{2\varepsilon_s\left(V_o + \phi_B - \varphi(y)\right)}{qN_A}} \tag{22.5}$$

and

$$W_D(y) = \sqrt{\frac{2\varepsilon_s\left(V_D + \phi_B - \varphi(y)\right)}{qN_A}} \tag{22.6}$$

where ϕ_B is the built-in potential across the pn^+ junction of the MOSFET. The hole channel current density of JFET, J_{JFET}, along the y-axis, is given by the following drift-based equation:

$$J_{JFET} = -qN_A\mu_p\frac{d\varphi(y)}{dy} \tag{22.7}$$

where μ_p is constant hole mobility. As shown in Figure 22.3d, the effective channel width of JFET, $W_{eff}(y)$, varies with W_S and W_D (see Eq. (22.8)). The maximal value of W_{eff} is equal to L_{eff} and the junction-gate voltage of JFET ($=V_D$) at $W_{eff}=0$ is called the pinch-off voltage, V_p, where $W_{eff}(y)$ is given by

$$W_{eff}(y) = L_{eff} - \eta\left(W_S(y) + W_D(y)\right) \tag{22.8}$$

where η is the fitting parameter. Although a simple estimation of $W_{eff}(y)$ gives $L_{eff} - \{W_S(y) + W_D(y)\}$, this overestimates the channel-width narrowing. Accordingly, we introduce fitting parameter η (<1) to adjust the depletion-region widths expanding from the source and drain junctions in the Si body. This is one way to take account of the geometrical effect of depletion region in a semiconductor body surrounded by insulators [31]. The complete hole channel current of JFET (I_{JFET}) is therefore given as [28]

$$\begin{aligned} I_{JFET} &= J_{JFET}W_{eff}(y)t_S \\ &= -qN_A\mu_p\frac{d\varphi(y)}{dy}\left\{L_{eff} - \eta\left(W_S(y) + W_D(y)\right)\right\}t_S \end{aligned} \tag{22.9}$$

where it is assumed that the local cross-section of the channel is $W_{eff}(y)t_s$ for simplicity. Eq. (22.9) is rewritten as

$$I_{JFET}dy = -qN_A\mu_p\left\{L_{eff} - \eta\left(W_S(y) + W_D(y)\right)\right\}t_s d\phi(y) \tag{22.10}$$

Substituting Eqs. (22.5) and (22.6) into Eq. (22.10) and integrating Eq. (22.10) from $y=0$ $(\varphi(y)=V_o)$ to $y=W_{eff}(\varphi(y)=0)$, we have

$$I_{JFET} = \frac{qN_A\mu_p t_s}{W_{eff}}\left[L_{eff}V_o + \frac{2}{3}K\phi_B^{3/2} + \frac{2}{3}K\left(V_D + \phi_B - V_o\right)^{3/2}\right.$$
$$\left. -\frac{2}{3}K\left(V_o + \phi_B\right)^{3/2} - \frac{2}{3}K\left(V_D + \phi_B\right)^{3/2}\right] \tag{22.11}$$

where $K = \eta\sqrt{2\varepsilon_s/qN_A}$. The potential of the MOSFET ad hoc source, V_o, is determined with the aid of current continuity.

22.2.3.3   Calculation of Channel Current of XCT-SOI MOSFET

We can calculate the channel current of the XCT-SOI MOSFET by connecting Eqs. (22.4) and (22.10); that is, $I_{MOSFET}=I_{JFET}$. This equation yields the solution of the potential of the MOSFET ad-hoc source, V_o, for a given V_G value and a given V_D value. However, it is intrinsically the transcendental equation on the parameter V_o and we must apply the numerical calculation technique when obtaining V_o values.

22.3   Results and Discussion

22.3.1   Measured Characteristics of XCT Devices

This subsection describes the measurements conducted on fabricated XCT devices. Nominal device parameters used in the fabrication are shown in Table 22.1. The partially depleted silicon-on-insulator metal oxide semiconductor field-effect transistor (PD SOI MOSFET) structure was designed by adopting a thick SOI layer so that the parasitic JFET would work well. The body doping level (N_A) of the SOI substrate as produced was 9×10^{14} cm^{-3} before device fabrication. Shallow ion-implantation for threshold-voltage control was employed in the fabrication process, so the following simulation assumes that $N_A=1.0\times10^{16}$ cm^{-3} and $\eta=0.85$. We evaluated the major features of an n-channel original SOI MOSFET with body contacts and an n-channel XCT SOI MOSFET composed of n-channel MOSFET and p-channel parasitic JFET in order to extract the physical parameters needed to reproduce the I_D-V_D characteristics of the XCT device. In addition, the measured I_D-V_D characteristics for the XCT device were compared to the results yielded by the device model described in section 22.2. Physical parameters used in the calculation are shown in Table 22.2; the value of N_A was adjusted so that the I_D-V_D characteristics yielded by the model reproduced the measured values In Table 22.2, mobility values follow conventional values [32].

Table 22.1 Nominal device parameters.

Device parameters	Values	(unit)
Nominal body doping, N_A (nMOS)	1.0×10^{16}	(cm^{-3})
Gate length, L_G	2	(µm)
Channel length, L_{eff}	1.5	(µm)
Gate width, W_G	10	(µm)
Channel width, W_{eff}	10	(µm)
Gate oxide thickness, t_{ox}	30	(nm)
SOI layer thickness, t_S	350	(nm)
BOX layer thickness, t_{BOX}	300	(nm)

$V_{TH} = 0.68$ V with the n$^+$ poly-Si gate.
Copyright (2011). Reprinted with permission from Elsevier from
Yasuhisa Omura, Yu Azuma, Yoshimasa Yoshioka, Kyota Fukuchi, and
Daishi Ino, "Proposal of preliminary device model and scaling scheme
of cross-current tetrode SOI MOSFET aiming at low-energy circuit
applications," Solid-State Electronics, Vol. 64, pp. 18–27, 2011.

Table 22.2 Physical parameters for calculations.

Physical parameters	Values	(unit)
Electron mobility for MOSFET, μ_n	600	(cm^2/V s)
Hole mobility for JFET, μ_p	450	(cm^2/V s)
Built-in potential, ϕ_B	0.9	(V)

(Reprinted from Publication Solid-State Electronics, Vol 64, Yasuhisa
Omura, Yu Azuma, Yoshimasa Yoshioka, Kyota Fukuchi, and Daishi
Ino, Proposal of Preliminary Device Model and Scaling Scheme of
Cross-Current Tetrode SOI MOSFET Aiming at Low-Energy
Circuit Applications, pp. 18–27, Copyright (2011), with permission
from Elsevier.)

Figure 22.4a shows the measured I_D-V_D curves of the original SOI n-channel MOSFET with body contacts connected to the source contact. One issue with the PD SOI MOSFET structure is, generally speaking, the floating-body effect, which happens due to the restoration of majority carriers in the Si body over time [1]. As the present device has body contacts, the floating-body effect and the kink effect are not seen in Figure 22.4a. The channel-length modulation effect is, however, slightly present in the saturation region of the drain current. Figure 22.4b shows the measured I_D-V_D curves of the n-channel XCT device. When a high drain voltage is applied to the XCT device, NDC appears in the saturation region of the drain current [10, 12]. It is also seen that NDC becomes significant as the gate voltage (V_G) rises. In this figure, symbols plot the simulation results yielded by the model described in section 22.2. The proposed model successfully reproduces the fundamental I_D-V_D characteristics of the XCT device. In other words, the advanced model proposed in section 22.2 is useful in conducting various device analyses. Figure 22.4c shows the measured I_D-V_G characteristics of the n-channel XCT-SOI MOSFET and the n-channel SOI MOSFET for comparison.

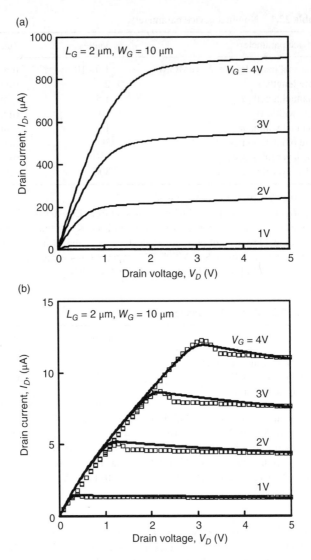

Figure 22.4 Experimentally derived $I_D\text{-}V_D$ and $I_D\text{-}V_G$ characteristics. (a) Original n-channel SOI MOSFET with body contacts connected to the source contact. (b) n-channel XCT device. (c) $I_D\text{-}V_G$ characteristics of SOI MOSFET and XCT-SOI MOSFET at V_D=0.1 V. Copyright (2011). Reprinted with permission from Elsevier from Yasuhisa Omura, Yu Azuma, Yoshimasa Yoshioka, Kyota Fukuchi, and Daishi Ino, "Proposal of preliminary device model and scaling scheme of cross-current tetrode SOI MOSFET aiming at low-energy circuit applications," Solid-State Electronics, Vol. 64, pp. 18–27, 2011.

The threshold voltage of the XCT device is 0.68 V and that of the SOI MOSFET is 0.45 V. The XCT device has higher threshold voltage than the original SOI MOSFET because the parasitic JFET eventually yields a negative body bias. It should be noted that the XCT-SOI MOSFET has smaller subthreshold swing (107 mV/dec) than the conventional SOI MOSFET (175 mV/dec). The smaller swing of the XCT is attributed to the suppression of short-channel

(c)

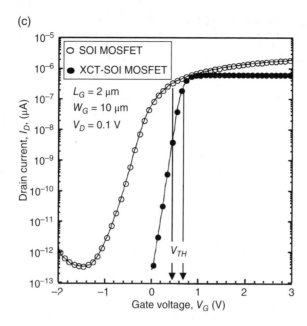

Figure 22.4   (Continued)

effects [33]. In the case of the conventional SOI MOSFET, the effective doping level around the bottom of the SOI body is less than $10^{15}\,\text{cm}^{-3}$. The lateral depletion layer widths from the source and drain junctions are wider than those near the top of the SOI body; this yields a slightly large drain-induced barrier lowering (DIBL) effect of the conventional SOI MOSFET. In addition, the XCT-SOI MOSFET does not exhibit any gate-induced drain leakage (GIDL) current, while the conventional SOI MOSFET clearly does. This is one of the advantages of the XCT-SOI MOSFET as is discussed later.

22.4   Design Guidelines

22.4.1   Drivability Control

The proposed model has been verified in broad outline, so here we will investigate the device characteristics that cannot be determined simply from the results of experiments; it is necessary to consider a device design method for circuit applications. As shown In Figure 22.4b, I_D takes its maximal value (I_{max}) at every V_G value because the XCT-SOI MOSFET demonstrates the NDC characteristic. Regarding this point, we already demonstrated the following [12]. A small gate length (L_G) yields significant NDC because the channel resistance of the parasitic JFET increases. However, when L_G is larger than a certain critical value, L_m, the NDC features of the I_D-V_D curves fade out and the fundamental characteristics are reduced to those of the usual MOSFET; the increase in channel width of the parasitic JFET (the channel length of the MOSFET) suppresses the increase in channel resistance of the parasitic JFET because the depletion-layer width of the pn junctions is much smaller than L_G (see Eq. (22.8)). In a similar way, we can get the I_{max} dependence on W_G (not shown here); the channel width at which the

Table 22.3 Scaling scheme for XCT SOI MOSFET.

Device parameters	N_A (cm^{-3})	L_{eff} (µm)	W_{eff} (µm)	t_{ox} (nm)	t_I (nm)	t_{BOX} (nm)	V_D (V)	V_{TH} (V)	SS (mV/dec)
Scaling scheme	k	k^{-1}	k^{-1}	$4/(3k)$	$k^{-1/3}$	$k^{-1/3}$	$k^{-1/2}$	—	—
$L_G=2\,\mu m$	5×10^{16}	1.5	10	30	350	300	5	0.68	107
$L_G=1\,\mu m$	1×10^{17}	0.75	5	20	278	238	3.5	0.58	103
$L_G=0.5\,\mu m$	2×10^{17}	0.38	2.5	10	221	189	2.5	0.33	93
$L_G=0.25\,\mu m$	4×10^{17}	0.2	1.3	5	170	150	1.8	0.15	83
$L_G=0.1\,\mu m$	2×10^{18}	0.075	0.5	2	129	111	1.1	0.10	76

n$^+$ poly-Si gate is assumed.

(Reprinted from Publication Solid-State Electronics, Vol 64, Yasuhisa Omura, Yu Azuma, Yoshimasa Yoshioka, Kyota Fukuchi, and Daishi Ino, Proposal of Preliminary Device Model and Scaling Scheme of Cross-Current Tetrode SOI MOSFET Aiming at Low-Energy Circuit Applications, pp. 18–27, Copyright (2011), with permission from Elsevier.)

drain current takes its maximal value (I_{max}) can be labeled W_m. When W_G is larger than W_m, the increase in JFET channel length (the channel width of the MOSFET) reduces I_{max} because the channel resistance of the parasitic JFET increases. Therefore, to make full use of the NDC attributes of the XCT device, we should select the condition of $L_G<L_m$ and/or $W_G>W_m$. Section 22.7 describes how L_m and W_m can be determined.

22.4.2 Scaling Issues

As the device dimensions considered above do not mirror the modern technology level, we must investigate the performance feasibility of scaled XCT devices; sub-100 nm long channel devices should be discussed. We start by noting that the model described above assumes constant mobility and does not take account of short-channel effects. When the scaled device model is investigated in detail, field-dependent mobility models for electrons and holes are needed to reproduce the measured results available reliably. For the purpose of achieving an analytical device model that can reproduce such scaled devices, we have to undertake very complicated mathematical operations. This is a future target that should be attempted after important aspects of XCT devices are demonstrated in detail. As the first step in this study, we performed 3-D semi-classical device simulations of scaled XCT devices, where the hydrodynamic transport model [34, 35] was assumed.

Here we consider five XCT SOI MOSFETs with different gate lengths; from $L_G=2\,\mu m$ ($L_{eff}=1.5\,\mu m$) to 100 nm (75 nm), where L_{eff} denotes the channel length. The fundamental scaling scheme examined in this chapter is shown in Table 22.3. It also shows the initial device parameters assumed. All device simulations assumed that the XCT device has abrupt source and drain junctions. The scaling scheme described in Table 22.3 covers gate oxide layer thickness (t_{ox}), SOI layer thickness (t_S), and BOX layer thickness (t_{BOX}). This scheme was derived from the results of many simulations. When this scheme is considered, we assume that the modern SOI substrate can be applied to XCT devices; that is, the application of a thin BOX layer as well as a thin SOI layer is assumed in order to develop the scaling scheme [36–38].

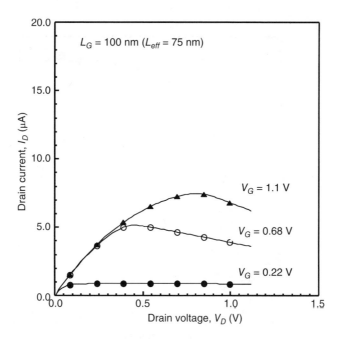

Figure 22.5 Calculated I_D-V_D characteristics of 100 nm long gate n-channel XCT device. Device characteristics are obtained by 3-D device simulations. Copyright (2011). Reprinted with permission from Elsevier from Yasuhisa Omura, Yu Azuma, Yoshimasa Yoshioka, Kyota Fukuchi, and Daishi Ino, "Proposal of preliminary device model and scaling scheme of cross-current tetrode SOI MOSFET aiming at low-energy circuit applications," Solid-State Electronics, Vol. 64, pp. 18–27, 2011.

Figure 22.5 plots simulated I_D-V_D characteristics of 100 nm long gate n-channel XCT device. These simulation results show that the scaling scheme yielded valid current-voltage characteristics. We discover that older simplified scaling schemes (e.g., constant field scaling and quasiconstant field scaling) fail to offer useful device characteristics (not shown here); that is, calculated I_D-V_D curves don't show the expected linear region followed by quasi-saturation and negative differential conductance.

Figure 22.6 shows 3-D simulation results of the I_D-V_G characteristics of various XCT devices (see Figure 22.6a) and conventional SOI MOSFETs (see Figure 22.6b); k denotes the scaling factor, and device parameters are given in Table 22.3. It should be noted that the subthreshold swing of XCT devices becomes sharp as the scaling is advanced, and that the GIDL current level is very low for $k=20$. Note that the subthreshold swing of the conventional SOI MOSFET degrades as scaling is advanced. It may also be seen that the threshold voltage of XCT devices is smoothly scaled in contrast to the conventional MOSFET, while the threshold voltage of conventional SOI MOSFETs rises. These aspects of XCT devices are quite desirable for low-energy designs. As is demonstrated in Ref. [27], XCT SOI CMOS circuits can suppress the energy dissipation of integrated circuits when the circuits don't demand very high-speed operations. Basically we think that one of high-κ materials should be applied to the gate insulator when a sharper subthreshold swing is requested in designing low-energy circuits.

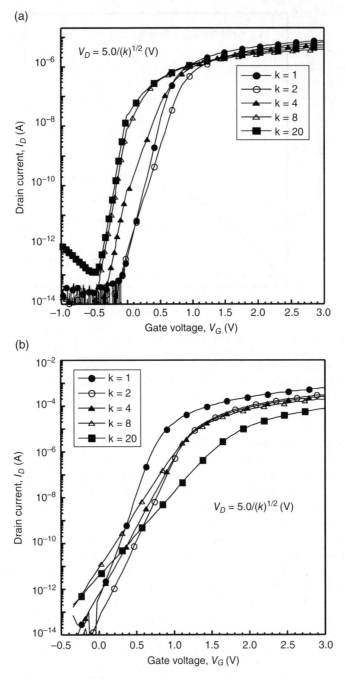

Figure 22.6 Calculated I_D-V_G characteristics of n-channel XCT devices and n-channel SOI MOSFET. Device characteristics obtained by 3-D device simulations. (a) n-channel XCT-SOI MOSFET and (b) n-channel SOI MOSFET. Copyright (2011). Reprinted with permission from Elsevier from Yasuhisa Omura, Yu Azuma, Yoshimasa Yoshioka, Kyota Fukuchi, and Daishi Ino, "Proposal of preliminary device model and scaling scheme of cross-current tetrode SOI MOSFET aiming at low-energy circuit applications," Solid-State Electronics, Vol. 64, pp. 18–27, 2011.

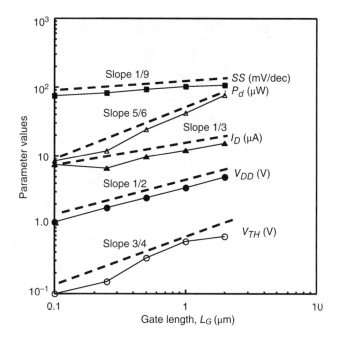

Figure 22.7 Scaling scheme proposed here. Fundamental data are extracted from device simulation results. Copyright (2011). Reprinted with permission from Elsevier from Yasuhisa Omura, Yu Azuma, Yoshimasa Yoshioka, Kyota Fukuchi, and Daishi Ino, "Proposal of preliminary device model and scaling scheme of cross-current tetrode SOI MOSFET aiming at low-energy circuit applications," Solid-State Electronics, Vol. 64, pp. 18–27, 2011.

Figure 22.7 reveals the simulated performance of scaled XCT SOI MOSFETs. The significant aspects of this scaling methodology are summarized below.

- V_{TH}/V_D falls with scaling, which is not a crucial drawback in terms of the power consumption because XCT SOI MOSFET drain current is also reduced. These aspects are not achieved by the conventional bulk and SOI MOSFETs.
- The subthreshold swing (SS) of the XCT-SOI MOSFET is also slowly scaled from 107 to 76 mV/dec when gate length (L_G) is scaled from 2 μm to 100 nm.
- Scaling rates of I_D and P_D are almost the same as those of the conventional bulk and SOI MOSFETs. Although this scaling scheme raises the power dissipation of the circuits per unit area, this is not a shortcoming because the overall power dissipation of the XCT SOI MOSFET is well suppressed by the device structure itself.

22.4.3 Potentiality of Low-Energy Operation of XCT CMOS Devices

Finally, preliminary circuit simulations of XCT CMOS devices are revealed in Figure 22.8, where the commercial SPice simulator [39] is used assuming the BSIM device model; the level-70 SOI MOSFET model (BSIMSOI 4.0) and the level-7 JFET model (TriQuint's original Model III) are assumed. In simulations, we assumed chains composed of 10 CMOS inverters

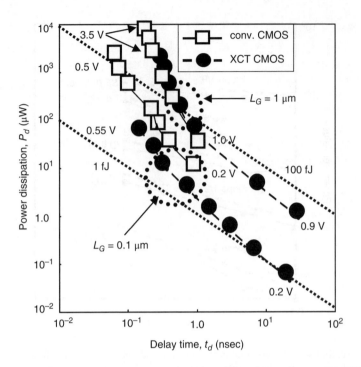

Figure 22.8 Dissipation power vs. delay time characteristics for various XCT-SOI CMOS and conventional SOI CMOS devices. Simulation results for $L_G = 1$ and $0.1\,\mu m$ are shown. Copyright (2011). Reprinted with permission from Elsevier from Yasuhisa Omura, Yu Azuma, Yoshimasa Yoshioka, Kyota Fukuchi, and Daishi Ino, "Proposal of preliminary device model and scaling scheme of cross-current tetrode SOI MOSFET aiming at low-energy circuit applications," Solid-State Electronics, Vol. 64, pp. 18–27, 2011.

based on device characteristics of fabricated devices [12]. Figure 22.8 shows dissipation power (P_D) versus delay time (t_D) characteristics. The delay time of the XCT SOI CMOS circuit is about ten times as long as that of the conventional SOI CMOS, and its power dissipation is about one-hundredth of that of the conventional SOI CMOS. The operation energy of XCT SOI CMOS is about one-tenth of that of the conventional SOI CMOS. The primary reason for such low-energy operation stems from the fact that the parasitic JFET works as the source follower component; details will be discussed elsewhere. The XCT SOI CMOS is thought to be very useful for the design of low-energy circuits for the following reasons:

- Suppression of short-channel effects improves the static noise margin of scaled static random-access memory (SRAM) memory cells.
- The power dissipation of the conventional SOI MOSFET is reduced by lowering the supply voltage. As a result, the low supply voltage degrades the signal-to-noise ratio of circuits. In the case of XCT-SOI MOSFET, however, the dissipation power is automatically reduced without a reduction of the supply voltage in comparison to the conventional SOI MOSFET having the same geometry.

22.5   Summary

This chapter proposed a physics-based model of the single-gate XCT SOI MOSFET aiming at the low energy-dissipation applications of XCT devices. Section 22.2 described features of the device structure and assumptions made. In analyzing the device characteristics, the n-channel XCT device was considered; that is, an intrinsic n-channel SOI MOSFET and a parasitic p-channel JFET were assumed. Several assumptions were made to simplify the calculations required by the drain current model proposed here. The model has two important points. The channel potential of JFET represents the body bias of the SOI MOSFET. The effective channel width of JFET varies as the depletion layers expand from the two n^+p junctions (source and drain junctions) of the MOSFET. For the n-channel XCT device, the entire drain current flows from the drain terminal of the MOSFET, via the MOSFET source and the source of the parasitic JFET, to the drain of the parasitic JFET.

We described the results of measurements on a fabricated XCT device and compared them to simulation results. Calculated I_D-V_D curves successfully reproduced the measured results of the n-channel XCT device including its NDC.

We also addressed a design methodology for the XCT-device aiming at low energy-dissipation circuit applications. It was revealed that the drain current and XCT device characteristics significantly depend on MOSFET gate length and width. It was shown how the performance of the XCT device can be controlled. Finally, a possible scaling scheme for the XCT MOSFET was considered for future sub-100 nm long gate XCT devices that should hold the feature of low energy dissipation. A couple of interesting results for the successful scaling scheme were summarized. Finally, we successfully demonstrated the potential of low-energy use of XCT CMOS devices.

22.6   Appendix: Calculation of MOSFET Channel Current

The resistance element, dR, of a small piece of the p-type Si body along the y axis is approximately defined by

$$dR = \frac{dy}{qN_A\mu_p\left\{L_{eff} - \eta\left(W_S(y) + W_D(y)\right)\right\}t_S} \tag{22.12}$$

where the carrier diffusion effect is neglected for simplicity. Assuming that the channel current along the y-axis has no generation-recombination component, the following relation is obtained:

$$d\varphi = -I_{JFET}\,dR. \tag{22.13}$$

Using Eqs. (22.10), (22.11), and (22.12), the local channel electric field of JFET is

$$\frac{d\varphi}{dy} = \left[-L_{eff}V_o - \frac{2}{3}K\phi_B^{3/2} - \frac{2}{3}K\left(V_D + \phi_B - V_o\right)^{3/2} + \frac{2}{3}K\left(V_o + \phi_B\right)^{3/2} + \frac{2}{3}K\left(V_D + \phi_B\right)^{3/2}\right] \Bigg/$$
$$W\left\{L_{eff} - \left(K\left(V_o + \phi_B - \varphi(y)\right)^{1/2} + K\left(V_D + \phi_B - \varphi(y)\right)^{1/2}\right)\right\}. \tag{22.14}$$

The electric field given by Eq. (22.13) is an explicit function of $\varphi(y)$, not coordinate y. As mentioned above, the integration in Eq. (22.9) is not completed yet. To simplify the integration of Eq. (22.9), we transform the integrand as follows:

$$I_{MOSFET} = \int_0^{W_{eff}} f(\varphi)\,dy = \int_{V_o}^0 f(\varphi)\frac{1}{\left(\dfrac{d\varphi}{dy}\right)}\,d\varphi \tag{22.15}$$

where $f(\varphi)$ is the integrand of Eq. (22.9). When Eq. (22.13) is substituted into Eq. (22.14), the expression for I_{MOSFET} is rewritten as

$$I_{MOSFET} = I_{M-i} + I_{M-ii} \tag{22.16}$$

where I_{M-i} and I_{M-ii} are defined as

$$
I_{M-i} = \frac{W_{eff}\,\mu_n}{L_{eff}} C_{ox} \int_{V_o}^0 d\varphi \left\{ \left(V_G - V_{FB} - \phi_s\right) - \frac{1}{2}\left(V_D + V_o\right)\right\}\left(V_D - V_o\right)
$$
$$
\times \left\{ L_{eff} - \left(K\left(V_o + \phi_B - \varphi(y)\right)^{1/2} + K\left(V_D + \phi_B - \varphi(y)\right)^{1/2}\right)\right\} \Big/
$$
$$
\left[-L_{eff}V_o - \frac{2}{3}K\phi_B^{3/2} - \frac{2}{3}K\left(V_D + \phi_B - V_o\right)^{3/2} + \frac{2}{3}K\left(V_o + \phi_B\right)^{3/2} + \frac{2}{3}K\left(V_D + \phi_B\right)^{3/2}\right] \tag{22.17}
$$

$$
I_{M-ii} = -\frac{W_{eff}\,\mu_n}{L_{eff}} \int_{V_o}^0 d\varphi \left\{ \frac{2}{3}D_A\left(\phi_s + V_D - \varphi(y)\right)^{3/2} - \frac{2}{3}D_A\left(\phi_s + V_o - \varphi(y)\right)^{3/2}\right\}
$$
$$
\times \left\{ L_{eff} - \left(K\left(V_o + \phi_B - \varphi(y)\right)^{1/2} + K\left(V_D + \phi_B - \varphi(y)\right)^{1/2}\right)\right\} \Big/
$$
$$
\left[-L_{eff}V_o - \frac{2}{3}K\phi_B^{3/2} - \frac{2}{3}K\left(V_D + \phi_B - V_o\right)^{3/2} + \frac{2}{3}K\left(V_o + \phi_B\right)^{3/2} + \frac{2}{3}K\left(V_D + \phi_B\right)^{3/2}\right] \tag{22.18}
$$

where $D_A = \eta\sqrt{2\varepsilon_s q N_A} = (qN_A)K$. The integration of I_{M-i} can be carried out yielding

$$I_{M-i} = \frac{W_{eff}\,\mu_n}{L_{eff}} C_{ox} \left\{ \left(V_G - V_{FB} - \phi_s\right) - \frac{1}{2}\left(V_D + V_o\right)\right\}\left(V_D - V_o\right) \tag{22.19}$$

Numerical calculation of I_{M-i} can be carried out when V_0 is determined from current continuity. We explain the termwise integration of I_{M-ii} in Eq. (22.15) below. Eq. (22.17) can be rewritten as

$$I_{M-ii} = I_{M-b-i} + I_{M-b-ii} + I_{M-b-iii} + I_{M-b-iv} + I_{M-b-v} \tag{22.20}$$

where the individual integration term is expressed as

$$I_{M-b-i} = \mu_n \frac{W_{eff}}{L_{eff}} \int_0^{V_o} d\varphi \left\{ \frac{2}{3} D_A \left(\phi_s + V_D - \varphi(y)\right)^{\frac{3}{2}} - \frac{2}{3} D_A \left(\phi_s + V_o - \varphi(y)\right)^{\frac{3}{2}} \right\} \frac{L_{eff}}{F_P} \quad (22.21)$$

$$I_{M-b-ii} = \mu_n \frac{W_{eff}}{L_{eff}} \int_0^{V_o} d\varphi \left\{ \frac{2}{3} D_A \left(\phi_s + V_D - \varphi(y)\right)^{\frac{3}{2}} \right\} \frac{K \left(V_o + \phi_B - \varphi(y)\right)^{\frac{1}{2}}}{F_P} \quad (22.22)$$

$$I_{M-b-iii} = \mu_n \frac{W}{L_{eff}} \int_0^{V_o} d\varphi \left\{ \frac{2}{3} D_A \left(\phi_s + V_D - \varphi(y)\right)^{\frac{3}{2}} \right\} \frac{K \left(V_D + \phi_B - \varphi(y)\right)^{\frac{1}{2}}}{F_P} \quad (22.23)$$

$$I_{M-b-iv} = -\mu_n \frac{W_{eff}}{L_{eff}} \int_0^{V_o} d\varphi \left\{ \frac{2}{3} D_A \left(\phi_s + V_o - \varphi(y)\right)^{\frac{3}{2}} \right\} \frac{K \left(V_o + \phi_B - \varphi(y)\right)^{\frac{1}{2}}}{F_P} \quad (22.24)$$

$$I_{M-b-v} = -\mu_n \frac{W}{L_{eff}} \int_0^{V_s} d\varphi \left\{ \frac{2}{3} D_A \left(\phi_s + V_o - \varphi(y)\right)^{\frac{3}{2}} \right\} \frac{K \left(V_D + \phi_B - \varphi(y)\right)^{\frac{1}{2}}}{F_P} \quad (22.25)$$

where the following expressions are assumed:

$$F_P = -L_{eff} V_o - \frac{2}{3} K \phi_B^{\frac{3}{2}} - \frac{2}{3} K P_{DBo}^{\frac{3}{2}} + \frac{2}{3} K P_{oB}^{\frac{3}{2}} + \frac{2}{3} K P_{DB}^{\frac{3}{2}} \quad (22.26)$$

$$P_{DBo} = \left(V_D + \phi_B - V_o\right) \quad (22.27)$$

$$P_{oB} = \left(V_o + \phi_B\right) \quad (22.28)$$

$$P_{DB} = \left(V_D + \phi_B\right) \quad (22.29)$$

When the above expressions are used, I_{M-b-i} is expressed as

$$I_{M-b-i} = \mu_n W_{eff} \times \frac{-\frac{4}{15} D_A P_{sDo}^{\frac{5}{2}} + \frac{4}{15} D_A P_{sD}^{\frac{5}{2}} + \frac{4}{15} D_A \varphi_s^{\frac{5}{2}} - \frac{4}{15} D_A P_{so}^{\frac{5}{2}}}{F_P} \quad (22.30)$$

where

$$P_{sDo} = \left(\phi_s + V_D - V_o\right) \quad (22.31)$$

$$P_{sD} = \left(\phi_s + V_D\right) \quad (22.32)$$

and

$$P_{so} = \left(\phi_s + V_o \right) \tag{22.33}$$

In performing the integrations of I_{M-b-ii} to I_{M-b-v}, the following formulae are useful:

$$I[m,n] = \int \left(px + q \right)^m \left(\sqrt{ax^2 + bx + c} \right)^n dx \tag{22.34}$$

$$
\begin{aligned}
I[1,1] &= \int \left(px + q \right) \left(\sqrt{ax^2 + bx + c} \right) dx \\
&= \frac{p}{3a} \left(ax^2 + bx + c \right)^{3/2} + \left(\frac{2aq - bp}{2a} \right) I[0,1]
\end{aligned}
\tag{22.35}
$$

$$
\begin{aligned}
I[0,1] &= \int \left(\sqrt{ax^2 + bx + c} \right) dx \\
&= \frac{2ax + b}{4a} \sqrt{ax^2 + bx + c} + \left(\frac{4ac - b^2}{8a} \right) I_F
\end{aligned}
\tag{22.36}
$$

$$
\begin{aligned}
I_F &= I[0,-1] = \int \left(\frac{1}{\sqrt{ax^2 + bx + c}} \right) dx \\
&= \begin{cases} \dfrac{1}{\sqrt{a}} \log \left| 2ax + b + 2\sqrt{a\left(ax^2 + bx + c \right)} \right| & (a > 0) \\[3mm] -\dfrac{1}{\sqrt{a}} \sin^{-1} \dfrac{2ax + b}{\sqrt{b^2 - 4ac}} & (a < 0) \end{cases}
\end{aligned}
\tag{22.37}
$$

In order to apply Eq. (22.33) to the integrations appearing in Eqs. (22.20)–(22.24), the following transformation is used for convenience.

$$\int \left(A - \varphi \right)^{3/2} \left(B - \varphi \right)^{1/2} d\varphi = \int \left(A - \varphi \right) \left(\sqrt{\varphi^2 - \left(A + B \right)\varphi + AB} \right) d\varphi \tag{22.38}$$

As a result, the total drain current of the MOSFET is expressed as

$$I_{MOSFET} = I_{M-i} + I_{M-b-i} + I_{M-b-ii} + I_{M-b-iii} + I_{M-b-iv} + I_{M-b-v} \tag{22.39}$$

All expressions related to the channel current of the MOSFET and parasitic JFET are derived from the assumptions described in section 22.2.2 as a function of the ad hoc source potential V_o. When we extract I_D-V_D characteristics of the XCT device from Eqs. (22.10) and (22.38), I_{MOSFET} must be equalized to I_{JFET} by varying the voltage V_o from 0 to V_D as indicated by current continuity

22.7   Appendix: Basic Condition for Drivability Control

As described in section 22.4.1, XCT device drivability is a function of the channel length and channel width of the original SOI MOSFET. When Eqs. (22.10) and (22.15) are assumed, the channel current is given by the following relation:

$$I_{MOSFET} = I_{JFET} \tag{22.40}$$

When Eq. (22.39) is differentiated against the channel length (L_{eff}), we have

$$L_{eff} = W_{eff} \sqrt{\frac{M2 - M1}{q N_A \mu_p t_S V_o}} \tag{22.41}$$

where

$$M1 = \mu_n C_{ox} \left\{ \left(V_G - V_{FB} - \phi_s \right) - \frac{1}{2} \left(V_D + V_o \right) \right\} \left(V_D - V_o \right) \tag{22.42}$$

$$M2 = \left(\frac{L_{eff}}{W_{eff}} \right) I_{M-ii} \tag{22.43}$$

Eq. (22.40) gives the channel length, L_m, that maximizes the channel current of the XCT device.

When Eq. (22.39) is differentiated against the channel width (W_{eff}), we have

$$W_{eff} = \sqrt{\frac{q N_A \mu_p t_S L_{eff} \left\{ L_{eff} V_o + M3 \right\}}{M2 - M1}} \tag{22.44}$$

where

$$M3 = \frac{2}{3} K \phi_B^{3/2} + \frac{2}{3} K \left(V_D + \phi_B - V_o \right)^{3/2} - \frac{2}{3} K \left(V_o + \phi_B \right)^{3/2} - \frac{2}{3} K \left(V_D + \phi_B \right)^{3/2} \tag{22.45}$$

Eq. (22.43) gives the channel width, W_m, that maximizes the channel current of the XCT device.

References

[1] J.-P. Colinge, "Silicon-on-Insulator Technology: Materials to VLSI," 3rd ed. (Kluwer Academic Publishers, 2004).

[2] J.-P. Colinge, "FinFETs and Other Multi-Gate Transistors" (Springer, 2008), Chapter 1.

[3] H.-S. P. Wong, K. K. Chan, and Y. Taur, "Self-Aligned (Top and Bottom) Double-Gate MOSFET with a 25 nm Thick Silicon Channel," Tech. Dig. IEEE Int. Electron Devices Meeting (Washington DC, Dec., 1997) pp. 427–430, 1997.

[4] B. Majkusiak, T. Jank, and J. Walczak, "Semiconductor thickness effects in the double-gate SOI MOSFET," IEEE Trans. Electron Devices, vol. 45, pp. 1127–1134, 1998.

[5] D. Hisamoto, W. Lee, J. Kedzierski, E. Anderson, H. Takeuchi, K. Asano, T. King, J. Bokor, and C. Hu, "A Folded-Channel MOSFET for Deep-Sub-tenth Micron Era," Tech. Dig. IEEE Int. Electron Devices Meeting (San Francisco, 1998) pp. 1032–1034, 1998.

[6] B. Doyle, B. Boyanov, S. Datta, M. Doczy, S. Hareland, B. Jin, J. Kavaieros, T. Linton, R. Rios, and R. Chau, "Tri-Gate Fully-Depleted CMOS Transistors: Fabrication, Design and Layout," Tech. Dig. Int. Symp. VLSI (Kyoto, 2003) pp. 133–134, 2003.

[7] B. J. Blalock, S. Cristoloveanu, B. M. Dufrene, F. Allibert, and M. Mojarradi, "The multiple-gate MOSFET-JFET transistor," *Int. J. High Speed Electron. Syst.*, vol. 12, pp. 511–520, 2002.

[8] J.-P. Colinge, "Multiple-gate SOI MOSFETs," *Solid State Electron.*, vol. 48, pp. 897–905, 2004.

[9] B. Dufrene, K. Akarvardar, S. Cristoloveanu, B. J. Blalock, P. Gentil, E. Kolawa, and M. M. Mojarradi, "Investigation of the four-gate action in G^4-FETs," *IEEE Trans. Electron Devices*, vol. 51, pp. 1931–1935, 2004.

[10] Y. Omura and K. Izumi, "High-Gain Cross-Current Tetrode MOSFET/SIMOX and Its Application," Ext. Abstr., 18th Int. Conf. Solid State Devices and Materials (Tokyo, 1986) p. 715, 1986.

[11] M. H. Gao, S. H. Wu, J. P. Colinge, C. Claeys, and G. Declerck, "Single Device Inverter Using SOI Cross-MOSFET's," Proc. IEEE Int. SOI Conf. (Colorado, Oct., 1991) pp. 138–139, 1991.

[12] Y. Azuma, Y. Yoshioka, and Y. Omura, "Cross-Current SOI MOSFET Model and Important Aspects of CMOS Operations," Ext. Abstr. Int. Conf. Solid State Devices and Materials (Tsukuba, Sep., 2007) pp. 460–461, 2007.

[13] M.-H. Gao, J.-P. Colinge, S.-H. Wu, and C. Claeys, "Improvement of Output Impedance in SOI MOSFETs," Proc. ESSDERC (Nottingham, 1990), pp. 445–448, 1990.

[14] Y. Omura and M. Nagase, "Abnormal threshold voltage dependence on gate length in ultrathin-film n-channel metal-oxide-semiconductor field-effect transistors (nMOSFET's) using separation by implanted oxygen (SIMOX) technology," *Jpn. J. Appl. Phys.*, vol. 35, pp. L304–L307, 1996.

[15] A. P. Chandrakasan, D. C. Day, D. F. Finchelstein, J. Kwong, Y. K. Ramadass, M. E. Sinangil, V. Sze, and N. Verma, "Technologies for ultradynamic voltage scaling," *Proc. IEEE*, vol. 98, pp. 191–214, 2010.

[16] D. Markovic, C. C. Wang, L. P. Alarcon, T.-T. Liu, and J. M. Rabaey, "Ultralow-power design in near-threshold region," *Proc. IEEE*, vol. 98, pp. 237–252, 2010.

[17] L. Chang, D. J. Frank, R. K. Montoye, S. J. Koester, B. L. Ji, P. W. Coteus, R. H. Dennard, and W. Haensch, "Practical strategies for power-efficient computing technologies," *Proc. IEEE*, vol. 98, pp. 215–236, 2010.

[18] R. G. Dreslinski, M. Wieckowski, D. Blaauw, D. Sylvester, and T. Mudge, "Near-threshold computing: reclaiming Moore's law through energy efficient integrated circuits," *Proc. IEEE*, vol. 98, pp. 253–266, 2010.

[19] H.-S. Lee, L. Brooks, and C. G. Sodini, "Zero-crossing-based ultra-low-power A/D converters," *Proc. IEEE*, vol. 98, pp. 315–332, 2010.

[20] S. A. Vitale, P. W. Wyatt, N. Checka, J. Kedzierski, and C. L. Keast, "FDSOI process technology for subthreshold-operation ultralow-power electronics," *Proc. IEEE*, vol. 98, pp. 333–342, 2010.

[21] M. Koyanagi, Y. Nakagawa, K. Lee, T. Nakamura, Y. Yamada, K. Inamura, K. Park, and K. Kurino, "Neuromorphic Vision Chip Fabricated Using Three-Dimensional Integration Technology," Proc. IEEE Int. Solid-State Circ. Conf. (ISSCC) (San Francisco, Feb., 2001), pp. 270–271, 2001.

[22] V. A. Gritsenko, "Design of SONOS Memory Transistor for Terabit Scale EEPROM," IEEE Conf. Electron Devices and Solid-State Circuits (Hong Kong, Dec., 2003), pp. 345–348.

[23] B. Otis and J. Rabaey, "Ultra-Low Power Wireless Technologies for Sensor Networks," (Springer, 2007).

[24] T. Ito, Y. Zhang, M. Matsumoto, and R. Maeda, "Wireless Sensor Network for Power Consumption Reduction in Information and Communication Systems," Proc. IEEE Sensors 2009 Conf., pp. 572–577, 2009.

[25] H. Kurino, Y. Nakagawa, T. Nakamura, Y. Yamada, K.-W. Lee, and M. Koyamagi, "Biologically inspired vision chip with three dimensional structure," *IEICE Trans. Electron.*, vol. E84-C, pp. 1717–1722, 2001.

[26] Y. Omura, "Cross-current silicon-on-insulator metal-oxide-semiconductor field-effect transistor and application to multiple voltage reference circuits," *Jpn. J. Appl. Phys.*, vol. 48, pp. 04C071–04C075, 2009.

[27] S. Tominaga and Y. Omura, "Sub-circuit SPICE Model of Cross-Current Tetrode (XCT) SOI MOSFET and Analysis of Low-power XCT CMOS Operation," Abstr. IEEE Int. Meet. Future Electron Devices, Kansai (IMFEDK), (Osaka, May, 2009), pp. 116–117, 2009.

[28] S. M. Sze, "Physics of Semiconductor Devices," 2nd ed. (Wiley Interscience, 1981), Chapters 6 and 8.

[29] K. Akarvardar, S. Cristoloveanu, and P. Gentil, "Analytical modeling of the two-dimensional potential distribution and threshold voltage of the SOI four-gate transistor," *IEEE Trans. Electron Devices*, vol. 53, pp. 2569–2577, 2006.

[30] K. Akarvardar, S. Cristoloveanu, P. Gentil, R. D. Schrimpf, and B. J. Blalock, "Depletion-all-around operation of the SOI four-gate transistor," *IEEE Trans. Electron Devices*, vol. 54, pp. 323–331, 2007.

[31] J.-P. Colinge (ed.), "FinFETs and Other Multi-Gate Transistors," ed. By (Springer, 2008) Chapter 4.

[32] S. M. Sze, "Physics of Semiconductor Devices," 2nd ed. (Wiley Interscience, 1981), p. 849 (Appendix G).

[33] Y. Omura, K. Fukuchi, D. Ino, and O. Hayashi, "Scaling scheme and performance perspective of cross-current tetrode (XCT) SOI MOSFET for future ultra-low power applications," *ECS Trans.*, vol. 35, no. 5, pp. 85–90, 2011.

[34] Synopsys Inc., "Sentaurus Device User Guide," Version A-2007, 2007, ftp://147.46.117.90/synopsys/manuals/PDFManual/data/sdevice_ug.pdf (accessed June 10, 2016).

[35] C. M. Snoden, "Introduction to Semiconductor Modeling," (World Scientific, 1987), Chapter 6.

[36] Y. Omura, S. Nakashima, K. Izumi, and T. Ishii, "0.1-µm-Gate, Ultrathin-Film CMOS Devices Using SIMOX Substrate with 80-nm-Thick Buried Oxide Layer," Tech. Dig. IEEE Int. Electron Devices Meeting (Washington DC, Dec., 1991) pp. 675–678, 1991.

[37] S. Yanagi, A. Nakakubo, and Y. Omura, "Proposal of a partial-ground-plane (PGP) silicon-on-insulator (SOI) MOSFET for deep sub-0.1-um channel regime," *IEEE Electron Device Lett.*, vol. 22, pp. 278–280, 2001.

[38] Y. Omura, "Silicon-on-insulator MOSFET structure for sub-100 nm channel regime and performance perspective," *J. Electrochem. Soc.*, vol.148, pp. G476–G479, 2001.

[39] Synopsys Inc., "HSpice Users Guide," 2008.

23

Low-Power Multivoltage Reference Circuit Using XCT-SOI MOSFET

23.1 Introduction

One of the editors (Omura) proposed the cross-current tetrode (XCT) silicon-on-insulator metal oxide semiconductor field-effect transistor (SOI MOSFET) and examined analog applications (see Figure 23.1) [1]. Although the scaling feasibility of XCT-like devices has been studied recently [2], the author thinks that XCT devices will yield new applications such as high-voltage devices and low-power static random-access memory (SRAM) cells with a high noise margin. In order to assess these applications in sufficient detail, device models are needed to perform circuit simulations. Since Azuma and coworkers have recently proposed an advanced device model for the cross-current tetrode metal oxide semiconductor field-effect transistor (XCT MOSFET) [3, 4], we are now able to study device applications.

In this chapter, we propose switched multiple-reference voltage supplier circuits for low-power SOI ICs because XCT MOSFET inherently shows a low drive current [1, 3, 4]. We examine their functions by circuit simulations (SPice) on the basis of measured XCT MOSFET parameters. The usefulness of the cross-current tetrode silicon-on-insulator metal oxide semiconductor field-effect transistor (XCT SOI MOSFET) is demonstrated.

23.2 Device Structure and Assumptions for Simulations

23.2.1 Device Structure and Features

The schematic device structure is shown in Figure 23.1. In an n-channel XCT device, the n-channel MOSFET, and p-channel junction-gate field-effect transistor (JFET) are self-merged and the electron current of the n-channel MOSFET is relayed to the hole current of the p-channel JFET in series; the drain current of XCT MOSFET is therefore inherently lower

MOS Devices for Low-Voltage and Low-Energy Applications, First Edition.
Yasuhisa Omura, Abhijit Mallik, and Naoto Matsuo.
© 2017 John Wiley & Sons Singapore Pte. Ltd. Published 2017 by John Wiley & Sons Singapore Pte. Ltd.

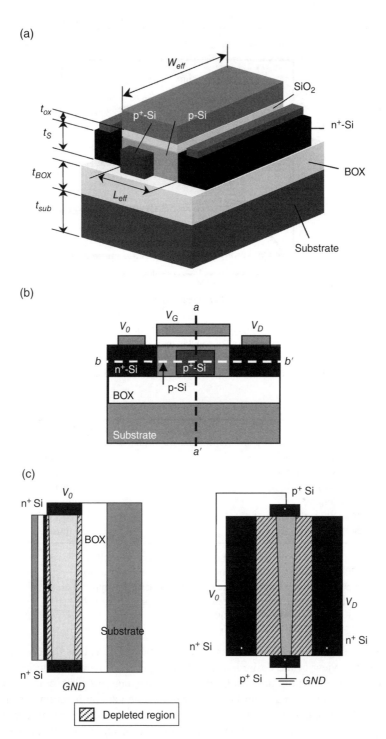

Figure 23.1 XCT SOI MOSFET (bird's-eye view and cross-sectional views). (a) Bird's-eye view of XCT device. (b) Cross-sectional view of device along MOSFET channel. (c) Cross-sectional view of device along a-a′ and b-b′. Reproduced by permission of the Japan Society of Applied Physics (2009). Yasuhisa Omura, "Cross-current silicon-on-insulator metal-oxide-semiconductor field-effect transistor and application to multiple voltage reference circuits," Jpn. J. Appl. Phys., Vol. 48, pp. 04C071-1–04C071-5, 2009.

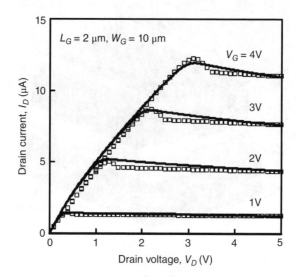

Figure 23.2 I_D-V_D characteristics of XCT device. Calculation results (symbols) are also shown for comparison [4]. Reproduced by permission of the Japan Society of Applied Physics (2009). Yasuhisa Omura, "Cross-current silicon-on-insulator metal-oxide-semiconductor field-effect transistor and application to multiple voltage reference circuits," Jpn. J. Appl. Phys., Vol. 48, pp. 04C071-1–04C071-5, 2009.

Table 23.1   Physical parameters of fabricated devices.

Parameters	Values
Nominal body doping, N_A	1×10^{16} (cm^{-3})
Gate width (nMOS), W_n	10 (μm)
Gate width (pMOS), W_p	20 (μm)
Gate length, L_G	2 (μm)
Gate oxide thickness, t_{ox}	30 (nm)
SOI layer thickness, t_S	350 (nm)

Reproduced by permission of the Japan Society of Applied Physics (2009). Yasuhisa Omura, "Cross-current silicon-on-insulator metal-oxide-semiconductor field-effect transistor and application to multiple voltage reference circuits," Jpn. J. Appl. Phys., Vol. 48, pp. 04C071-1–04C071-5, 2009.

than that of the original MOSFET without the parasitic JFET. This is the crucial drawback from the point of view of the desired "high-speed" operation. However, in this chapter, we demonstrate that the apparent drawback may be an advantage for some circuit applications. In XCT MOSFET, source and drain junctions work as the gate pn-junctions of the parasitic p-channel JFET. When the drain voltage rises, the JFET reaches the pinch-off condition of its channel region (the body region of MOSFET). After the pinch-off of the channel, the drain conductance decreases as the drain voltage rises. Therefore, the XCT device offers negative differential conductance in the saturation region of drain current, as shown in Figure 23.2 [1, 3]; device parameters of fabricated devices are summarized in Table 23.1. In Figure 23.2,

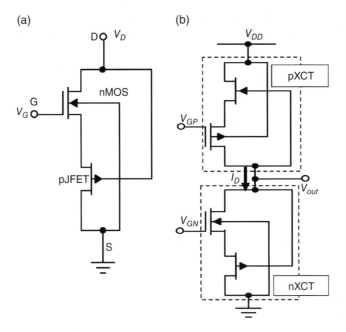

Figure 23.3 Equivalent circuit model for nXCT device and XCT CMOS device. (a) Fundamental equivalent circuit. (b) CMOS circuit composed of nXCT and pXCT devices. Reproduced by permission of the Japan Society of Applied Physics (2009). Yasuhisa Omura, "Cross-current silicon-on-insulator metal-oxide-semiconductor field-effect transistor and application to multiple voltage reference circuits," Jpn. J. Appl. Phys., Vol. 48, pp. 04C071-1–04C071-5, 2009.

symbols show results obtained from the device model proposed by Azuma and coworkers [3, 4]. Since the XCT device has active body contacts from the p-channel JFET, the body-floating effect is eliminated automatically, which is a merit of XCT MOSFET.

23.2.2 Assumptions for Simulations

In circuit simulations, the equivalent circuit for the n-channel cross-current tetrode (nXCT) device is shown in Figure 23.3a, where the BSIM device model of level 27 is assumed. We examined whether the equivalent circuit reproduces the I_D-V_D characteristics of the XCT device. The SPICE simulation parameters were extracted semi-manually from experimental results using the device parameters shown in Table 23.1. Actual parameters are shown below for n-channel XCT MOSFET:

```
*****Subcircuit of XCT SOI MOSFET
.SUBCKT NXCT 11 12 10
MXCT 11 12 13 10 NMOS L=2U W=10U AD=1.05E-12 AS=1.05E-12
  PS=4.4E-6 PD=4.4E-6
JXCT 13 11 10 PJFET L=10U W=2U
.ENDS NXCT
```

```
*****Device Model for SOI MOSFET
.model NMOS NMOS (LEVEL=27 SOSLEV=1lmin=0.5u lmax=100u
  wmin=0.5u wmax=500u $model selector
+ ld=0.1u wd=.15u xl=0 xw=0 acm=3 hdif=hdif rsh=25 rs=40 rd=40
  $resistors
+ ldif=0.1u $junction cap (ACM=3 (h9007 only) allows diode on
  gate edge
+ cj=0 cjsw=0 cgate=0.4e-9 mjsw=0.33 php=0.6
+ js=0 jsw=1e-9 n=1 vnds=.5 nds=1 $junction leakage
+ bex=-1.5 tcv=2m $temperature
+ tox=7.0e-8 capop=7 xqc=.4 meto=0.08u $gate cap
+ alpha=0.1 vcr=18 $impact ionization
+ vto=0.6 phi=1 gamma=1 $threshold
+ eta=10 xj=0.1u $threshold
+ wic=3 n0=1.6 nd0=0 weff=2e-6 leff=1e-6 $subthreshold
+ vth=0.745 vsb=0.35 zn0=190 cox=3.453e-4 $subthreshold
+ uo=330 theta=1m $dc mobility
+ vmax=100k kappa=0 $dc saturation
+ cgdo=3.1e-10 cgso=3.1e-10 ld=0.6u)

******* Device Model for JFET
.model PJFET PJF LEVEL=2 (VTO=-8.5, BETA=5.5E-6 IS=1.0E-13,
  RD=90k, RS=90k, LAMBDA=1.0E-6
+ CGS=5P, CGD=1P, CAPOP=1, ALPHA=2 ACM=1
+ LAM1=0.0 $Channel length modulation gate voltage parameter
+ ND=0.0 NG=0.0 $Subthreshold factor)
.end
```

23.3   Proposal for Voltage Reference Circuits and Simulation Results

23.3.1   Two-Reference Voltage Circuit

We first propose two reference-voltage circuits, each with one cross-current tetrode metal oxide-semiconductor (XCT MOS) complementary metal oxide semiconductor(CMOS) device. In Figure 23.3b, the XCT devices are shown as equivalent circuits. This is a simple CMOS configuration except for the independent gate terminals of the nXCT device and the pXCT device. It is assumed that $V_{DD}=5V$ and the gate voltage of pXCT device (V_{GP}) is basically constant; the threshold voltages are +0.6 V for nXCT and −0.6 V for pXCT.

The concept of switching the reference voltage is shown in Figure 23.4, where simulated time sequences of the gate voltage of the nXCT device (V_{GN}), output voltage (V_{OUT}), and V_{DD} are shown. Switching mechanisms are shown schematically in Figure 23.5, where the four figures correspond to notations "(a)" to "(d)" in the time evolution shown in Figure 23.4.

At $V_{DD}=0V$ and $V_{GP}=3V$, V_{GN} first rises from 0 to 2V. When V_{DD} rises from 0 to 5V (which takes several microseconds), V_{OUT} rises from 0 to 1V (V_L), where V_L means the target reference voltage. This process is shown in Figure 23.5a. Next, at $V_{DD}=5V$, V_{GN} falls to 0V and V_{OUT} rises to 5V. This process is shown in Figure 23.5b. After that, V_{GN} rises from 0 to 2V again,

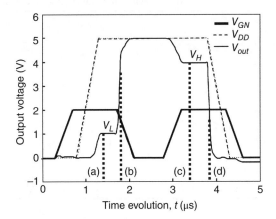

Figure 23.4 Time evolution of input and output signals. Reproduced by permission of the Japan Society of Applied Physics (2009). Yasuhisa Omura, "Cross-current silicon-on-insulator metal-oxide-semiconductor field-effect transistor and application to multiple voltage reference circuits," Jpn. J. Appl. Phys., Vol. 48, pp. 04C071-1–04C071-5, 2009.

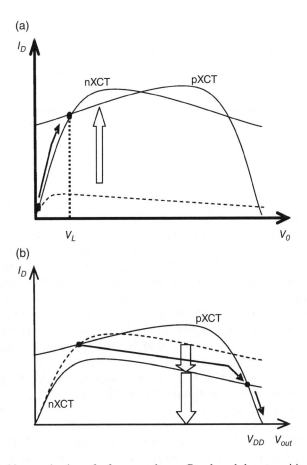

Figure 23.5 Switching mechanism of reference voltages. Panels a–d show transition of states of output level. Reproduced by permission of the Japan Society of Applied Physics (2009). Yasuhisa Omura, "Cross-current silicon-on-insulator metal-oxide-semiconductor field-effect transistor and application to multiple voltage reference circuits," Jpn. J. Appl. Phys., Vol. 48, pp. 04C071-1–04C071-5, 2009.

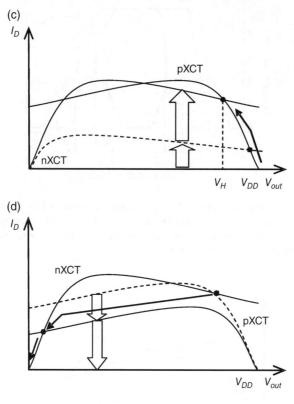

Figure 23.5   (Continued)

which leads V_{OUT} to the other target reference voltage (V_H) – in this case, V_H=4V. This process is shown in Figure 23.5c. When V_{GP} falls to 0V at this stage, V_{OUT} also falls to almost 0V, as shown in Figure 23.5d.

We note that the sequence shown in Figure 23.4 is quite inconvenient. Our solution is a more practical control method. Figure 23.6 shows an example of switching V_{OUT} ($V_L \leftrightarrow V_H$). In the operation shown in Figure 23.6, a falling trigger is applied to V_{GP} and a rising trigger is applied to V_{GN}. At the initial stage, a V_{DD} of 5V is supplied to the circuit initially, whereupon V_{GN} rises to 2V, and finally V_{GP} rises to 3V; so, V_{OUT} holds V_L at the initial stage. In Figure 23.6, at t=0.4μs, V_{GP} decreases by 0.8V briefly and then rebounds, followed by a high V_{OUT} (V_H); this process is illustrated in Figure 23.7a,b. The crossing point of current curves of nXCT and pXCT devices moves from V_L to a certain high level following the sequence shown in Figure 23.7a, and it finally returns to V_H when V_{GP} returns to 3V as shown in Figure 23.7b. At the next stage, V_{GN} rises by 0.8V at t=0.8μs and returns to 2V; this trigger forces V_{OUT} to decrease to V_L. This sequence is illustrated in Figure 23.7c,d.

In the simulations, it is assumed that the target V_{OUT} values are 1, 2, 3, and/or 4V for an example to demonstrate the potential of cross-current tetrode complementary metal oxide semiconductor (XCT CMOS) devices. In order to realize such results, I_D-V_D characteristics of n-channel cross-current tetrode silicon-on-insulator metal oxide semiconductor field-effect

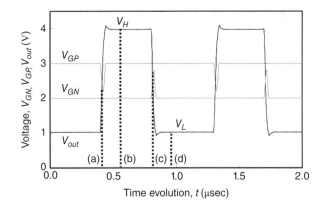

Figure 23.6   Time evolution of V_{GN}, V_{GP}, and V_{OUT}. Reproduced by permission of the Japan Society of Applied Physics (2009). Yasuhisa Omura, "Cross-current silicon-on-insulator metal-oxide-semiconductor field-effect transistor and application to multiple voltage reference circuits," Jpn. J. Appl. Phys., Vol. 48, pp. 04C071-1–04C071-5, 2009.

transistor (n-XCT SOI MOSFET) and p-channel cross-current tetrode silicon-on-insulator metal oxide semiconductor field-effect transistor (p-XCT SOI MOSFET) were calculated by variously changing the gate width of devices; as suggested in the discussion, the output resistance of n- and p-XCT devices plays an important role in obtaining target V_{OUT}. Here, it is assumed that input gate voltages (V_{GN} and V_{GP}) are 2 and 3 V. However, selection of possible input gate voltage is not so limited; when V_{GN} and/or V_{GP} must be changed, the gate width should be changed so that the resistance of the device is the desired value. In the reference voltage circuit, the accuracy and stability of output voltage are often discussed. In the present circuit, it is anticipated that accuracy is primarily ruled by the definition of channel length, channel width, body-doping profile, threshold voltage, and carrier mobility. These parameters are usually well controlled in fabrication processes, so the final accuracy depends on device design rules for the device fabrication. On the other hand, the stability of output voltage is not so easily anticipated. XCT MOSFET consists of MOSFET and JFET, as shown in Figure 23.3a; when the temperature increases, the channel conductance of MOSFET increases because of threshold voltage lowering, and that of JFET falls because of mobility reduction. Therefore, the temperature drift of output conductance of the XCT device is automatically suppressed. This may be an advantage of the device.

In practical circuit operations, pulse triggers are utilized in order to change output voltage. When such pulse triggers are used, the rising time and falling time of a pulse should be longer than the possible response time of the XCT device; in the present case these values are of the order of nanoseconds. The effective pulse height can be determined from comparison of possible currents of the n-XCT device and p-XCT device, as suggested in Figure 23.5b,d. In Figure 23.5b, the lowering of V_{GN} reduces the current of the n-XCT device. The effective extent of decrease in V_{GN} must be determined so that the current equality point at a lower drain voltage of the n-XCT device is lost. A similar determination must be made in the case of Figure 23.5d.

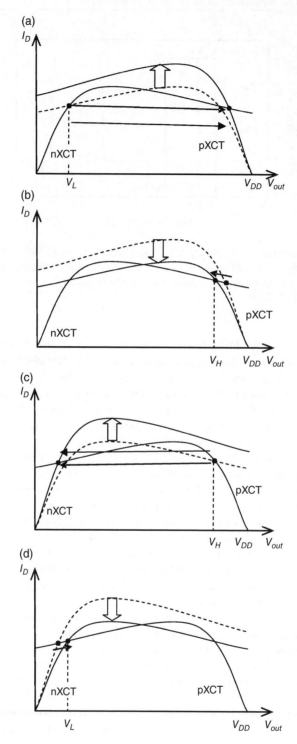

Figure 23.7 Switching mechanism of reference voltages. Panels a–d show transition of states of output level. Reproduced by permission of the Japan Society of Applied Physics (2009). Yasuhisa Omura, "Cross-current silicon-on-insulator metal-oxide-semiconductor field-effect transistor and application to multiple voltage reference circuits," Jpn. J. Appl. Phys., Vol. 48, pp. 04C071-1–04C071-5, 2009.

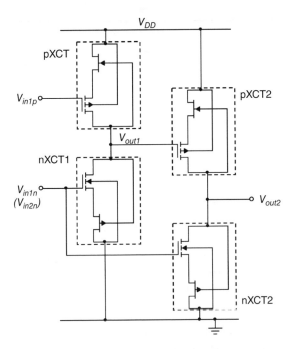

Figure 23.8 Three-reference-voltage circuit. Reproduced by permission of the Japan Society of Applied Physics (2009). Yasuhisa Omura, "Cross-current silicon-on-insulator metal-oxide-semiconductor field-effect transistor and application to multiple voltage reference circuits," Jpn. J. Appl. Phys., Vol. 48, pp. 04C071-1–04C071-5, 2009.

23.3.2   Three-Reference Voltage Circuit

We also propose a three-reference-voltage circuit for practical convenience; its circuit diagram is illustrated in Figure 23.8; the circuit is composed of two XCT CMOS devices that are connected sequentially. The time evolution of signals (V_{IN1n} and V_{IN1p}) and V_{OUT2} is shown in Figure 23.9. The basics of the operation mechanism are the same as those illustrated in Figure 23.7.

Thus, multiple-reference voltage circuits can be easily realized with XCT CMOS devices as described above. The through-current of the XCT CMOS is very low [3, 4], so the power dissipation of the proposed circuits is quite low. These considerations strongly suggest that the XCT CMOS is promising as one of the key devices for future SOI ICs. The characteristic control and design guidelines of the three-reference voltage circuit are identical to those of the two-reference voltage circuit described in the previous section.

23.4   Summary

In this chapter, we introduced important aspects of the XCT SOI MOSFET and proposed its application to practical multiple-reference voltage suppliers. The fundamental functions of the proposed reference voltage supplier circuits were examined by circuit simulations (SPICE) on the basis of measured XCT MOSFET parameters. The drivability of XCT SOI MOSFET is

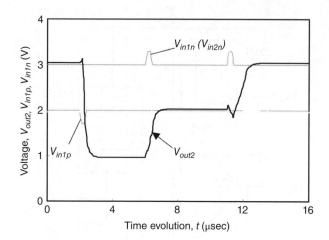

Figure 23.9 Time evolution of V_{IN1p}, V_{IN1n}, and V_{OUT2}. Reproduced by permission of the Japan Society of Applied Physics (2009). Yasuhisa Omura, "Cross-current silicon-on-insulator metal-oxide-semiconductor field-effect transistor and application to multiple voltage reference circuits," Jpn. J. Appl. Phys., Vol. 48, pp. 04C071-1–04C071-5, 2009.

lower than that of the original SOI MOSFET without the parasitic JFET. However, it has been shown that the apparent drawback of XCT SOI MOSFET produces an advantage of low-power operation of some circuit applications. Consequently, in this chapter, we successfully demonstrated the usefulness of XCT SOI MOSFETs for integrated circuits.

References

[1] Y. Omura and K. Izumi, "High-Gain Cross-Current Tetrode MOSFET/SIMOX and Its Application," Ext. Abstr. 18th Int. Conf. Solid State Devices and Materials (Tokyo, 1986) pp. 715–716, 1986.
[2] B. J. Blalock, S. Cristoloveanu, B. M. Dufrene, F. Allibert, and M. Mojarradi, "The multiple-gate MOSFET-JFET transistor," Int. J. High Speed Electron. Syst., vol. 12, pp. 511–520, 2002.
[3] Y. Azuma, Y. Yoshioka, and Y. Omura, "Analytical Modeling of Cross-Current SOI (Silicon-on-Insulator) MOSFET," 2007 Int. Meeting Future Electron Devices, Kansai (IMFEDK) (Osaka, 2007) pp. 29–30, 2007.
[4] Y. Azuma, Y. Yoshioka, and Y. Omura, "Cross-Current SOI MOSFET Model and Important Aspects of CMOS Operations," Ext. Abstr. Int. Conf. Solid State Devices and Materials (Tsukuba, Sept., 2007) pp. 460–461, 2007.

24

Low-Energy Operation Mechanisms for XCT-SOI CMOS Devices: Prospects for a Sub-20 nm Regime

24.1 Introduction

Since the 1990s, various families of fully depleted silicon-on-insulator metal oxide semiconductor field-effect transistors (SOI MOSFETs) have been proposed and extensively studied [1, 2] due to their various merits in terms of device scaling (high drivability, steep swing, less short-channel effects, smaller foot print, etc.). One of the authors (Omura) proposed the cross-current tetrode-silicon-on-insulator metal oxide semiconductor field-effect transistor (XCT-SOI MOSFET) for analog applications (see Figure 24.1), in 1986, on the basis of partially depleted SOI MOSFET technology [3], and conducted experiments to evaluate the fundamental aspects of the device. While the scaling feasibility of cross-current tetrode (XCT)-like devices has been studied recently [5, 6], we expect that cross-current tetrode-silicon-on-insulator (XCT-SOI) devices will yield new applications, such as medical implants, which demand low-energy operation with a high noise margin. In order to examine those applications in detail, device models have already been proposed to perform circuit simulations [7, 8]. However, modeling an XCT-SOI device is not so easy due to the three-dimensionality of its operations [3, 6]. Recently, our laboratory group examined the device model proposed in [7], and its usefulness was comprehensively revealed [9]. In addition, a mechanism analysis demonstrated the potential of scaled cross-current tetrode complementary metal oxide semiconductor (XCT CMOS) devices for extremely low-energy operations [4].

This chapter considers the dynamic and standby power dissipation characteristics of the sub-30 nm long gate XCT-SOI MOSFET. We start by analyzing the low-energy operation of cross-current tetrode-silicon-on-insulator complementary metal oxide semiconductor

MOS Devices for Low-Voltage and Low-Energy Applications, First Edition.
Yasuhisa Omura, Abhijit Mallik, and Naoto Matsuo.

(a)

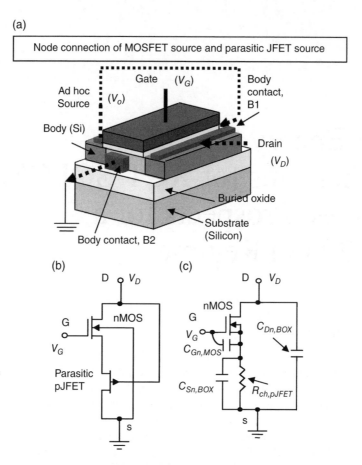

Figure 24.1   XCT-SOI MOSFET (bird's-eye view and equivalent circuit models). (a) Bird's-eye view of XCT-SOI device with parameter definitions and current flow. Broken arrows reveal the current flow; (b) basic equivalent circuit model of n-XCT-SOI MOSFET [3], and (c) AC circuit model of n-XCT-SOI MOSFET [4]. Yasuhisa Omura and Daiki Sato, "Mechanisms of low-energy operation of XCT-SOI CMOS devices–prospect of sub-20-nm regime," J. Low Power Electron. Appl., vol. 4, no. 1, pp. 15–25, 2014; doi:10.3390/jlpea4010015.

(XCT-SOI CMOS) circuits; the model proposed here strongly suggests that the "source potential floating effect (SPFE)" substantially reduces the operation power consumption. The advantages of the design methodology are also elucidated. This chapter also addresses the reality of sub-20 nm long gate XCT-SOI devices and a scaling scheme to suppress standby power consumption for future low-energy applications.

24.2   Device Structure and Assumptions for Modeling

The schematic device structure is shown in Figure 24.1a. In an n-channel XCT-SOI device, the n-channel SOI MOSFET, and p-channel junction-gate field-effect transistor (JFET) are self merged and the electron current of the n-channel metal oxide semiconductor field-effect

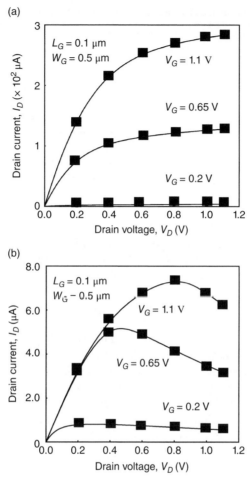

Figure 24.2 Simulated I_D-V_D characteristics. (a) Original 0.1-μm-gate SOI MOSFET without parasitic JFET, and (b) nXCT-SOI MOSFET. Yasuhisa Omura and Daiki Sato, "Mechanisms of low-energy operation of XCT-SOI CMOS devices – prospect of sub-20-nm regime," J. Low Power Electron. Appl., vol. 4, no. 1, pp. 15–25, 2014; doi:10.3390/jlpea4010015.

transistor (nMOSFET) is relayed to the hole current of the p-channel junction-gate field-effect transistor (pJFET) in series; the broken arrows reveal the current flow by the node connection of the MOSFET source and the JFET source.

As an example, we show the fundamental I_D-V_D characteristics of a 0.1-μm-long gate partially-depleted nMOSFET without the parasitic pJFET – see Figure 24.2a –following 3-D device simulations [10]. Device parameters are summarized in Table 24.1. As the XCT-SOI device has two active body contacts (*B1* and *B2*) with the SOI MOSFET, the conventional "quasi-static body-floating effect" [1] is eliminated automatically. Details of the device's operation have been described in a previous paper [9]. The nMOSFET shows slight short-channel effects due to channel length modulation. Figure 24.2b shows the I_D-V_D characteristics

Table 24.1 Scaling scheme and device parameters for sub-100 nm long gate XCT-SOI MOSFET.

Parameters	N_A (cm^{-3})	L_G (nm)	EOT (nm)	t_S (nm)	t_{BOX} (nm)	V_D (V)	$V_{TH}{}^a$ (V)
Scaling scheme	$k^{4/3}$	$1/k$	$4/3\,k$	$1/k^{1/3}$	$1/k^{1/3}$	$1/k^{1/2}$	–
$L_G = 100$ nm	2.0×10^{18}	100	2.0	129	111	1.00	0.20
$L_G = 75$ nm	4.0×10^{18}	75	1.5	117	100	0.97	0.16
$L_G = 20$ nm	2.3×10^{19}	20	0.4	75	65	0.50	−0.06
$L_G = 15$ nm	3.4×10^{19}	15	0.3	69	59	0.43	−0.10

^{a} n$^+$ poly-Si gate is assumed.
Yasuhisa Omura and Daiki Sato, "Mechanisms of low-energy operation of XCT-SOI CMOS devices – prospect of sub-20-nm regime," J. Low Power Electron. Appl., vol. 4, no. 1, pp. 15–25, 2014; doi:10.3390/jlpea4010015.

of an n-channel XCT-SOI MOSFET as determined from 3-D device simulations. It is seen that the XCT-SOI device has a lower I_D than the MOSFET because of the series resistance of the parasitic pJFET. In addition, the XCT-SOI device shows the negative-differential conductance (NDC) in the saturation region of the drain current [3, 7, 8]. On the other hand, the short-channel effects (SCEs) of the original MOSFET are sufficiently suppressed in the XCT device [9]; subthreshold swing is 76 mV/dec at $V_D = 1$ V. This is one of great advantages of XCT devices.

In support of XCT-SOI MOSFET modeling, we have already proposed the equivalent circuit (for quasistatic analysis), shown in Figure 24.1b. The basic feasibility of this model has been examined by circuit simulations [6, 8]. Low-energy operation mechanisms were also examined by a simplified analysis, based on the model shown in Figure 24.1c [4], where $C_{Gn,MOS}$, $C_{Sn,BOX}$, $C_{Dn,BOX}$, and $R_{ch,pJFET}$ denote the gate-to-source capacitance of SOI MOSFET, the source-to-substrate capacitance of SOI MOSFET, the drain-to-substrate capacitance of SOI MOSFET, and the source-to-drain resistance of the parasitic pJFET, respectively. In the case of p-channel XCT-SOI MOSFET, we label them as $C_{Gp,MOS}$, $C_{Sp,BOX}$, $C_{Dp,BOX}$, and $R_{ch,nJFET}$, respectively.

24.3   Circuit Simulation Results of SOI CMOS and XCT-SOI CMOS

Here, we advance the discussion to better understand XCT-CMOS EXOR circuit features. We assume an OR-NAND type EXOR circuit, which consists of four CMOS inverters (standard layout). We concentrate the discussion on the energy ratio (ER) of CMOS-EXOR circuits, where energy ratio is defined as the energy dissipated by the XCT-CMOS EXOR over that of the comparable conventional SOI-CMOS EXOR. Calculation results, based on HSpice [10] simulation results, are shown in Figure 24.3.

First, it is seen that the *ER* value of 1 μm long gate devices is almost unity regardless of the V_{DD} value. This occurs for the following reasons. The energy dissipation of conventional devices, evaluated by the P_d-t_d product, is not a function of V_{DD} due to the simple recognition of the metal oxide semiconductor (MOS) gate capacitor's charging and discharging operations. It is anticipated that the XCT-CMOS, with a 1 μm-long gate, follows this principle. In the case of 0.1 μm-long gate devices, on the other hand, the *ER* value rapidly falls as V_{DD} rises. As this is a very interesting result and somewhat mysterious. We discuss below a possible mechanism based on physics.

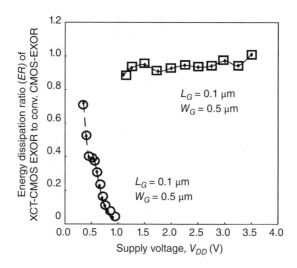

Figure 24.3 Simulation results for the energy ratio (*ER*) of the energy dissipation of a 1.0 μm long gate CMOS EXOR and a 100 nm long gate CMOS EXOR. The energy ratio is defined by the energy dissipated by the XCT-EXOR divided by that of the conventional CMOS EXOR. Yasuhisa Omura and Daiki Sato, "Mechanisms of low-energy operation of XCT-SOI CMOS devices – prospect of sub-20-nm regime," J. Low Power Electron. Appl., vol. 4, no. 1, pp. 15–25, 2014; doi:10.3390/jlpea4010015.

Frequency-dependent *ER* is defined by [4]:

$$ER(\omega) = \frac{C_{\text{Gn,XCT}} + C_{\text{Gp,XCT}} + C_{\text{Dn,BOX}} + C_{\text{Dp,BOX}}}{C_{\text{Gn,MOS}} + C_{\text{Gp,MOS}} + C_{\text{Dn,BOX}} + C_{\text{Dp,BOX}}} \tag{24.1}$$

where $C_{\text{Gn,XCT}}$ and $C_{\text{Gp,XCT}}$ denote the effective gate capacitance of nXCT-SOI MOSFET and pXCT-SOI MOSFET, respectively. Here, for simplicity, we do not take account of the depletion layer beneath the buried oxide layer. These capacitances are calculated using the equivalent circuit model shown in Figure 24.1c as:

$$C_{\text{Gn,XCT}} = \frac{C_{\text{Gn,MOS}}\left(1 + \omega^2 C_{\text{Sn,BOX}}^2 R_{\text{ch,PJFET}}^2\right)}{1 + \omega^2 C_{\text{Sn,BOX}} R_{\text{ch,PJFET}}^2 \left(C_{\text{Gn,MOS}} + C_{\text{Sn,BOX}}\right)} \tag{24.2}$$

$$C_{\text{Gp,XCT}} = \frac{C_{\text{Gp,MOS}}\left(1 + \omega^2 C_{\text{Sp,BOX}}^2 R_{\text{ch,nJFET}}^2\right)}{1 + \omega^2 C_{\text{Sp,BOX}} R_{\text{ch,nJFET}}^2 \left(C_{\text{Gp,MOS}} + C_{\text{Sp,BOX}}\right)} \tag{24.3}$$

At the low-frequency limit ($\omega \to 0$), $C_{\text{Gn,XCT}}$ and $C_{\text{Gp,XCT}}$ are reduced to $C_{\text{Gn,MOS}}$ and $C_{\text{Gp,MOS}}$, respectively, as expected. At the high-frequency limit, however, we have:

$$C_{\text{Gn,XCT}} = \frac{C_{\text{Gn,MOS}} C_{\text{Sn,BOX}}}{C_{\text{Gn,MOS}} + C_{\text{Sn,BOX}}} \tag{24.4}$$

$$C_{\text{Gp,XCT}} = \frac{C_{\text{Gp,MOS}} C_{\text{Sp,BOX}}}{C_{\text{Gp,MOS}} + C_{\text{Sp,BOX}}} \tag{24.5}$$

Generally speaking, reducing the gate length (L_G) raises the operation frequency at the same supply voltage. The *ER* rises when the scaling is enhanced; the scaling scheme assumed here [6, 9] slightly reduces $C_{Sn,BOX} R_{ch,pJFET}$ and $C_{Sp,BOX} R_{ch,nJFET}$ as the scaling is advanced. For $f < 1/(C_{Sn,BOX} R_{ch,pJFET})$ and $f < 1/(C_{Sp,BOX} R_{ch,nJFET})$, the roles of $C_{Sn,BOX}$ and $C_{Sp,BOX}$ are lost; that is, we have $C_{Gn,XCT} \sim C_{Gn,MOS}$ and $C_{Gp,XCT} \sim C_{Gp,MOS}$. This is equivalent to the low-frequency limit. As a result, the *ER* value approaches unity; the intrinsic advantage of the XCT-SOI CMOS is lost. For $f > 1/(C_{Sn,BOX} R_{ch,pJFET})$ and $f > 1/(C_{Sp,BOX} R_{ch,nJFET})$, on the other hand, we have $C_{Gn,XCT} < C_{Gn,MOS}$ and $C_{Gp,XCT} < C_{Gp,MOS}$. Device operation approaches the high-frequency limit. In this case, the *ER* value decreases as the frequency rises. The *ER* value approaches zero as the supply voltage rises, shown in Figure 24.3. It is anticipated that the depletion layer beneath the buried oxide layer reduces the parasitic capacitance of source diffusion. In order to achieve a small *ER* value, therefore, we have to increase the effective channel resistance of the parasitic JFET when scaling is advanced.

Calculation results of the *ER*, defined by Eqs. (24.1) and (24.2 and b), are shown in Figure 24.4. The model clearly predicts that high-frequency drive will reduce XCT-SOI CMOS power consumption drastically. This stems from the SPFE obtained by the model

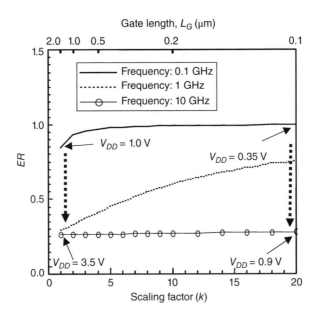

Figure 24.4 Dissipated energy ratio as a function of the scaling factor. It is assumed that the energy dissipation consists only of the charging and discharging processes of the gate capacitor. The scaling scheme is described in [6]. The dotted arrows reveal the lowest to highest range of supply voltage applicable to 1 μm long gate CMOS (for $k = 1$) and 0.1 μm long gate CMOS ($k = 20$). Yasuhisa Omura and Daiki Sato, "Mechanisms of low-energy operation of XCT-SOI CMOS devices – prospect of sub-20-nm regime," J. Low Power Electron. Appl., vol. 4, no. 1, pp. 15–25, 2014; doi.10.3390/jlpea4010015.

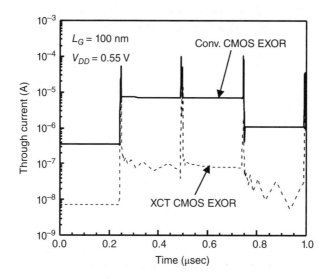

Figure 24.5 Simulation results of through-current of CMOS EXOR for $L_G = 100$ nm. Arrows show standby power reduction. Yasuhisa Omura and Daiki Sato, "Mechanisms of low-energy operation of XCT-SOI CMOS devices – prospect of sub-20-nm regime," J. Low Power Electron. Appl., vol. 4, no. 1, pp. 15–25, 2014; doi:10.3390/jlpea4010015.

shown in Figure 24.1c. Consequently, we can conclude that the SPFE of the source diffusion of the SOI MOSFET plays an important role in reducing the energy dissipated by XCT-SOI CMOS devices

Figure 24.5 shows the simulation results of the time-dependent through current of the conventional SOI CMOS EXOR chain circuit and the XCT-SOI CMOS EXOR chain circuit, both under dynamic operation. It is seen that the standby power of the XCT-SOI CMOS EXOR chain circuit is about two orders lower than that of the conventional SOI CMOS EXOR chain circuit. This suggests that logic circuits composed of XCT-SOI CMOS can offer drastically lower standby energy dissipation as well as lower switching-energy dissipation.

24.4 Further Scaling Potential of XCT-SOI MOSFET

We now investigate the scaling potential of the XCT-SOI CMOS. The fundamental scaling scheme of the XCT-SOI MOSFET has already been studied [6, 9] and a 100 nm long gate XCT-SOI CMOS has been realized. However, body doping is apt to rise to 10^{19} cm^{-3} as the XCT-SOI MOSFET is inherently a partially depleted silicon-on-insulator (SOI) device. We investigated how well the XCT-SOI MOSFET can be scaled down to realize a 15 nm long gate. Using 3-D device simulations [10], we simulated the performance of 20 and 15 nm long gate XCT-SOI MOSFETs under the scaled bias condition, where we assumed abrupt source and drain junctions, for simplicity, in the scaling scheme proposed recently [11]. As the previous scheme [6, 9] cannot be applied to the sub-100 nm regime, we restructured the scheme (see Table 24.1) so that devices work well [11].

Here, simulated I_D-V_D characteristics of the 15 nm long gate XCT-SOI MOSFET, with device parameters, shown in Table 24.1, shown in Figure 24.6; the device has abrupt source

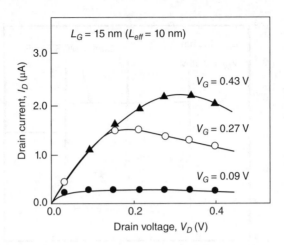

Figure 24.6   Simulation results of I_D-V_D characteristics of 15 nm long gate XCT-SOI MOSFET. L_{eff} denotes the effective channel length. Yasuhisa Omura and Daiki Sato, "Mechanisms of low-energy operation of XCT-SOI CMOS devices – prospect of sub-20-nm regime," J. Low Power Electron. Appl., vol. 4, no. 1, pp. 15–25, 2014; doi:10.3390/jlpea4010015.

and drain junctions. The device shows the negative differential conductance in the saturation region, which is expected.

Simulated subthreshold characteristics of the 15 nm long gate XCT-SOI MOSFET and the 20 nm long gate XCT-SOIMOSFET are compared in Figure 24.7; both devices have abrupt junctions. It is seen in Figure 24.7 that the subthreshold swing (SS) is 77 mV/dec, for the 20 nm long gate device, and 81 mV/dec, for the 15 nm long gate device. Short-channel effects are well suppressed. However, the gate-induced drain-leakage (GIDL) current level of the 15 nm long gate XCT-SOI device is somewhat high (~10^{-9} A/μm) as the device simulation assumes an abrupt junction for simplicity.

In response, we introduced a 10 nm long graded doping region for source and drain junctions as a "new" scaling scheme, as shown in Figure 24.8. Simulated I_D-V_G characteristics of the 15 nm long gate XCT-SOI MOSFET are shown in Figure 24.9. Drain current characteristics of 15 nm long gate XCT-SOI devices with the abrupt and graded junctions are compared. It should be noted that the subthreshold swing is 65 mV/dec for the 15 nm long gate device with the graded junction. Short-channel effects are well suppressed in the 15 nm long gate XCT-SOI MOSFET. The GIDL current level is lowered to 200 pA/μm. The most noticeable result is the improved drivability of the device with the graded junction; this is due to the remarkable improvement in carrier velocity over the whole device region, as shown in Figure 24.10 [12], and the reduction of the channel resistance of the parasitic pJFET [11].

24.5   Performance Expected from the Scaled XCT-SOI MOSFET

This section comprehensively examines the potential of the scaled XCT-SOI MOSFET. Calculated values of the on current (I_{ON}), the off current (I_{OFF}), the intrinsic switching time ($C_G V_D / I_{ON}$), the switching energy (E_{sw}), and the standby power (P_{stby}), are summarized in

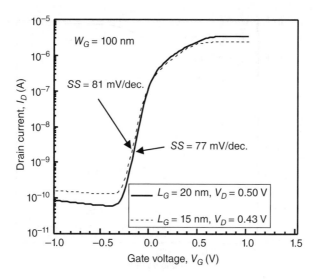

Figure 24.7 Simulation results of I_D-V_G characteristics of 20 nm long and 15 nm long gate XCT-SOI MOSFET. It is assumed that the devices have abrupt source and drain junctions. Yasuhisa Omura and Daiki Sato, "Mechanisms of low-energy operation of XCT-SOI CMOS devices – prospect of sub-20-nm regime," J. Low Power Electron. Appl., vol. 4, no. 1, pp. 15–25, 2014; doi:10.3390/jlpea4010015.

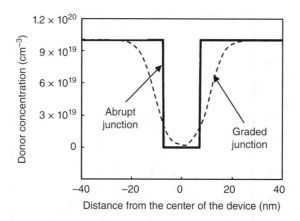

Figure 24.8 Doping profiles from the source to drain. The doping profile with a graded junction is compared to that with an abrupt junction. Yasuhisa Omura and Daiki Sato, "Mechanisms of low-energy operation of XCT-SOI CMOS devices – prospect of sub-20-nm regime," J. Low Power Electron. Appl., vol. 4, no. 1, pp. 15–25, 2014; doi:10.3390/jlpea4010015.

Table 24.2, where it is assumed that the threshold voltage (V_{TH}) is 0.2 V and $W_G/L_G = 5$, regardless of scale. Here, we assume that $C_{G,XCT} = C_{G,MOS}$; in other words, SPFE is not assumed in Table 24.2. From Table 24.2, we can identify the following.

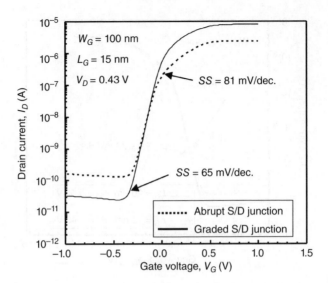

Figure 24.9   Simulation results of I_D-V_G characteristics of 15 nm long gate XCT-SOI MOSFET. Impact of doping profile on I_D-V_G characteristics is compared. Yasuhisa Omura and Daiki Sato, "Mechanisms of low-energy operation of XCT-SOI CMOS devices—prospect of sub-20-nm regime," J. Low Power Electron. Appl., vol. 4, no. 1, pp. 15–25, 2014; doi:10.3390/jlpea4010015.

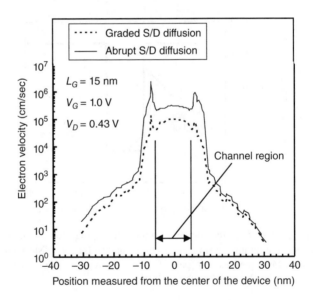

Figure 24.10   Simulation results of carrier velocity along the channel for the 15 nm long gate XCT-SOI MOSFET. Impact of doping profile on velocity profile is compared. Yasuhisa Omura and Daiki Sato, "Mechanisms of low-energy operation of XCT-SOI CMOS devices—prospect of sub-20-nm regime," J. Low Power Electron. Appl., vol. 4, no. 1, pp. 15–25, 2014; doi:10.3390/jlpea4010015.

Table 24.2 Performance perspectives of scaled XCT-SOI MOSFET.

L_G (nm)/W_G (nm)		I_{ON} (A)	I_{OFF} (A)	$C_G V_D/I_{ON}$ (s)[a]	E_{sw} (J)[a]	P_{stby} (W)[a]
100/500		5.4×10^{-6}	2.5×10^{-9}	1.7×10^{-10}	5.1×10^{-16}	2.8×10^{-9}
75/375		1.0×10^{-5}	1.3×10^{-9}	1.6×10^{-11}	3.0×10^{-16}	1.3×10^{-9}
20/100	Abrupt S/D	1.7×10^{-6}	7.6×10^{-9}	4.9×10^{-11}	2.1×10^{-17}	3.8×10^{-9}
	Graded S/D	2.2×10^{-6}	2.4×10^{-9}	3.8×10^{-11}	2.1×10^{-17}	1.9×10^{-9}
15/75	Abrupt S/D	5.2×10^{-7}	2.1×10^{-10}	1.4×10^{-10}	1.2×10^{-17}	9.0×10^{-11}
	Graded S/D	4.4×10^{-6}	1.9×10^{-9}	1.6×10^{-11}	1.2×10^{-17}	8.2×10^{-10}

[a] It is assumed that $C_{G,XCT} = C_{G,MOS}$.
Yasuhisa Omura and Daiki Sato, "Mechanisms of low-energy operation of XCT-SOI CMOS devices – prospect of sub-20-nm regime," J. Low Power Electron. Appl., vol. 4, no. 1, pp. 15–25, 2014; doi:10.3390/jlpea4010015.

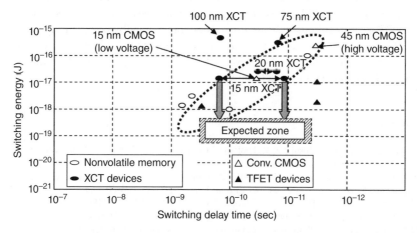

Figure 24.11 Comprehensive examination of scaled XCT-SOI MOSFET based on the ITRS roadmap 2011. Expected zone is estimated by assuming the SPFE, based on the model shown in Figure 24.1c. Yasuhisa Omura and Daiki Sato, "Mechanisms of low-energy operation of XCT-SOI CMOS devices – prospect of sub-20-nm regime," J. Low Power Electron. Appl., vol. 4, no. 1, pp. 15–25, 2014; doi:10.3390/jlpea4010015.

- The graded source/drain junction improves the drivability;
- The graded source/drain junction does not always reduce the off-current;
- The intrinsic switching time is degraded as the scaling is advanced if an abrupt junction is assumed. However, it can be drastically improved by using a graded junction due to the improved drivability;
- The switching energy will be further reduced when SPFE is taken into account.

In order to evaluate the above aspects of the scaled XCT-SOI MOSFET, we start by referring to the ITRS roadmap 2011 (the International Technology Roadmap for Semiconductors) [13]. Figure 24.10 shows the projected switching performance suggested by the ITRS roadmap

2011; the calculation results of various XCT-SOI devices are also plotted. Simple calculation results reveal that 15 nm long gate XCT-SOI devices without SPFE are identical to the non-volatile memory device group. However, it is worthwhile noting that the 15 nm long gate XCT-SOI device is expected to lie in the zone below the nonvolatile memory zone; SPFE should provide XCT-SOI devices with much better performance ("expected zone") than that shown in Figure 24.11. As the thermal energy is $\sim 4 \times 10^{-21}$ J at 300 K, it is not yet a crucial concern for the "expected zone."

24.6　Summary

We demonstrated the low-energy operation of XCT-SOI CMOS devices scaled down to 15 nm and analyzed the key underlying mechanism. It was shown that the source-follower like operation of the XCT-SOI MOSFET dynamically reduces the effective input capacitance and, thus, the energy dissipated by XCT-SOI devices. This operation should be called the "source potential floating effect SPFE." It is predicted, based on a physics-based model, that realizing SPFE in XCT-CMOS circuits will significantly suppress standby power dissipation.

It was also suggested that such aspects are still available in the sub-30 nm long gate regime. Therefore, we can state that XCT-SOI CMOS devices are very promising for future extremely low-energy circuits that suit medical implant applications.

References

[1] J.-P. Colinge, "Silicon-On-Insulator Technology: Materials to VLSI," 3rd ed., Dordrecht, the Netherlands: Kluwer Academic Publishers, 2004, pp. 151–277.

[2] J.-P. Colinge (ed.), "FinFETs and Other Multi-Gate Transistors," (Springer, 2008), Chapter 1, pp. 4–28.

[3] Y. Omura and K. Izumi, "*High-Gain Cross-Current Tetrode MOSFET/SIMOX and Its Application*," Ext. Abstr. 18th Int. Conf. Solid State Devices and Materials (Tokyo, 1986) pp. 715–716, 1986.

[4] Y. Omura and D. Ino, "*Definite Feature of Low-Energy Operation of Scaled Cross-Current Tetrode (XCTY) SOI CMOS Circuit*," Proc. 17th Workshop on Synthesis and System Integration of Mixed Information Technologies (SASIMI 2012), Beppu, Japan, 2012, pp. 355–360, 2012.

[5] B. J. Blalock, S. Cristoloveanu, B. M. Dufrene, F. Allibert, and M. Mojarradi, "The multiple-gate MOSFET-JFET transistor," *Int. J. High Speed Electron. Syst.*, vol. 12, pp. 511–520, 2002.

[6] Y. Omura, K. Fukuchi, D. Ino, and O. Hayashi, "Scaling scheme and performance perspective of cross-current tetrode (XCT) SOI MOSFET for future ultra-low power applications," *ECS Trans.*, vol. 35, pp. 85–90, 2011.

[7] Y. Azuma, Y. Yoshioka, and Y. Omura, "*Cross-Current SOI MOSFET Model and Important Aspects of CMOS Operations*," Proc. Int. Conf. Solid State Devices and Materials (Tsukuba, Japan, 2007) pp. 460–461, 2007.

[8] Y. Omura, "Cross-current silicon-on-insulator metal-oxide semiconductor field-effect transistor and application to multiple voltage reference circuits," *Jpn. J. Appl. Phys.*, vol. 48, pp. 04C07–04C11, 2009.

[9] Y. Omura, Y. Azuma, Y. Yoshioka, K. Fukuchi, and D. Ino, "Proposal of preliminary device model and scaling scheme of cross-current tetrode silicon-on-insulator metal-oxide-semiconductor field-effect transistor aiming at low-energy circuit applications," *Solid-State Electron.*, vol. 64, pp. 18–27, 2011.

[10] Synopsys Inc. *Sentaurus Operations Manual*; Synopsys: Mountain View, CA, 2008.

[11] D. Sato and Y. Omura, "*Scaling Scheme Prospect of XCT-SOI MOSFET Aiming at Medical Implant Applications Showing Long Lifetime with a Small Battery*," Proc. IEEE Int. Meet. Future of Electron Devices, Kansai (IMFEDK), Suita, Japan, 2013, pp. 38–39, 2013.

[12] Y. Omura, S. Nakano, and O. Hayashi, "Gate-field engineering and source/drain diffusion engineering for high-performance Si wire GAA MOSFET and low-power strategy in sub-30-nm-channel regime," *IEEE Trans. Nanotechnol.*, vol. 10, pp. 715–726, 2011.

[13] ITRS Roadmap, http://www.itrs2.net/itrs-reports.html;#_blank, see "2011 ITRS"(accessed June 12, 2016).

Part VI

QUANTUM EFFECTS AND APPLICATIONS – 1

25

Overview

Quantum effects have been studied extensively in the field of compound semiconductor materials because many compound semiconductor materials demonstrate the direct interband transition applicable to optoelectronics [1]. The molecular beam epitaxial technique has greatly assisted such applications by making it possible to propose and demonstrate nonconventional semiconductor devices. Light-emitting diodes (LEDs) [2], semiconductor lasers [2], and resonant transistors [3] showing differential negative conductance properties [4] have been examined and developed.

The feature size of Si metal oxide semiconductor field-effect transistors (MOSFETs) reached 100 nm or so at the research level in the 1990s [5], and silicon-on-insulator (SOI) substrates with sub-100 nm thick ultrathin Si films were created for laboratory testing [6]. This encouraged the study of Si device technologies that used quantum effects [7–10] because they were seen as breakthroughs that could overwhelm various technical barriers.

The primary focus of this part is to address the question of "Are quantum effects useful in the development of low-energy Si devices?" The following chapters answer in the affirmative by describing some advanced studies that introduce high-performance devices with the possibility of high drivability; their important aspects are described. Although the initial attempts at high drivability were not successful, this does not mean that such devices are meaningless. The following text will introduce the latest works that elucidate the possibility of realizing the device performance needed in the twenty-first century.

References

[1] B. E. A. Saleh and M. C. Teich, "Fundamentals of Photonics," (John Wiley & Sons, Inc., 1991).
[2] A. A. Bergh and P. J. Dean, "Light-Emitting Diodes," (Clarendon, 1976).
[3] R. Tsu and L. Esaki, "Tunneling in a finite superlattice," *Appl. Phys. Lett.*, vol. 22, pp. 562–564, 1973.

MOS Devices for Low-Voltage and Low-Energy Applications, First Edition.
Yasuhisa Omura, Abhijit Mallik, and Naoto Matsuo.
© 2017 John Wiley & Sons Singapore Pte. Ltd. Published 2017 by John Wiley & Sons Singapore Pte. Ltd.

[4] L. Esaki and P. J. Stiles, "New type of negative resistance in barrier tunneling," *Phys. Rev. Lett.*, vol. 16, pp. 1108, 1966.

[5] G. A. Sai-Halasz, M. R. Wordeman, D. P. Kern, S. Rishton, and E. Ganin, "High transconductance and velocity overshoot in NMOS devices at the 0.1-μm gate-length level," *IEEE Electron Device Lett.*, vol. 9, pp. 464–466, 1988.

[6] Y. Omura, S. Nakashima, K. Izumi, and T. Ishii, "0.1-μm-Gate, Ultrathin-Film CMOS Devices Using SIMOX Substrate with 80-nm-Thick Buried Oxide Layer," 1991 IEEE Int. Electron Devices Meeting, Tech. Dig., pp. 675–678, 1991.

[7] Y. Omura, S. Horiguchi, M. Tabe, and K. Kishi, "Quantum-mechanical effects on the threshold voltage of ultrathin-SOI nMOSFET's," *IEEE Electron Device Lett.*, vol. 14, no. 12, pp. 569–571, 1993.

[8] Y. Takahashi, M. Nagase, H. Namatsu, K. Kurihara, K. Iwadate, Y. Nakajima, S. Horiguchi, K. Murase, and M. Tabe, "Conductance Oscillations of a Si Single Electron Transistor at Room Temperature," Ext. Abstr. IEEE 1994 Int. Electron Devices Meeting (San Francisco, Dec., 1994), pp. 938–940, 1994.

[9] Y. Omura, K. Kurihara, Y. Takahashi, T. Ishiyama, Y. Nakajima, and K. Izumi, "50-nm channel nMOSFET/SIMOX with an ultrathin 2- or 6-nm thick silicon layer and their significant features of operations," *IEEE Electron Device Lett.*, vol. 18, no. 5, pp. 190–193, 1997.

[10] Y. Omura, "Negative conductance properties in extremely thin silicon-on-insulator insulated-gate pn-junction devices (silicon-on-insulator surface tunnel transistor)," *Jpn. J. Appl. Phys.*, vol. 35, no. 11A, pp. L1401–L1403, 1996.

26

Si Resonant Tunneling MOS Transistor

26.1 Introduction

The channel length of metal oxide semiconductor field-effect transistorss (MOSFETs) has been decreasing with increasing circuit integration. MOSFETs with a variety of structures, such as Fin MOSFETs [1], double-gate MOSFETs [2], and wire MOSFETs [3] have been proposed. For gate lengths of less than 30 nm, the ballistic transport of electrons has also been observed [4]. The physical limit of the channel length for MOSFETs is a quantum mechanical length: the direct tunneling (DT) of electrons from the source to the drain occurs via the band gap without applying a voltage to the gate. From the observed subthreshold current at low temperatures for an electrically variable shallow junction (EJ)-MOSFET [5], the physical limit of the channel length is considered to be approximately 8 nm. To improve an increase in the electric field at the channel edge near the drain electrode, a new metal oxide semicon-ductor transistor (MOST) with thin dielectric films at both edges of the channel, named a tunneling metal oxide semiconductor transistor (T-MOST), was proposed [6]. It was found that the electric field at the channel edge near the drain for the T-MOST is less than that for a conventional MOST. In addition, a fabrication method was proposed for the T-MOST using the super self-aligned metal oxide-semiconductor field-effect transistor (SSA MOSFET) [7]. In the case of a channel with a quantum mechanical length, resonant tunneling (RT) appears in the T-MOST, and to realize low-power and high-speed characteristics, a Si-resonant tunneling metal oxide semiconductor transistor (SRTMOST) with the same structure as the MOST was pro-posed [8]. The resonant tunnel enables the transistor operation [9]. The other role of the double barriers of the SRTMOST is to suppress the direct tunnel from the source to the drain, and therefore extend the physical limit of conventional MOSFETs [10]. Because the barrier height between the dielectric film and the Si substrate influences the operation of the SRTMOST [11], it is important for the barrier height to be fixed to the optimum value. In addition, a

MOS Devices for Low-Voltage and Low-Energy Applications, First Edition.
Yasuhisa Omura, Abhijit Mallik, and Naoto Matsuo.
© 2017 John Wiley & Sons Singapore Pte. Ltd. Published 2017 by John Wiley & Sons Singapore Pte. Ltd.

tunneling barrier junction (TBJ) MOSFET on a silicon-on-insulator (SOI) structure was discussed from the viewpoint of the ultimate device structure [12].

This chapter describes the configuration of the SRTMOST, including the structure, the electrostatic potential, the operation principle, and the subthreshold characteristics. Second, the device performance of the SRTMOST including its transistor characteristics and its application to logic circuits are explained.

26.2   Configuration of SRTMOST

26.2.1   Structure and Electrostatic Potential

Figure 26.1a–c show the tunneling through the double barriers, a cross-sectional view along the channel, and the electrostatic potential of the SRTMOST, respectively. Using the transfer

matrices θ_1 and θ_2, $\begin{pmatrix} a_1 \\ b_1 \end{pmatrix} = \theta_1 \begin{pmatrix} a_2 \\ b_2 \end{pmatrix}$ and $\begin{pmatrix} a_2 \\ b_2 \end{pmatrix} = \theta_2 \begin{pmatrix} 1 \\ 0 \end{pmatrix}$ are defined. From these formulas, the

following formula is obtained:

$$\begin{pmatrix} a_1 \\ b_1 \end{pmatrix} = \begin{pmatrix} \exp\{i(\alpha_1 + \alpha_2)\}\left[\cosh(S_1)\cosh(S_2) + \sinh(S_1)\sinh(S_2)\exp\{i(\beta_1 - \beta_2 - \alpha_1 - \alpha_2)\}\right] \\ \exp\{i(\alpha_1 - \alpha_2)\}\left[\sinh(S_1)\cosh(S_2) + \cosh(S_1)\sinh(S_2)\exp\{i(\beta_1 - \beta_2 - \alpha_1 - \alpha_2)\}\right] \end{pmatrix}$$

(26.1)

here, a_1 and a_2 refer to the incident waves and b_1 and b_2 refer to the reflected waves. S, α and β denote the reflection by the potential, the total phase shift at the potential, and the asymmetry of each barrier, respectively. From Eq. (26.1), $a_1 = \exp\{i(\alpha_1 + \alpha_2)\}\cosh(S_1 - S_2)$ and $b_1 = \exp\{i(\alpha_2 - \beta_1)\}\sinh(S_1 - S_2)$. It was found that $T * T$ becomes small if the difference between S_1 and S_2 becomes large. The increase in the difference between S_1 and S_2 means that the asymmetry of the double barriers becomes marked. The asymmetry of the oxide double barriers is evaluated by calculating the value of $S_1 - S_2$, where S_1 and S_2 are the reflections by the oxide potentials of both sides, respectively. $S_1 - S_2$ is given as

$$S_1 - S_2 = \frac{2\pi}{h}(2m*)^{1/2}\left\{\int_{w_0}^{w_1}(|E - \varphi_1|)^{1/2}dx - \int_{w_2}^{w_3}(|E - \varphi_2|)^{1/2}dx\right\}$$

(26.2)

where E and h are the electron energy and Planck's constant, respectively [13].

The channel is formed at the gate-oxide/p-Si interface and is indicated by dashed lines in Figure 26.1b. The electrons in the channel are confined to a limited area by the double barriers and the potential barrier that generates the surface quantization, in which they become one-dimensional (1D) electron gas. The resonant tunnel occurs through the 1D electron gas. Although the transconductance g_m is considered to be reduced by the high resistivity of the direct tunnel current flowing through each double barrier, it is actually compensated for by the resonant tunnel, and the degradation of g_m does not occur. The double barriers are formed asymmetrically so as to make the resonant tunneling remarkable. The dielectric film is thicker

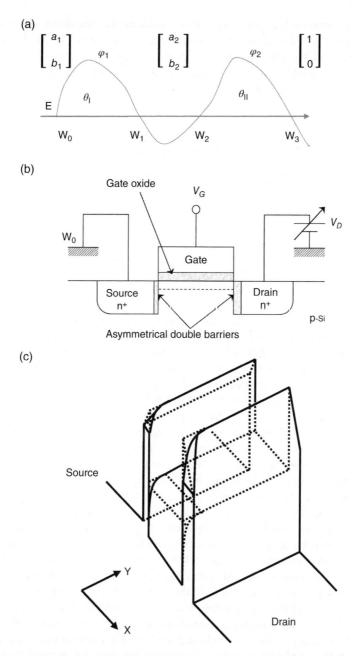

Figure 26.1 (a) Tunneling through the double barriers. Reprinted with permission from Elsevier from N. Matsuo, T. Miura, H. Fujiwara and T. Miyoshi, "Effects of electric field concentration on resonant tunneling of asymmetric oxide-nitride-oxide films," Solid-State Electronics, vol. 43, pp. 653–657. Copyright 1999. (b) Cross-sectional view along the channel. (c) Electrostatic potential of the SRTMOST. N. Matsuo, T. Miura, J. Yamauchi, H. Hamada and T. Miyoshi, "Simulation of Si resonant tunneling MOS transistor-transconductance and subthreshold swing," IEICE Trans.Electron., vol. 82, pp. 131–133, 1999. (Copyright 1999 IEICE, permission No. RA0011)

on the drain side than on the source side. It is considered that the direct tunnel of electrons from the source to the drain occurs for a quantum mechanical channel length if double barriers are not formed, and the phenomenon determines the physical limit of the conventional MOSFET. However, the physical limit is improved by decreasing the transmission coefficient of electrons from the source to the drain by the formation of thin dielectric films at both edges of the channel.

26.2.2   Operation Principle and Subthreshold Characteristics

Figure 26.2a,b show the energy diagram of the MOS cross section and a schematic I_D-V_G characteristic to explain the operation principle of the SRTMOST. Quantized energy levels are formed at the gate-oxide/Si interface, which are given by

$$E_n = \frac{\pi^2 \hbar^2}{\left(2m_{Si}^* L_{eff}^2\right)} n_x^2 + \frac{\left(\pi \hbar q F_G\right)^{2/3} n_y^{2/3}}{\left(2m_{Si}^*\right)^{1/3}} \tag{26.3}$$

here; m_{Si}^*, L_{eff}, $\hbar$, n_x, n_y, and F_G are the effective electron mass in the Si, the channel length, Planck's constant, and the quantum numbers for the double barrier, for the surface quantization, and the electric field of Si at the interface, respectively. The quantized energy levels E_1, E_2, and E_3 are 0.068, 0.099, and 0.126 eV assuming $F_G = 10^5$ V/cm and $L_{eff} = 10$ nm, respectively. The differences between these energy levels approximately correspond to the thermal energy at room temperature. The SRTMOST operates under depletion and inversion conditions in the range $0 \leq \varphi_s \leq 2\varphi_f$. In the case of strong inversion, nonresonant currents flow. After the first peak of the drain current I_D is generated through resonance by the double barriers and surface quantization, it flows continuously via the energy level by surface quantization. The resonant peaks cannot be observed clearly because of the small difference between the energy levels. At the threshold voltage V_{TH}, the drain current due to the strong inversion of the channel becomes dominant as shown in Figure 26.2b.

Figure 26.3a–d shows the potential diagram used for the calculations, the definitions of the specific voltages V_{D1} and V_{D2}, and the schematic I_D-V_D characteristics of the n-channel Si resonant tunneling metal oxide semiconductor transistor (n-SRTMOST), respectively.

The transmission coefficient from the source to the drain is calculated by solving the 1D Schrödinger equation numerically using the effective mass approximation [14]. V_{D1} and V_{D2} are the voltages where the Fermi level E_f of the source is equal to the minimum energy level of the quantum well, and E_f is equal to the ground energy level E_1. The drain current I_{D1} at the drain voltage V_{D1} is calculated using an analytical function of the direct tunnel current based on the Wentzel, Kramers, Brillouin (WKB) method [15]. The resonant tunnel current I_{D2} at the drain voltage of V_{D2} is calculated by Eq. (26.4) assuming the delta function for the transmission coefficient at the resonant peak [16].

$$\begin{aligned} I_{D2} &= q^2 V_{D2} m^*_{Si} \times \Delta E / 2\pi^2 \hbar^3 \int_0^{E_f} \delta\left(E_f - E_\perp + qV_a/2 - E_o\right) dE_\perp \\ &= q^2 V_{D2} m^*_{Si} \times \Delta E / 2\pi^2 \hbar^3 \end{aligned} \tag{26.4}$$

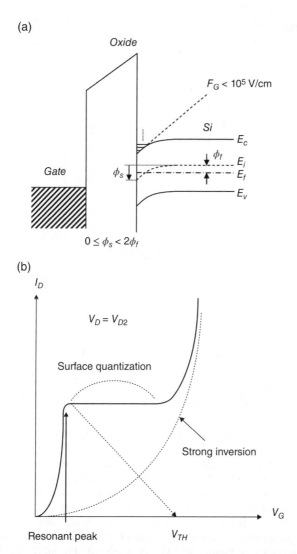

Figure 26.2 (a) Energy diagram of MOS cross section. Reprinted, with permission from Elsevier, from N. Matsuo, Y. Kitagawa, Y. Takami, J. Yamauchi, H. Hamada and T. Miyoshi, "Influence of off-set energy on the electrical characteristics of Si resonant tunneling MOST (SRTMOST)," Superlattices and Microstructures. vol. 28, pp. 407–412. Copyright 2000. (b) Schematic I_D-V_G characteristic of the SRTMOST.

Although the drain currents are assumed to remain constant at voltages larger than V_{D2}, strictly speaking they increase gradually with the drain voltage because of the density of state in the 1D channel. However, the rate of increase in the drain current after the resonance is small, and an increase in the drain current due to strong inversion does not occur at the present drain voltages. Therefore, the assumption of a constant drain current does not affect the output characteristics of

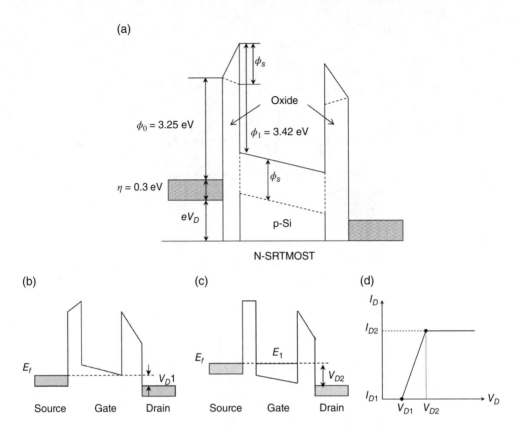

Figure 26.3   (a) Potential diagram used for the calculations. N. Matsuo, Y. Kitagawa, Y. Takami, J. Yamauchi, H. Hamada and T. Miyoshi, "Influence of off-set energy on the electrical characteristics of Si resonant tunneling MOST(SRTMOST)," Superlattices and Microstructures, vol. 28, pp. 407–412, 2000. License number is 3670680248296. (b) Definitions of specific voltages V_{D1} Reproduced by permission of the Institute of Electronics, Information, and Communication Engineers 2002. N. Matsuo, Y. Takami, T. Nozaki and H. Hamada, Silicon Resonant Tunneling Metal-Oxide-Semiconductor Transistor for Sub-0.1um Era, IEICE Trans. Electronics, vol. 85, pp. 1086–1090. Copyright © 2002 IEICE. (c) Definitions of specific voltages V_{D2}. Reproduced by permission of the Institute of Electronics, Information, and Communication Engineers (2002). N. Matsuo, Y. Takami, T. Nozaki and H. Hamada, Silicon Resonant Tunneling Metal-Oxide-Semiconductor Transistor for Sub-0.1um Era, IEICE Trans. Electronics, vol. 85, pp. 1086–1090. Copyright © 2002 IEICE. (d) Schematic I_D-V_D characteristics of n-SRTMOST. Reproduced by permission of the Institute of Electronics, Information, and Communication Engineers (2002). N. Matsuo, Y. Takami, T. Nozaki and H. Hamada, Silicon Resonant Tunneling Metal-Oxide-Semiconductor Transistor for Sub-0.1um Era, IEICE Trans. Electronics, vol. 85, pp. 1086–1090. Copyright © 2002 IEICE.

the inverter circuit, and the conventional subthreshold characteristic is considered to be achieved. The transconductance g_m and subthreshold swing (SS) are calculated as

$$g_m = d\left(I_{D,RT}\right)/dV_G$$
$$\approx I_{D,RT}/\left(\Delta\varphi_{S,P-P}/2\right)$$

(26.5)

$$SS = d(V_G) / d \log I_{D,RT}$$
$$\approx (\Delta\varphi_{S,P-P} / 2) / d \log I_{D,RT} \tag{26.6}$$

where ΔE is the uncertainty of the resonant tunnel energy. ΔE is given by the full width at half maximum (FWHM) of the resonant tunnel peak for the relationship between the transmission coefficient and the gate voltage, and the magnitude of ΔE is 10^{-5} to 10^{-3} eV. The transition time from the source to the drain is calculated using the uncertainty principle as $\tau \cong \hbar/\Delta E$, where τ and $\hbar$ are the lifetime of electrons in the potential well, which is equal to the transition time, and the reduced Planck's constant, respectively. $\Delta\phi_{S,P-P}$ is the difference between the neighboring resonant peaks for the relationship between T^*T and φ_s. The characteristics of the inverter circuit, which is constructed of the SRTMOST and conventional p-MOSFET in series, is examined. The channel width W_{eff} of the SRTMOST is 300 nm.

26.3 Device Performance of SRTMOST

26.3.1 Transistor Characteristics of SRTMOST

Figure 26.4a and b show the relationships between the transmission coefficient T^*T and the surface potential φ_S of an SRTMOST with a channel length of 10 nm and between the transconductance g_m, SS and channel length, assuming the thicknesses of the asymmetrical double barriers to be 2 and 3 nm.

It was found that g_m maintains a constant value of approximately 0.04 mS for channel lengths of 5–30 nm and that SS is 10 and 20 meV/dec. for channel lengths of 10 and 5 nm, respectively. SS has a smaller value than the theoretical value of 60 meV/dec. This advantageous result is due to the resonant tunnel effect.

Figure 26.5a–d respectively show the calculated transmission coefficient as a function of the surface potential φ_S for offset energy φ_o from 0.3 to 3.25 eV, the drain current $I_{D,RT}$ as a function of φ_o, the gate-off current density as a function of the drain voltage for various φ_o, and the transition time τ_e as a function of φ_o, with a channel length of 10 or 5 nm and thicknesses of the asymmetrical double barriers of 0.4 and 0.7 nm.

The peak-to-peak energy $\Delta\varphi_{S,P-P}$ and peak energy width ΔE increase as φ_o decreases. The actual peak-to-peak energy is smaller than the calculated value because of the nonconsideration of the surface quantization at the potential of the dielectric film/Si interface. The peak-to-peak energy is approximately $k_B T$ at room temperature considering the surface quantization as discussed in section 26.2.2. It is expected that the same subthreshold characteristics as those of the SRTMOST with the asymmetrical double barriers of 2 and 3 nm thicknesses are realized.

Next, the switching characteristics are discussed. $I_{D,RT}$ increases as φ_o decreases for the channel lengths of 10 and 5 nm. From the finding that the gate-off current densities for $\phi_o = 0.3$ eV and $\phi_o = 0.6$ eV are larger than those of the conventional MOSFET at low voltages, the optimum offset potential is approximately 1.0 eV and the resonant tunnel peak current was found to be 3×10^{11} A/m² for φ_o of 1.0 eV. Therefore, the current ratio under the gate-on and gate-off conditions is approximately 10^{25}. Switching characteristics superior to those of the conventional MOSFET were obtained for the SRTMOST. Considering the trend of the supply voltage, the drain voltage will be 0.5–1.0 V for a gate length of approximately

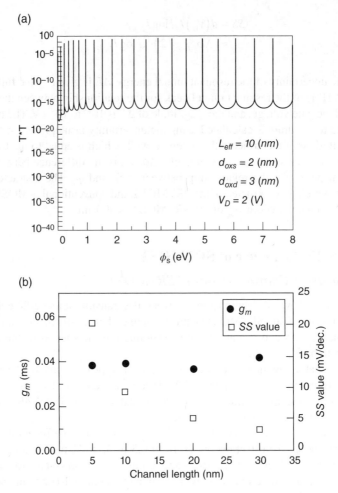

Figure 26.4 (a) Relationships between the transmission coefficient $T*T$ and the surface potential ϕ_S of a SRTMOST with a channel length of 10 nm. Reproduced by permission of the Institute of Electronics, Information, and Communication Engineers (1999), N. Matsuo, T. Miura, J. Yamauchi, H. Hamada and T. Miyoshi, "Simulation of Si resonant tunneling MOS transistor-transconductance and subthreshold swing," IEICE Trans. Electronics, vol. 82, pp. 131–133. Copyright © 1999 IEICE. (b) Relationships between the transconductance g_m, subthreshold swing SS and channel length of a SRTMOST assuming the thicknesses of the asymmetrical double barriers to be 2 and 3 nm. Reproduced by permission of the Institute of Electronics, Information, and Communication Engineers (1999), N. Matsuo, T. Miura, J. Yamauchi, H. Hamada and T. Miyoshi, "Simulation of Si resonant tunneling MOS transistor-transconductance and subthreshold swing," IEICE Trans. Electronics, vol. 82, pp. 131–133. Copyright © 1999 IEICE.

10 nm, and it is inferred that a conventional MOSFET with a channel length of less than 10 nm cannot be realized without double barriers. The transition time τ_e under the operation condition is also the crucial parameter as the transmission coefficient. τ_e from the source to the drain is comparable with that for the conventional MOSFET for φ_o of 1.0 eV. Although the present

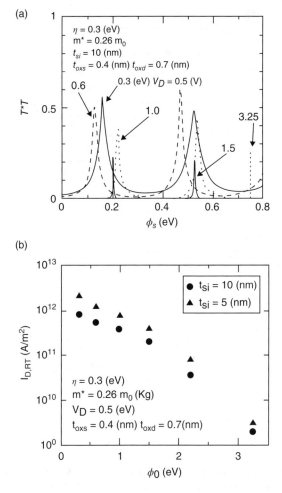

Figure 26.5 (a) Calculated transmission coefficient as a function of the surface potential ϕ_s for offset energy ϕ_o from 0.3 to 3.25 eV. Reprinted, with permission from Elsevier, from N. Matsuo, Y. Kitagawa, Y. Takami, J. Yamauchi, H. Hamada, and T. Miyoshi, "Influence of off-set energy on the electrical characteristics of Si resonant tunneling MOST (SRTMOST)," Superlattices and Microstructures. vol. 28, pp. 407–412. Copyright 2000. (b) Drain current $I_{D,RT}$ as a function of ϕ_o. Reproduced by permission of the Institute of Electronics, Information, and Communication Engineers (2001). N. Matsuo, Y. Kitagawa, Y. Takami, and H. Hamada, "Study of barrier-height of si resonant tunneling MOS transistor (SRTMOST) – relationship between barrier-height and gate length." IEICE Trans. Electronics, vol. 84, pp. 326–327. Copyright 2001.

calculation is performed for a channel length of 10 nm, SRTMOST with a channel length smaller than 10 nm can operate. Regarding the subthreshold characteristics, the following phenomenon will occur. The ground energy level, which is a function of the distance between the double barriers, becomes large with decreasing channel length. However, the energy level due to surface quantization remains constant despite the decrease in the channel length. This results in an increase in the threshold gate voltage. The gate-off current density increases

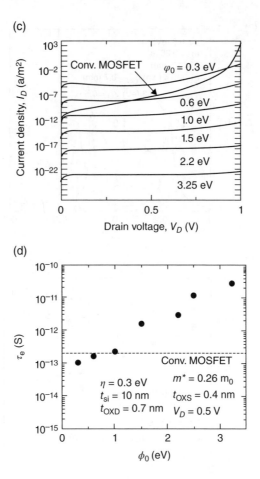

Figure 26.5 (Continued) (c) Gate-off current density as a function of the drain voltage for various offset energies. Reprinted, with permission from Elsevier, from N. Matsuo, Y. Kitagawa, Y. Takami, J. Yamauchi, H. Hamada, and T. Miyoshi, "Influence of off-set energy on the electrical characteristics of Si resonant tunneling MOST (SRTMOST)," Superlattices and Microstructures. vol. 28, pp. 407–412. Copyright 2000. (d) Transition time τ_e as a function of the offset energy ϕ_o, with channel length of 10 or 5 nm and thicknesses of the asymmetrical double barriers of 0.4 and 0.7 nm. Reprinted, with permission from Elsevier, from N. Matsuo, Y. Kitagawa, Y. Takami, J. Yamauchi, H. Hamada, and T. Miyoshi, "Influence of off-set energy on the electrical characteristics of Si resonant tunneling MOST (SRTMOST)," Superlattices and Microstructures. vol. 28, pp. 407–412. Copyright 2000.

because of the increase in the transmission coefficient. The transition time becomes small with decreasing channel length. Therefore, the optimum offset energy φ_o is approximately 1.0–1.5 eV for a channel length of 5–10 nm.

26.3.2 Logic Circuit Using SRTMOST

This section discusses a logic circuit using the SRTMOST [17, 18]. Figure 26.6 shows the relationship between the output voltage V_{out} and the input voltage V_{in}. The inset shows

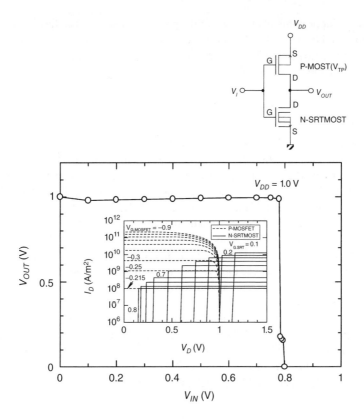

Figure 26.6 Relationship between the output voltage V_{OUT} and the input voltage V_{IN}. The inset shows the relationship between the drain current and the drain voltage of the n-SRTMOST and p-MOSFET at a supply voltage V_{DD} of 1.0 V for the inverter circuit. Reproduced by permission of the Institute of Electronics, Information, and Communication Engineers (2002). N. Matsuo, Y. Takami, T. Nozaki and H. Hamada, "Silicon Resonant Tunneling Metal-Oxide-Semiconductor Transistor for Sub-0.1um Era," IEICE Trans. Electronics, vol. 85, pp. 1086–1090. Copyright © 2002 IEICE.

the relationship between the drain current and the drain voltage of an n-SRTMOST and a p-MOSFET at a supply voltage V_{DD} of 1.0 V for an inverter circuit constructed from a p-MOSFET and an n-SRTMOST. Here, the gate length L_G and gate width W_G of the p-MOSFET are assumed to be 0.1 μm. The Fermi energy η, m^*_{Si}, the barrier height of the Si/SiO$_2$ φ_o, the channel length L_{eff} of the n-SRTMOST, the thickness of the double-barrier at the source side t_{OXS} and that at the drain side t_{OXD} are assumed to be 0.3 eV, 0.26m$_0$, 3.26 eV, 10, 0.4, and 0.7 nm, respectively. The threshold voltages of the p-MOS and n-MOS are −0.2 and 0.8 V, respectively. The pull-up transistor, p-MOSFET, and the pull-down transistor, n-SRTMOST, perform at an input voltage V_{in}. The current for the point located at the intersection of the corresponding load lines flows through the p-MOSFET and the n-SRTMOST, and the drain voltage equals the output. For V_{in} larger than 0.8 V, V_{out} is nearly equal to 0 V because the channel of the n-SRTMOST is subjected to strong inversion. Thus, it was found that the output characteristics using the p-MOSFET and n-SRTMOST were nearly the same as those using a conventional complementary metal oxide semiconductor (CMOS) inverter. There is a plateau at V_{out} of approximately 0.15 V. For the present transistor size, the plateau is small and difficult to control. However, the plateau area depends on the transistor size. This phenomenon shows that the present logic circuit constructed of an n-SRTMOST and a conventional p-MOSFET is theoretically suitable for use as a three-valued output circuit.

26.4   Summary

The device performance of an SRTMOST has been discussed with regard to its transistor characteristics and its application to logic circuits by assuming the MOSFET structure and its electrostatic potential where the double barriers are fabricated at both ends of the channel. It was found that g_m takes a constant value of approximately 0.04 mS for channel lengths of 5–30 nm and that SS is 10 and 20 meV/dec. for channel lengths of 10 and 5 nm, respectively, regardless of the thicknesses of the asymmetrical double barriers. SS has a smaller value than the theoretical value of 60 meV/dec. This advantageous feature is due to the resonant tunneling effect. The switching characteristics of the SRTMOST are expected to be superior to those of the conventional MOSFET because the current ratio under the gate-on and gate-off conditions is approximately 10^{25}. It was inferred that a conventional MOSFET with a channel length of less than 10 nm cannot be realized without double barriers by considering the drain leakage current under the gate-off condition. The offset energy of the double barrier/Si must be approximately 1.0–1.5 eV for a channel length of less than 10 nm. The feasibility of using the SRTMOST as a multivalued logic circuit was shown. The SRTMOST is a potential candidate for realizing a breakthrough in conventional MOSFETs, enabling the realization of next-generation devices with a channel length of less than 10 nm.

References

[1] D. Hisamoto, W. C. Lee, J. Kedziershi, E. Anderson, H. Takeuchi, K. Asano, T. J. King, J. Bokov, and C. Hu, "A Folded-Channel MOSFET for Deep-Sub-Tenth Micron Era," IEEE IEDM Tech. Dig., pp. 1032–1034, 1998.

[2] H. S. P. Wong, D. J. Frank, and P. M. Solomon, "Device Design Consideration for Double-Gate, Ground-Plane, and Single-Gated Ultra-Thin SOI MOSFET's at the 25 nm Channel Length Generation," IEEE IEDM Tech. Dig., pp. 407–410, 1998.

[3] W. J. Skocpol, L. D. Jackel, R. E. Howard, P. M. Mankiewich, and D. M. Tennant, "Quantum transport in narrow MOSFET channels," Surf. Sci., vol. 170, pp. 1–13, 1986.

[4] G. Timp, J. Bude, K. K. Bourdelle, J. Garno, A. Ghetti, H. Gossmann, M. Green, G. Forsyth, Y. Kim, R. Kleiman, F. Klemens, A. Kornblit, C. Lochstampton, W. Mansfield, S. Moccio, T. Sorsch, D. M. Tennant, W. Timp, and R. Tung, "The Ballistic Nano-Transistor," IEEE IEDM Tech. Dig., pp. 55–58, 1999.

[5] H. Kawaura, T. Sakamoto, and T. Baba, "Transport Properties in Sub-10-nm-gate EJ-MOSFETs," Ext. Abstr. 1999 Int. Conf. Solid State Devices and Materials, 1999, pp. 20–21, 1999.

[6] N. Matsuo, T. Miura, H. Hamada, and T. Miyoshi, "Application of tunneling effect for thin dielectric film to MOS transistor," IEICE Trans. Electron., vol. 81, pp. 266–267, 1998.

[7] T. Sakai, K. Murakami, and T. Kawachi, "SSA MOSFET of a New Device Structure with nm Self-Aligned Contact," Ext. Abstr. (61st Autumn Meeting, 2000); The Japan Society of Applied Physics, 3P-K-9, p. 672, 2000.

[8] N. Matsuo, T. Miura, J. Yamauchi, H. Hamada, and T. Miyoshi, "Simulation of Si resonant tunneling MOS transistor-transconductance and subthreshold swing," IEICE Trans. Electron., vol. 82, pp. 131–133, 1999.

[9] N. Matsuo, Y. Kitagawa, Y. Takami, and H. Hamada, "Study of barrier-height of Si resonant tunneling MOS transistor (SRTMOST) – Relationship between barrier-height and gate length," IEICE Trans. Electron., vol. 84, pp. 326–327, 2001.

[10] N. Matsuo, J. Yamauchi, Y. Kitagawa, H. Hamada, T. Miura, and T. Miyoshi, "Extension of physical limit of conventional metal-oxide-semiconductor transistor by double barriers formed at the channel edges," *Jpn. J. Appl. Phys.*, vol. 39, pp. 3850–3853, 2000.

[11] N. Matsuo, Y. Kitagawa, Y. Takami, J. Yamauchi, H. Hamada, and T. Miyoshi, "Influence of off-set energy on the electrical characteristics of Si resonant tunneling MOST (SRTMOST)," *Superlattices and Microstructures*, vol. 28, pp. 407–412, 2000.

[12] Y. Omura, "A Tunneling-Barrier Junction SOI MOSFET with a Suppressed Short-Channel Effect for the Ultimate Device Structure," Abstr. 2001 IEEE Silicon Nanoelectron. Workshop, pp. 58–59, 2001.

[13] N. Matsuo, T. Miura, H. Fujiwara, and T. Miyoshi, "Effects of electric field concentration on resonant tunneling of asymmetric oxide-nitride-oxide films," *Solid-State Electron.*, vol. 43, pp. 653–657, 1999.

[14] N. Matsuo, H. Fujiwara, and T. Miyoshi, "Electron tunneling through thin oxide-nitride-oxide films under electric field concentration," *Solid-State Electron.*, vol. 42, pp. 441–446, 1998.

[15] N. Matsuo, T. Miura, A. Urakami, and T. Miyoshi, "Analysis of direct tunneling of thin SiO_2 film," *Jpn. J. Appl. Phys.*, vol. 38, pp. 3967–3971, 1999.

[16] D. K. Ferry, "Quantum Mechanics – An Introduction for Device Physicists and Electrical Engineers," (Institute of Physics Publishing, 1995).

[17] N. Matsuo, Y. Takami, T. Nozaki, and H. Hamada, "Silicon resonant tunneling metal-oxide-semiconductor transistor for sub-0.1 um era," *IEICE Trans. Electron.*, vol. 85, pp. 1086–1090, 2002.

[18] N. Matsuo, A. Yamamoto, and Y. Kitamon, "Feasibility of Si resonant tunneling MOS transistor (SRTMOST) to new multi-valued logic circuit," *IEICE Trans. Electron.*, vol. 86, pp. 1370–1371, 2003.

27

Tunneling Dielectric Thin-Film Transistor

27.1 Introduction

With the increase in the pixel density of liquid crystal displays (LCDs) and organic light emitting diode (OLED) displays, the number of thin-film transistors (TFTs) used in these displays has increased and the TFTs have also been downsized. To realize high speed, high brightness, and low power consumption for these displays, a decrease in the gate-off current of the TFTs is one of the key issues [1, 2]. The gate-off current is due to grain boundaries or in-grain defects, such as point defects or dislocations. Although attempts to decrease the numbers of defects have been made by enlarging the grain size [3] or terminating the dangling bonds with hydrogen [4], it is difficult to remove these defects completely. To solve this problem, a new transistor called a tunneling dielectric thin-film transistor (TDTFT) [5, 6] has been proposed and designed. The TDTFT employs nonlinear resistors connected in series to the source and drain electrodes. By using the tunneling dielectric film as a resistance, it is theoretically possible to decrease the gate-off current without the deterioration of subthreshold swing. A bottom-gate TDTFT was fabricated using a silicon-on-insulator (SOI) substrate and its electrical characteristics were examined [7]. Furthermore, the conduction mechanism of electrons flowing in the channel of the TDTFT was examined in detail, to discuss the application of TDTFTs as the switching or driver devices in flat-panel displays [8–11].

As one of the phenomena of TFT degradation, the hump phenomenon has been examined intensively [12–16]. The hump phenomenon was analyzed with regard to donor-like traps in the ZnO channel [13], the stability of indium gallium zinc oxide (IGZO) TFTs for various channel widths under the high-gate and drain bias stresses [14], anomalous stress-induced hump in amorphous indium gallium zinc oxide thin-film transistors (IGZO TFTs) [15] and anomalous hump in gate current after negative-bias temperature instability in HfO$_2$/metal gate

MOS Devices for Low-Voltage and Low-Energy Applications, First Edition.
Yasuhisa Omura, Abhijit Mallik, and Naoto Matsuo.
© 2017 John Wiley & Sons Singapore Pte. Ltd. Published 2017 by John Wiley & Sons Singapore Pte. Ltd.

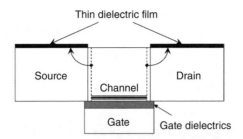

Figure 27.1 Cross-sectional structure of the TDTFT.

p-channel metal oxide semiconductor field-effect transistors (MOSFETs) [16]. It was also clarified that the TDTFT suppressed the hump phenomenon [17].

In this chapter, we first describe the configuration of the fabricated TDTFT using SiN_x as a tunneling dielectric film and compare its electrical characteristics with those of the conventional TFT. Secondly, the temperature dependences of the drain current and the transconductance g_m are given and the conduction mechanism at low temperatures is discussed. Thirdly, the temperature dependence of the drain current at high temperatures is also given. From these results, the dominant conduction process, quantum tunneling, or thermal activation at high temperatures, is clarified. Fourthly, it is shown that the TDTFT suppresses the hump phenomenon and short-channel effect.

27.2 Fundamental Device Structure

Figure 27.1 shows the cross-sectional structure of the TDTFT fabricated in this study. The TDTFT has the novel feature of thin dielectric films at both ends of the channel area. The position of the dielectric film is shown by the dashed lines. However, it was actually fabricated on the source and drain as shown by the shaded area. The gate-off currents of a conventional TFT depend on the density of crystal defects, which are located in the channel area, therefore large off-currents are observed. For the TDTFT, the gate-off current is equal to the tunneling current via the thin dielectric films, and therefore the off-currents are expected to be less than those of a conventional TFT.

27.3 Experiment

27.3.1 Experimental Method

Figure 27.2a and b shows a schematic cross-sectional view of the bottom-gate-type TDTFT and an optical micrograph of the fabricated TDTFT. Tunnel diodes are fabricated directly onto a poly-Si layer as a nonlinear resistor. In the experiment, an n$^+$ Si wafer was used as a gate electrode. A field oxide and gate oxide with thicknesses of 280 and 66 nm, respectively, were formed in turn by thermal oxidation. A poly-Si film with a thickness of 53 nm was deposited by low-pressure chemical vapor deposition (LPCVD) at 635 °C. A source and drain (S/D) were doped by arsenic ion implantation at a dose of 10^{15} cm^{-2}. To prevent electric field concentration at the corner of the source and drain, thermal annealing was carried out at 850 °C

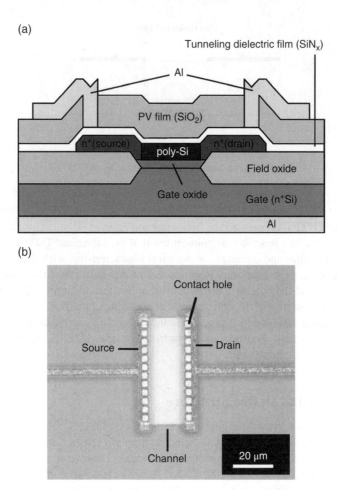

Figure 27.2 (a) Cross-sectional view of fabricated bottom-gate-type TDTFT. Reproduced by permission of the Institute of Electronics, Information, and Communication Engineers (2007). N. Matsuo, A. Fukushima, K. Ohkura, A. Heya, and S. Yokoyama, "Fabrication of tunneling dielectric thin-film transistor with very thin SiN$_x$ films onto source and drain," IEICE Electron Express (on-line journal), vol. 4, pp. 442–447. Copyright © 2007 IEICE. (b) Optical micrograph. Reproduced by permission of the Institute of Electronics, Information, and Communication Engineers (2007). N. Matsuo, A. Fukushima, K. Ohkura, A. Heya, and S. Yokoyama, "Fabrication of tunneling dielectric thin-film transistor with very thin SiN$_x$ films onto source and drain," IEICE Electron Express (on-line journal), vol. 4, pp. 442–447. Copyright © 2007 IEICE.

in an oxygen atmosphere. For the tunneling dielectric film, a 1.7-nm thick SiN$_x$ film was deposited at 750 °C by LPCVD. The thickness of the SiN$_x$ was measured using a spectroscopic ellipsometer. To clarify whether the SiNx film was formed successfully, the direct tunneling (DT) current was measured for the Al electrode/SiNx film/n$^+$ Si structure, and the Al/SiNx film/poly-Si structure was also observed by cross-section transmission electron microscopy (TEM) and energy-dispersive X-ray fluorescence spectrometry (EDX). SiO$_2$ passivation (PV)

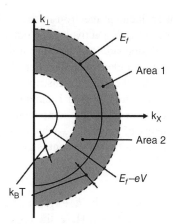

Figure 27.3 Fermi energy surface of k-space. Reproduced from Takahiro Kobayashi, Naoto Matsuo, Yasuhisa Omura, Shin Yokoyama, Akira Heya, "Electrical conduction mechanism in the high temperatures for thin-film transistor utilizing tunnel effect," J. Electron Devices, vol. 20, pp. 1733–1739, 2014 (Euro-Mediterranean Institute for Sustainable Development).

film with a thickness of 620 nm was deposited by atmospheric-pressure chemical vapor deposition (APCVD). Contact holes were etched by wet etching. Aluminum (Al) layers were formed by sputtering. The thicknesses of the Al layers on the front and back sides were 800 and 400 nm, respectively. Hydrogen annealing was performed at 400 °C. The channel length L_{eff} and channel width W_{eff} were 10 and 50 μm, respectively. A conventional TFT was also simultaneously fabricated by the same process. Low-temperature measurement of the I_D-V_G characteristic was performed at 20, 40, 100, 150, and 300 K using a BCT-4R, and high-temperature measurements were performed at 293, 353, 453, and 623 K using an AC-450H Hot Chuck Controller. The g_m value was calculated from the I_D-V_G characteristics.

27.3.2   Calculation Method

The tunnel current is given by [11, 18]:

$$J_{nt} = \frac{4\pi m^* q}{h^3} \int_{E_{fm2}}^{E_{max}} \left[\int_0^\infty \left(f_m - f_s \right) dE_t \right] \times \exp\left[-\gamma_n \left(\overline{\psi}_m - E_{fm1} - E_x \right)^{1/2} \right] dE_x \quad (27.1)$$

where J_{nt}, q, h, m^*, E_x, E_t, E_{fm1}, E_{fm2}, f_m, f_s, and $\overline{\psi}_m$ are the total tunnel current, the electric charge, Planck's constant, the effective mass of an electron, the energy component of an incident electron in the X direction, the energy component of an incident electron normal to the X direction, the Fermi energies of Al and poly-Si, the Fermi-Dirac distribution functions (FD functions) of Al and poly-Si, and the average energy height of the tunnel barrier, respectively. In the present experiment, electrons are emitted from the Al electrode to the poly-Si via SiN$_x$ film by direct tunnel.

Figure 27.3 shows the Fermi energy surface of the *k*-space for the source side of the TDTFT, by which the electron distribution contributing to the direct tunnel is shown. Here, E_f, V, k_B, k_x, $k_\perp$, and T are the Fermi energy of the Al on the emission side, the applied voltage, the Boltzmann constant, the wave numbers of an incident electron in the X direction and normal to the X direction and the absolute temperature, respectively. The outer solid line denoted by E_f is the Fermi surface of the Al on the emission side and the inner solid line denoted by $(E_f\text{-}qV)$ is the Fermi surface of the poly-Si on the incident side. The Fermi energy surface is spread over a radial distance of k_BT, and therefore the electron distribution that contributes to the direct tunnel is shown by the dark areas denoted by area 1 and area 2. By considering the temperature, the total direct tunnel current is divided into components J_{nt1}, which corresponds to area 1 and is larger than the Fermi energy of Al, and J_{nt2}, which corresponds to area 2 smaller than the Fermi energy of Al. The total direct tunnel current is given by

$$J_{nt} = J_{nt1} + J_{nt2},\tag{27.2}$$

By assuming the total electron energy to be $E_f + k_BT$ in area 1, the ranges of E_X and E_t are $E_f \sim E_f + k_BT$ and $0 \sim E_f + k_BT - E_X$, respectively. Here, the upper limit is replaced by ∞ to simplify the integral calculation. In addition, assuming the total electron energy to be E_f in area 2, the ranges of E_X and E_t are $E_f - eV + k_BT \sim E_f$ and $0 \sim E_f - E_X$ respectively.

Here, the band diagrams are used to calculate J_{nt1} and J_{nt2}. Figure 27.4a and b shows the band diagrams of the TDTFT before and after the application of gate voltage at a high temperature. For the Al layer and n$^+$ doping layer, the relationships between the electron density and the electron energy are shown. J_{nt1} and J_{nt2} are obtained by integrating Eq. (27.1) as follows.

When evaluating J_{nt1}, $f_m = \exp[-\beta(E_t + E_x - E_{fm1})]$ and $f_S = \exp[-\beta(E_t + E_x - E_{fm2})]$. Here, E_t and E_X are integrated from 0 to ∞ and from E_{fm1} to ∞, respectively. J_{nt1} is given by

$$J_{nt1} = 4\pi m^* q \left(k_B T\right)^2 /h^3 \times \left[1 - \gamma_n / 2\beta \bar{\psi}_m^{1/2}\right]^{-1} \times \exp\left(-\gamma_n \bar{\psi}_m^{1/2}\right) \times \left(1 - \exp\left(\beta V_{app}\right)\right)\tag{27.3}$$

where $\gamma_n = 4\pi(2m^*)^{1/2} t_{SiN} /h$

When evaluating J_{nt2}, $f_m = 1$ and $f_S = \exp[-\beta(E_t + E_x - E_{fm2})]$. Here, E_t and E_X are integrated from 0 to $E_{fm1} - E_X$ and E_{fm2} to E_{fm1}, respectively. J_{nt2} is given by

$$J_{nt2} = \frac{4\pi m^* q}{h^3} \times \frac{2\left(E_{fm1} - E_{fm2}\right)\bar{\psi}_m^{1/2}}{3\gamma_n} \times \exp\left(-\gamma_n \bar{\psi}_m^{1/2}\right)$$

$$- \frac{4\pi m^* q}{h^3} \times \frac{2\left(E_{fm1} - E_{fm2}\right)\left(\bar{\psi}_m + E_{fm1} - E_{fm2}\right)^{1/2}}{3\gamma_n} \times \exp\left[-\gamma_n \left(\bar{\psi}_m + E_{fm1} - E_{fm2}\right)^{1/2}\right]$$

$$\tag{27.4}$$

where t_{SiN} is the thickness of the SiN$_X$ film and β and qV_{app} are given by $(k_BT)^{-1}$ and $E_{fm1} - E_{fm2}$, respectively.

The voltage applied to the tunneling dielectric film is calculated as

$$V_{tun} = R_{tun} \times I_{TD},\tag{27.5}$$

where

$$R_{tun} = \frac{1}{2}\left(\frac{V_D}{I_{TD}} - \frac{V_D}{I_{C_0}}\right).\tag{27.6}$$

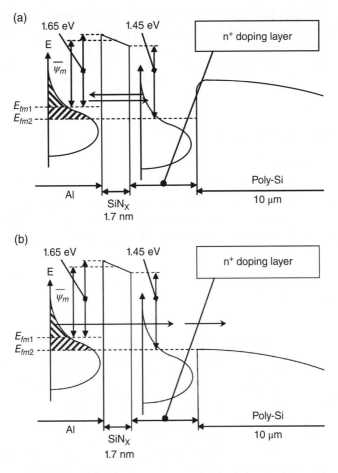

Figure 27.4 Band diagrams of TDTFT. (a) Band diagrams of TDTFT at a high temperature without application of bias voltage. Reproduced from Takahiro Kobayashi, Naoto Matsuo, Yasuhisa Omura, Shin Yokoyama, Akira Heya, "Electrical conduction mechanism in the high temperatures for thin-film transistor utilizing tunnel effect," J. Electron Devices, vol. 20, pp. 1733–1739, 2014 (Euro-Mediterranean Institute for Sustainable Development). (b) Band diagrams of TDTFT at a high temperature with application of bias voltage. Reproduced from Takahiro Kobayashi, Naoto Matsuo, Yasuhisa Omura, Shin Yokoyama, Akira Heya, "Electrical conduction mechanism in the high temperatures for thin-film transistor utilizing tunnel effect," J. Electron Devices, vol. 20, pp. 1733–1739, 2014 (Euro-Mediterranean Institute for Sustainable Development).

here, R_{tun}, I_{TD}, I_{Co}, V_{D}, and V_{tun} are the resistance of the tunneling dielectric film, the drain current of the TDTFT, the drain current of the conventional TFT, the drain voltage and the voltage applied to the tunneling dielectric film, respectively. When the drain voltage is 2.5 V, the voltages applied to the tunneling dielectric film at 293 and 353 K are 8.0×10^{-2} and 9.8×10^{-2} V, respectively. The total drain current of the TDTFT was calculated by multiplying J_{nt} by the tunneling probability of the tunneling dielectric film on the drain side. The energy barrier between the doping layer and the poly-Si is formed as shown in Figure 27.4a because the channel is not formed before the gate voltage is applied. When the drain voltage is applied, electrons are emitted from the Al layer to the n^+ doping layer by direct tunnel via the SiN_x film, and the electrons are reflected by the energy barrier. However, when the gate voltage is applied, electrons are emitted from the Al layer to the channel by direct tunnel via the SiN_x film, as shown in Figure 27.4b, because the energy barrier disappears upon gate voltage application.

27.4   Results and Discussion

27.4.1   Evaluation of SiN_x Film

Figure 27.5a–c shows the measured direct tunnel currents and direct tunnel current calculated by the Wentzel, Kramers, Brillouin (WKB) approximation method using Eq. (27.7) [18, 19], a cross-sectional transmission electron microscope (TEM) image of Al/SiNx/poly-Si and contact area between Al and the source/drain measured by EDX.

$$J_t = \frac{q}{2\pi h(\beta S)^2} \bar{\varphi} \exp\left(-A\sqrt{\bar{\varphi}}\right) - \frac{q}{2\pi h(\beta S)^2}(\bar{\varphi}+qV)\exp\left(-A\sqrt{\bar{\varphi}+qV}\right) \qquad (27.7)$$

where $A=(4\pi\beta S/h)(2m^*)^{1/2}$ and q, h, V, β, $\bar{\varphi}$, S, and m^* are the electronic charge, Planck's constant, the applied voltage, a correction factor, the average potential of the barrier ($\bar{\varphi} = \varphi_0 - qV/2$), the barrier thickness and the effective mass of electrons in the SiN_x film, respectively. φ_0 is the barrier height of the SiN_x film. $\beta=0.99$, $S=1.7$ nm, $m^*=0.31 m_0$ and $\varphi_0=2$ eV were set. This equation reproduces the direct tunnel current exactly except for in the early stage of the direct tunnel. In the voltage range of approximately −2.5 to −4 V, the slope of the calculated direct tunnel current coincided with that of the measured data. Therefore, the current flowing in the SiN_x film was confirmed to be the direct tunnel current. Here, hysteresis was observed in the early stage for the current-bias relation. This is considered to be due to the electron traps in the SiN_x film [20]. Although the exact thickness of the SiN_x film could not be obtained from the image shown in Figure 27.5b because it was sandwiched between polycrystalline films, it was confirmed that a uniform SiN_x film was formed at the Al/SiN_x/poly-Si structure. The result of the EDX measurement down to the bottom interface at the contact holes indicated the existence of N and Si atoms at the Al/poly-Si interface.

27.4.2   Characteristics of the TDTFT

Figure 27.6a–c shows the I_D-V_D characteristics of the conventional TFT and TDTFT, and the I_D-V_G characteristics of the conventional TFT at $V_D=0.1$ V and the TDTFT at $V_D=0.1$ and 1.0 V. The contact area is the total overlapping area between the SiN_x film and Al layer. The drain current I_D of the TDTFT was less than that of the conventional TFT. It was confirmed that the SiN_x film at the bottom of the contact holes serves as a tunneling resistance. The gate-off current of the TDTFT at $V_D=0.1$ V was less than 1/10 that of the conventional TFT at $V_D=0.1$ V. It was clarified that the tunnel effect limited the increase in the gate-off current. I_D for the TDTFT at $V_D=1.0$ V was approximately one order of magnitude larger than that at

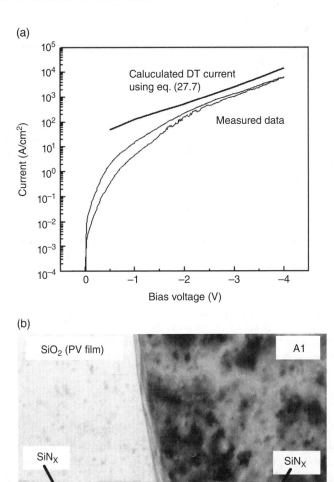

Figure 27.5 (a) DT currents measured and calculated by the WKB approximation method using Eq. (27.6). Reproduced by permission of the Institute of Electronics, Information, and Communication Engineers (2007), N. Matsuo, A. Fukushima, K. Ohkura, A. Heya, and S. Yokoyama, "Fabrication of tunneling dielectric thin-film transistor with very thin SiN$_x$ films onto source and drain," IEICE Electron. Express (on-line journal), vol. 4, pp. 442–447. Copyright © 2007 IEICE. (b) Cross-section TEM image of Al/SiNx/poly-Si. Reproduced by permission of the Institute of Electronics, Information, and Communication Engineers (2007), N. Matsuo, A. Fukushima, K. Ohkura, A. Heya, and S. Yokoyama, "Fabrication of tunneling dielectric thin-film transistor with very thin SiN$_x$ films onto source and drain," IEICE Electron. Express (on-line journal), vol. 4, pp. 442–447. Copyright © 2007 IEICE.

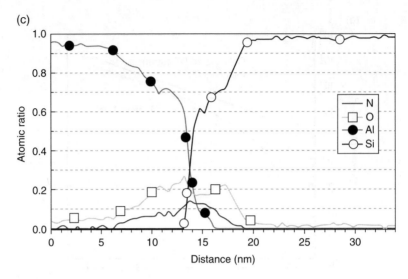

Figure 27.5 (Continued) (c) Atomic ratio in the vicinity of contact area between Al and source/drain measured by EDX. Reproduced by permission of the Institute of Electronics, Information, and Communication Engineers (2011), T. Kobayashi, N. Matsuo, T. Tochio, K. Ohkura, Y. Omura, S. Yokoyama and A. Heya, "Temperature dependence of drain currents of thin-film transistor with very thin SiN$_x$ film formed at source and drain region," IEICE Trans. Electronics (Japanese edition), vol. 94, no. 3, pp. 79–87, Copyright © 2011 IEICE.

$V_D=0.1\,$V. Although it is difficult to obtain the nonlinear behavior for $V_D=0.1\,$V from Figure 27.5a, the TFT performance for $V_D=1.0\,$V is clearly nonlinear. For other electrical characteristics, however, the desirable results were not obtained. The threshold voltages of the TDTFT and conventional TFT are 16.5 and 11.3 V and the subthreshold swings of the TDTFT and conventional TFT are 2.14 and 1.15 V/dec, respectively. The maximum trans-conductances g_m of the TDTFT and conventional TFT are 1.15×10^{-7} and 2.27×10^{-7} S, respectively. The subthreshold swing (SS) and g_m for the TDTFT were inferior to those of the conventional TFT. These phenomena are due to the tunneling resistance in the contact area, which has a close relationship with the thickness of the SiN$_x$ film. It was found by a simple calculation [5, 6] that the tunneling resistance is larger than the channel resistance in the present case. Further thinning of the SiN$_x$ film down to 0.5 nm and the use of a material such as TiO$_2$, with a lower barrier height than SiN$_x$ will improve the electric characteristics of the TDTFT. The ratio of the on-current to the off-current of the TDTFT is the same as that of the conventional TFT because the transmission coefficient of the electrons through the SiN$_x$ film is constant and independent of the DT current. To improve the on/off ratio, first, the reduction of the electric field perpendicular to the interface between the channel and the gate oxide by the use of a double SOI structure [6] is recommended. Second, the formation of asymmetrical tunnel dielectric film will limit the decrease in the on-current and also decrease the off-current. We have obtained preliminary data indicating that having the tunnel oxide only on the drain side improves the on/off ratio compared with that of the conventional TFT.

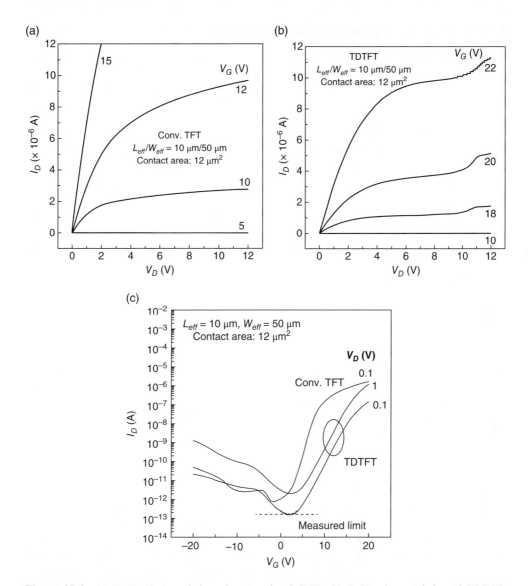

Figure 27.6 (a) I_D-V_D characteristics of conventional TFT. (b) I_D-V_D characteristics of TDTFT. Reproduced by permission of the Institute of Electronics, Information, and Communication Engineers (2007), N. Matsuo, A. Fukushima, K. Ohkura, A. Heya, and S. Yokoyama, "Fabrication of tunneling dielectric thin-film transistor with very thin SiN$_x$ films onto source and drain," IEICE Electron. Express (on-line journal), vol. 4, pp. 442–447. Copyright © 2007 IEICE. (c) I_D-V_G characteristics of conventional TFT at V_D=0.1 V and TDTFT at V_D=0.1 and 1.0 V. Reproduced by permission of the Institute of Electronics, Information, and Communication Engineers (2007), N. Matsuo, A. Fukushima, K. Ohkura, A. Heya, and S. Yokoyama, "Fabrication of tunneling dielectric thin-film transistor with very thin SiN$_x$ films onto source and drain," IEICE Electron. Express (on-line journal), vol. 4, pp. 442–447. Copyright © 2007 IEICE.

27.4.3 TFT Performance at Low Temperatures

Figure 27.7a and b shows the I_D-V_G characteristics of the conventional TFT and TDTFT at 20, 40, 100, 150, and 300 K, respectively. It was found that the drain currents of both TFTs increase as the temperature increases. A shoulder was observed in the I_D-V_G characteristic of the conventional TFT but not in that of the TDTFT. This is due to the increase in the tunnel resistance at the source and drain.

Figure 27.8a and b shows the relationship between g_m and the reciprocal of the measured temperature 1000/T for the TDTFT and conventional TFT at 20, 40, 100, 150, and 300 K and that for the conventional TFT at 293, 353, 453, and 623 K. g_m for the TDTFT was constant in the range of 20–150 K. This strongly indicates that direct tunneling via the SiN_x film dominates the drain current of the TDTFT from approximately 20 to 150 K. g_m for the conventional TFT increased with the temperature in the range 20–150 K. The dominant process is electron trapping in the poly-Si film. g_m for both TFTs also increased when temperature was increased from 150 to 300 K. The thermal activation process of the conventional TFT is attributed to the Poole–Frenkel mechanism.

The Poole–Frenkel current J_{PF} is given by:

$$\ln\left(\frac{J_{PF}}{E_N}\right) = \ln C - \left(\frac{q}{k_B T}\right)\left(\varphi_t - \sqrt{\frac{q}{\pi \varepsilon_0 \varepsilon_{Si}}} \cdot \sqrt{E_N}\right) \tag{27.8}$$

where φ_t, E_N, ε_0, ε_{si}, q, k_B, and C are the barrier height of the trap level, the electric field in SiN_x, the dielectric constant of free space, the dynamic dielectric constant of Si, the electronic charge, the Boltzmann constant, and a constant of integration. Calculated values of φ_{t1} for the conventional TFT in the low temperatures are shown in Table 27.1. φ_{t2} for the conventional TFT at high temperatures was calculated from Figure 27.8b to be 0.15 eV.

27.4.4 TFT Performance at High Temperatures

Figure 27.9a and b shows the I_D-V_G characteristics of the TDTFT and conventional TFT, respectively. The slope in the subthreshold region of the TDTFT is smaller than that for the conventional TFT. This is due to the tunnel resistivity of the SiN_x film formed in the source/drain area of the TDTFT. Although the leakage currents of the TDTFT and the conventional TFT are similar, the off-state voltage of the TDTFT is slightly larger than that of the conventional TFT. This is considered to be due to the difference in the Fermi level for the poly-Si serving as the active layer between the TDTFT and the conventional TFT. As the Fermi level of the TDTFT is lower than that of the conventional TFT, the leakage current of the TDTFT at $V_G=0$ V for the TDTFT is larger. Therefore, the leakage current of the TDTFT is smaller than that of the conventional TFT because of the use of poly-Si with an identified Fermi level [7]. Here, the origin of the carriers in the subthreshold region that are generated by increasing the temperature is discussed. The carriers are excited from a defect level of 0.15 eV [10] below the band edge for T = 293–453 K and from the valence band for temperatures of above 623 K in the intrinsic semiconductor region. The conduction of the excited carriers in the channel is dominated by the tunnel emission via the SiNx film on the source/drain side as discussed later.

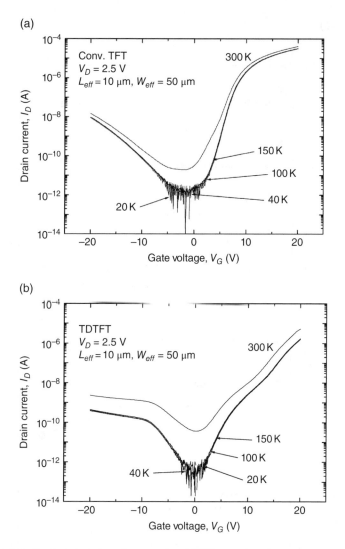

Figure 27.7 (a) I_D-V_G characteristics of conventional TFT at various low temperatures. Reproduced by permission of the Institute of Electronics, Information, and Communication Engineers (2011), T. Kobayashi, N. Matsuo, T. Tochio, K. Ohkura, Y. Omura, S. Yokoyama and A. Heya, "Temperature dependence of drain currents of thin-film transistor with very thin SiN$_x$ film formed at source and drain region," IEICE Trans. Electronics (Japanese edition), vol. 94, no. 3, pp. 79–87. Copyright © 2011 IEICE. (b) I_D-V_G characteristics of TDTFT at various low temperatures. Reproduced by permission of the Institute of Electronics, Information, and Communication Engineers (2011), T. Kobayashi, N. Matsuo, T. Tochio, K. Ohkura, Y. Omura, S. Yokoyama and A. Heya, "Temperature dependence of drain currents of thin-film transistor with very thin SiN$_x$ film formed at source and drain region," IEICE Trans. Electronics (Japanese edition), vol. 94, no. 3, pp. 79–87, Copyright © 2011 IEICE.

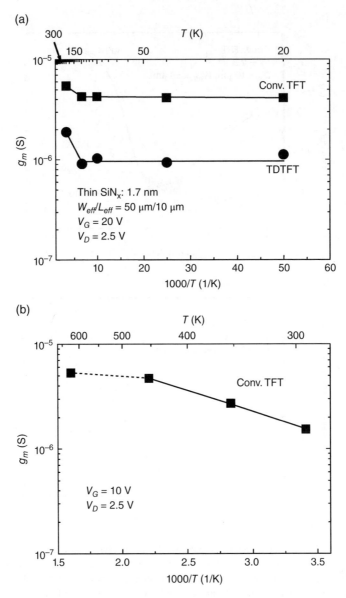

Figure 27.8 (a) Relationship between g_m and the reciprocal of the absolute temperature, $1000/T$, for the TDTFT and conventional TFT at 20, 40, 100, 150, and 300 K. Reproduced by permission of the Institute of Electronics, Information, and Communication Engineers (2011), T. Kobayashi, N. Matsuo, T. Tochio, K. Ohkura, Y. Omura, S. Yokoyama and A. Heya, "Temperature dependence of drain currents of thin-film transistor with very thin SiN$_x$ film formed at source and drain region," IEICE Trans. Electronics (Japanese edition), vol. 94, no. 3, pp. 79–87. Copyright © 2011 IEICE.(b) Relationship between g_m and the reciprocal of the absolute temperature, $1000/T$, for the conventional TFT at 293, 353, 453, and 623 K. Reproduced by permission of the Institute of Electronics, Information, and Communication Engineers (2011), T. Kobayashi, N. Matsuo, T. Tochio, K. Ohkura, Y. Omura, S. Yokoyama and A. Heya, "Temperature dependence of drain currents of thin-film transistor with very thin SiN$_x$ film formed at source and drain region," IEICE Trans. Electronics (Japanese edition), vol. 94, no. 3, pp. 79–87. Copyright © 2011 IEICE.

Table 27.1 Calculated trap energy φ_{tl} of conventional TFT at low temperatures.

	Temperature range (K)	φ_t (eV)	
		$V_D = 1\,V$	$V_D = 2.5\,V$
Conventional TFT	20–150	0.009	0.014
	150–300	0.01	0.020

It was finally confirmed that the TDTFT shows a distinct characteristic owing to the tunnel effect of the SiN$_x$ film and that it is an improvement on the conventional TFT. Next, we analyzed the conduction of the TDTFT at high temperatures using Eq. (27.2).

Figure 27.10 shows the relationship between the calculated and measured drain currents of the TDTFT. The inset is an enlarged view around the threshold voltage in Figure 27.9 for the temperatures of 293 and 353 K. The threshold voltage for each temperature was determined by interpolation of the I_D-V_G curve. The drain current was defined as the current at the threshold voltage for each temperature. The conditions used in the direct tunnel current simulation to obtain the results in Figure 27.10 were a SiN$_x$ film thickness and effective mass of $t_{SiN} = 1.3$ nm and $m^*/m_0 = 0.25$, respectively. When the temperature increases from 293 to 353 K, the conduction band edge of Si is decreased by 0.02 eV because of the decrease in the band gap with increasing temperature [21]. The band gap of the SiN$_x$ film is 4.4 eV [22]. At 293 K, the calculated direct tunnel current is 7.4×10^{-7} A, which is very close to the measured value of 7.5×10^{-7} A. Moreover, at 353 K, it is 1.1×10^{-6} A, which is very close to the measured value of 1.2×10^{-6} A. Furthermore, a maximum temperature at which the direct tunnel is dominant exists between 350 and 400 K according to the calculated results. In addition, the values of J_{nt1} at 293 and 353 K are 1.3×10^{-7} and 2.1×10^{-7} A, and those of J_{nt2} at 293 and 353 K are 6.1×10^{-7} and 8.5×10^{-7} A, respectively. These data strongly indicate that the dominant conduction mechanism of the TDTFT is the direct tunnel. The tail area of the FD function, being larger than the Al Fermi energy, has a strong effect on the direct tunnel. Although the thickness of the SiN$_x$ film was assumed to be 1.3 nm, this thickness is within the measurement error of spectroscopic ellipsometer. The effective mass was calculated by extrapolation of the calculated effective mass as a function of the oxide thickness [19]. As a result, the effective mass of the SiN$_x$ film was obtained as 0.25 in the present experiment. In the case of a thick film of nm-order thickness, an effective mass of 0.42 is used in the two-band model [23, 24]. However, in the case of a thin film, the effective mass decreases to 0.33 ± 0.08, in accordance with the one-band model [25]. In the present calculation, the effective mass for the one-band model was used because of the low thickness of the SiN$_x$ film.

The drain current cannot be controlled for gate voltage application at higher temperatures (T > 453 K) according to Figure 27.9 and Figure 27.10. The reason for this is as follows. The change in the electron carrier concentration in the channel upon the gate voltage application is small because many electron carriers have already been generated for nongate application at

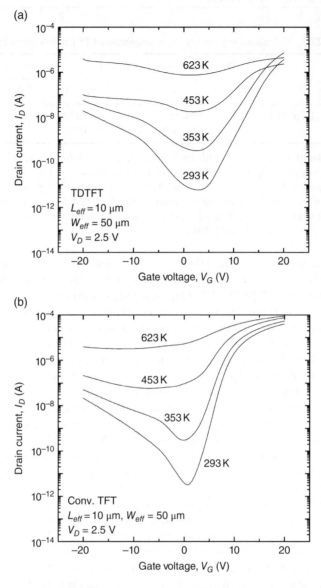

Figure 27.9 (a) I_D-V_G characteristics of conventional TFT at high temperatures. Reproduced from Takahiro Kobayashi, Naoto Matsuo, Yasuhisa Omura, Shin Yokoyama, Akira Heya, "Electrical conduction mechanism in the high temperatures for thin-film transistor utilizing tunnel effect," J. Electron Devices, vol. 20, pp. 1733–1739, 2014 (Euro-Mediterranean Institute for Sustainable Development). Reprinted with permission from T. Kobayashi, N. Matsuo, A. Heya, Y. Omura, and S. Yokoyama, "Electrical characteristic in the high temperatures for thin-film transistor with very thin SiN$_x$ film formed at source and drain region, IEEE International Meeting for Future of Electron Devices, Kansai (May, 2010), pp. 78–79. (b) I_D-V_G characteristics of TDTFT at high temperatures. Reproduced from Takahiro Kobayashi, Naoto Matsuo, Yasuhisa Omura, Shin Yokoyama, Akira Heya, Electrical Conduction Mechanism in the High Temperatures for Thin-Film Transistor Utilizing Tunnel Effect, J. Electron Devices, vol. 20, pp. 1733–1739, 2014 (Euro-Mediterranean Institute for Sustainable Development). Copyright(c) 2010 IEEE. Reprinted with permission from T. Kobayashi, N. Matsuo, A. Heya, Y. Omura, and S. Yokoyama, "Electrical characteristic in the high temperatures for thin-film transistor with very thin SiN$_x$ film formed at source and drain region, IEEE International Meeting for Future of Electron Devices, Kansai (May, 2010), pp. 78–79.

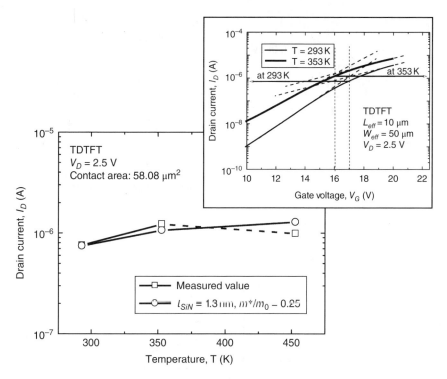

Figure 27.10 Relationship between calculated and measured drain current of TDTFT under various conditions. Reproduced from Takahiro Kobayashi, Naoto Matsuo, Yasuhisa Omura, Shin Yokoyama, Akira Heya, "Electrical conduction mechanism in the high temperatures for thin-film transistor utilizing tunnel effect," J. Electron Devices, vol. 20, pp. 1733–1739, 2014. (Euro-Mediterranean Institute for Sustainable Development.)

high temperatures. Thus, it is concluded that the conduction mechanism is direct tunnel from 293 to 353 K and that the tail area of the FD function being larger than the Fermi energy of Al strongly affects the conduction mechanism of the TDTFT. The direct tunnel from 293 to 353 K for the TDTFT is different from that at low temperatures, from 20 to 150 K [10].

Figure 27.11a–c shows the conduction mechanism of the conventional TFT and the TDTFT at low and high temperature. For the conventional TFT, the electron conduction is dominated by the Poole–Frenkel mechanism through the trap sites with a φ_{t1} at low temperature and that of φ_{t2} at high temperature. For the TDTFT at low temperatures, electrons are emitted from the Fermi level of Al to the conduction band of poly-Si by direct tunnel via the SiN$_x$ films. The shallow electron trap level is approximately 0.01 eV from the conduction band edge of poly-Si. At low temperatures, the relaxation time of the trapped electrons is considered to be small. Therefore, the direct tunnel becomes dominant. At high temperatures, the thermally activated electrons in Al are emitted to the poly-Si via the SiN$_x$ film by direct tunnel. The conduction mechanism of the TDTFT at high temperatures is not the simple activation process but the direct tunnel via the SiN$_x$ film.

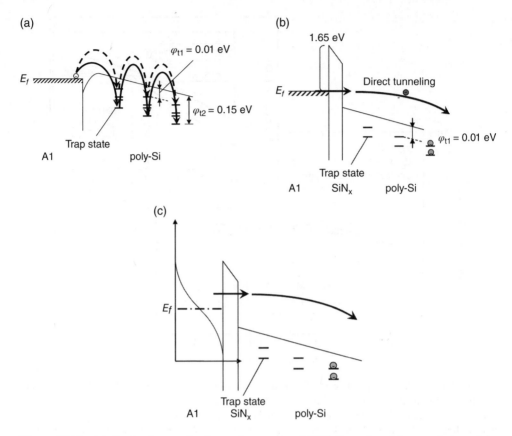

Figure 27.11 (a) Conduction mechanisms of conventional TFT. Reproduced by permission of the Institute of Electronics, Information, and Communication Engineers (2011), T. Kobayashi, N. Matsuo, T. Tochio, K. Ohkura, Y. Omura, S. Yokoyama and A. Heya, "Temperature dependence of drain currents of thin-film transistor with very thin SiN$_x$ film formed at source and drain region," IEICE Trans. Electronics (Japanese edition), vol. 94, no. 3, pp. 79–87. Copyright © 2011 IEICE. (b) Conduction mechanisms of TDTFT at low-temperatures. Reproduced by permission of the Institute of Electronics, Information, and Communication Engineers (2011), T. Kobayashi, N. Matsuo, T. Tochio, K. Ohkura, Y. Omura, S. Yokoyama and A. Heya, "Temperature dependence of drain currents of thin-film transistor with very thin SiN$_x$ film formed at source and drain region," IEICE Trans. Electronics (Japanese edition), vol. 94, no. 3, pp. 79–87. Copyright © 2011 IEICE. (c) Conduction mechanisms of TDTFT at high temperatures.

27.4.5 Suppression of the Hump Effect by the TDTFT

Figure 27.12 shows the I_D-V_G characteristics of the conventional TFT for channel lengths L_{eff} of 10, 1, and 0.5 μm at drain voltages of 1.0–3.0 V in steps of 0.5 V. A hump is clearly observed from the beginning of the subthreshold region to the gate voltage, where the slope of the subthreshold current changes for $L_{eff} = 10$ μm. For a large V_G, the current flowing in the bulk becomes larger than that flowing at the edge. In addition, the hump becomes more pronounced at a large drain voltage. Although a characteristic similar to the hump phenomenon, named the

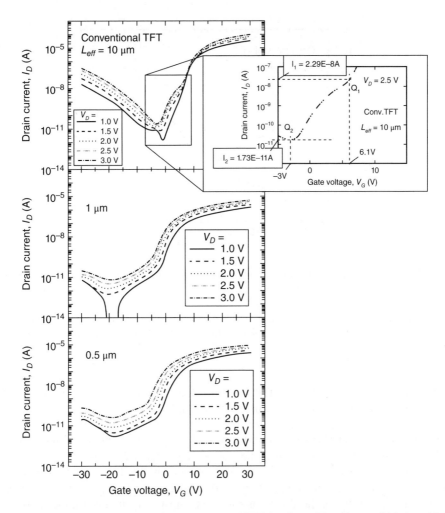

Figure 27.12 I_D-V_G characteristics of the conventional TFT for channel lengths L_{eff} of 10, 1, and 0.5 μm at drain voltages of 1.0–3.0 V. Reproduced by permission of the Institute of Electronics, Information, and Communication Engineers (2014), T. Kobayashi, N. Matsuo, A. Heya, and S. Yokoyama, "Improvement of hump phenomenon of thin-film transistor by SiNx film," IEICE Trans. Electron., vol. E97, no. 11, pp. 1112–1116. Copyright © 2014 IEICE.

quasihump phenomenon, was observed for L_{eff}=1 and 0.5 μm, this was not the real hump phenomenon. The I_D-V_G curves for small channel lengths are shifted to the left compared with those for large channel lengths. This phenomenon is due to the short channel effect. The hump phenomenon is reproducible and has been observed in other TFTs at different positions.

Figure 27.13 shows the I_D-V_G characteristics of another conventional TFT in another region of the same substrate as that of the TFT of Figure 27.12 for L_{eff}=10, 1.0, and 0.5 μm at a drain voltage of 2.5 V. The hump or quasihump phenomenon is pronounced at a large drain voltage and short channel length. The gate-off current of the TFT with L_{eff}=10 μm is larger than those

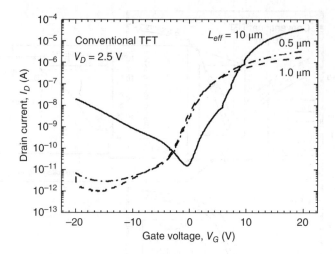

Figure 27.13 I_D-V_G characteristics of conventional TFT in another region of the same substrate as the TFT of Figure 27.12 for the channel lengths L_{eff} of 10, 1.0, and 0.5 μm at a drain voltage of 2.5 V. Reproduced by permission of the Institute of Electronics, Information, and Communication Engineers (2014). T. Kobayashi, N. Matsuo, A. Heya, and S. Yokoyama, "Improvement of hump phenomenon of thin-film transistor by SiNx Film," IEICE Trans. Electron., vol. E97, no. 11, pp. 1112–1116, Copyright © 2014 IEICE.

of the TFTs with L_{eff} = 1 and 0.5 μm. This is mainly due to the difference in the channel width as indicated by the ratios W_{eff}/L_{eff} = 50 μm/10 μm, 1/1, and 1/0.5.

Figure 27.14 shows the I_D-V_G characteristics of the TDTFT for L_{eff} of 10, 1.0, and 0.5 μm at drain voltages of 1.0–3.0 V in steps of 0.5 V. For L_{eff} of 10 μm, the hump was faintly observed. For L_{eff} of 1.0 μm, the quasihump phenomenon shown in Figure 27.12 was not observed. For L_{eff} of 0.5 μm, the quasihump phenomenon was also not observed at V_D of 1.0 V, although it was faintly observed at the beginning of the subthreshold region for V_D of 1.5–3.0 V. Here, the variation of V_{TH} with V_D for the TDTFT is small in comparison with that for the conventional TFT. It is considered that this is related to the impact ionization. The electric field in the channel for the conventional TFT is larger than that for the TDTFT owing to the voltage drop at the SiN$_X$ film. This effect induces the impact ionization in the channel in the vicinity of the drain. Next, the mechanism of hump suppression by this structure is discussed.

Figure 27.15a–c shows the layout of the lithography mask of the TDTFT or conventional TFT for W_{eff}/L_{eff} = 50 μm/10 μm, a schematic cross-sectional view of the edge area in the poly-Si island where the TFT with the SiN$_X$ film is fabricated and the hump suppression model. This model indicates the following assumptions: (i) The electrons in the SiN$_X$ film, that is, the fixed charges, block the carriers flowing into the channel from the source by Coulomb repulsion. (ii) The charge density per unit area, referred to as the fixed-charge density, of the SiN$_X$ film is equal to that of the carriers in the edge channel when the hump phenomenon occurs for a conventional TFT. Figure 27.15b shows the cross-sectional area of the TDTFT cut at the solid line in Figure 27.15a, where the source and drain are aligned perpendicular to the paper surface. Electron parasitic channels 1–4 are formed at the edge of the islands owing to the electric field concentration for the conventional TFT as shown by the dashed lines in

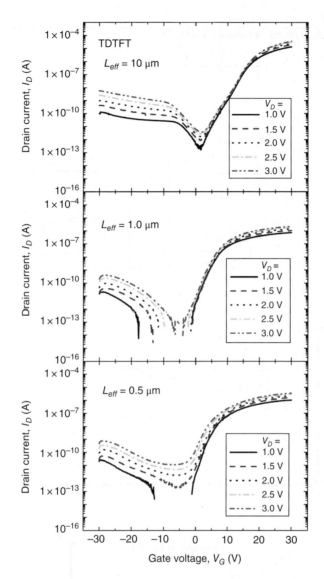

Figure 27.14 I_D-V_G characteristics of the TDTFT for L_{eff} of 10, 1.0, and 0.5 μm at the drain voltages of 1.0–3.0 V. Reproduced by permission of the Institute of Electronics, Information, and Communication Engineers (2014), T. Kobayashi, N. Matsuo, A. Heya, and S. Yokoyama, "Improvement of hump phenomenon of thin-film transistor by SiNx film," IEICE Trans. Electron., vol. E97, no. 11, pp. 1112–1116. Copyright © 2014 IEICE.

Figure 27.15b. The electric field around the edge is approximately 3–6×10^5 V/cm because the electric field at the interface between the poly-Si and the local oxidation of silicon (LOCOS) SiO_2 film was calculated to be 2×10^5 V/cm at V_G of 5 V and the electric field becomes concentrated at the convex edge. By considering the calculated electric field at the convex edge of the poly-Si [26], the coefficient of the field concentration at the convex edge

(a)

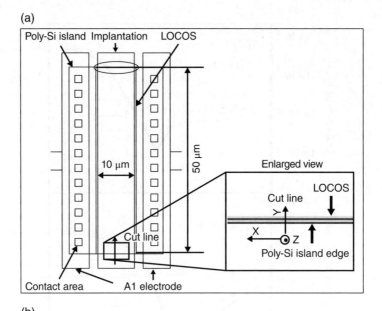

(b)

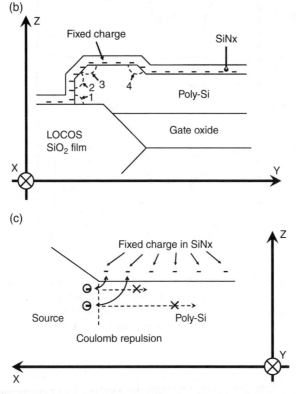

(c)

Figure 27.15 (a) Layout of the lithography mask of the TDTFT or conventional TFT for the $W_{eff}/L_{eff} = 50\,\mu m/10\,\mu m$. Reproduced by permission of the Institute of Electronics, Information, and Communication Engineers (2014), T. Kobayashi, N. Matsuo, A. Heya, and S. Yokoyama, "Improvement of hump phenomenon of thin-film transistor by SiNx film," IEICE Trans. Electron., vol. E97, no. 11, pp. 1112–1116. Copyright © 2014 IEICE.

may become 1.5–3 for the present dimensions of the convex edge. Therefore, the carrier concentration increases at the edge of the island-shape active layer owing to the electric field concentration in the case of a tapered profile. The parasitic channels are formed in the area indicated by the ellipse marked in the poly-Si island mask of Figure 27.15a. They are also formed in the opposite edge area. However, the parasitic channels are not formed by Coulomb repulsion in the TDTFT as shown in Figure 27.15c, because the poly-Si island is covered with a SiN_x film that has negative fixed charges in it [7]. As a consequence, the TDTFT suppresses the hump phenomenon. The quantity of the fixed charge in the SiN_x film was estimated in order to verify that the hump suppression is closely related to the fixed charge in the SiNx film. It was calculated at V_D of 2.5 V because the saturated region starts at this value as shown approximately in Figure 27.12. In Figure 27.12, $dQ_1/dt = I_1$ and $dQ_2/dt = I_2$ were obtained by assuming the numbers of charges in the channel, Q_1 and Q_2. The total number of charges, Q_{total}, flowing at the edge during the hump phenomenon, is simply calculated using

$\int d(Q_1 - Q_2) = \int_{t_1}^{t_2} (I_1 - I_2)\, dt$. Here, I_1 and I_2 are 2.29×10^{-8} and 1.73×10^{-11} A, respectively.

The total charge flowing at the edge during the hump phenomenon was found to be 8.33×10^{-8} C. This was transformed to the charge density per unit area, which is the number of electrical charges per unit area as a ratio of the total numbers of charges flowing at the edges and is assumed to be equal to the fixed charge density of the SiN_x film. The total area of the region where the carriers flow is calculated as $S = \dfrac{W_{total} \cdot V_D \cdot \mu \cdot t}{L_{eff}}$, where W_{total}, V_D, μ, and t are the total width of the edge where carriers are flowing, the drain voltage, the electron mobility, and the total holding time, respectively. The total width is 6×10^{-8} m. The total holding time was calculated using $t = t_{hold} \cdot (V_{G2} - V_{G1})/V_{step}$, where t_{hold} and V_{step} are the holding time and the voltage step in the gate voltage application, respectively. μ was calculated from the I_D-V_G characteristic, where V_{G1} and V_{G2} are −3 and 6.1 V, respectively. The total area of the edge region where the carriers flow was found to be 1.43×10^{-5} m². Therefore, the trap density in the SiN_x film was simply calculated as $N_t = Q_{total}/qS$, where N_t and q are the trap density and electron charge, respectively. The trap density in the SiN_x film was found to be 3.64×10^{12} cm⁻². This value is consistent with the reported values of 2.65×10^{12} to 1.3×10^{13} cm⁻² [27, 28]. Here, the formation of parasitic channel 4 is closely related to the fabrication process. By smoothing the LOCOS SiO_2 film surface, the projection of the poly-Si disappears. Therefore, the area, where field concentration occurs, decreases. In addition, if the number of fixed charges in the SiNx film increases, it is considered that the faintly observed hump for L_{eff} of 10 and 0.5 μm will be removed entirely.

Figure 27.15 (Continued) (b) Schematic cross-sectional view of the edge area in the poly-Si island where the TFT with SiN_x film is fabricated. The arrow indicates the region where the electric field concentration occurs. Reproduced by permission of the Institute of Electronics, Information, and Communication Engineers (2014), T. Kobayashi, N. Matsuo, A. Heya, and S. Yokoyama, "Improvement of hump phenomenon of thin-film transistor by SiNx film," IEICE Trans. Electron., vol. E97, no. 11, pp. 1112–1116. Copyright © 2014 IEICE. (c) Hump suppression model. The two-directional arrows show the Coulomb repulsion between the fixed charge in SiNx and the electron in the source. Reproduced by permission of the Institute of Electronics, Information, and Communication Engineers (2014), T. Kobayashi, N. Matsuo, A. Heya, and S. Yokoyama, "Improvement of hump phenomenon of thin-film transistor by SiNx film," IEICE Trans. Electron., vol. E97, no. 11, pp. 1112–1116. Copyright © 2014 IEICE.

27.5 Summary

- A TDTFT having a 1.7 nm thick SiNx film as a tunneling dielectric film on the source and drain was fabricated for the first time. The gate-off current of the TDTFT was decreased to less than 1/10 that of a conventional TFT. It was confirmed that a SiN_x film at the bottom of the contact hole serves as an effective tunneling resistance. The TFT performance under a nonlinear regime was also obtained for $V_D = 1.0$ V. However, the subthreshold swing and g_m were degraded. The reason for these phenomena is related to the thickness of the present SiN_x film. Further thinning of the film and the use of materials with a lower barrier height than SiN_x will improve the TDTFT performance.
- The conduction mechanism of the TDTFT with a 1.7 nm thick SiN_x layer at low temperatures was investigated. Direct tunneling via the SiN_x film dominates the drain current of the TDTFT from approximately 20 to 150 K. For the conventional TFT, the electron conduction was dominated by the PF mechanism through a trap site with a 0.01 eV at low temperatures and one with a 0.15 eV at high temperatures.
- The conduction mechanism of the TDTFT at high temperatures from 293 to 623 K was investigated by comparing with the results of a simulation of direct tunnel. The simulation of direct tunnel via the SiN_x film was performed dividing the electron distribution contributing to the direct tunnel into two areas. The direct tunnel was considered in areas larger than the Fermi energy affected by the temperature and smaller than the Fermi energy unaffected by the temperature. As a result, when the thickness of the SiN_x film and the effective mass were 1.3 nm and 0.25, respectively, the relationship between the drain current and the temperature reproduced the experimental results very closely. These results strongly indicate that the dominant conduction mechanism of the TDTFT at higher temperatures is direct tunnel. In addition, a tail area larger than the Fermi energy of Al has a strong effect on the direct tunnel. The reason why the electron effective mass of 0.25 was used for the present simulation is that the effective mass decreases in accordance with the one-band model for a thin dielectric film.
- It was clarified that the SiN_x film with a thickness of 1.7 nm, which was formed at the interface between the poly-Si source/drain and the Al layer, suppressed the hump phenomenon of TFT with a channel length of 10 μm. The mechanism of the hump suppression by this structure was discussed. It is considered that the fixed charges in the SiN_x film suppressed the formation of parasitic channels at the poly-Si edge. The trap density in the SiN_x film was estimated to be 3.64×10^{12} cm^{-2}. This value is consistent with the reported values of 2.65×10^{12} to 1.3×10^{13} cm^{-2}. It is considered that the fixed charges in the SiN_x film suppressed the formation of parasitic channels at the poly-Si edge by Coulomb repulsion.

References

[1] H. Hamada, H. Abe, and Y. Miyai, "Development of high-performance poly-Si TFTs and improvement of image characteristics for high-definition LCD light-valves," *IEICE Trans. Electron. (Japanese edition)*, vol. J84-C, no. 2, pp. 65–75, 2001.

[2] T. Noguchi, J. Y. Kwon, J. S. Jung, J. M. Kim, K. B. Park, H. Lim, D. Y. Kim, H. S. Cho, X. X. Zhang, H. X. Yin, and W. X. Xianyu, "Low Temperature Process for Advanced Si TFT Technology," Dig. Int. Workshop on Active-Matrix Liquid-Crystal Displays, (Kanazawa, Japan, July, 2005) TFT1-1, pp. 281–284, 2005.

[3] N. Sasaki, A. Hara, F. Takeuchi, K. Suga, M. Takei, K. Yoshino, and M. Senda, "A new low-temperature poly-Si TFT technology realizing mobility above 500 cm²/Vs by using CW laser lateral crystallization (CLC)," *IEICE Trans. Electron. (Japanese edition)*, vol. J85-C, no. 8, pp. 601–608, 2002.

[4] H. Watakabe, Y. Tsunoda, T. Sameshima, and M. Kimura, "Characterization of Polycrystalline Silicon Thin Film Transistor," Dig. Int. Workshop on Active-Matrix Liquid-Crystal Displays (Kanazawa, Japan, July, 2001) TFT1-3, pp. 45–48, 2001.

[5] N. Matsuo, H. Kihara, A. Yamamoto, N. Kawamoto, and H. Hamada, "Proposal and examination of new type of TFT with tunneling dielectric film at both ends of channel fabrication area," *IEICE Trans. Electron. (Japanese edition)*, vol. J86-C, no. 10, pp. 1070–1078, 2003.

[6] N. Matsuo, N. Kawamoto, A. Yamamoto, and A. Heya, "Application of tunneling dielectric TFT on double SOI to AM-OLED drive circuit," *IEICE Trans. Electron. (Japanese edition)*, vol. J89-C, no. 6, pp. 409–415, 2006.

[7] N. Matsuo, A. Fukushima, K. Ohkura, A. Heya, and S. Yokoyama, "Fabrication of tunneling dielectric thin-film transistor with very thin SiNx films onto source and drain," *IEICE Electron. Express* (on-line journal), vol. 4, pp. 442–447, 2007.

[8] T. Kobayashi, A. Fukushima, N. Matsuo, A. Heya, K. Ohkura, S. Yokoyama, and Y. Omura, "Electrical Characteristic in the Low Temperature for Thin-Film Transistor with Very Thin SiNx Film Formed at Source and Drain Region," IEEE International Meeting for Future of Electron Devices, Kansai (April, 2009), pp. 88–89, 2009.

[9] T. Kobayashi, N. Matsuo, A. Heya, Y. Omura, and S. Yokoyama, "Electrical Characteristic in the High Temperatures for Thin-Film Transistor with Very Thin SiN$_x$ Film Formed at Source and Drain Region," IEEE International Meeting for Future of Electron Devices, Kansai (May, 2010), pp. 78–79, 2010.

[10] T. Kobayashi, N. Matsuo, T. Tochio, K. Ohkura, Y. Omura, S. Yokoyama, and A. Heya, "Temperature dependence of drain currents of thin-film transistor with very thin SiNx film formed at source and drain region," *IEICE Trans. Electron. (Japanese edition)*, vol. 94, no. 3, pp. 79–87, 2011.

[11] T. Kobayashi, N. Matsuo, Y. Omura, S. Yokoyama, and A. Heya, "Electrical conduction mechanism in the high temperatures for thin-film transistor utilizing tunnel effects," *J. Electron Devices*, vol. 20, pp. 1733–1739, 2014.

[12] S. Kim, T. Y. Oh, J. Y. Yang, M. S. Yang, and I. J. Chung, "Analysis of the hump characteristics in poly-Si thin film transistor," *ECS Trans.*, vol. 31, no. 8, pp. 63–67, 2006.

[13] M. Furuta, Y. Kamada, M. Kimura, T. Hiramatsu, T. Matsuda, H. Furata, C. Li, S. Fujita, and T. Hirao, "Analysis of hump characteristics in thin-film transistors with ZnO channels deposited by sputtering at various oxygen partial pressures," *IEEE Electron Device Lett.*, vol. 31, no. 11, pp. 1257–1259, 2010.

[14] S.-H. Choi and M.-K. Han, "Effect of channel widths on negative shift of threshold voltage, including stress-induced hump phenomenon in InGaZnO thin-film transistors under high-gate and drain bias stress," *Appl. Phys. Lett.*, vol. 100, no. 4, pp. 043503–043506, 2012.

[15] Y.-M. Kim, K.-S. Jeong, H.-J. Yun, S.-D. Yang, S.-Y. Lee, Y.-C. Kim, J.-K. Jeong, H.-D. Lee, and G.-W. Lee, "Investigation of zinc interstitial ions as the origin of anomalous stress-induced hump in amorphous indium gallium zinc oxide thin film transistors," *Appl. Phys. Lett.*, vol. 102, no. 17, pp. 173502–173505, 2013.

[16] S.-H. Ho, T.-C. Chang, C.-W. Wu, W.-H. Lo, C.-E. Chen *et al.*, "Investigation of an anomalous hump in gate current after negative-bias temperature-instability in HfO2/metal gate p-channel metal-oxide- semiconductor field-effect transistors," *Appl. Phys. Lett.*, vol. 102, no. 1, pp. 012103–012106, 2013.

[17] T. Kobayashi, N. Matsuo, A. Heya, and S. Yokoyama, "Improvement of hump phenomenon of thin-film transistor by SiNx film," *IEICE Trans. Electron.*, vol. E97, no. 11, pp. 1112–1116, 2014.

[18] J. G. Simmons, "Generalized formula for the electric tunnel effect between similar electrodes separated by a thin insulating film," *J. Appl. Phys.*, vol. 34, no. 6, pp. 1793–1803, 1963.

[19] N. Matsuo, J. Yamauchi, Y. Kitagwa, and T. Miyoshi, "Analysis of effective mass of tunneling electron through thin SiO$_2$ film by WKB method," *IEICE Trans. Electron. (Japanese edition)*, C, vol. 83, no. 6, pp. 577–580, 2000.

[20] H. Yamamoto, H. Iwasawa, and A. Sasaki, "Discharging process by multiple tunneling in MNOS structures," *IEEE Electron Device Lett.*, vol. 2, pp. 21–23, 1981.

[21] A. S. Grove, "Physics and Technology of Semiconductor Devices," (John Wiley & Sons, Inc., 1967), p. 104.

[22] H. Kondo, S. Zaima, M. Hori, A. Sakai, and M. Ogawa, "Initial stage of processes and energy bandgap formation in nitridation of silicon surface using nitrogen radicals," *J. Vac. Soc. Jpn.*, vol. 50, no. 11, pp. 665–671, 2007.

[23] B. Brar, G. D. Wilk, and A. C. Seabaugh, "Direct extraction of the electron tunneling effective mass in ultrathin SiO$_2$," *Appl. Phys. Lett.*, vol. 69, pp. 2728–2730, 1996.

[24] E. O. Kane and E. I. Blount, "Interband Tunneling," pp. 79–91, in E. Burstein and S. Lundqvist, "Tunneling Phenomena in Solids," (Plenum Press, 969).

[25] M. Hiroshima, T. Yasudaka, S. Miyazaki, and M. Hirose, "Electron tunnleing through ultrathin gate oxide formed on hydrogen-terminated Si(100) surfaces," *Jpn. J. Appl. Phys.*, vol. 33, no. 1B, pp. 295–398, 1994.

[26] N. Matsuo, H. Fujiwara, T. Miyoshi, and T. Koyanagi, "Numerical analysis for conduction mechanism of thin oxide-nitride-oxide films on rough poly-Si," *IEEE Electron Device Lett.*, vol. 17, no. 2, pp. 56–58, 1996.

[27] M. Nishihara, S. Fujita, and A. Sasaki "Temperature dependence of charge transfer in metal-nitride-semiconductor diode structure," *Jpn. J. Appl. Phys.*, vol. 20, no. 10, pp. 1975–1976, 1981.

[28] E. Suzuki and Y. Hayashi, "Carrier conduction and trapping in metal-nitride oxide-semiconductor structures," *J. Appl. Phys.*, vol. 53, no. 12, pp. 8880–8885, 1982.

28

Proposal for a Tunnel-Barrier Junction (TBJ) MOSFET

28.1 Introduction

Extensive research efforts have been directed towards the development of design guidelines for the next generation of metal oxide semiconductor field-effect transistors (MOSFETs) and other devices, in order to realize high-performance integrated circuits (ICs). In particular, the silicon-on-insulator (SOI) device structure, including the double-gate SOI structure, is attracting interest because it offers a lower short-channel effect (SCE) and the possibility of quantum effect devices [1–3].

In the 2000s, researchers proposed or demonstrated several notable devices [4–6] from the viewpoint of using explicit quantum-mechanical effects. An editor of this book (Omura) proposed a device structure on which future generations of MOSFET devices should be based [4, 7]. This kind of device is attracting attention [8, 9] because its high barrier suppresses the unwanted thermal current across the junction or between the source and drain terminals [10].

This chapter demonstrates its advantages over the scaled MOSFET by simple simulations and discusses possible applications by assuming that the channel current in the Si body is ruled by the drift current.

28.2 Device Structure and Model

The device structure proposed here, a tunneling-barrier junction (TBJ) MOSFET on an SOI substrate, is shown in Figure 28.1a. Here an accumulation mode MOSFET is taken as an example. It has TBJs at the edges of the metal source and drain electrodes. It looks like a single electron transistor [2] and also a double-barrier resonant tunneling transistor [5].

MOS Devices for Low-Voltage and Low-Energy Applications, First Edition.
Yasuhisa Omura, Abhijit Mallik, and Naoto Matsuo.
© 2017 John Wiley & Sons Singapore Pte. Ltd. Published 2017 by John Wiley & Sons Singapore Pte. Ltd.

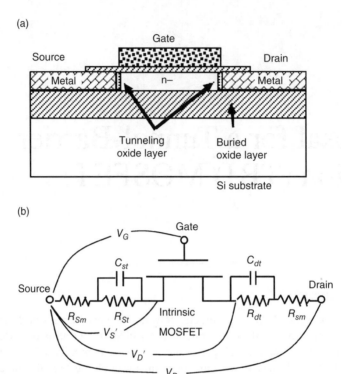

Figure 28.1 Schematic of device cross-section. (a) Schematic device structure, and (b) equivalent circuit of TBJ MOSFET. Reproduced by permission of The Japan Society of Applied Physics (2001). Yasuhisa Omura, "A Tunneling-Barrier Junction SOI MOSFET with a Suppressed Short-Channel Effect for the Ultimate Device Structure," 2001 Silicon Nanoelectronics Workshop (Kyoto, 2001), pp. 58–59.

This chapter, as a first step, examines only the direct tunneling current through thin oxide layers at the source and the drain; layer thicknesses are denoted by t_{st} and t_{dt}, respectively.

The equivalent circuit for the tunneling-barrier junction metal oxide semiconductor field-effect transistor (TBJ MOSFET) is shown in Figure 28.1b. The source and drain regions feature the parallel elements of the effective capacitance of the thin oxide layers ($C_{st} = \varepsilon_{ox}/t_{st}$ and $C_{dt} = \varepsilon_{ox}/t_{dt}$) and the tunneling resistance (R_{st} and R_{dt}). R_{sm} and R_{dm} are the resistances of the metal contacts at the source and drain junctions, respectively. Since the gate oxide layer thickness (t_{ox}) is much smaller than the buried oxide layer thickness (t_{BOX}) and the gate voltage (V_G) is low, the influence of the effective substrate bias on channel potential (V) is neglected here for simplicity.

28.3 Calculation Results

The following simulations assume the direct tunneling mechanism for modeling the current across the source and drain junctions, and that the work function difference between the metal and the silicon is very small. From this, tunneling currents R_{st} and R_{dt} are estimated for each applied voltage, where the resistances are assumed to be the same. We have to solve Eq. (28.1) by means of iteration:

$$
\begin{aligned}
I_D &= \left(\frac{W_{eff}}{L_{eff}}\right)\mu_n \left\{qN_D t_S \left[V_D - \left(R_{st} + R_{dt} + R_{sm} + R_{dm}\right)I_D\right]\right\} \\
&+ \left(\frac{W_{eff}}{L_{eff}}\right)\mu_n C_{ox} \left\{V_G - V_{FB} - (1/2)\left[V_D - \left(R_{dt} + R_{dm}\right)I_D\right]\right\}\left[V_D - \left(R_{dt} + R_{dm}\right)I_D\right] \\
&- \left(\frac{W_{eff}}{L_{eff}}\right)\mu_n C_{ox} \left\{V_G - V_{FB} - (1/2)\left(R_{st} + R_{sm}\right)I_D\right\}\left(R_{st} + R_{sm}\right)I_D
\end{aligned}
\tag{28.1}
$$

where V_{FB} is the flat-band voltage, μ_n is the electron mobility, and C_{ox} is the gate oxide capacitance per unit area. This equation is made by equating the metal oxide semiconductor (MOS) channel current (drift current is assumed) to the tunnel current across the source and drain tunnel barrier junctions. Typical simulation results of the drain current (I_D) versus the drain voltage (V_D) characteristics are shown in Figure 28.2a. Here, an n-type poly-Si gate structure is assumed. The TBJ MOSFET shows the usual MOSFET-like I_D vs. V_D characteristics.

In Figure 28.2a, I_D is insensitive to the channel length (L_{eff}) within the range 20 to 5 nm, to the body doping concentration (N_D) within the range 10^{18} to 10^{14} cm^{-3}, and to the gate oxide layer thickness (t_{ox}) within the range 10 to 2.5 nm. I_D is sensitive to t_{st} and t_{dt} within the range 1 to 0.3 nm, as is shown in Figure 28.2b. These features are reasonable given the high tunneling resistances of the source and drain junctions. Significant issues related to device performance are discussed below.

As shown in Figure 28.2b, I_D is of the order of 10^{-4} A (10^{-5} A for $W_{eff}/L_{eff}=1$), where W_{eff} is the channel width. This value is much smaller than those of usual MOSFETs, but much larger than those of single-electron transistors ($\sim 10^{-9}$ A). However, we must consider the case that the ultimate device structure would have a thin-SOI structure with a silicon layer of 10 nm or less. In that case, the total parasitic resistance of the 10 nm thick source and drain regions would be about 30 kΩ for silicon, or at least 1 kΩ, even for partial-silicide. As the supply voltage must be limited to below 1 V in conventional MOSFETs with sub-50 nm channel length to prevent direct tunneling between pn junctions of source and drain, its drain current will be much less than 10^{-4} A for $W_{eff}/L_{eff}=1$.

Here, it is assumed that a metal Schottky junction structure is not realistic because the drain depletion layer enhances the SCE. Figure 28.2b indicates that I_D of the TBJ MOSFET easily reaches the order of 10^{-3} A (10^{-4} A for $W_{eff}/L_{eff}=1$) by means of device design techniques such as thickness control of the tunneling oxide layer, which suggests that the drivability of TBJ MOSFET will exceed that of the ultimate structure of the conventional MOSFET.

The reality of such a thin tunneling oxide layer has been examined by using the partial-pressure oxidation technique; the tunneling current density of 2 nm thick partial-pressure oxide, formed by the rapid-thermal process, can be 100 times that of the usual oxide layer [11, 12]. The advantage of the partial-pressure oxidation is that it yields a lower tunneling resistance despite its thicker oxide layer, which is important from the viewpoint of thickness control.

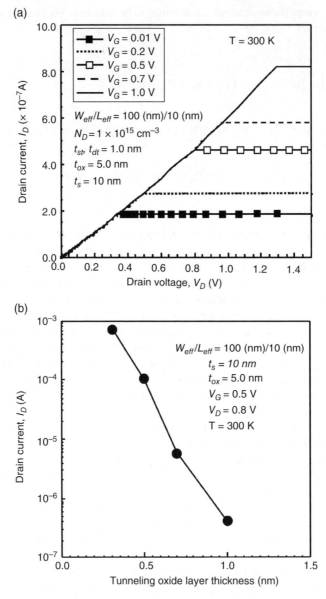

Figure 28.2 Drain current characteristics of TBJ MOSFET. (a) Calculated I_D vs. V_D characteristics of TBJ MOSFET with a 10 nm long channel and (b) drain current dependence on thickness of SiO_2 film. Reproduced by permission of The Japan Society of Applied Physics (2001). Yasuhisa Omura, "A Tunneling-Barrier Junction SOI MOSFET with a Suppressed Short-Channel Effect for the Ultimate Device Structure," 2001 Silicon Nanoelectronics Workshop (Kyoto, 2001), pp. 58–59.

28.4   Summary

This chapter proposed the TBJ MOSFET on the SOI substrate, and its fundamental characteristic was elucidated by simplified calculations. Detail is discussed in Chapter 30.

References

[1] Y. Nakajima, Y. Takahashi, S. Horiguchi, K. Iwadate, H. Namatsu, K. Kurihara, and M. Tabe, "Quantized conductance of a silicon wire fabricated by separation-by-implanted-oxygen technology," *Jpn. J. Appl. Phys.*, vol. 34, pp. 1309–1314, 1995.

[2] Y. Takahashi, M. Nagase, H. Namatsu, K. Kurihara, K. Iwadate, Y. Nakajima, S. Horiguchi, K. Murase, and M. Tabe, "Conductance Oscillations of a Si Single Electron Transistor at Room Temperature," Ext. Abstr. IEEE 1994 Int. Electron Devices Meeting (San Francisco, Dec., 1994) pp. 938–940, 1994.

[3] Y. Omura, S. Horiguchi, M. Tabe, and K. Kishi, "Quantum-mechanical effects on the threshold voltage of ultrathin-SOI nMOSFETs," *IEEE Electron Device Lett.*, vol. 14, no. 12, pp. 569–571, 1993.

[4] Y. Omura and M. Shoji, Jpn. Pat. Pend., No. H4-207001, 1992.

[5] H. Namatsu, S. Horiguchi, Y. Takahashi, M. Nagase, and K. Kurihara, "Fabrication of $SiO_2/Si/SiO_2$ double barrier diodes using two-dimensional Si structures," *Jpn. J. Appl. Phys.*, vol. 36, pp. 3669–3674, 1997.

[6] N. Matsuo, T. Miura, H. Hamada, and T. Miyoshi, "Application of tunneling effect for thin dielectric film to MOS transistor," *Trans. IEICE C*, vol. J-81-C-II, pp. 266–267, 1998 (in Japanese).

[7] Y. Omura, "Quantum-Effect Devices on SOI Substrates with an Ultrathin Silicon Layer," Proc. of NATO Advanced Research Workshop (Ukraina, Oct., 1998, Kluwer Academic Publisher) pp. 257–268, 1999.

[8] K. Nakazato, R. J. Blaikie, J. R. A. Cleaver, and H. Ahmed, "Single-electron memory," *Electron. Lett*, vol. 29, pp. 384–385, 1993.

[9] M. Wagner, H. Mizuta, and K. Nakazato, "A Fast Three-Dimensional MC Simulator for Tunneling Diodes," 2000 Int. Conf. Simulation of Semiconductor Processes and Devices (SISPAD), pp. 31–33, 2000.

[10] K. Morita, K. Morimoto, H. Sorada, K. Araki, K. Yuki, M. Niwa, T. Uenoyama, and K. Ohnaka, "Si Interband Tunnel Diode through a Thin Oxide with a Degenerate poly-Si Electrode," 1997 International Workshop on Quantum Function Devices, pp. 175–176, 1997.

[11] H. Nakatsuji, Y. Kotani, and Y. Omura, "Significant Aspects of Direct Tunneling Current in Thin-Oxide Films Formed by Partial-Pressure Rapid-Thermal Oxidation," Abstr. of 1999 Silicon Nanoelectron. Workshop (IEEE and JSAP, Kyoto, June, 1999) pp. 70–71, 1999.

[12] K. Nanjo and Y. Omura, "Consideration on theoretical model to describe growth process of thin silicon oxide films," *Recent Res. Dev. Electrochem.* (Transworld Research Network), vol. 8, pp. 1–24, 2005.

29

Performance Prediction of SOI Tunneling-Barrier-Junction MOSFET

29.1 Introduction

In order to develop devices with a high performance, device dimensions are being reduced to below the sub-100 nm scale. Reduction of channel length, however, leads to a drastic increase in the off-leakage current of conventional metal oxide semiconductor field-effect transistors (MOSFETs) because of short-channel effects (SCEs) [1]. Short-channel effects including an increment of subthreshold swing are important issues for device miniaturization [1]. A silicon-on-insulator (SOI) structure moderately suppresses the degradation of device performance due to the above influences. For SOI devices, off-leakage current resulting from the drain-induced barrier lowering (DIBL) is generally decreased because a thin buried-oxide layer shuts the lateral penetration of the drain-induced electrical field [2]; the substrate terminates most of the electric field flux from the drain.

However, in conventional short-channel SOI devices having pn junctions or hi-lo junctions, an energy barrier to suppress the off-leakage current is inevitably lowered by the built-in and external potentials across source and drain junctions. In the case of high-temperature operations, the off-leakage current should be quite notable because electrons in source electrodes have a higher thermal energy and also intrinsic carrier concentration increases exponentially with temperature. The tunneling-barrier junction (TBJ) MOSFET has been proposed as a device that can suppress various off-leakage current processes as well as short-channel effects [3, 4]; the drain current characteristics are briefly estimated on the basis of the simplified equivalent circuit model. However, features of operation characteristics have not yet been studied in detail.

In this chapter the transport characteristics of tunneling-barrier junction metal oxide semiconductor field-effect transistor (TBJ MOSFET) are discussed. Under the assumption of

MOS Devices for Low-Voltage and Low-Energy Applications, First Edition.
Yasuhisa Omura, Abhijit Mallik, and Naoto Matsuo.
© 2017 John Wiley & Sons Singapore Pte. Ltd. Published 2017 by John Wiley & Sons Singapore Pte. Ltd.

ballistic transport in the device, we simulate device characteristics with the aid of a commercial two-dimensional device simulator [5]. The transmission probability of electrons between the source and drain is calculated by the transfer matrix method [6]. We also optimize the device structure to increase the drive current for practical use. We also discuss low-temperature operation and threshold voltage fluctuation.

29.2   Simulation Model

The schematic cross-sectional view of TBJ MOSFET is shown in Figure 29.1a. The gate oxide thickness and the SOI layer thickness of the simulated devices are 5 and 10 nm, respectively. The silicon layer thickness is chosen to avoid quantum mechanical influences perpendicular

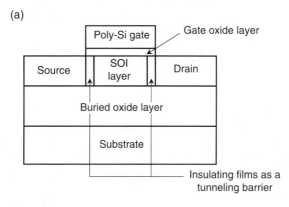

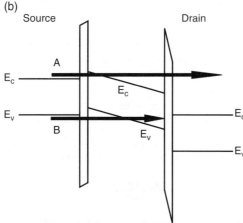

Figure 29.1   Device structure and band scheme for transport phenomena. (a) Schematic device structure. (b) Band scheme for transport phenomena. Here, the tunneling current component denoted by label "A" is calculated. Reproduced by permission of the Japan Society of Applied Physics (2003). Hidehiko Nakajima, Akira Kawamura, Kenji Komiya, and Yasuhisa Omura, "Simulation models for silicon-on-insulator tunneling-barrier-junction metal-oxide-semiconductor field-effect transistor and performance perspective," Jpn. J. Appl. Phys., vol. 42, pp. 1206–1211, 2003.

to the gate oxide layer/silicon layer interface [7]; the difference in quantized energy levels is much less than the thermal energy. The thick gate oxide film is selected to suppress gate leakage current. The 100 nm thick buried oxide layer is assumed to suppress short channel effects by enhancing the SOI-to-substrate electric field. The doping level of the n-type SOI layer and n-type substrate is 1×10^{15} cm^{-3}. The gate length (L_G) is 10 nm. In simulations, the tunneling insulator thickness ranges from 1.0 to 0.3 nm, and its material is assumed to be SiO$_2$ or Ta$_2$O$_5$. Therefore, the effective channel length (L_{eff}) ranges from 8 to 9.4 nm. Matsuo *et al.* simulated the characteristics of a similar device with a 50 nm long channel [8], although the proposed device is not an SOI structure. When the channel is long, the drain current oscillates as a function of gate bias in a resonant tunneling manner as discussed in [7]. However, such oscillation characteristics are usually undesirable because a flat current characteristic is used in many circuits. Thus, we assume a very short channel in order to avoid the oscillation characteristics; the short channel length makes the energy difference of quantized levels wide. The most significant feature of the operation characteristics of TBJ MOSFET is direct tunneling across the thin insulators between the source/drain diffusion and the channel region. As these tunneling layers are expected to suppress the undesired thermal current, off-leakage current must be significantly decreased in comparison to that of conventional devices.

Simulation flow of drain current estimation is shown in Figure 29.2. First, the internal potential profile is calculated by using the commercial 2-D device simulator [5]. As the simulation does not reflect the tunneling current across the tunneling barrier junctions at this step, the calculated potential profile seems to be slightly impractical. However, the drain current (I_D) level is very low, as will be shown later. Thus, the calculated potential is sufficiently practical in this study. At the second step, transmission probability of electrons from the source to the drain electrode is calculated by the transfer matrix method using the potential profile obtained by the two-dimensional device simulator. The internal potential profile inducing the short-channel effect is determined by the two-dimensional device simulator with

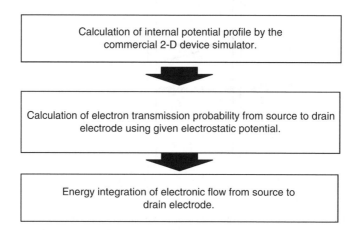

Figure 29.2 Simulation flow of drain current estimation. Reproduced by permission of the Japan Society of Applied Physics (2003). Hidehiko Nakajima, Akira Kawamura, Kenji Komiya, and Yasuhisa Omura, "Simulation models for silicon-on-insulator tunneling-barrier-junction metal-oxide-semiconductor field-effect transistor and performance perspective," Jpn. J. Appl. Phys., vol. 42, pp. 1206–1211, 2003.

the drift-diffusion model because of a low current level, so the short-channel effect or its suppression is successfully reflected on the drain current simulation results. When the buried oxide layer is thick, the drain potential easily penetrates into the buried oxide later, which enhances the short-channel effects [9]. In the case of a thin buried oxide layer, the drain-induced electric field almost terminates the substrate [10]. These physics are stringently reflected on the simulated potential profile. Finally, the drain current (I_D) is estimated by energy integration of electronic flow. The drain current (I_D) is given by [11]

$$
\begin{aligned}
I_D &= \frac{2qW}{h}\int dz \int \frac{d^2k_t}{(2\pi)^2}\int dE_l D\big(E_l,V(V_G,V_D)\big)\big[f_{FD}(E_l+E_t)-f_{FD}(E_l+E_t+qV_D)\big] \\
&= \frac{4\pi qm^* k_B TW}{h^3}\int_0^{t_{si}}\int_{E_{min}}^{E_{max}} dE_l D\big(E_l,V(V_G,V_D)\big)\Bigg[\ln\left\{\left(1+\exp\left(\frac{E_F-E_l}{k_B T}\right)\right)\Big/\left(1+\exp\left(\frac{E_F-E_{max}}{k_B T}\right)\right)\right\} \\
&\quad -\ln\left\{\left(1+\exp\left(\frac{E_F-E_l-qV_D}{k_B T}\right)\right)\Big/\left(1+\exp\left(\frac{E_F-E_{max}-qV_D}{k_B T}\right)\right)\right\}\Bigg]
\end{aligned} \tag{29.1}
$$

where q is the elementary charge of electrons, h is the Planck's constant, k_t is the transverse wave number component of tunneling electrons at the source/drain junction, E_l is the longitudinal energy component of tunneling electrons at the source/drain junction, E_t is the transverse energy component of tunneling electrons at the source/drain junction, $D(E_l, V(V_G,V_D))$ is the transmission probability of electrons between the source and the drain, $V(V_G,V_D)$ is the potential, V_G is the gate voltage, V_D is the drain voltage, $f_{FD}(E_l+E_t)$ and $f_{FD}(E_l+E_t+qV_D)$ are the Fermi–Dirac distribution functions, m^* is the density-of-state effective mass of electrons in silicon, E_{max} and E_{min} are the top and the bottom values in energy integration, E_F is the Fermi level energy at the source electrode, k_B is the Boltzmann's constant, T is the absolute temperature, W is the gate width, and the z-axis is perpendicular to the gate oxide/SOI layer interface.

In the present simulation, it is assumed that the drain voltage (V_D) is at most 1 V. Thus, the source-to-drain electron tunneling through the valence band of the SOI layer, denoted by label "B" in Figure 29.1b, is not taken into account; the tunneling current component denoted by label "A" is calculated by Eq. (29.1). It is also assumed that there is no gate tunneling current because the gate oxide is sufficiently thick. In addition, it is assumed that there is no scattering event of electrons traveling in the SOI layer; all electrons show ballistic conduction. As the channel length of the present device is on the sub-10 nm scale and it is considered that the mean free path of electrons is a few nanometers long [12], we think that the assumption of no scattering event in the SOI layer is valid. Energy levels of the conduction band in the channel region are quantized because the channel is very short, so distinct resonant levels should be manifested there. Figure 29.3a shows how the transmission probability, $D(E_l, V_G)$, depends on the longitudinal energy component of electrons at the source region; E_l is measured from the conduction band bottom of the source electrode, and $E_F - E_c = 6$ meV. Simulations are carried out at V_G of 1 V and V_D of 1 V. Labels $E_{l,1}$, $E_{l,2}$, $E_{l,3}$, and $E_{l,4}$ indicate resonant levels. The energy band diagram along the channel and resonant levels at V_G of 1 V and V_D of 1 V are both shown in Figure 29.3b. They are extracted from the simulation results of the potential profile.

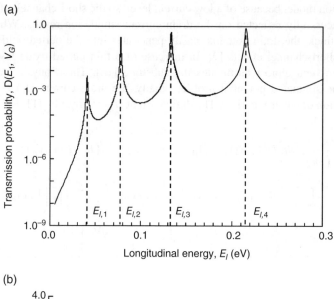

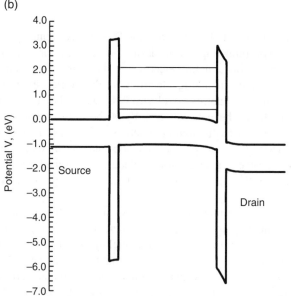

Figure 29.3 Dependence of transmission probability, $D(E_l, V(V_G, V_D))$, on longitudinal energy of electrons, E_l, at the source electrode. (a) $D(E_l, V(V_G, V_D))$ versus E_l characteristics at V_G of 1 V and V_D of 1 V. (b) Band diagram along the channel extracted from the simulated potential profile and resonant levels at V_G of 1 V and V_D of 1 V. The potential in eV is measured from the conduction band bottom of the source electrode. Reproduced by permission of the Japan Society of Applied Physics (2003). Hidehiko Nakajima, Akira Kawamura, Kenji Komiya, and Yasuhisa Omura, "Simulation models for silicon-on-insulator tunneling-barrier-junction metal-oxide-semiconductor field-effect transistor and performance perspective," Jpn. J. Appl. Phys., vol. 42, pp. 1206–1211, 2003.

It can be seen in Figure 29.3a and b that the lowest resonant level is much higher than the Fermi level of the source electrode. Thus, the oscillation of drain current is not observed in the I_D versus V_G characteristics as shown later because very few electrons flow through the resonant levels. Drain current begins to flow saliently when the Fermi level of the source electrode exceeds the conduction band bottom of the channel region at a certain high V_G. In this study, we calculate I_D on the basis of the quantum wave mechanics. The influence of scattering-limited conduction in the SOI layer is not discussed here.

29.3 Simulation Results and Discussion

29.3.1 Fundamental Properties of TBJ MOSFET

In this section, the fundamental current-voltage characteristics of TBJ MOSFET are discussed in detail. It is assumed here that the source/drain electrode material is n+ Si, tunneling barrier material is SiO_2, and tunneling barrier is 1 nm thick. In Figure 29.4a and b, I_D-V_G and I_D-V_D characteristics of TBJ MOSFET are shown, respectively, where V_G is the gate voltage. Figure 29.4a shows that the subthreshold (off-leakage) current of TBJ MOSFET is much lower than that of conventional MOSFET as expected. Transmission probability of electrons is much smaller than unity when there is a potential barrier along the electron conduction path, which results in suppression of off-leakage current. Drive current (on-current) is, however, also decreased simultaneously as shown in Figure 29.4a.

The transmission probability is calculated by the transfer matrix method, so the wave nature is precisely reflected on to electron conduction and resonant levels appear in the SOI layer. In resonant tunneling transistors, conduction current usually oscillates as a function of applied voltage. However, such current oscillation is not observed in the TBJ MOSFET. As the channel confined with two barriers is so short, the lowest resonant level is quite far from the conduction band bottom. When V_G is positive, the accumulation layer of electrons is firm at the gate oxide layer/silicon layer interface. In this case, the surface potential in the channel region is small and hardly depends on V_G. The incremental electron density is proportional to V_G, so the drain current increases linearly with V_G. Thus, the lowest resonant level does not exceed the Fermi level of the source electrode in the range of simulated gate voltages; in other words, major electrons in the source electrode travel to the drain electrode without passing through the resonant level. In Figure 29.4, the subthreshold swing (SS) is 70 mV/dec at V_D of 1 V, and both the swing and the drain current (I_D) slightly depend on V_D even in the 10 nm long channel device. The g_m value of the device is 2.4×10^{-10} S/µm at V_D of 1 V. TBJ MOSFET has an excellent switching property and reveals a smaller short-channel effect, although its drivability should be improved.

29.3.2 Optimization of Device Parameters and Materials

As shown in the previous section, the off-leakage current of TBJ MOSFET is suppressed in comparison to that of conventional silicon-on-insulator metal oxide semiconductor field-effect transistors (SOI MOSFETs) [13] as expected. However, drive current is also reduced simultaneously. Thus, this section considers how to increase drive current with low off-leakage current.

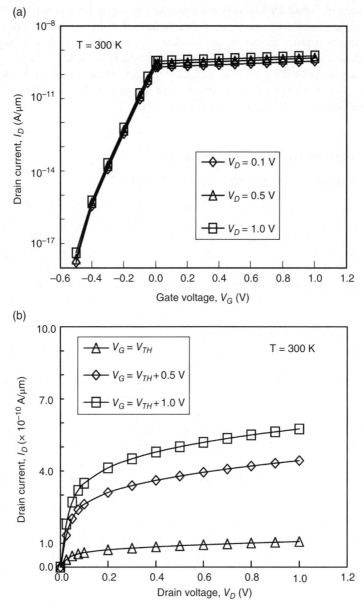

Figure 29.4 Transport characteristics of TBJ MOSFET. It is assumed that the source/drain electrode is n+ Si, and that the tunneling barrier is a 1 nm thick SiO_2 film. (a) I_D–V_G characteristics. (b) I_D–V_D characteristics. Reproduced by permission of the Japan Society of Applied Physics (2003). Hidehiko Nakajima, Akira Kawamura, Kenji Komiya, and Yasuhisa Omura, "Simulation models for silicon-on-insulator tunneling-barrier-junction metal-oxide-semiconductor field-effect transistor and performance perspective," Jpn. J. Appl. Phys., vol. 42, pp. 1206–1211, 2003.

29.3.2.1   Influence of Source/Drain Electrode Material

Figure 29.5a and b shows simulation results of I_D-V_G and I_D-V_D characteristics for aluminum or an n$^+$ Si source/drain electrode when the tunneling insulator is SiO$_2$ and its thickness is 1 nm. The drive current of the device with aluminum source/drain electrodes is larger than that of the device with n$^+$ Si source/drain electrodes. This can be explained by considering the on-state band diagram. Since the work function of n$^+$ Si with the impurity density of 10^{20} cm^{-3} is slightly larger than that of aluminum [2], the Fermi level of the aluminum source electrode is closer to the first resonant level at low gate voltages beyond the threshold. The aluminum source/drain electrode raises the electron transmission probability between the source and the drain electrode at low gate voltages.

In addition, the energy range of incident electrons at the source electrode should be noted. For an n$^+$ Si electrode, the energy range of incident electrons is of the order of millielectron volts. For an aluminum electrode, however, the energy range is of the order of tens of millielectron volts. As a consequence, the drive current for the aluminum electrode is larger than that for the n$^+$ Si electrode, as seen in Figure 29.5. However, the modern fabrication technology often uses Si material. Thus, we use n$^+$ Si as a source/drain electrode material hereafter.

29.3.2.2   Influence of Tunneling Barrier Thickness

Drive current is proportional to transmission probability between the source and the drain electrodes. Thus, in order to step up the drive current, the transmission probability should be increased somehow. Thinning the barrier is one way to improve the drive current. Reduction of tunneling barrier thickness exponentially raises the transmission probability, so this is a powerful method to achieve our purpose.

Figure 29.6a and b shows I_D-V_G and I_D-V_D characteristics when the tunneling barrier material is SiO$_2$ and source/drain electrode material is n$^+$ Si. The tunneling barriers are 1 or 0.3 nm thick. In Figure 29.6, the drive current for the 0.3 nm thick tunneling barrier is considerably larger than that for 1 nm thick tunneling barrier. The drive current of the device with a 0.3 nm thick tunneling barrier, however, is still two orders smaller with identical bias conditions than that of the conventional MOS device [13]. Thus, further improvement of device structure is required to apply TBJ MOSFETs to practical modern circuitry.

Usually, the film thickness of 0.3 nm is not a realistic range because it is a monolayer. Although a rectangular potential barrier is assumed in the simulation for simplicity, generally speaking, the assumption is not acceptable because the bulky band structure is not formed in such a thin film; we cannot easily determine the potential barrier of the film. Thus, this simulation has been carried out to estimate the possible drain current level. In the following sections, this is the aim of simulations. In practice, the monolayer oxide film can be fabricated inside a molecular beam epitaxy (MBE) system [14]; a monocrystalline silicon film has been successfully formed on the thin oxide film.

29.3.2.3   Influence of Tunneling Barrier Material

In the previous sections, a couple of methods to increase the drive current have been discussed. However, the drain current of the device is ruled by the tunneling effect across the barrier, which seems to suggest the limitation of improvement of drivability. Thus, we try to lower the barrier height for electrons; this should lead to the increase in drive current and also to the

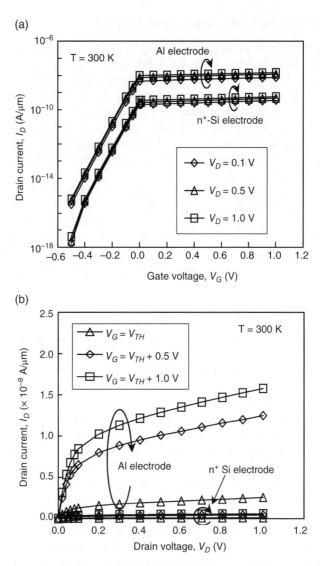

Figure 29.5 Transport characteristics of TBJ MOSFET. It is assumed that the source/drain electrode is n⁺ Si or aluminum, and that the tunneling barrier is a 1 nm thick SiO_2 film. (a) I_D–V_G characteristics. (b) I_D–V_D characteristics. Reproduced by permission of the Japan Society of Applied Physics (2003). Hidehiko Nakajima, Akira Kawamura, Kenji Komiya, and Yasuhisa Omura, "Simulation models for silicon-on-insulator tunneling-barrier-junction metal-oxide-semiconductor field-effect transistor and performance perspective," Jpn. J. Appl. Phys., vol. 42, pp. 1206–1211, 2003.

increase in undesired thermal current. Here, we replace a SiO_2 layer as a high barrier material with a Ta_2O_5 film as a low barrier material. The material parameters of Ta_2O_5 are as follows: the permittivity is $25\varepsilon_0$ F/m [15], the barrier height for electrons is $0.3\,eV$ [15] and the effective mass of tunneling electrons is $0.3\,m_0$, where ε_0 is the permittivity in vacuum and m_0 is the rest mass of an electron.

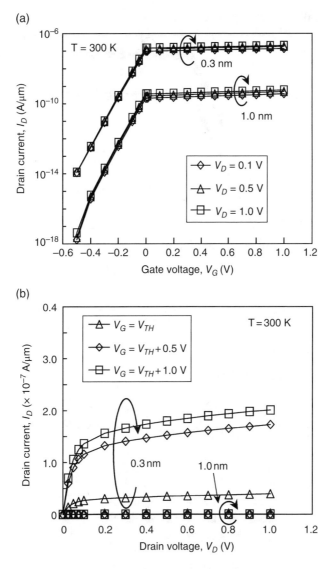

Figure 29.6   Transport characteristics of TBJ MOSFET. It is assumed that the source/drain electrode is n⁺ Si, and that the tunneling barrier is a 0.3 nm or 1 nm thick SiO_2 film. (a) I_D–V_G characteristics. (b) I_D–V_D characteristics. Reproduced by permission of the Japan Society of Applied Physics (2003). Hidehiko Nakajima, Akira Kawamura, Kenji Komiya, and Yasuhisa Omura, "Simulation models for silicon-on-insulator tunneling-barrier-junction metal-oxide-semiconductor field-effect transistor and performance perspective," Jpn. J. Appl. Phys., vol. 42, pp. 1206–1211, 2003.

Figure 29.7a and b shows I_D-V_G and I_D-V_D characteristics, where a 0.3 nm thick SiO_2 tunneling barrier is used, and the source/drain electrode material is n⁺ Si; the tunneling barrier material is SiO_2 or Ta_2O_5. The drive current of the device with a 0.3 nm thick Ta_2O_5 barrier is about one-order larger than that of the device with a 0.3 nm thick SiO_2 barrier. Moreover,

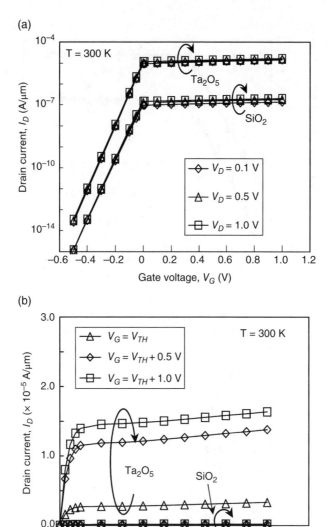

Figure 29.7 Transport characteristics of TBJ MOSFET. It is assumed that the source/drain electrode is n⁺ Si, and that the 0.3 nm thick tunneling barrier is a SiO_2 film or a Ta_2O_5 film. (a) I_D–V_G characteristics. (b) I_D–V_D characteristics. Reproduced by permission of the Japan Society of Applied Physics (2003). Hidehiko Nakajima, Akira Kawamura, Kenji Komiya, and Yasuhisa Omura, "Simulation models for silicon-on-insulator tunneling-barrier-junction metal-oxide-semiconductor field-effect transistor and performance perspective," Jpn. J. Appl. Phys., vol. 42, pp. 1206–1211, 2003.

subthreshold swing is 60 mV/dec at V_D of 1 V when a Ta_2O_5 barrier is used. This swing value is close to the theoretical limit of conventional metal oxide semiconductor (MOS) devices. Thus, a TBJ MOSFET with a Ta_2O_5 film as a tunneling barrier has excellent switching characteristics.

In this study, we have optimized the device structure to obtain a high drive current and a low off-leakage current. Although the off-leakage current of TBJ MOSFET is more suppressed than that of conventional devices as expected, the drive current of TBJ MOSFET does not yet reach the level of conventional devices. We consider that one of the reasons results from the no-scattering conduction in the SOI layer: that is, the electron conduction in the SOI layer is stringently ruled by wave mechanics. Thus, the transmission probability of incidental electrons having an energy below the resonant level in the SOI layer is very small. The drive current is still low because many electrons cannot pass through the resonant level at the assumed gate and drain voltages.

Therefore, a significant way to increase the drive current is to generate many allowed conductive "channels" in the SOI layer; in other words, inelastic scattering processes in the SOI layer yield numerous "channels" of conductive electrons. However, this is not discussed here because the purpose of this chapter is to demonstrate the fundamental performance of TBJ MOSFET.

29.3.2.4 Low-Temperature Characteristics

In this section, the characteristics of TBJ MOSFET at low temperatures are discussed. Figure 29.8a and b shows I_D-V_G and I_D-V_D characteristics at $T=1$ and $300\,K$ at V_D of $1\,V$, respectively. It is assumed that the tunneling barrier material is Ta_2O_5, that the tunneling barrier is $0.3\,nm$ thick, and that the source/drain electrode material is n^+ Si. As shown in Figure 29.8, the drain current at $T=1\,K$ is seven orders smaller than that at $T=300\,K$. Electrons have a very low thermal energy at $1\,K$, so almost no electrons have an energy value higher than the Fermi level on the basis of the Fermi–Dirac distribution function.

The transmission probability of major electrons is much smaller than that at room temperature because the Fermi level in the source electrode is still far from the lowest resonant level. Thus, it is considered that drain current is quite low at $1\,K$. Figure 29.8 also shows that the drain current does not saturate, but increases with the gate bias at $1\,K$. This is due to the fact that the transmission probability of electrons increases abruptly when the Fermi level exceeds the lowest resonant level.

29.3.2.5 Impact of Impurity Density Variation in SOI Layer

As is often discussed for conventional short-channel MOS devices, local variation of impurity density in the channel region affects the transport characteristics, such as threshold voltage and conductance [16]. In conventional MOS devices, the threshold voltage shows a high sensitivity to the impurity density variation, which is an issue for future integrated circuits.

We simulated how the impurity density in the SOI layer changes the threshold voltage. Impurity density was varied from 10^{15} to $10^{20}\,cm^{-3}$ for n-type and p-type silicon, respectively. As a result, a notable threshold voltage dependence on impurity density in SOI layer was not observed; it was at most $1\,mV$. Since the Fermi level of the SOI layer is pinned at a certain potential level due to the work function difference between the SOI layer and the source/drain electrodes, which resembles a punch-through phenomenon. When $V_G \sim 0\,V$, the conduction band edge at the gate oxide/SOI layer interface goes under the Fermi level

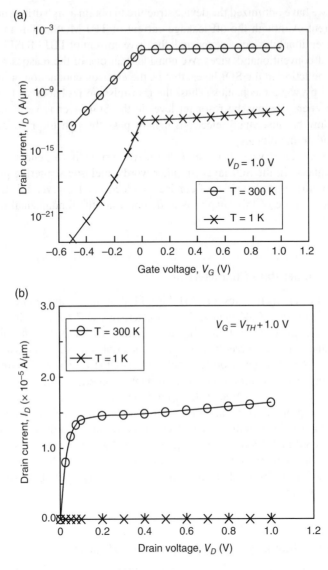

Figure 29.8 Transport characteristics of TBJ MOSFET. It is assumed that the source/drain electrode is n⁺ Si, and that the tunneling barrier is a 0.3 nm thick Ta_2O_5 film. Simulated temperature is 1 or 300 K. (a) I_D–V_G characteristics. (b) I_D–V_D characteristics. Reproduced by permission of the Japan Society of Applied Physics (2003). Hidehiko Nakajima, Akira Kawamura, Kenji Komiya, and Yasuhisa Omura, "Simulation models for silicon-on-insulator tunneling-barrier-junction metal-oxide-semiconductor field-effect transistor and performance perspective," Jpn. J. Appl. Phys., vol. 42, pp. 1206–1211, 2003.

in the source electrode; thus, the threshold voltage (V_{TH}) is almost equal to 0 V. In all the simulations presented in this chapter, the threshold voltage is about 0 V regardless of the device parameters, which is also due to the work function difference between the SOI layer and the source/drain electrodes.

29.4 Summary

We simulated room-temperature drain current characteristics of TBJ MOSFET on the basis of wave mechanics. It was demonstrated that subthreshold swing and off-leakage current of the device were much smaller than those of modern MOSFETs; short-channel effects were sufficiently suppressed. However, the tunneling barrier limited the drive current simultaneously. By possible optimization of the device parameters, a considerably high drive current was obtained when the tunneling barrier material was a 0.3 nm thick Ta_2O_5 film and an aluminum source/drain electrode was used. Although the drive current was still small for practical use, even after possible optimization of device parameters, it can be expected that the drain current should be improved by introducing inelastic scattering processes in the SOI layer. This is an issue for future discussion.

References

[1] J. P. Colinge, "Silicon-on-Inculator Technology: Materials to VLSI," 2nd ed., (Kluwer Academic Publishers, 1997).

[2] S. M. Sze, "Physics of Semiconductor Devices," 2nd ed., (John Wiley & Sons, Inc., 1981).

[3] Y. Omura, "Quantum Effect Devices on SOI Substrates with an Ultrathin Silicon Layer," in P. L. F. Hemment (ed.), "Perspective, Science and Technology for Novel SOI Devices," (Kluwer Academic Publishers, 2000), pp. 257–268.

[4] Y. Omura, "A Tunneling-Barrier Junction MOSFET on SOI Substrates with a Suppressed Short-Channel Effect for the Ultimate Device Structure," 10th Int. Symp. on SOI Technol. Dev. (The Electrochem. Soc., Washington DC, 2001) PV2001-3, pp. 451–456, 2001.

[5] Synopsys, "Sentaurus Device User Guide," Version A-2007, 2007, ftp://147.46.117.90/synopsys/manuals/PDFManual/data/sdevice_ug.pdf (accessed June 10, 2016).

[6] E. Merzbacher, "Quantum Mechanics," 3rd ed. (John Wiley & Sons, Inc., 1998).

[7] Y. Omura, S. Horiguchi, M. Tabe, and K. Kishi, "Quantum-mechanical effects on the threshold voltage of ultrathin-SOI nMOSFET's," *IEEE Electron Device Lett.*, vol. 14, no. 12, pp. 569–571, 1993.

[8] N. Matsuo, J. Yamaguchi, Y. Kitagawa, H. Hamada, T. Miura, and T. Miyoshi, "Extension of physical limit of conventional metal-oxide-semiconductor transistor by double barriers formed at the channel edges," *Jpn. J. Appl. Phys.*, vol. 39, pp. 3850–3853, 2000.

[9] Y. Omura, S. Nakashima, K. Izumi, and T. Ishii, "0.1-μm-Gate, Ultrathin-Film CMOS Devices Using SIMOX Substrate with 80-nm-Thick Buried Oxide Layer," 1991 IEEE Int. Electron Devices Meeting, Tech. Dig., pp. 675–678, 1991.

[10] S. Yanagi, A. Nakakubo, and Y. Omura, "Proposal of a partial-ground-plane (PGP) silicon-on-insulator (SOI) MOSFET for deep sub-0.1-um channel regime," *IEEE Electron Device Lett.*, vol. 22, no. 6, pp. 278–280, 2001.

[11] D. K. Ferry, "Quantum Mechanics: An Introduction for Device Physists and Electrical Engineers," (IOP Publishing, 1995).

[12] M. Wagne0072, H. Mizuta, and K. Nakazato, "A Fast Three-Dimensional MC Simulator for Tunneling Diodes," 2000 Int. Conf. Simulation of Semiconductor Processes and Devices (SISPAD), pp. 31–33, 2000.

[13] H. V. Meer and K. D. Meyer, "The spacer/replacer concept: a viable route for sub-100 nm ultrathin-film fully-depleted SOI CMOS," *IEEE Electron Device Lett.*, vol. 23, pp. 46–48, 2002.

[14] T. J. Maloney, "MBE growth and characterization of single crystal silicon oxides on (111) and (100) silicon," *J. Vac. Sci. Technol., B*, vol. 1, pp. 773–777, 1983.

[15] J. Robertson, "Band offsets of wide-band-gap oxides and implications for future electronic devices," *J. Vac. Sci. Technol., B*, vol. 18, pp. 1785–1791, 2000.

[16] M. J. Sherony, L. T. Su, J. E. Chung, and D. A. Antoniadis, "Reduction of threshold voltage sensitivity in SOI MOSFET's," *IEEE Electron Device Lett.*, vol. 16, pp. 100–102, 1995.

30

Physics-Based Model for TBJ-MOSFETs and High-Frequency Performance Prospects

30.1 Introduction

Extensive research efforts have been directed towards the development of design guidelines for the next generation of metal oxide semiconductor field-effect transistors (MOSFETs) and other devices [1–3]. In particular, the silicon-on-insulator (SOI) device structure, including the double-gate SOI structure, is attracting interest because it offers a lower short-channel effect (SCE) [4] and the possibility of investigating quantum effect devices (QEDs) [5–7], which have not yet been comprehensively studied as hardware.

In the 2000s, several notable devices were proposed or demonstrated [8–10] from the viewpoint of applying explicit quantum-mechanical effects [11]. This area is quite important and seems close to achieving a significant breakthrough. One of editors (Omura) proposed a device structure on which future generations of MOSFET devices should be based [8, 12]. One of its important features is that it employs a tunneling-barrier junction (TBJ) structure, with thin oxide layers at the source and drain junctions. This kind of device is attracting attention [13, 14] because its high barrier suppresses the unwanted thermal current [15].

This chapter demonstrates its advantages over the scaled MOSFET by simple simulations, and addresses its potential for high-frequency performance, which will make many attractive applications possible.

MOS Devices for Low-Voltage and Low-Energy Applications, First Edition.
Yasuhisa Omura, Abhijit Mallik, and Naoto Matsuo.
© 2017 John Wiley & Sons Singapore Pte. Ltd. Published 2017 by John Wiley & Sons Singapore Pte. Ltd.

30.2 Device Structure and Device Model for Simulations

The device structure proposed here, the tunneling-barrier junction metal oxide semiconductor field-effect transistor (TBJ MOSFET) with an SOI substrate, is shown in Figure 30.1a. Here an accumulation mode MOSFET is taken as an example. It has TBJs at the edges of the metal source and drain electrodes. It looks like a single electron transistor [6] and also a double-barrier resonant tunneling transistor [9]. This chapter, as a first step, examines only the direct tunneling current through the thin oxide layer (the thickness is denoted by t_{st} and t_{dt}, respectively) at the source and the drain, not the overall operation.

The equivalent circuit for the TBJ MOSFET is shown in Figure 30.1b. The source and drain regions feature the parallel elements of the effective capacitance of the thin oxide layers (they are defined as $C_{st} = \varepsilon_{ox}/t_{st}$ and $C_{dt} = \varepsilon_{ox}/t_{dt}$, respectively, where ε_{ox} is the permittivity of the oxide layer) and the tunneling resistances, denoted by R_{st} and R_{dt}, respectively. R_{sm} and R_{dm} are the resistances of the metal contacts at the source and drain junctions. When the gate bias and the drain bias are applied to the device, as shown in Figure 30.1b, the channel current in the intrinsic MOSFET region, I_D, is expressed using the charge-sheet approximation as

$$I_D = W_{eff}\mu_n \left[qN_D t_s + C_{ox}\left(V_G - V_S' - V_{FB} - V\right)\right]\frac{dV}{dx} \qquad (30.1)$$

(a)

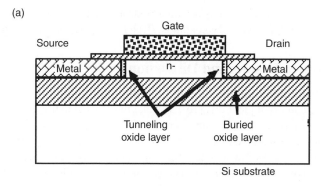

(b)

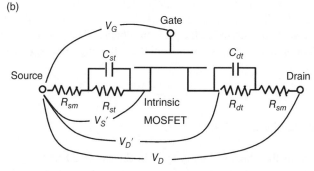

Figure 30.1 Device structure of TBJ MOSFET. The device was fabricated on the surface of an SOI substrate. (a) Schematic of device cross-section and (b) equivalent dc circuit model of TBJ MOSFET and the bias configuration. Reprinted with permission from the Electrochemical Society Proceedings, 2001–2003, 451 (2001). Copyright 2001, the Electrochemical Society.

where W_{eff} is the channel width, μ_n is the electron mobility, N_D is the doping concentration in the SOI layer, t_s is the silicon layer thickness, C_{ox} is the gate oxide capacitance per unit area, V_G is the gate voltage, V_{FB} is the flat-band voltage, V_S' is the voltage drop across the source junction, and V is the channel potential measured from the source-side edge. The gate oxide layer thickness (t_{ox}) is much smaller than the buried-oxide layer thickness (t_{BOX}) and V_G is low, so the influence of the effective substrate bias on channel potential (V) is neglected here for simplicity. From the current continuity condition, the drain current I_D is expressed using the conventional algebra for the analysis of MOSFET devices:

$$I_D = \left(\frac{W_{eff}}{L_{eff}}\right)\mu_n\left\{qN_Dt_S\left[V_D - \left(R_{st} + R_{dt} + R_{sm} + R_{dm}\right)I_D\right]\right\}$$

$$+ \left(\frac{W_{eff}}{L_{eff}}\right)\mu_nC_{ox}\left\{V_G - V_{FB} - (1/2)\left[V_D - \left(R_{dt} + R_{dm}\right)I_D\right]\right\}\left[V_D - \left(R_{dt} + R_{dm}\right)I_D\right]$$

$$- \left(\frac{W_{eff}}{L_{eff}}\right)\mu_nC_{ox}\left\{V_G - V_{FB} - (1/2)\left(R_{st} + R_{sm}\right)I_D\right\}\left(R_{st} + R_{sm}\right)I_D \qquad (30.2)$$

where L_{eff} is the channel length. As this is a transcendental equation, we must iterate it in order to obtain the drain current I_D for every bias condition. R_{st} and R_{dt} depend on t_{st}, t_{dt} and the junction's area.

30.3   Simulation Results and Discussion

The following simulations assume the direct tunneling mechanism for modeling the current across the source and drain junctions [16], and that the work function difference between the metal and the silicon is very small. The electron effective mass (m^*) during tunneling is taken as $0.3m_0$ [17] where m_0 is the electron rest mass. From this tunneling current, R_{st} and R_{dt} are estimated for each applied voltage, where the resistances are assumed to be the same. We have to solve Eq. (30.2) by means of iteration. Typical simulation results of I_D vs. V_D characteristics are shown in Figure 30.2a. Here, an n-type poly-Si gate structure is assumed. Device parameters are summarized in Table 30.1. The TBJ MOSFET shows the usual MOSFET-like I_D-V_D characteristics as we demonstrated in the previous chapter. The pinch-off condition (i.e., saturation condition of drain current) is given from the following relation (see Figure 30.1).

$$V_D' = V_S' \qquad (30.3)$$

The significant results are as follows.

- I_D is insensitive to the channel length (L_{eff}) in the range 20 to 5 nm, to the body doping level (N_D) in the range 10^{18} to 10^{14} cm^{-3}, and to the gate oxide layer thickness (t_{ox}) in the range 10 to 2.5 nm.
- Naturally, I_D is proportional to the silicon layer thickness (t_s) and the channel width (W_{eff}) because of the increase in junction contact areas.
- I_D is sensitive to the tunneling oxide layer thickness (t_{st} and t_{dt}) in the range 1.0 to 0.3 nm, which is shown in Figure 30.2b.

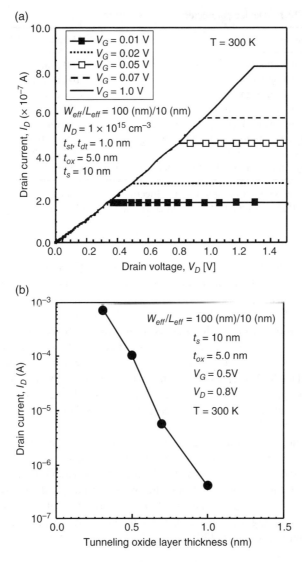

Figure 30.2   Drain current characteristics of TBJ MOSFET. (a) Drain current (I_D) dependence on drain voltage (V_D) at 300 K for five-different gate voltages (V_G). The silicon layer is 10 nm thick. (b) Drain current dependence on tunneling oxide layer thickness (t_{st}, t_{dt}). Reprinted with permission from the Electrochemical Society Proceedings, 2001–2003, 451 (2001). Copyright 2001, the Electrochemical Society.

These features are reasonable because of the high tunneling resistances of source and drain junctions. The following discusses significant issues related to device performance.

30.3.1   Current Drivability

Now, as shown in Figure 30.2b, I_D is of the order of 10^{-4} A (10^{-5} A for $W_{eff}/L_{eff}=1$). This value is much smaller than those of usual MOSFETs, but much larger than that of the single-electron transistor (~10^{-9} A). However, we must consider that the ultimate device structure would have

Table 30.1   Device parameters.

Parameters	Values
Channel width (W_{eff})	100 (nm)
Channel length (L_{eff})	10 (nm)
Gate oxide layer thickness (t_{ox})	5.0 (nm)
Silicon layer thickness (t_{s})	10 (nm)
Tunneling oxide layer thickness (t_{st}, t_{dt})	1.0 (nm)
Doping concentration of the Si body (N_{D})	1×10^{15} (cm^{-3}) (n type)
Electron mobility (μ_{n})	500 (cm^{2}/V s)
Doping concentration of the gate poly-Si (N_{G})	4×10^{20} (cm^{-3}) (n type)

Reprinted with permission from the Electrochemical Society Proceedings, 2001–2003, 451 (2001). Copyright 2001, the Electrochemical Society.

a thin-SOI structure with a silicon layer of 10 nm or less [2, 4, 11]. In that case, the total parasitic resistance of the 10 nm thick source and drain regions is about 30 kΩ for silicon [18], or at least 1 kΩ, even for partial silicide. The supply voltage must be limited below 1 V in conventional MOSFETs with a sub-50 nm channel length so as to prevent direct tunneling between pn junctions of source and drain, so its drain current will become much less than 10^{-4} A for $W_{eff}/L_{eff} = 1$. Here, it is assumed that a metal Schottky junction structure is not realistic because the drain depletion layer enhances the short-channel effect. Figure 30.2b indicates that the I_D of the TBJ MOSFET easily reaches the order of 10^{-3} A (10^{-4} A for $W_{eff}/L_{eff} = 1$) by means of device design techniques including thickness control of tunneling oxide layer, which suggests that the drivability of the TBJ MOSFET will exceed that of the ultimate structure of the conventional MOSFET.

The reality of such a thin tunneling oxide layer has been examined by using the partial-pressure oxidation technique. Gate leakage current characteristics of metal oxide semiconductor (MOS) capacitors with 2.2 nm thick SiO$_2$ film are shown in Figure 30.3. The tunneling current density of 2.2 nm thick partial-pressure oxide, formed by the rapid-thermal process, can be 100 times that of the usual oxide layer [19]. The advantage of partial-pressure oxidation is to realize lower tunneling resistance despite the thicker oxide layer, which is important from the viewpoint of thickness control [20].

30.3.2   Threshold Voltage Issue

When the tunneling resistances (R_{st} and R_{dt}) are rather large, the effective substrate bias due to V_s' is reflected on the threshold voltage (V_{TH}), resulting in the increase in threshold voltage (V_{TH}); we have $\Delta V_{TH} \sim (t_{ox}/t_{BOX})V_s'$. On the other hand, when the resistances are reduced, the device should have the threshold voltage of the intrinsic MOSFET. We aim here to achieve low tunneling resistance, so the threshold voltage of the TBJ MOSFET can be designed with the conventional guidelines. One advantage of the TBJ MOSFET is that a higher doping level does not limit the supply voltage, in part because the device has no pn junction.

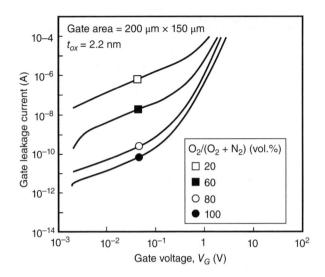

Figure 30.3 Gate leakage current dependence on gate voltage in various MOS capacitors with 2.2 nm thick SiO_2 film. SiO_2 films are formed by the partial-pressure oxidation technique. Reprinted with permission from the Electrochemical Society Proceedings, 2001–2003, 451 (2001). Copyright 2001, the Electrochemical Society.

30.3.3 Subthreshold Characteristics

Another factor that may influence device performance is subthreshold operation. This does not appear to be a serious problem because it is expected that the internal resistance of the intrinsic MOSFET is as high as that of the tunneling barrier. Actually, the sub-threshold characteristics of a similar device have already been simulated numerically [14]. One result is that the subthreshold slope is about 60 mV/dec, as demonstrated in the previous chapter.

30.3.4 Radio-Frequency Characteristics

To consider dynamic operation, the equivalent circuit of the TBJ MOSFET is shown for the common-source configuration in Figure 30.4, where the source terminal and the substrate are both grounded. The TBJ MOSFET was originally an SOI device, so the equivalent circuit includes the channel-to-substrate capacitance ($C_{csub(i)}$). The area of the intrinsic MOSFET area is delineated by the dotted line, and the label "(i)" means the "intrinsic value." The resistances of metal source and drain regions are neglected for simplicity.

The tunneling resistance (R_{st}) reduces the transconductance and drain conductance, which limits drivability. However, at the same time, the tunneling junction capacitance reduces the gate-to-drain capacitance, which improves the gain-band product. The maximum operating frequency (f_m) is given by

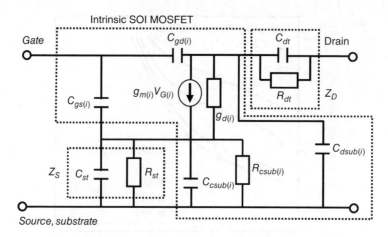

Figure 30.4 Small-signal equivalent circuit model for TBJ MOSFET. The source terminal and the substrate are grounded. Z_S and Z_D denote the component contributing to the impedance of source and drain junctions, respectively. Reprinted with permission from the Electrochemical Society Proceedings, 2001–2003, 451 (2001). Copyright 2001, the Electrochemical Society.

$$f_m = \frac{g_m}{2\pi C_{in}} \tag{30.4}$$

where g_m and C_{in} are the actual transconductance and the input capacitance, respectively. From Figure 30.4, they are given, at angular frequency ω, by

$$g_m = \frac{g_{m(i)}}{1 + g_{d(i)}\left\{ R_{st} + R_{dt} + \dfrac{1+\omega^2\tau^2}{C_{csub(i)}\,\omega^2\tau} \right\}} \tag{30.5}$$

$$C_{in} = \frac{1}{\dfrac{1}{C_{gs(i)}} + \dfrac{1}{C_{st} + \dfrac{C_{csub(i)}}{1+\omega^2\tau^2}}} \tag{30.6}$$

with the time constant τ $(=R_{csub(i)}C_{csub(i)})$. Thus, f_m is rewritten as

$$f_m = \frac{f_{m(i)}\left\{ 1 + C_{gs(i)}\left[C_{st} + \dfrac{C_{csub(i)}}{1+\omega^2\tau^2} \right] \right\}}{1 + g_{d(i)}\left\{ R_{st} + R_{dt} + \dfrac{1+\omega^2\tau^2}{C_{csub(i)}\,\omega^2\tau} \right\}} \tag{30.7}$$

$$f_{m(i)} = \frac{g_{m(i)}}{2\pi C_{gs(i)}} \tag{30.8}$$

where $f_{m(i)}$ is the intrinsic value of f_m. As $L_{eff} \sim t_S$ and $t_{st} \sim t_{ox}$ in the case considered here, f_m can be reduced to

$$f_m = \frac{f_{m(i)}}{1 + g_{d(i)} \left\{ R_{st} + R_{dt} + \dfrac{1 + \omega^2 \tau^2}{C_{csub(i)} \omega^2 \tau} \right\}} \tag{30.9}$$

Therefore, when R_{st} and R_{dt} are at most $1\,k\Omega$, the denominator of Eq. (30.9) is almost unity at $\omega \gg 1/\tau$ (~100 GHz) and $f_m \sim f_{m(i)}$. This situation can be realized for t_{st}, $t_{dt} < 0.3\,nm$ as shown in Figure 30.2b. Thus, TBJ MOSFET devices fully realize the merits of the SOI structure and so would be applicable to low-power and high-frequency circuits. As a result, the TBJ MOSFET should be seen as an intermediate step between the conventional MOSFET and the future single-electron device.

30.4 Summary

In this chapter, a new device structure, the TBJ MOSFET with SOI substrate, has been examined in detail as a candidate for the ultimate device structure. The device uses direct tunneling at source and drain junctions. The TBJ MOSFET offers superior drivability (compared to conventional devices) if device scale is very small because its tunneling resistances can be extremely small. In addition, it has been shown that the TBJ MOSFET is likely to offer better high-frequency performance.

The TBJ MOSFET appears to be a promising device for application to extremely low-power integrated circuits in the future.

References

[1] R. H. Dennard, F. H. Geansslen, H.-N. Yu, V. L. Rideout, E. Bassous, and A. R. LeBlanc, "Design of ion-implanted MOSFET's with very small physical dimensions," *IEEE J. Solid-State Circuits*, vol. SC-9, pp. 256–268, 1974.
[2] J. R. Brews, W. Fichtner, E. H. Hicollian, and S. M. Sze, "Generalized guide for MOSFET miniaturization," *IEEE Electron Device Lett.*, vol. EDL-1, pp. 2–4, 1980.
[3] C. Fiegna, H. Iwai, T. Wada, T. Saito, E. Sangiorgi, and B. Ricco, "A New Scaling Methodology for the 0.1-0.025 μm MOSFET," Tech. Dig. 1992 Symp. on VLSI Technology, pp. 33–34, 1992.
[4] D. J. Frank, S. E. Laux, and M. V. Fischetti, "Monte Carlo Simulation of a 30 nm Dual-Gate MOSFET: How Short Can Si Go?" Ext. Abstr. 1992 IEEE Int. Electron Devices Meeting, pp. 553–556, 1992.
[5] Y. Nakajima, Y. Takahashi, S. Horiguchi, K. Iwadate, H. Namatsu, K. Kurihara, and M. Tabe, "Quantized Conductance of a Si Wire Fabricated Using SIMOX Technology," Ext. Abstr. 1994 Int. Conf. Solid State Devices and Materials, pp. 538–540, 1994.
[6] Y. Takahashi, M. Nagase, H. Namatsu, K. Kurihara, K. Iwadate, Y. Nakajima, S. Horiguchi, K. Murase, and M. Tabe, "Conductance Oscillations of a Si Single Electron Transistor at Room Temperature," Ext. Abstr. IEEE Int. Electron Devices Meeting, pp. 938–940, 1994.

[7] Y. Omura, S. Horiguchi, M. Tabe, and K. Kishi, "Quantum-mechanical effects on the threshold voltage of ultrathin-SOI nMOSFETs," *IEEE Electron Devices*, vol. 14, pp. 569–571, 1993.

[8] Y. Omura and M. Shoji, Jpn. Pat. Pend., No. H4-207001, 1992.

[9] H. Namatsu, S. Horiguchi, Y. Takahashi, M. Nagase, and K. Kurihara, "Fabrication of $SiO_2/Si/SiO_2$ double barrier diodes using two dimensional Si structures," *Jpn. J. Appl. Phys.*, vol. 36, pp. 3669–3674, 1997.

[10] N. Matsuo, T. Miura, H. Hamada, and T. Miyoshi, "Application of tunneling effect for thin dielectric film to MOS transistor," *IEICE*, vol. J-81-C-1l, pp. 266–267, 1998 (in Japanese).

[11] Y. Omura, "Features of ultimately miniaturized MOSFETs/SOI: a new stage in device physics and design concepts," *IEICE Trans. Electron.*, vol. E80-C, pp. 394–406, 1997.

[12] Y. Omura, "Quantum-Effect Devices on SOI Substrates with an Ultrathin Silicon Layer," Proc. of NATO Advanced Research Workshop (Ukraina, Oct., 1998, Kluwer Academic Publisher) pp. 257–268, 1999.

[13] K. Nakazato, R. J. Blaikie, J. R. A. Cleaver, and H. Ahmed, "Single-*electron* memory," *Electron. Lett*, vol. 29, pp. 384–385, 1993.

[14] M. Wagner, H. Mizuta, and K. Nakazato, "A Fast Three-Dimensional MC Simulator for Tunneling Diodes," 2000 Int. Conf. Simulation of Semiconductor Processes and Devices (SISPAD), pp. 31–33, 2000.

[15] K. Morita, K. Morimoto, H. Sorada, K. Araki, K. Yuki, M. Niwa, T. Uenoyama, and K. Ohnaka, "Si Interband Tunnel Diode through a Thin Oxide with a Degenerate poly-Si Electrode," 1997 International Workshop on Quantum Function Devices, pp. 175–176, 1997.

[16] J. G. Simmons, "Generalized formula for the electric tunnel effect between similar electrodes separated by a thin insulating film," *J. Appl. Phys.*, vol. 34, pp. 1793–1803, 1963.

[17] M. Hirose, J. L. Alay, T. Yoshida, and S. Miyazaki, "Electronic density of states at the ultrathin SiO2/Si interfaces," Proc. of the 3rd Int. Symp. on the Physics and Chemistry of SiO_2 and the SiO_2-Si interface, vol. 96-1, pp. 485–496, 1996.

[18] M. Ono, M. Saito, T. Yoshitomi, C. Fiegna, T. Ohguro, and H. Iwai, "Sub-50 nm Gate Length n-MOSFETs with L0 nm Phosphorus Source and Drain Junctions," Ext. Abstr. IEEE Int. Electron Devices Meeting, pp. 119–122, 1993.

[19] H. Nakatsuji, Y. Kotani, and Y. Omura, "Significant Aspects of Direct Tunneling Current in Thin-Oxide Films Formed by Partial-Pressure Rapid-Thermal Oxidation," Abstr. of 1999 Silicon Nanoelectron. Workshop (IEEE & JSAP, Kyoto, June, 1999) pp. 70–71, 1999.

[20] K. Nanjo and Y. Omura, "Consideration on theoretical model to describe growth process of thin silicon oxide films," *Recent Res. Dev. Electrochem.* (Transworld Research Network), vol. 8, pp. 1–24, 2005.

31

Low-Power High-Temperature-Operation-Tolerant (HTOT) SOI MOSFET

31.1 Introduction

The long-term goal in integrated circuits is to lower the dimensionality of metal oxide semiconductor (MOS) transistors in order to increase the function density and also the speed of extremely large-scale silicon-integrated circuits [1]. The necessity of the silicon-on-insulator (SOI) MOSFET is clear given its merits of high-speed operation and low-power operation with fewer short-channel effects [2]. However, its off-leakage current is significant, even in thin SOI MOSFETs in the sub-100 nm regime [3]. Omura recently proposed the tunneling barrier junction (TBJ) SOI MOSFET that offers suppressed off-leakage current [4, 5]. It has been shown that the TBJ SOI MOSFET suffers from low drive current if used at low temperatures as intended [5].

It has been demonstrated, however, that the thin SOI MOSFET is a promising device for applications that work at 300 °C [1]. Its off-leakage current is still a serious problem and prevents its use at higher temperatures. When analyzing high-temperature operation it is anticipated that we will not need full quantum-mechanical simulations even for a thin SOI MOSFET because the influence of various carrier scattering events on the transport in the channel region is crucial; so-called *thermalization* is dominant in the Si material.

This study applies the semi-classical transport model to assess the feasibility of the SOI MOSFET in achieving high-temperature operation. This chapter introduces the high-temperature-operation-tolerant (HTOT) SOI MOSFET and shows preliminary simulation results of its characteristics. A commercial 2-D device simulator [6] is used to simulate the drain current characteristics throughout the study.

MOS Devices for Low-Voltage and Low-Energy Applications, First Edition.
Yasuhisa Omura, Abhijit Mallik, and Naoto Matsuo.
© 2017 John Wiley & Sons Singapore Pte. Ltd. Published 2017 by John Wiley & Sons Singapore Pte. Ltd.

31.2 Device Structure and Simulations

A schematic of the HTOT SOI MOSFET is shown in Figure 31.1a. The device has an n^+-Si gate, a thin n-type body, two thin p-type bodies, and two local-thin Si regions; it is assumed that the top SOI layer surface has a (001) orientation. The gate oxide layer is 5 nm thick, the buried oxide layer is 100 nm thick, the thin n-type Si body, and thin p-type Si body are 10 nm thick, and two local thin Si regions are 1 nm or 2 nm thick; the local thin Si regions are 2 nm long. n^+-Si source and drain diffusion regions are 10 nm thick and their doping concentrations are 1×10^{20} cm^{-3}; the junction is assumed to be abrupt for simplicity. P-type body has a doping concentration of 3×10^{17} cm^{-3} and n-type body has a doping concentration of 1×10^{15} cm^{-3}. Gate length (L_G) is 100 nm (or 60 nm) and gate width (W_G) is 1 µm; the n-type Si region is typically 10 nm long and the p-type Si region is typically 40 nm long (20 nm long in some cases); this dimension is selected to suppress the short-channel effect. The Si layer is very thin

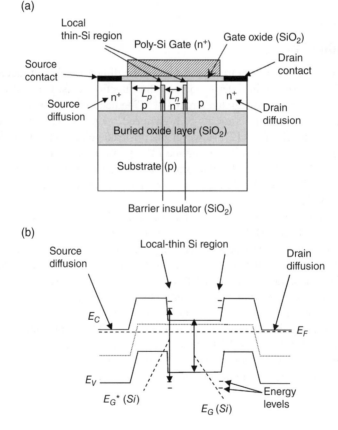

Figure 31.1 Schematic device structure of HTOT SOI MOSFET. (a) Schematic view of device. (b) Schematic band structure from source to drain at $V_D = 0$ V. E_G^* is the effective bandgap energy of ultra-thin Si regions and $E_G^* > E_G$. Yasuhisa Omura, "Proposal of high-temperature-operation tolerant SOI MOSFET and preliminary study on device performance evaluation," Active and Passive Electronic Components, vol. 2011, Article ID 850481, 2011. doi:10.1155/2011/850481

in the two local-thin regions, so the energy levels in these regions are distinctly quantized. The schematic band structure at $V_D=0\,\mathrm{V}$ and $V_G=V_{FB}$ is shown in Figure 31.1b and the effective bandgap energy (E_G^*) of the local thin Si regions is larger than the nominal bandgap energy of bulk Si (E_G); its theoretical expression is given as [7]

$$E_G^* = E_{n1} - E_{p1} \tag{31.1}$$

where E_{n1} is the ground-state level energy of confined electrons in the conduction band and E_{p1} is the ground-state level energy of confined holes in the valence band. In calculating E_G^*, the temperature dependence of intrinsic bandgap energy ($E_G(T)$) of Si is taken into account [8].

When the confinement is along the z-axis (normal to (001) surface), the ground-state energy level of twofold X-valleys (E_{n1}) in the local thin Si body is given by

$$E_{n1} - E_C = \frac{\pi^2 \hbar^2}{2m_{z,2-fold} t_{S,thin}^2} \tag{31.2}$$

where $m_{z,2\text{-}fold}$ is the effective mass of electrons for the twofold X-valley ($=m_l m_0=0.92m_0$) and $t_{S,thin}$ is the thickness of local-thin Si region ($=2$ or $1\,\mathrm{nm}$). $E_{n1}-E_C$ is about $0.1\,\mathrm{eV}$ ($\sim 1200\,\mathrm{K}$) when the local-thin Si region is 2 nm thick and about $0.2\,\mathrm{eV}$ when the local-thin Si region is 1 nm thick [9]; this suggests that the following consideration based on quantum mechanics is acceptable because the maximal operation temperature assumed is $700\,\mathrm{K}$.

As it is assumed that the device works at high temperature, it is expected that semiclassical analysis can be used in the simulations, where we basically assume semiclassical hydrodynamic transport in both thin and thick bodies and the thermionic emission model [10] is introduced to calculate the transport through the local thin Si regions using the conventional hetero-junction model. This approximation is valid except for the degenerate semiconductor. Therefore, it is expected that, at high temperatures with $V_D=1\,\mathrm{V}$, the effective energy barrier of the 2 nm long local thin Si region enhances thermionic conduction rather than electron tunneling. Accordingly, we apply the thermionic emission model in the present simulations. Mobility models for carrier transport comply with the following physics: the Masseti model for doping dependent mobility [11], the Lombardi model for mobility degradation at the Si/SiO$_2$ interface [12], and the Canali model for mobility degradation due to velocity saturation [13]. The present consideration focuses on the subthreshold characteristics because the increase in the off-current is crucial for such devices at high temperature. The mobility models primarily rule the on-current, not the off-current. Therefore, it is anticipated that the mobility models assumed here do not influence the present consideration significantly.

In addition, as n-type and p-type Si regions are 10 nm thick, discreteness of electronic states is not so crucial for the semiclassical analysis. Thus, to develop an overall consideration of the transport characteristics of the HTOT SOI MOSFET, it can be concluded that the semiclassical analysis is sufficiently verifiable. In the simulation, therefore, we replace the default parameters for the ultrathin Si region with a new set of physical parameters, where the bandgap energy and the effective electron affinity are revised following Eqs. (31.1) and (31.2). When <011> confinement is applied to the HTOT SOI MOSFET, a 3.5 nm thick local-thin Si region yields an approximately 0.1 eV high barrier to the conduction electrons; <111> confinement yields an identical result.

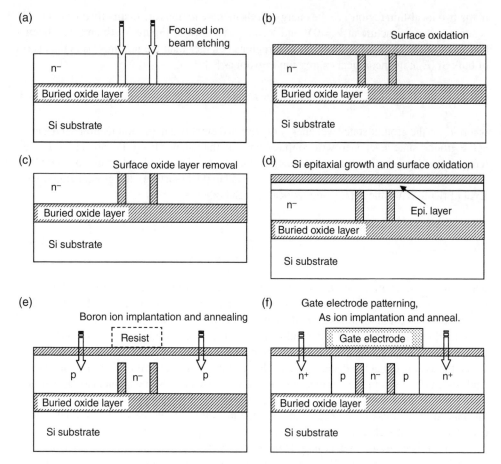

Figure 31.2 Possible fabrication process for HTOT SOI MOSFET. (a) Focused ion beam etching, (b) surface oxidation, (c) removal of surface oxide layer, (d) epitaxial growth and surface oxidation, (e) boron ion implantation and annealing, and (f) gate electrode deposition and etching. Yasuhisa Omura, "Proposal of high-temperature-operation tolerant SOI MOSFET and preliminary study on device performance evaluation," Active and Passive Electronic Components, vol. 2011, Article ID 850481, 2011, doi:10.1155/2011/850481.

Possible fabrication process of the HTOT SOI MOSFET is introduced in Figure 31.2. The major processing steps are given below.

1. Shallow and narrow trenches are formed on the *n*-type SOI layer by the focused ion beam etching technique.
2. The surface is oxidized in a furnace tube, resulting in separation of the central *n*-type body.
3. The surface oxide layer is removed.
4. A thin crystalline Si layer is deposited epitaxially and a surface oxide layer is formed as the gate insulator.

5. After the n-type body region is covered by resist, p-type body regions are formed by boron ion implantation and the substrate is annealed.
6. Gate poly-Si is deposited and the gate electrode pattern is formed as ion implantation is performed to form n^+ source and drain regions.

The fabrication process mentioned above requires challenging techniques, and simulations to predict device characteristics must also cover atomic-scale physics. The local thin Si layer has countable atomic layers. Regarding such a thin Si layer, several articles consider the impact of Si-layer thickness on transport properties [14–19]. These articles predict the following:

- Four atomic Si layers can roughly hold a bulk band structure [14, 15].
- The effective mass values of conduction band electrons increase as the Si layer is thinned [16–19].

The second point [16–19] suggests that the effective barrier height of the local thin Si layer may be overestimated. In that case, as suggested later, we should take a longer local thin Si region to have better characteristics at a high temperature.

31.3   Results and Discussion

31.3.1   Room-Temperature Characteristics

In the HTOT SOI MOSFET, most carriers cannot tunnel through the insulators but can pass through the local-thin Si regions between gate oxide and barrier insulators; the barrier insulator acts as a hard barrier. Since the local thin Si region is very narrow along the surface channel region, distinct energy quantization should be assumed in the local thin Si region even at high temperatures because $k_B T < E_{n1} - E_C$. Therefore, the bandgap of the Si region between the gate oxide layer and the barrier insulator is effectively widened which reduces the total drain current. In high-temperature environments the thermal energy of some carriers can exceed the ground-state level ($E_{n1} - E_C$ or $E_V - E_{p1}$) in the channel. Thus it can be expected that both the drive current and off-current of a HTOT SOI MOSFET operating at high temperatures will be larger than those of a TBJ MOSFET [4, 5], while its subthreshold swing at high temperatures is superior to that of the conventional SOI MOSFET.

Figure 31.3 shows I_D-V_G characteristics of a HTOT SOI MOSFET with a 40 nm long p-type region (L_p) at 300 K for various V_D values; it is assumed that the HTOT SOI MOSFET has 1 nm thick local-thin Si regions on both sides of the n-type Si body. It shows that the HTOT SOI MOSFET has a subthreshold swing value of ~70 mV/dec and ON-current of about 350 μA/μm at $V_G = 1$ V.

I_D-V_D characteristics of the HTOT SOI MOSFET at 300 K for various V_G conditions are shown in Figure 31.4 for $L_p = 40$ nm. It is also assumed that the HTOT SOI MOSFET has 1 nm thick local-thin Si regions on both sides of the n-type Si body. In order to consider conduction mechanisms, the schematic band diagram of the HTOT SOI MOSFET is shown in Figure 31.5 for $V_D \gg 0$ V; arrows indicate carrier-flow paths in the energy space. Two characteristic behaviors are considered: (i) the superlinear increase in the drain current stems from the non-ohmic conduction through the local thin Si regions at the source side, and (ii) the negative differential conductance at $V_G = V_{TH}$ stems from the high impedance created by the local-thin Si regions at the drain side.

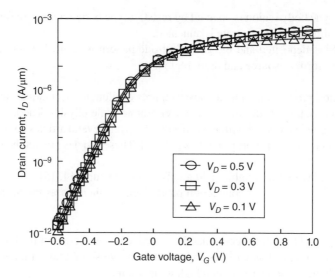

Figure 31.3 I_D-V_G characteristics of HTOT SOI MOSFET at 300 K for various drain voltage conditions. The device has a 40 nm long p-type region (L_p); L_G = 100 nm. The device has 1 nm thick local thin Si regions at both sides of n-type Si body. Yasuhisa Omura, "Proposal of high-temperature-operation tolerant SOI MOSFET and preliminary study on device performance evaluation," Active and Passive Electronic Components, vol. 2011, Article ID 850481, 2011, doi:10.1155/2011/850481.

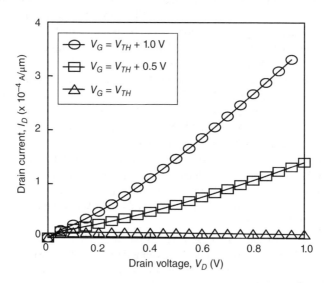

Figure 31.4 I_D-V_G characteristics of HTOT SOI MOSFET for various gate voltage conditions at 300 K. The device has a 40 nm long p-type region (L_p); L_G = 100 nm. The device has 1 nm thick local-thin Si regions at both sides of n-type Si body. Yasuhisa Omura, "Proposal of high-temperature-operation tolerant SOI MOSFET and preliminary study on device performance evaluation," Active and Passive Electronic Components, vol. 2011, Article ID 850481, 2011, doi:10.1155/2011/850481.

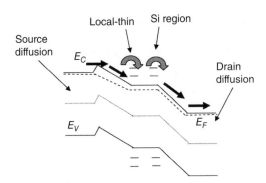

Figure 31.5 Schematic band diagram of HTOT SOI MOSFET from source to drain at $V_D > 0\,V$. Yasuhisa Omura, "Proposal of high-temperature-operation tolerant SOI MOSFET and preliminary study on device performance evaluation," Active and Passive Electronic Components, vol. 2011, Article ID 850481, 2011, doi:10.1155/2011/850481.

With regard to the I_D-V_G characteristics of the TBJ MOSFET [4, 5] under the assumption of full tunneling and ballistic transport, the drain current curve shows a kink at around the threshold voltage (V_{TH}) and the drain current is almost constant for $V_G > V_{TH}$. In Figure 31.4, however, the drain current of the HTOT SOI MOSFET smoothly increases around $V_G = V_{TH}$ and it increases monotonously for $V_G > V_{TH}$. This means that the drain current of the HTOT SOI MOSFET is ruled by the semiclassical mechanism.

31.3.2 High-Temperature Characteristics

This section discusses in detail the I-V characteristics of the HTOT MOSFET at temperatures ranging from 300 to 700 K. The temperature dependencies of the I_D-V_G characteristics of the conventional SOI MOSFET with a 10 nm thick p-type body and the HTOT SOI MOSFET with 1 nm thick local-thin Si regions at the edges of 10 nm thick n-type body at T = 300, 500, and 700 K are shown in Figure 31.6 and Figure 31.7, respectively. The subthreshold swing at each temperature is also indicated in the figures. Little difference in subthreshold swing between the conventional SOI MOSFET and the HTOT SOI MOSFET is seen at $T = 300\,K$ in Figure 31.6 and Figure 31.7. At 700 K, however, the difference in subthreshold swing is 40 mV/dec; the advantage of subthreshold swing of the HTOT SOI MOSFET is about 5%. The HTOT SOI MOSFET operates safely at 700 K with no thermal instability because of its expanded effective bandgap [7]. Thus the HTOT MOSFET is somewhat superior to the conventional SOI MOSFET in high-temperature operation. When the threshold voltage (V_{TH}) is set to 0.3 V at the drain voltage (V_D) of 0.1 V at 700 K, the drain current of the HTOT SOI MOSFET has an on/off dynamic range of about 1.7.

Next, device operations are discussed for the case where the energy levels in the 1 nm thick local thin Si region are quantized. When energy levels are quantized in the local thin Si region, the effective conduction band bottom rises to the ground state energy level; consequently, the bandgap is effectively widened. When the local thin Si region is 1 nm wide, the ground state energy level is higher by 0.26 eV than the conduction band bottom. In the following, I_D-V_G

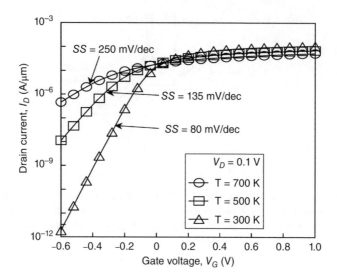

Figure 31.6 I_D-V_G characteristics of the conventional SOI MOSFET at V_D=0.1 V for various temperature conditions. The device has a 10 nm thick SOI body. Yasuhisa Omura, "Proposal of high-temperature-operation tolerant SOI MOSFET and preliminary study on device performance evaluation," Active and Passive Electronic Components, vol. 2011, Article ID 850481, 2011, doi:10.1155/2011/850481.

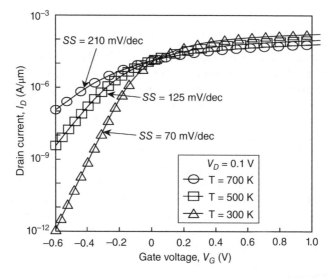

Figure 31.7 I_D-V_G characteristics of HTOT SOI MOSFET at V_D=0.1 V for various temperature conditions. The device has a 40 nm long p-type region (L_p). It has 1 nm thick local-thin Si regions at both sides of n-type Si body. Yasuhisa Omura, "Proposal of high-temperature-operation tolerant SOI MOSFET and preliminary study on device performance evaluation," Active and Passive Electronic Components, vol. 2011, Article ID 850481, 2011, doi:10.1155/2011/850481.

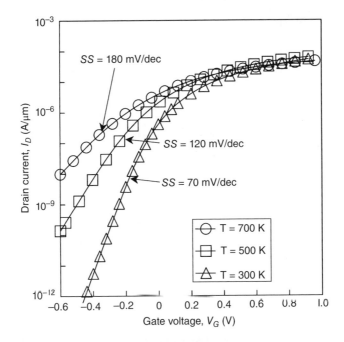

Figure 31.8 I_D–V_G characteristics of HTOT SOI MOSFET with 2 nm thick and 10 nm long local-thin Si regions at V_D=0.1 V for various temperature conditions; L_p=20 nm and L_G=60 nm. It is assumed that the ground-state level of the conduction band of the local-thin Si region is higher by 0.1 eV than the conduction band bottom. Yasuhisa Omura, "Proposal of high-temperature-operation tolerant SOI MOSFET and preliminary study on device performance evaluation," Active and Passive Electronic Components, vol. 2011, Article ID 850481, 2011, doi:10.1155/2011/850481.

characteristics are simulated for two cases; the ground state energy level is higher by 0.1 or 0.2 eV than the conduction band bottom. In addition, the thickness of the "hard barrier" insulator adjacent to the local-thin Si regions is changed to 10 nm in order to clarify the influence of long and narrow conduction paths on overall carrier transport. In this case, L_n = 10 nm and L_p = 20 nm.

I_D–V_G characteristics depending on temperature (300–700 K) for the HTOT SOI MOSFET at V_D=0.1 V are shown in Figure 31.8 and Figure 31.9 for L_p=20 nm; in Figure 31.8, it is assumed that the ground state energy level of the local-thin Si region is higher by 0.1 eV than the conduction band bottom, and in Figure 31.9, it is assumed to be higher by 0.2 eV. The following points are found in Figure 31.8 and Figure 31.9: (i) Subthreshold swing values are insensitive to the width of the local thin Si regions because subthreshold conduction is inherently similar to the thermionic process. (ii) Drain current at V_G=1 V is sensitive to the width of the local thin Si regions as expected because the long and narrow conduction path reduces channel conductivity. When energy levels of the conduction band in the local thin Si region are discretely quantized, the ground state level in the local thin Si regions should be higher than the conduction band bottom of the surface-inverted p-type region. (iii) The leakage current at V_G=−0.6 V is sensitive to the width of the local thin Si regions because the electron density in the conduction band strongly depends on the "effective bandgap energy" ($E_G{}^*$).

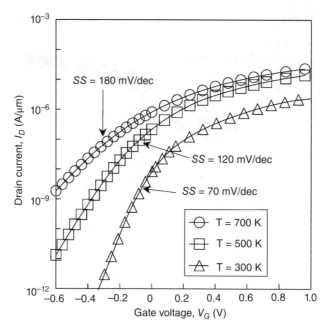

Figure 31.9 I_D–V_G characteristics of HTOT SOI MOSFET with 1 nm thick and 10 nm long local thin Si regions at V_D=0.1 V for various temperature conditions; L_p=20 nm and L_G=60 nm. It is assumed that the ground-state level of the conduction band of the local-thin Si region is higher by 0.2 eV than the conduction band bottom. Yasuhisa Omura, "Proposal of high-temperature-operation tolerant SOI MOSFET and preliminary study on device performance evaluation," Active and Passive Electronic Components, vol. 2011, Article ID 850481, 2011, doi:10.1155/2011/850481.

As the bandgap energy of the local thin Si regions is larger than that of bulk Si, the intrinsic carrier density value of the local thin Si regions ($n_i{}^*$) should be lower than that of the bulk Si (n_i) as expected by Eq. (31.3) [7]:

$$ n_i^* (T) = \left\{ \left(D_{osn} D_{osp} \right)^{1/2} k_B T / t_S \right\} \exp\left[-\frac{E_G^*}{2 k_B T} \right] \tag{31.3} $$

where D_{osn} and D_{osp} are the density of states of two-dimensional conduction-band electrons and the density of states of two-dimensional valence band holes, respectively. The density of states is a function of the effective mass; for simplicity, it is assumed the effective mass is independent of temperature. On the other hand, it is assumed that Debye length (L_D) and the intrinsic bandgap energy (E_G) is a function of temperature [8]. Equation (31.3) is quite valid for high temperatures because it was derived on the basis of high-temperature approximations. The Si conduction band presents the electrons with an effective barrier height of ($E_G{}^*-E_G$)/2 and the thermionic emission current is controlled by the barrier of ($E_G{}^*-E_G$)/2. This barrier suppresses the subthreshold leakage current at high temperatures. Therefore, the thickness and length of the local-thin Si regions is a design issue and depends on the implementation demands.

Simulated threshold voltage (V_{TH}) and subthreshold swing (SS) at drain voltage of 0.1 V are summarized in Figure 31.10. It is assumed that the SOI layer has a (001) Si surface. The HTOT SOI MOSFET is compared to the conventional SOI MOSFET with a 10 nm thick SOI layer. It is assumed in Figure 31.10a that the local thin Si regions are 2 nm thick and 2 nm long. At room temperature, both devices show almost identical characteristics; only the threshold voltage is slightly different. However, the HTOT SOI MOSFET exhibits much lower performance degradation than the conventional SOI MOSFET; $dV_{TH}/dT = -0.6$ mV/K and $dS/dT = 0.3$ mV/dec/K for the HTOT SOI MOSFET. When threshold voltage is set to 0.3 V at 700 K, the present HTOT SOI MOSFET has superior off-leakage, by a factor of 3, to the conventional SOI MOSFET. The threshold voltage of the HTOT SOI MOSFET is higher than that of the SOI MOSFET by about 0.2 V; this is slightly larger than $(E_{nl} - E_C)/q$ because electrons contributing to the threshold current should have an average energy slightly larger than $(E_{nl} - E_C)$ at the threshold [20].

In Figure 31.10b, simulated threshold voltage (V_{TH}) and SS are shown at the drain voltage of 0.1 V for 1 nm thick local thin Si regions; the HTOT SOI MOSFET is compared to the conventional SOI MOSFET with a 10 nm thick SOI layer. The results demonstrate that the HTOT SOI MOSFET has outstanding performance: $dV_{TH}/dT = 0.0$ mV/K and $dS/dT = 0.25$ mV/dec/K. Subthreshold swing of the HTOT SOI MOSFET is about 178 mV/dec at 700 K (427 °C). It should be noted that the HTOT SOI MOSFET with 1 nm thick local thin Si regions is almost insensitive to temperature for $T < 700$ K (427 °C). The mechanism is the same as that described previously.

Finally, the I_{ON} vs. I_{OFF} characteristics of the HTOT SOI MOSFET are shown in Figure 31.11 at $V_D = 0.1$ V. Slopes of curves range from -5 to -8. The reduction in on-current values at high temperatures is due to mobility degradation, and the drastic increase in off-current at high temperatures is due to the thermionic emission process. It is seen that the difference in local thin Si region thickness primarily impacts the on-current value (I_{ON}), and that the off-current value (I_{OFF}) is not so sensitive to the local-thin Si region thickness; these are important aspects of the HTOT SOI MOSFET from the viewpoint of device design.

31.4 Summary

This chapter proposed the HTOT SOI MOSFET and demonstrated its characteristics with preliminary device simulations.

The HTOT SOI MOSFET has local-thin Si regions and operates safely at 700 K with no thermal instability because of its expanded effective band gap. A HTOT SOI MOSFET with 2 nm thick local thin body regions exhibits much lower performance degradation than the conventional SOI MOSFET; $dV_{TH}/dT = -0.6$ mV/K and $dS/dT = 0.3$ mV/dec/K for the HTOT SOI MOSFET. The subthreshold swing of the HTOT SOI MOSFET is about 180 mV/dec at 700 K (427 °C). The threshold voltage of the HTOT SOI MOSFET is higher than that of SOI MOSFET by about 0.2 V because the local thin Si regions offer an expanded effective band gap.

A HTOT SOI MOSFET with 1 nm thick local thin Si regions shows outstanding performance: $dV_{TH}/dT = 0.0$ mV/K and $dS/dT = 0.25$ mV/dec/K. Subthreshold swing of the HTOT SOI MOSFET is about 178 mV/dec at 700 K (427 °C). The HTOT SOI MOSFET is therefore a promising device for future high-temperature applications when its device parameters are appropriately optimized.

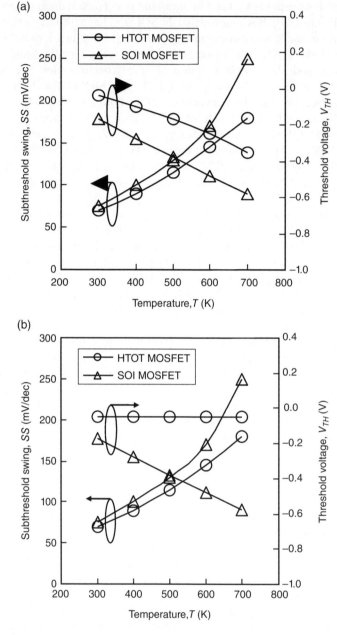

Figure 31.10 Temperature dependence of subthreshold swing and threshold voltage: HTOT SOI MOSFET compared to conventional SOI MOSFET with a 10 nm thick SOI layer. (a) Characteristics of HTOT SOI MOSFET with 2 nm thick local thin body. (b) Characteristics of HTOT SOI MOSFET with 1 nm thick local thin body. Yasuhisa Omura, "Proposal of high-temperature-operation tolerant SOI MOSFET and preliminary study on device performance evaluation," *Active and Passive Electronic Components*, vol. 2011, Article ID 850481, 2011, doi:10.1155/2011/850481.

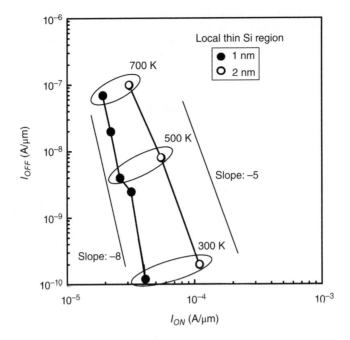

Figure 31.11 I_{ON} vs. I_{OFF} characteristics of HTOT SOI MOSFET at $V_D = 0.1\,V$. It is assumed that $V_{TH} = 0.3\,V$. I_{ON} is defined at $V_G = V_{TH} + 0.7\,V$, and I_{OFF} is defined at $V_G = V_{TH} - 0.3\,V$. Yasuhisa Omura, "Proposal of high-temperature-operation tolerant SOI MOSFET and preliminary study on device performance evaluation," Active and Passive Electronic Components, vol. 2011, Article ID 850481, 2011, doi:10.1155/2011/850481.

References

[1] J.-P. Colinge, "Silicon-on-Insulator Technology: Materials to VLSIs," 2nd ed. (Kluwer Academic Publishers, 1997).

[2] Y. Omura, "Silicon-on-Insulator (SOI) MOSFET Structure for Sub-50-nm Channel Regime," Proc. 10th Int. Symp. on SOI Technol. Dev. (The Electrochem. Soc., March, 2001) PV2001-3, pp. 205–210, 2001.

[3] H. Nakajima, S. Yanagi, K. Komiya, and Y. Omura, "Off-leakage and drive current characteristics of sub-100-nm SOI MOSFETs and impact of quantum tunnel current," *IEEE Trans. Electron Devices*, vol. 49, pp. 1775–1782, 2002.

[4] Y. Omura, "A Tunneling-Barrier Junction MOSFET on SOI Substrates with a Suppressed Short-Channel Effect for the Ultimate Device Structure," Proc. 10th Int. Symp. on SOI Technol. Dev. (The Electrochem. Soc., March, 2001) PV2001-3, pp. 451–456, 2001.

[5] H. Nakajima, A. Kawamura, K. Komiya, and Y. Omura, "Simulation models for silicon-on-insulator tunneling-barrier-junction metal-oxide-semiconductor field-effect transistor and performance perspective," *Jpn. J. Appl. Phys.*, vol. 42, no. 3, pp. 1206–1211, 2003.

[6] Synopsys Inc., "Sentaurus Device User Guide," Version A-2007, 2007, ftp://147.46.117.90/synopsys/manuals/PDFManual/data/sdevice_ug.pdf (accessed June 10, 2016).

[7] Y. Omura, T. Ishiyama, M. Shoji, and K. Izumi, "Quantum Mechanical Transport Characteristics in Ultimately Miniaturized MOSFETs/SIMOX," Proc. 10th Int. Symp. on SOI Technol. Dev. (The Electrochem. Soc., March, 1996) PV96-3, pp. 199–205, 1996.

[8] S. M. Sze, "Physics of Semiconductor Devices," 2nd ed. (John Wiley & Sons, Inc., 1981), p. 15.

[9] Y. Omura, S. Horiguchi, M. Tabe, and K. Kishi, "Quantum-mechanical effects on the threshold voltage of ultrathin-SOI nMOSFETs," *IEEE Electron Device Lett.*, vol. 14, pp. 569–571, 1993.

[10] D. Schroeder, "Modeling of Interface Carrier Transport for Device Simulations,"(Springer, 1994).

[11] G. Masetti, M. Severi, and S. Solmi, "Modeling of carrier mobility against carrier concentration in arsenic-, phosphorus-, and boron-doped silicon," *IEEE Trans. Electron Devices*, vol. ED-30, pp. 764–769, 1983.

[12] C. Lombardi, S. Manzini, A. Saporito, and M. Vanzi, "A physically based mobility model for numerical simulation of nonplanar devices," *IEEE Trans. Comput. Aided Des.*, vol.7, pp. 1164–1171, 1988.

[13] C. Canali, G. Majni, R. Minder, and G. Ottaviani, "Electron and hole drift velocity measurements in silicon and their empirical relation to electric field and temperature," *IEEE Trans. Electron Devices*, vol. ED-22, pp. 1045–1047, 1975.

[14] S. Wakui, J. Nakamura, and A. Natori, "Atomic scale dielectric constant near the SiO2/Si(001) interface," *J. Vac. Sci. Technol.*, vol. B26, pp. 1579–1584, 2008.

[15] H. Kageshima and A. Fujiwara, "First-principles study on inversion layer properties of double-gate atomically thin silicon channels," *Appl. Phys. Lett.*, vol. 93, pp. 043516-1–043516-3, 2008.

[16] K. Nehari, N. Cavassilas, J. L. Autran, M. Bescond, D. Munteanu, and M. Lannoo, "Influence of band structure on electron ballistic transport in silicon nanowire MOSFET's: an atomistic study," *Solid State Electron.*, vol. 50, pp. 716–721, 2006.

[17] P. V. Sushko and A. L. Shluger, "Electronic structure of insulator-confined ultra-thin Si channels," *Microelectron. Eng.*, vol. 84, pp. 2043–2046, 2007.

[18] M. D. Michielis, D. Esseni, Y. L. Tsang, P. Palestri, L. Selmi, A. G. O'Neill, and S. Chattopadhyay, "A semiana-lytical description of the hole band structure in inversion layers for the physically based modeling of pMOS transistors," *IEEE Trans. Electron Devices*, vol. 54, pp. 2164–2173, 2007.

[19] Y. Omura, "Extension of analytical model for conduction band non-parabolicity to transport analysis of nano-scale metal-oxide-semiconductor field-effect transistor," *J. Appl. Phys.*, vol. 105, pp. 014310–014317, 2009.

[20] Y. Tahara and Y. Omura, "Empirical quantitative modeling of threshold voltage of sub-50-nm double-gate SOI MOSFETs," *Jpn. J. Appl. Phys.*, vol. 45, pp. 3074–3078, 2006.

Part VII

QUANTUM EFFECTS AND APPLICATIONS – 2

32

Overview of Tunnel Field-Effect Transistor

In recent years, the device dimensions in complementary metal oxidesemiconductor (CMOS) technology has reached the nanometer level, where power management has become the most crucial issue for further scaling of this technology. Due to dimension scaling, severe short-channel effects such as drain-induced barrier lowering result in a substantial increase in the leakage current. Furthermore, the fundamental physical limit of $(k_B T/q)$. $\ln 10$ (~60 mV/dec at room temperature) on the subthreshold swing (SS) of a conventional metal oxide semiconductor field-effect transistor (MOSFET) poses a major roadblock for further scaling of its power-supply voltage.

The devices that are demonstrated to have SS of sub-60 mV/dec at room temperature are tunnel field-effect transistors (TFETs) and impact ionization MOSFETs (I-MOSs). An I-MOS structure consists of a p-i-n diode where the intrinsic region is partially gated so that a very high electric field exists in the nongated portion of the i-region when the device is in the ON-state leading to avalanche breakdown [1]. The impact ionization process helps the I-MOS to achieve a very small SS and high I_{ON} [2]. Experimental *SS* values of about 4 and 9 mV/dec have been reported [3] for n- and p-channel devices, respectively. The drawbacks of an I-MOS include: (i) it is not scalable due to the presence of a gated and an ungated region between the source and drain; (ii) it has a reliability problem, as the hot carriers, created by impact ionization, can go into the gate oxide, and (iii) the difficulty of low-voltage operation of the device as high voltages are required to induce breakdown.

Due to the above problems of an I-MOS, the novel and simple device structure of a TFET has become a strong contender to replace the present state-of-the-art MOSFET for further extending Moore's law [4–12]. The device structures of an n-channel MOSFET and an n-channel TFET look very similar except: (i) the highly doped n-region of the MOSFET is replaced by a highly doped p-region in a TFET, and (ii) either an intrinsic or a lightly doped n- or p-type substrate (channel) may be used for a TFET instead of the lightly doped p-type substrate (channel)

MOS Devices for Low-Voltage and Low-Energy Applications, First Edition.
Yasuhisa Omura, Abhijit Mallik, and Naoto Matsuo.
© 2017 John Wiley & Sons Singapore Pte. Ltd. Published 2017 by John Wiley & Sons Singapore Pte. Ltd.

for the MOSFET. The highly doped p-region, the intermediate lightly doped region, and the highly doped n-region act, respectively, as the source, the channel, and the drain. An insulated gate similar to a gate for a MOSFET is placed over the channel. The device may be thought of as a gated p-i-n diode that exploits a band-to-band tunneling (BTBT) mechanism to overcome the SS limitation of a conventional MOSFET. For the device operation of an n-TFET, a positive V_{DS} is applied. The mechanism of current flow in this device is based on the BTBT of electrons from the valance band of the source to the conduction band of the channel region. In the absence of a positive gate bias V_{GS}, no tunneling takes place, because the top of the valance band of the source region E_{Vp} is below the bottom of the conduction band of the channel region E_{Ci}, and hence, the device is in the OFF-state. However, when a positive V_{GS} is applied, the energy bands of the channel region are pushed down. At some value of V_{GS}, E_{Vp} aligns with E_{Ci}. If V_{GS} is further increased, the tunneling of electrons from the valance band of the source to the conduction band of the channel takes place, resulting in a drain current. The simulated energy band diagram is shown in Figure 32.1 in the OFF- and ON-state of a typical n-channel TFET.

Both the theoretical and experimental results show that SS can be much lower than 60 mV/dec for a TFET [9–11]. Other advantages of a TFET include: (i) compatibility with CMOS process; (ii) higher transconductance due to exponential variation of the output (drain) current with input (gate) voltage, and (iii) higher immunity to random discrete dopant-induced threshold voltage variability. Despite the advantages, the TFET was not seriously considered as a future CMOS-compatible technology until recently, due to its low I_{ON}. In the recent past, researchers from across the world have tried to overcome the ON-state current limitation of a TFET by improved device design. Bhuwalka *et al.* [9, 13] reported that a TFET with a delta layer of SiGe at the edge of the source improves both I_{ON} and SS. The ON-state current has been reported to improve further by gate work function engineering [13]. Toh *et al.* [14] proposed a double-gate (DG) TFET with a SiGe source that overcomes much of the ON-state current limitation of a single-gate TFET. Boucart and Ionescu [15] reported that a DG TFET with optimum silicon thickness and a high-κ gate dielectric results in higher I_{ON}. Vertical n-channel and p-channel TFETs based on heterostructure silicon/intrinsic-SiGe channel layers have been reported to have higher ON-state currents and lower subthreshold swings and threshold voltages [16].

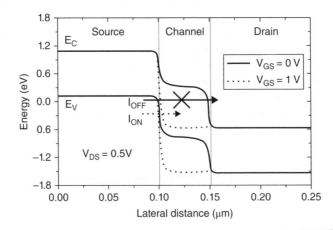

Figure 32.1 Simulated energy band diagram in the OFF- and ON-state of a typical n-channel TFET.

References

[1] E.-H. Toh, G. Wang, L. Chan, G.-Q. Lo, G. Samudra, and Y.-C. Yeo, "Strain and materials engineering for the I-MOS transistor with an elevated impact-ionization region," *IEEE Trans. Electron Devices*, vol. 54, no. 10, pp. 2778–2785, 2007.

[2] K. Gopalakrishnan, P. B. Griffin, and J. D. Plummer, "I-MOS: A Novel Semiconductor Device with Subthreshold Slope Lower than kT/q," IEDM Tech. Dig., 2002, pp. 289–292, 2002.

[3] W. Choi, J. Song, J. Lee, Y. Park, and B.-G. Park, "70-nm Impact-Ionization Metal-Oxide-Semiconductor (I-MOS) Devices Integrated with Tunneling Field Effect Transistors (TFETs)," IEDM Tech. Dig., 2005, pp. 955–958, 2005.

[4] W. M. Reddick and G. A. J. Amaratunga, "Silicon surface tunnel transistor," *Appl. Phys. Lett.*, vol. 67, no. 4, pp. 494–497, 1995.

[5] J. Koga and A. Toriumi, "Three-terminal silicon surface junction tunneling device for room temperature operation," *IEEE Electron Device Lett.*, vol. 20, no. 10, pp. 529–531, 1999.

[6] C. Aydin, A. Zaslavsky, S. Luryi, S. Cristoloveanu, D. Mariolle, D. Fraboulet, and S. Deleonibus, "Lateral interband tunneling transistor in silicon-on-insulator," *Appl. Phys. Lett.*, vol. 84, no. 10, pp. 1780–1782, 2004.

[7] W. Hansch, C. Fink, J. Schulze, and I. Eisele, "A vertical MOS-gated Esaki tunneling transistor in silicon," *Thin Solid Films*, vol. 369, no. 1/2, pp. 387–389, 2000.

[8] W. Hansch, P. Borthen, J. Schulze, C. Fink, T. Sulima, and I. Eisele, "Performance improvement in vertical surface tunneling transistors by a boron surface phase," *Jpn. J. Appl. Phys.*, vol. 40, no. 5A, pp. 3131–3136, 2001.

[9] K. K. Bhuwalka, J. Schulze, and I. Eisele, "Performance enhancement of vertical tunnel field-effect transistor with SiGe in the δp+ layer," *Jpn. J. Appl. Phys.*, vol. 43, no. 7A, pp. 4073–4078, 2004

[10] P.-F. Wang, K. Hilsenbeck, T. Nirschl, M. Oswald, C. Stepper, M. Weiss, D. Schmitt-Landsiedel, and W. Hansch, "Complementary tunneling transistor for low power applications," *Solid State Electron.*, vol. 48, no. 12, pp. 2281–2286, 2004.

[11] W. Y. Choi, B.-G. Park, J. D. Lee, and T.-J. K. Liu, "Tunneling field-effect transistors (TFETs) with subthreshold swing (SS) less than 60 mV/dec," *IEEE Electron Device Lett.*, vol. 28, no. 8, pp. 743–745, 2007.

[12] K. K. Bhuwalka, S. Sedlmaier, A. Ludsteck, C. Tolksdorf, J. Schulze, and I. Eisele, "Vertical tunnel field-effect transistor," *IEEE Trans. Electron Devices*, vol. 51, no. 2, pp. 279–282, 2004.

[13] K. Bhuwalka, J. Schulze, and I. Eisele, "Scaling the vertical tunnel FET with tunnel bandgap modulation and gate workfunction engineering," *IEEE Trans. Electron Devices*, vol. 52, no. 5, pp. 909–917, 2005.

[14] E.-H. Toh, G. H. Wang, L. Chan, D. Sylvester, C.-H. Heng, G. S. Samudra, and Y.-C. Lee, "Device design and scalability of a double-gate tunneling field-effect transistor with silicon-germanium source," *Jpn. J. Appl. Phys.*, vol. 47, no. 4, pp. 2593–2597, 2008.

[15] K. Boucart and A. M. Ionescu, "Double-gate tunnel FET with high-κ gate dielectric," *IEEE Trans. Electron Devices*, vol. 54, no. 7, pp. 1725–1733, 2007.

[16] Y. Khatami and K. Banerjee, "Steep subthreshold slope n- and p-type tunnel-FET devices for low-power and energy-efficient digital circuits," *IEEE Trans. Electron Devices*, vol. 56, no. 11, pp. 2752–2761, 2009.

33

Impact of a Spacer Dielectric and a Gate Overlap/Underlap on the Device Performance of a Tunnel Field-Effect Transistor

33.1 Introduction

The impact of using a high-κ dielectric as a gate insulator is different for a conventional metal oxide semiconductor field-effect transistor (MOSFET) and a tunnel field-effect transistor (TFET). The fringing field arising out of the high-κ gate dielectric causes fringing-induced barrier lowering [1], which is known to cause the performance of both the conventional MOSFET and the FinFET to deteriorate [2, 3], whereas the same has been reported in [4] to enhance I_{ON} of a TFET because of the different current injection mechanism in this device. On the other hand, although the use of a low-κ dielectric as a spacer is beneficial for a conventional MOSFET [5], a high-κ spacer in a FinFET with undoped underlap regions enhances its I_{ON} considerably [6]. Degradation of the device performance has been reported in [4] for a TFET with a high-κ spacer, as compared with that without a spacer. It has been reported recently [7] that a double-gate tunnel field-effect transistor (DG TFET) with a low-κ spacer and a high-κ gate dielectric improves I_{ON} by a factor of 3.8, as compared with that with a high-κ spacer and a high-κ gate dielectric. On the other hand, it is reported in [8] that a high-κ spacer causes an improvement both in I_{ON} and subthreshold swing (SS) for a heterojunction TFET with a SiGe source, which seems to have a conflict with that reported in [4, 7]. It is neither clear why such a conflict arises nor a detailed study on the effects of the spacer on the device performance of a TFET is available in the literature. In this chapter, a comprehensive investigation of the role of a spacer on the device performance of a TFET is reported. The impact of a gate overlap/underlap on the device's performance is also investigated.

MOS Devices for Low-Voltage and Low-Energy Applications, First Edition.
Yasuhisa Omura, Abhijit Mallik, and Naoto Matsuo.
© 2017 John Wiley & Sons Singapore Pte. Ltd. Published 2017 by John Wiley & Sons Singapore Pte. Ltd.

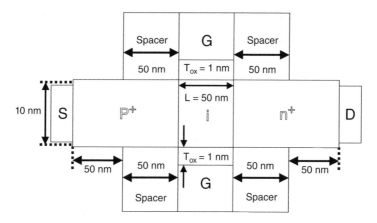

Figure 33.1 Device structure for a DG n-channel TFET. Copyright © 2011 IEEE. Reprinted, with permission, from A. Chattopadhyay and A. Mallik, "Impact of a spacer dielectric and a gate overlap/ underlap on the device performance of a tunnel field-effect transistor," IEEE Trans. Electron Devices, vol. 58, no. 3, pp. 677–683, 2011.

33.2 Device Structure and Simulation

Two-dimensional device simulations are carried out for the double gate (DG) silicon TFET structure, as shown in Figure 33.1, using the Version 5.11.24.C Silvaco ATLAS device simulator. Unless otherwise mentioned, the device's dimensions, as shown in Figure 33.1, are used for simulations. A metallurgical channel length L of 50 nm is used. The equivalent oxide thickness (EOT) of the gate dielectric is 1 nm. Gate leakage is neglected in the simulations. The source and drain contacts are made of aluminum, and the gate contact is made of a metal for which the work function is 4.5 eV. Unless otherwise mentioned, the source and the drain are made of highly doped (1×10^{20} atoms/cm³) p-type and n-type regions, respectively. The intermediate channel region is made of a moderately doped (1×10^{17} atoms/cm³) n-type layer. A uniform doping profile is used for all the regions, namely the source, the channel, and the drain, with an abrupt profile at the interfaces as applicable between them. The results presented here are obtained by using a nonlocal band-to-band tunneling (BTBT) model combined with a band-gap-narrowing model. Although the use of a classical drift-diffusion model or an abrupt doping profile in the device may result in some disagreement in the current levels, as obtained by the simulations with their experimental counterparts, this does not have much impact on the findings as the focus of this work is not on the exact values of currents but more on the general trends and relative results due to variation in different device parameters. It has been verified that the use of a quantum density-gradient model in the simulation has a very little impact on the results presented here.

33.3 Results and Discussion

33.3.1 Effects of Variation in the Spacer Dielectric Constant

The device performance is first studied for different values of the dielectric constant κ of the spacer while keeping its width fixed at 50 nm for the device structure shown in Figure 33.1. Hafnium oxide of 1 nm EOT is used as the gate insulator. The κ values of the spacer used

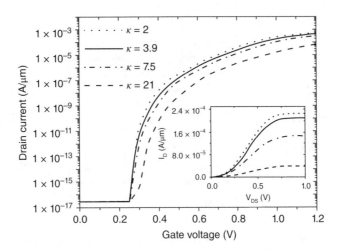

Figure 33.2 Transfer characteristics for $V_{DS}=1\,\text{V}$ for different values of κ of the spacer dielectric. The inset shows the output characteristics for a gate overdrive voltage of 0.5 V. Hafnium oxide of 1 nm EOT is used as the gate dielectric. Copyright © 2011 IEEE. Reprinted, with permission, from A. Chattopadhyay and A. Mallik, "Impact of a spacer dielectric and a gate overlap/underlap on the device performance of a tunnel field-effect transistor," IEEE Trans. Electron Devices, vol. 58, no. 3, pp. 677–683, 2011.

are 2, 3.9, 7.5, and 21, which may correspond to the interlayer dielectric, silicon dioxide, silicon nitride, and hafnium oxide, respectively. The transfer characteristics for different values of κ of the spacer are shown in Figure 33.2. In addition to an increase in the threshold voltage V_T, it is observed in Figure 33.2 that an increase in the κ value results in corresponding degradation of the device's performance in terms of both I_{ON} and SS, similar to that reported in [7]. The value of V_T is extracted using the constant current ($10^{-7}\,\text{A/}\mu\text{m}$) definition and found to be 0.45, 0.48, 0.5, and 0.6 V, respectively, for $\kappa=2$, 3.9, 7.5, and 21. The output characteristics for different values of κ are shown in the inset in Figure 33.2 for a gate overdrive voltage ($V_{GS}-V_T$) of 0.5 V from which it can be clearly seen that a decrease in the κ value of the spacer makes a good improvement in I_{ON} for the same gate overdrive voltage. It may be noted that, although a symmetric device structure is used for the simulations, only the n-type branch is shown in the characteristics for simplicity.

In order to gain an insight into the above observations, the simulated energy band diagram around the tunneling junction of the device biased at $V_{GS}=V_{DS}=1\,\text{V}$ at a distance of 1 nm below the top oxide-semiconductor interface is plotted in Figure 33.3 for different values of κ of the spacer. For a TFET, it may be recalled that the electrons from the valance band of the source, which is henceforth referred to as the *tunnel source*, tunnel into the conduction band of the adjacent (normally the channel) region, which is henceforth referred to as the *tunnel destination*. The potential barrier formed by the band gap of the semiconductor across which the tunneling of electrons takes place is henceforth referred to as the *tunneling junction*. The tunneling junction for the device in Figure 33.1 lies very close to the metallurgical source-channel junction, as can be verified in Figure 33.3. It is evident in Figure 33.3 that an increase in the κ value of the spacer results in more band lowering of the tunnel source region. For a given value of κ, the band lowering is nearer to the tunneling junction, and the same decreases as one

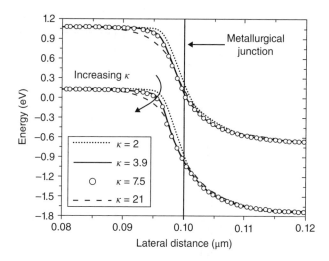

Figure 33.3 Simulated energy band diagram near the tunneling junction at a depth of 1 nm from the oxide-semiconductor interface as a function of the κ value of the spacer for the device in Figure 33.1 biased at $V_{GS} = V_{DS} = 1$ V. Copyright © 2011 IEEE. Reprinted, with permission, from A. Chattopadhyay and A. Mallik, "Impact of a spacer dielectric and a gate overlap/underlap on the device performance of a tunnel field-effect transistor," IEEE Trans. Electron Devices, vol. 58, no. 3, pp. 677–683, 2011.

moves away from the tunneling junction toward the source. The fringing field in the device is shown in Figure 33.4 for two different values of κ of the spacer. It is evident in Figure 33.4 that the fringing field in the source near the gate edge is larger for a device with a high-κ spacer, as compared with that with a low-κ spacer, which confirms that the band lowering, as seen in Figure 33.3, is indeed due to the fringing field arising out of the spacer. This band lowering results in the formation of depletion zones in the source near the gate edge, as reported in [7]. It has been suggested in [7] that due to the formation of depletion zones near the surface, the holes are pushed away from the surface toward the body. This causes carrier tunneling to occur in the body instead of near the surface that degrades I_{ON} when a high-κ spacer is used. It may be noted, however, that the current in a TFET depends upon the tunneling of the electrons from the valence band of the source to the conduction band of the channel and not upon the hole concentration in the source. Again, the device current depends upon the tunnel width and the electric field across the tunneling junction [9], and this has nothing to do with the hole concentration in the source. The fringing field arising out of the spacer dielectric that causes band lowering, as observed in Figure 33.3, also results in a variation both in the minimum tunnel width and the maximum electric field at the tunneling junction, as can be seen in Figure 33.5. The minimum value of the tunnel width and maximum value of the electric field across the tunneling junction in Figure 33.5 are extracted for $V_{GS} = V_{DS} = 1$ V at two different locations of the channel in the vertical direction: one at a depth of 1 nm below the top oxide-semiconductor interface and the other in the middle of the channel. It is evident in Figure 33.5 that the tunnel width and the electric field are, respectively, lower and higher near the surface, as compared with that in the middle of the channel for a low value of κ of the spacer. A crossover is observed for $\kappa \sim 6$ above which the tunnel width becomes lower, and the electric field becomes higher in the middle, as compared with that near the surface. This indicates that the

(a)

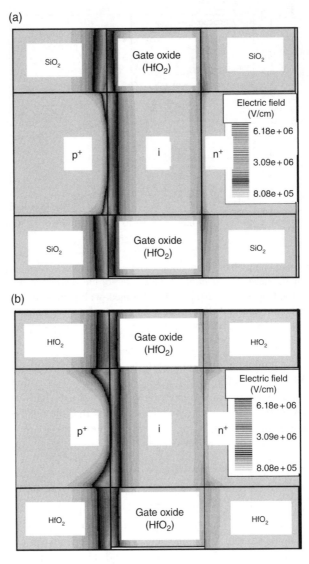

(b)

Figure 33.4 Fringing field plot for a TFET biased at a gate overdrive voltage of 0.5 V and $V_{DS} = 1$ V for two different κ values of the spacer. (a) $\kappa = 3.9$ and (b) $\kappa = 21$. Copyright © 2011 IEEE. Reprinted, with permission, from A. Chattopadhyay and A. Mallik, "Impact of a spacer dielectric and a gate overlap/underlap on the device performance of a tunnel field-effect transistor," IEEE Trans. Electron Devices, vol. 58, no. 3, pp. 677–683, 2011.

tunneling occurs near the surface and at the middle (bulk) of the device, respectively, for the low (<6) and high (>6) values of κ of the spacer. It may also be seen in Figure 33.5 that the minimum tunnel width is less for a low-κ spacer, as compared with that for a high-κ spacer, which explains why degradation of the device performance occurs due to the increasing fringing field arising out of the spacer when its κ value is increased, as observed in Figure 33.2.

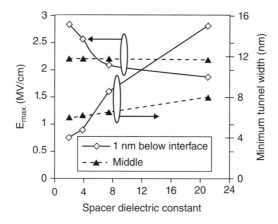

Figure 33.5 Minimum tunnel width and maximum electric field across the tunneling junction for $V_{GS} = V_{DS} = 1\,V$ for different κ values of the spacer for the device in Figure 33.1 at two different locations of the channel in the vertical direction: one at a depth of 1 nm below the top oxide-semiconductor interface and the other at the middle of the channel. Copyright © 2011 IEEE. Reprinted, with permission, from A. Chattopadhyay and A. Mallik, "Impact of a spacer dielectric and a gate overlap/underlap on the device performance of a tunnel field-effect transistor," IEEE Trans. Electron Devices, vol. 58, no. 3, pp. 677–683, 2011.

The spacer κ value dependence of the TFET characteristics has been also studied for each of the following cases: when larger silicon body thickness (>10 nm) is used and when HfO$_2$ is replaced by 1 nm thick SiO$_2$ as the gate dielectric. It was found for both cases that the trends of the spacer κ value dependence of the device characteristics are very similar to that observed in Figure 33.2, although the impact of varying the κ value of the spacer is slightly less when either the silicon body thickness is increased or SiO$_2$ is used as the gate dielectric instead of HfO$_2$.

A high-κ spacer in a heterojunction TFET with a SiGe source is reported in [8] to improve both I_{ON} and SS, which seems to conflict not only with the findings mentioned above but also with those reported in [4, 7]. To resolve this issue, the typical band diagram in the ON-state, indicating the relative positions of the tunnel source, the tunnel destination, and the tunnel path, is shown in Figure 33.6 for a heterojunction TFET with a Si$_{0.6}$Ge$_{0.4}$ source. Both the tunnel source and the tunnel destination for such a device are located in the narrow-band-gap source region, as evident in Figure 33.6, which is also consistent with that reported in [8, 10]. A high-κ spacer in such a device therefore causes more band lowering of the tunnel destination adjacent to the gate edge, resulting in a reduction in the tunnel width and an increase in the electric field across the tunneling junction. This, in turn, causes an improvement in the device's performance.

33.3.2 Effects of Variation in the Spacer Width

To study the effects of variation in the spacer width, it is varied from 0 to 50 nm in steps of 10 nm for a fixed value of κ (corresponding to HfO$_2$) of the spacer dielectric. The device structure and all other parameters are kept the same as in the previous case, with 1 nm SiO$_2$ as the gate dielectric. It is very clear from the transfer characteristics in Figure 33.7 that V_T increases,

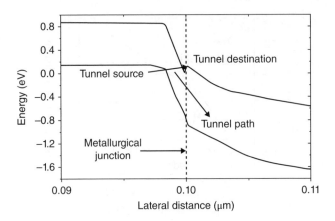

Figure 33.6 Simulated energy band diagram near the tunneling junction at a depth of 1 nm from the oxide-semiconductor interface for a heterojunction TFET with a SiGe source biased at $V_{GS} = V_{DS} = 1$ V. Copyright © 2011 IEEE. Reprinted, with permission, from A. Chattopadhyay and A. Mallik, "Impact of a spacer dielectric and a gate overlap/underlap on the device performance of a tunnel field-effect transistor," IEEE Trans. Electron Devices, vol. 58, no. 3, pp. 677–683, 2011.

and the device performance degrades with an increasing width of the spacer. The V_T values are found to be 0.45, 0.535, 0.545, 0.57, 0.59, and 0.595 V, respectively, for a spacer width of 0, 10, 20, 30, 40, and 50 nm. The output characteristics for different widths of the spacer are shown in the inset in Figure 33.7 for a constant gate overdrive voltage of 0.5 V, from which it is clear that an increase in the spacer width results in a corresponding degradation of the device performance, although the dependence becomes weak for relatively large widths (≥30 nm). It is also observed in Figure 33.7 that a 0 nm spacer or no spacer results in a better performance, which is consistent with that reported in [4]. An increase in the width increases the coupling between the gate metal and the source through the spacer, thereby causing degradation in the device performance. After a certain width, the coupling does not increase significantly for further increases in width because of the physical distance, making the device performance less sensitive for the larger widths (≥30 nm). Similar trends have been observed for a variation of the spacer width when HfO$_2$ is replaced by SiO$_2$ as the spacer dielectric, although the impact is much smaller in the case of SiO$_2$.

33.3.3 Effects of Variation in the Source Doping Concentration

To investigate the impact of a varying source doping concentration on the spacer dependence of the device performance the device structure in Figure 33.1 with 1 nm thick SiO$_2$ as the gate dielectric and a 50 nm wide spacer is simulated for three different values of the source doping concentration as 3×10^{20}, 1×10^{20}, and 5×10^{19} atoms/cm^3. For each of the source doping concentrations, the transfer characteristic is plotted in Figure 33.8 for two different κ values of the spacer, corresponding to SiO$_2$ and HfO$_2$. In addition to a change in V_T, as expected, it is observed in Figure 33.8 that variation in the source doping concentration results in a difference in the κ value dependence of the device characteristics. The dependence of the device characteristics on the κ value of a spacer decreases with an increasing source doping concentration.

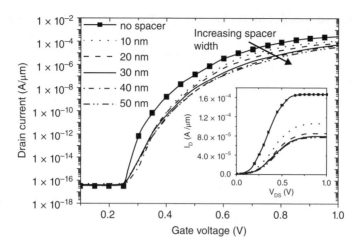

Figure 33.7 Transfer characteristics for $V_{DS}=1\,\text{V}$ for different spacer widths. The inset shows the output characteristics for a gate overdrive voltage of 0.5 V. Hafnium oxide ($\kappa=21$) as the spacer and 1 nm thick silicon dioxide as the gate dielectric are used in this case. Copyright © 2011 IEEE. Reprinted, with permission, from A. Chattopadhyay and A. Mallik, "Impact of a spacer dielectric and a gate overlap/underlap on the device performance of a tunnel field-effect transistor," IEEE Trans. Electron Devices, vol. 58, no. 3, pp. 677–683, 2011.

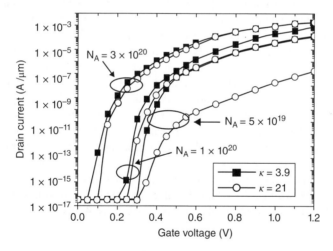

Figure 33.8 Transfer characteristics showing the impact of different source doping concentrations on the spacer κ value dependence of the device characteristics. Copyright © 2011 IEEE. Reprinted, with permission, from A. Chattopadhyay and A. Mallik, "Impact of a spacer dielectric and a gate overlap/underlap on the device performance of a tunnel field-effect transistor," IEEE Trans. Electron Devices, vol. 58, no. 3, pp. 677–683, 2011.

Higher source doping causes less depletion of the source and, hence, less band lowering of the tunnel source. As a result, less impact of the spacer dielectric on the device characteristics is observed for a higher source doping concentration. It may also be noted that the device with a source doping concentration of 3×10^{20} atoms/cm^3 shows an improved SS for the high-κ

spacer, as compared with that for the low-κ spacer. This is probably due to band narrowing of the source for such high doping concentrations, which has, to some extent, a similar effect to the use of a lower band-gap material such as SiGe in the source, as discussed earlier. Although the effect of donor deactivation due to a dielectric mismatch between the conducting channel and its dielectric surroundings [11] is not considered in this chapter, the impact of the spacer on the device characteristics for a given doping concentration is likely to increase due to this effect.

33.3.4   Effects of a Gate-Source Overlap

In order to study the impact of a gate-source overlap on the spacer dependence of the performance, the device simulation is done for a TFET structure, which is very similar to that shown in Figure 33.1, except with a gate-source overlap of 5 nm. The transfer characteristics of this device for two different κ values of the spacer corresponding to SiO_2 and HfO_2 are compared with that of a device without any gate-source overlap, that is, for which the gate is exactly aligned with the metallurgical source-channel junction. Except the overlap, all other device parameters for both the devices are kept the same. Silicon dioxide of 1 nm thickness is used as the gate insulator, and a doping concentration of 1×10^{20} atoms/cm^3 is used for the source. It is observed from the transfer characteristics in Figure 33.9 that the spacer dependence of the device characteristics is somewhat reduced for the device with a gate-source overlap, as compared with that without an overlap. Due to the overlap, the fringing field arising from the spacer dielectric cannot significantly influence the band structure in the part of the source that is covered by the gate. It may be noted that the band lowering of this part of the source adjacent to the tunneling junction mainly influences the tunnel width and, hence, the device current. As the spacer covers the rest of the source, which is away from the tunneling junction, it has less influence on the tunneling parameters. As a result, a reduced dependence of the spacer on the device characteristics is observed for the overlap devices in Figure 33.9.

It also may be noted in Figure 33.9 that, for a given spacer, the device current is significantly degraded for the overlap devices, as compared with that without an overlap. The applied gate bias causes the band lowering of the tunnel destination that reduces the tunnel width, making the device-current gate bias dependent, as discussed earlier. For the overlap devices, the applied gate bias also causes the band lowering of the part of the source that is covered by the gate. Such band lowering of the tunnel source adjacent to the tunneling junction results in degradation of the device current for the overlap devices, as observed in Figure 33.9.

33.3.5   Effects of a Gate-Channel Underlap

The transfer characteristics for two different devices with 5 and 10 nm gate channel underlaps are compared with that for a device without any underlap (i.e., for which the gate is exactly aligned with the metallurgical source-channel junction) in Figure 33.10. For all the devices, the transfer characteristics are shown for two different κ values of the spacer to see the impact of the gate underlap on the spacer dependence of the device characteristics. All other device parameters are kept the same as that in the previous case. It is observed in Figure 33.10 that, although a high-κ spacer causes the device performance for a device without an underlap to deteriorate, it actually helps improve the performance when there is a

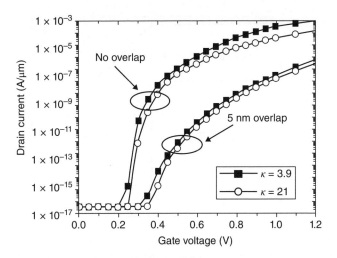

Figure 33.9 Transfer characteristics for $V_{DS} = 1\,\mathrm{V}$ showing the impact of gate-source overlap on the device characteristics. Copyright © 2011 IEEE. Reprinted, with permission, from A. Chattopadhyay and A. Mallik, "Impact of a spacer dielectric and a gate overlap/underlap on the device performance of a tunnel field-effect transistor," IEEE Trans. Electron Devices, vol. 58, no. 3, pp. 677–683, 2011.

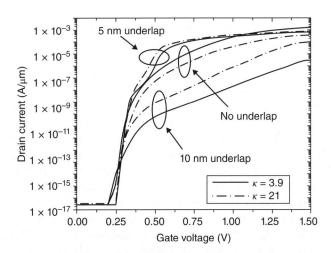

Figure 33.10 Transfer characteristics for $V_{DS} = 1\,\mathrm{V}$ showing the impact of a gate-channel underlap on the device characteristics. Copyright © 2011 IEEE. Reprinted, with permission, from A. Chattopadhyay and A. Mallik, "Impact of a spacer dielectric and a gate overlap/underlap on the device performance of a tunnel field-effect transistor," IEEE Trans. Electron Devices, vol. 58, no. 3, pp. 677–683, 2011.

gate underlap in the device. The amount of improvement for the high-κ spacer over the low-κ spacer for the gate underlap devices increases with an increasing width of the underlap region. Figure 33.11 shows the energy band diagram for a 10 nm underlap device biased at $V_{GS} = 0.5\,\mathrm{V}$ and $V_{DS} = 1\,\mathrm{V}$ for two different κ values of the spacer. It is evident in Figure 33.11 that a high-κ spacer causes more band lowering of the tunnel destination as compared with

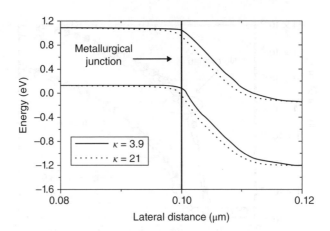

Figure 33.11 Simulated energy band diagram near the tunneling junction at a depth of 1 nm from the oxide-semiconductor interface for a 10 nm underlap device biased at $V_{GS}=0.5$ V and $V_{DS}=1$ V for two different κ values of the spacer. Copyright © 2011 IEEE. Reprinted, with permission, from A. Chattopadhyay and A. Mallik, "Impact of a spacer dielectric and a gate overlap/underlap on the device performance of a tunnel field-effect transistor," IEEE Trans. Electron Devices, vol. 58, no. 3, pp. 677–683, 2011.

that of the tunnel source, resulting in less tunnel width for such an underlap structure. Hence, an underlap device with a high-κ spacer shows a better performance, as observed in Figure 33.10.

It is also interesting to observe in Figure 33.10 that a small gate-channel underlap of 5 nm improves the device performance for a given κ value of the spacer, particularly for $V_{GS}<1$ V. A relatively large gate underlap of 10 nm, however, degrades the device performance drastically. In order to provide an insight into this, the energy band diagram is shown in Figure 33.12 for the 0, 5, and 10 nm gate underlap devices with the same spacer dielectric $\kappa=3.9$ for $V_{GS}=0.5$ V and $V_{DS}=1$ V. For the device without an underlap, it is clear in Figure 33.12 that the applied gate bias causes band lowering not only for the channel region that lies exactly underneath the gate but also for a part of the source, which is adjacent to the gate edge. The band lowering of the tunnel source increases the tunnel width and hence degrades the device performance to some extent. For a relatively small gate underlap of 5 nm, no such gate-bias-induced band lowering of the tunnel source is observed in Figure 33.12, which in turn results in a relatively small tunnel width and, hence, improved device performance, as compared with the previous case. No such gate-bias induced band lowering of the tunnel source is also observed for the device with a relatively large gate underlap of 10 nm. For such devices, however, the applied gate bias can induce significant band lowering of the channel region, which is away from the tunneling junction, as can be verified in Figure 33.12. As a result, the tunnel width becomes significantly larger, which drastically degrades the device performance for the 10 nm overlap devices, as observed in Figure 33.10.

A high-κ spacer would also result in higher fringe capacitance, which, in TFETs, can be significantly higher than that in MOSFETs due to enhanced Miller effects [12]. This could nullify some of the advantages of higher I_{ON} for the device with a small gate underlap and a high-κ spacer.

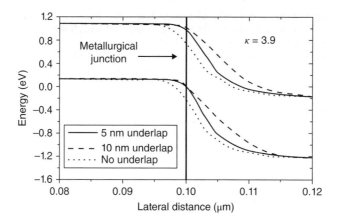

Figure 33.12 Simulated energy band diagram near the tunneling junction at a depth of 1 nm from the oxide-semiconductor interface for the 0, 5, and 10 nm underlap gate devices biased at $V_{GS}=0.5$ V and $V_{DS}=1$ V with the same spacer dielectric (k=3.9). Copyright © 2011 IEEE. Reprinted, with permission, from A. Chattopadhyay and A. Mallik, "Impact of a spacer dielectric and a gate overlap/underlap on the device performance of a tunnel field-effect transistor," IEEE Trans. Electron Devices, vol. 58, no. 3, pp. 677–683, 2011.

33.4 Summary

A systematic investigation of the impact of the spacer dielectric on the device performance of a TFET was made. It was found that a low-κ dielectric as a spacer with minimum possible width produced the best device performance. High source doping or a gate-source overlap reduces the spacer dependence of the device characteristics. A gate underlap structure with a high-κ spacer in it shows improved performance. For a given spacer, although a gate overlap or a relatively large gate underlap degrades the device performance, a small gate underlap shows an improvement in it.

References

[1] G. C.-F. Yeap, S. Krishnan, and M.-R. Lin, "Fringing-induced barrier lowering (FIBL) in sub-100 nm MOSFETs with high-κ gate dielectrics," *Electron. Lett*, vol. 34, no. 11, pp. 1150–1152, 1998.

[2] B. Cheng, M. Cao, R. Rao, A. Inani, P. V. Voorde, W. M. Greene, J. M. C. Stork, Z. Yu, P. M. Zeitzoff, and J. C. S. Woo, "The impact of high-κ gate dielectrics and metal gate electrodes on sub-100 nm MOSFETs," *IEEE Trans. Electron Devices*, vol. 46, no. 7, pp. 1537–1544, 1999.

[3] C. R. Manoj and V. R. Rao, "Impact of high-κ gate dielectrics on the device and circuit performance of nanoscale FinFETs," *IEEE Electron Device Lett.*, vol. 28, no. 4, pp. 295–297, 2007.

[4] M. Schlosser, K. K. Bhuwalka, M. Sauter, T. Zilbauer, T. Sulima, and I. Eisele, "Fringing-induced drain current improvement in the tunnel field-effect transistor with high-κ gate dielectrics," *IEEE Trans. Electron Devices*, vol. 56, no. 1, pp. 100–108, 2009.

[5] N. R. Mohapatra, M. P. Desai, S. G. Narendra, and V. R. Rao, "The effect of high-κ gate dielectrics on deep submicrometer CMOS device and circuit performance," *IEEE Trans. Electron Devices*, vol. 49, no. 5, pp. 826–831, 2002.

[6] A. B. Sachid, C. R. Manoj, D. K. Sharma, and V. R. Rao, "Gate fringe-induced barrier lowering in underlap FinFET structures and its optimization," *IEEE Electron Device Lett.*, vol. 29, no. 1, pp. 128–130, 2008.

[7] C. Anghel, P. Chilagani, A. Amara, and A. Vladimirescu, "Tunnel field effect transistor with increased ON current, low-κ spacer and high-κ dielectric," *Appl. Phys. Lett.*, vol. 96, no. 12, pp. 122104-1–122104-3, 2010.

[8] H. G. Virani and A. Kottantharayil, "Optimization of Hetero Junction n-Channel Tunnel FET with High-κ Spacers," Proc. 2nd Int. Workshop Electron Devices Semicond. Technol., 2009, pp. 1–6, 2009.

[9] E. O. Kane, "Zener tunneling in semiconductors," *J. Phys. Chem. Solids*, vol. 12, no. 2, pp. 181–188, 1960.

[10] E.-H. Toh, G. H. Wang, L. Chan, G. Samudra, and Y.-C. Yeo, "Device physics and guiding principles for the design of double-gate tunneling field effect transistor with silicon–germanium source heterojunction," *Appl. Phys. Lett.*, vol. 91, no. 24, pp. 243505-1–243505-3, 2007.

[11] M. T. Björk, H. Schmid, J. Knoch, H. Riel, and W. Riess, "Donor deactivation in silicon nanostructures," *Nat. Nanotechnol.*, vol. 4, no. 2, pp. 103–107, 2009.

[12] S. Mookerjea, R. Krishnan, S. Datta, and V. Narayanan, "Effective capacitance and drive current for tunnel FET (TFET) CV/I estimation," *IEEE Trans. Electron Devices*, vol. 56, no. 9, pp. 2092–2098, 2009.

34

The Impact of a Fringing Field on the Device Performance of a P-Channel Tunnel Field-Effect Transistor with a High-κ Gate Dielectric

34.1 Introduction

It is well known that the fringing field arising out of a high-κ dielectric causes fringing-induced barrier lowering (FIBL) [1]. Such FIBL has been reported to improve the device performance of an n-channel tunnel field-effect transistor (n-TFET) [2]. However, when used as a spacer for an n-TFET, it causes the device performance [2–4] to deteriorate. Very little is, however, reported in the literature on the impact of either a high-κ gate dielectric or a spacer on the device performance of a p-channel tunnel field-effect transistor (p-TFET). In this chapter, a comprehensive investigation on the impact of varying the dielectric constant κ of both the gate dielectric and the spacer on the device performance of a p-TFET is reported. The effects of varying the source doping concentration on the gate dielectric dependence of the device characteristics for a p-TFET is reported. The impact of scaling on the gate dielectric dependence of the device characteristics is also investigated.

34.2 Device Structure and Simulation

Two-dimensional device simulations are done for the double gate (DG) silicon p-TFET structure, as shown in Figure 34.1, using the Version 5.11.24.C Silvaco ATLAS device simulator. Unless otherwise mentioned, the device dimensions, as shown in Figure 34.1, are used for the simulations. Gate leakage is neglected in the simulations. The source and drain

MOS Devices for Low-Voltage and Low-Energy Applications, First Edition.
Yasuhisa Omura, Abhijit Mallik, and Naoto Matsuo.
© 2017 John Wiley & Sons Singapore Pte. Ltd. Published 2017 by John Wiley & Sons Singapore Pte. Ltd.

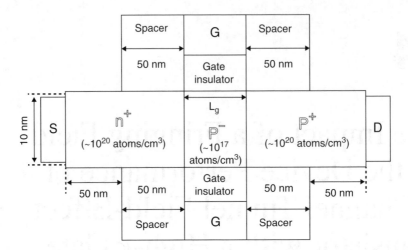

Figure 34.1 Device structure for a DG p-TFET. Copyright © 2011 IEEE. Reprinted, with permission, from A. Mallik and A. Chattopadhyay, "The impact of fringing field on the device performance of a p-channel tunnel field-effect transistor with a high-κ gate dielectric," IEEE Trans. Electron Devices, vol. 59, no. 2, pp. 277–282, 2012.

contacts are made of aluminum, and the gate contact is made of a metal, for which the work function is 5.41 eV. Unless otherwise mentioned, the source and the drain are made of highly doped (1×10^{20} atoms/cm^3) n- and p-type regions, respectively. Gaussian doping profile with a doping gradient of 2 nm/dec is used for both the source and the drain. The intermediate channel region is made of a moderately doped (1×10^{17} atoms/cm^3) p-type layer. The simulated doping profile and the location of the metallurgical source-channel junction are shown in Figure 34.2. The results presented here are obtained by using a nonlocal band-to-band tunneling (BTBT) model combined with a band-gap-narrowing model and a quantum density-gradient model.

34.3 Results and Discussion

34.3.1 Effects of Variation in the Gate Dielectric Constant

The purpose of using a high-κ gate dielectric in complementary metal oxide semiconductor (CMOS) technology is to achieve lower equivalent oxide thickness (EOT) of the gate dielectric without enhancing the gate leakage. This is achieved by increasing the κ value of the gate dielectric while keeping its physical thickness constant. In doing so, a high-κ dielectric induces a higher electrical field at the surface due to the increased oxide capacitance. The impact of such a higher electric field at the surface is investigated by keeping the physical thickness of the gate dielectric constant, as in [3, 5] for an n-TFET. On the other hand, the fringing field arising out of a high-κ gate dielectric causes FIBL [1]. Although such FIBL is known to degrade the performance of a conventional metal oxide semiconductor field effect transistor (MOSFET) [1, 6], it actually helps improve the performance of an n-TFET [2]. The impact of such fringing field arising out of a high-κ gate dielectric in a p-TFET is studied by keeping the EOT of the gate dielectric constant, as in [2] for an n-TFET.

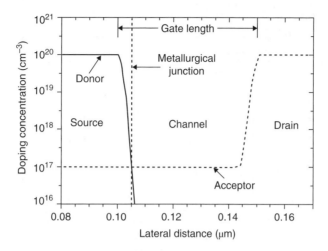

Figure 34.2 Simulated doping profile and location of the metallurgical junction for the device, as shown in Figure 34.1. Copyright © 2011 IEEE. Reprinted, with permission, from A. Mallik and A. Chattopadhyay, "The impact of fringing field on the device performance of a p-channel tunnel field-effect transistor with a high-κ gate dielectric," IEEE Trans. Electron Devices, vol. 59, no. 2, pp. 277–282, 2012.

34.3.1.1 Impact of Fringing Field (Constant EOT)

The impact of the fringing field arising out of a high-κ gate dielectric by keeping the EOT of the gate dielectric constant is first investigated. The effects of varying the κ value of the gate dielectric on the transfer characteristics at drain-to-source voltage $V_{DS} = -1$ V of the p-TFET, as shown in Figure 34.1, with a drawn gate length $L_G = 50$ nm and EOT $= 2$ nm, are shown in Figure 34.3a. No spacer is used in this case. The κ values of the gate dielectric used are 3.9, 7.5, 21, 50, 100, and 200. It may be noted that, although a symmetric device structure is used for the simulations, only the p-type branch is shown in Figure 34.3a for simplicity. The values of threshold voltage V_T and I_{ON} for different κ values of the gate dielectric are extracted from the transfer characteristics in Figure 34.3a and plotted in Figure 34.3b as a function of the κ value. In this chapter, the constant current (10^{-7} A/μm) method is used for extracting V_T, and I_{ON} is assumed to be the value of drain current I_D corresponding to a gate-to-source voltage V_{GS} swing of -1 V from the point where the current changes from p-i-n leakage to tunneling. It is observed in Figure 34.3 that an increase in the κ value, except from $\kappa = 100$ to 200, results in corresponding degradation of device performance in terms of both V_T and I_{ON}. An improvement in both V_T and I_{ON} is observed for $\kappa = 200$, compared with that for $\kappa = 100$. This is in contrast with that reported for an n-TFET, where an increase in the κ value of the gate dielectric results in improvement of the device performance [2], except when κ is relatively low. When κ is increased from 3.9 to 21, an improvement and deterioration in I_D have been reported in [2] for relatively low and high values of V_{GS}, respectively. It has been verified that the contrast is not due to the difference in the type (n-channel/p-channel) of TFET structures but is due to the difference in the doping profile. The characteristics reported in [2] are for an n-channel TFET with abrupt doping profile, whereas a practical doping profile with a doping gradient of 2 nm/dec is used for the p-channel TFET used in this study.

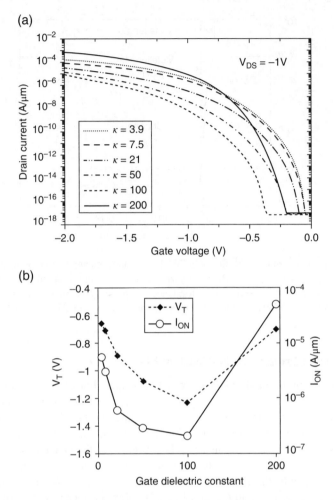

Figure 34.3 Impact of varying the κ value of the gate dielectric for the device in Figure 34.1 with $L_G = 50$ nm and EOT $= 2$ nm but without a spacer. (a) Transfer characteristics at $V_{DS} = -1$ V and (b) variation of V_T and I_{ON} with the κ value. Copyright © 2011 IEEE. Reprinted, with permission, from A. Mallik and A. Chattopadhyay, "The impact of fringing field on the device performance of a p-channel tunnel field-effect transistor with a high-κ gate dielectric," IEEE Trans. Electron Devices, vol. 59, no. 2, pp. 277–282, 2012.

It has been also verified that the qualitative results presented in this chapter for I_{ON} do not change when the same is extracted at a constant gate overdrive voltage $V_{GT} = V_{GS} - V_T$.

The simulated electron energy band diagram for the p-TFET biased at $V_{GS} = V_{DS} = -1$ V at a distance of 1 nm below the top oxide-semiconductor interface is shown in Figure 34.4 for different κ values of the gate dielectric. The band diagram is shown only around the tunneling junction. In a p-TFET, the BTBT of holes occurs from the source to the channel, resulting in the drain current. This is equivalent to the BTBT of electrons from the valance band of the channel, which is henceforth referred to as the *tunnel source*, into the conduction band of the source,

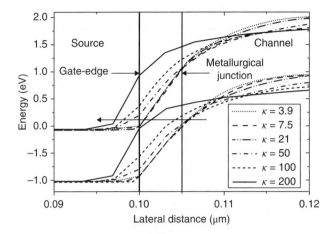

Figure 34.4 Simulated energy band diagram near the tunneling junction at a depth of 1 nm from the oxide-semiconductor interface as a function of the κ value of the gate dielectric at $V_{GS}=V_{DS}=-1\,V$. Copyright © 2011 IEEE. Reprinted, with permission, from A. Mallik and A. Chattopadhyay, "The impact of fringing field on the device performance of a p-channel tunnel field-effect transistor with a high-κ gate dielectric," IEEE Trans. Electron Devices, vol. 59, no. 2, pp. 277–282, 2012.

which is henceforth referred to as the *tunnel destination*. The potential barrier formed by the band gap of the semiconductor across which the tunneling takes place is henceforth referred to as the *tunneling junction*. The fringing field arising out of a high-κ dielectric causes fringing-induced barrier lowering for an n-channel device [1]. In contrast, it is observed in Figure 34.4 that the bands are pushed up in energy when the κ value of the gate dielectric is increased in the case of a p-TFET, which is simply due to the use of opposite polarity (negative) of the gate bias in such devices. The fringing field in the device is shown in Figure 34.5 for two different κ values of the gate dielectric, corresponding to SiO_2 and HfO_2. It is observed in Figure 34.5 that the fringing field in the source near the gate edge is larger for a device with a high-κ gate dielectric, compared with that with a low-κ dielectric. As a result, as evident in Figure 34.4, the impact of the fringing field is found to be larger near the gate edge than the rest of the device. It is also evident in Figure 34.4 that, when the κ value is increased from 3.9 to 100, the amount by which the energy bands are pushed up, which is henceforth referred to as *the impact of fringing field*, becomes larger for the tunnel destination than that for the tunnel source. This results in a larger value of the minimum tunnel width and a correspondingly smaller value of the maximum electric field across the tunneling junction, when the κ value of the gate dielectric is increased from 3.9 to 100, as can be verified in Figure 34.6. This, in turn, causes degradation of the device performance for increasing the κ value of the gate dielectric up to $\kappa=100$, as observed in Figure 34.3. It is also noticed in Figure 34.4 that an increase in the κ value results in a corresponding shift in the location of the tunneling junction toward the source. Due to a significant shift in the location of the tunneling junction when the κ value is increased from 100 to 200, the impact of the fringing field, which is always at its maximum at the gate edge, becomes larger at the tunnel source than that at the tunnel destination. As a result, an improvement in the device performance is observed in Figure 34.3 when the κ value is increased from 100 to 200.

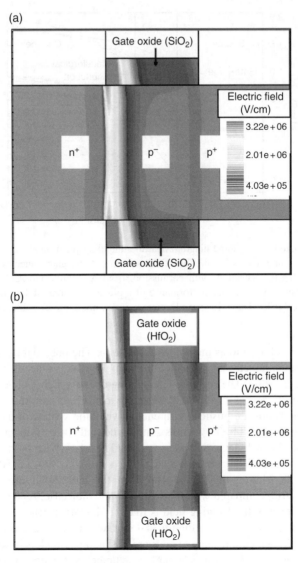

Figure 34.5 Fringing field plot for the p-TFET biased at $V_{GS} = V_{DS} = -1\,\mathrm{V}$ for two different κ values of the gate dielectric of EOT = 2 nm. (a) $\kappa = 3.9$ and (b) $\kappa = 21$. Copyright © 2011 IEEE. Reprinted, with permission, from A. Mallik and A. Chattopadhyay, "The impact of fringing field on the device performance of a p-channel tunnel field-effect transistor with a high-κ gate dielectric," IEEE Trans. Electron Devices, vol. 59, no. 2, pp. 277–282, 2012.

34.3.1.2   Impact of Varying the Source Doping

The impact of varying the source doping concentration on the gate-dielectric dependence of the device performance is now investigated. The device structure, as shown in Figure 34.1, with $L_G = 50\,\mathrm{nm}$ and EOT = 2 nm but without a spacer, as in the previous case, is simulated for two different values of the source doping concentration as 3×10^{20} and 5×10^{19} atoms/cm³.

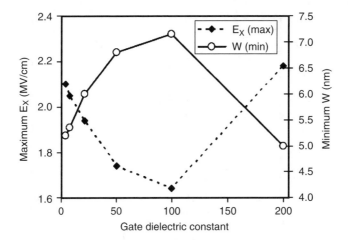

Figure 34.6 Minimum tunnel width and maximum electric field across the tunneling junction for the p-TFET biased at $V_{GS}=V_{DS}=-1$ V for different κ values of the gate dielectric of the same EOT. Copyright © 2011 IEEE. Reprinted, with permission, from A. Mallik and A. Chattopadhyay, "The impact of fringing field on the device performance of a p-channel tunnel field-effect transistor with a high-κ gate dielectric," IEEE Trans. Electron Devices, vol. 59, no. 2, pp. 277–282, 2012.

For each of the source doping concentrations, the transfer characteristic is plotted in Figure 34.7 for two different κ values of the gate dielectric, corresponding to SiO_2 and HfO_2. The corresponding energy band diagrams at $V_{GS}=V_{DS}=-1$ V are shown in Figure 34.8. For a given value of κ, it is observed in Figure 34.7 that an increase in the source doping concentration results in corresponding improvement in the device performance, as expected. This is simply due to fact that a relatively higher source doping not only causes more band lowering of the tunnel destination but it also reduces the source depletion, and, hence, less depletion width, as evident in Figure 34.8. As a result, the minimum value of the tunnel width is reduced, and the maximum electric field is increased across the tunneling junction. It is, however, interesting to note that, in Figure 34.7, a relatively lower source doping reduces the gate-dielectric dependence of the device performance, which is in contrast with that intuitively expected. This can be attributed to the combined influence of two factors. First, a decrease in the source doping concentration results in corresponding shift in the location of the tunneling junction toward the source, which is due to larger source depletion for lower source doping, as evident in Figure 34.8. Second, a larger impact of the fringing field arising out of the gate dielectric is observed at the gate edge than the rest of the device, as can also be verified in Figure 34.8. As a result, for a device with higher source doping, the impact of the fringing field across the tunneling junction is found to be much larger at the tunnel destination, compared with that at the tunnel source. This, in turn, results in significant widening of the tunnel, thereby degrading the device performance drastically. In contrast, due to a shift in the location of the tunneling junction toward the source for a device with lower source doping, the impact of the fringing field on the tunnel source partly compensates its impact on the tunnel destination. As a result, the gate-dielectric dependence of the device characteristics is somewhat reduced when a relatively lower source doping concentration is used.

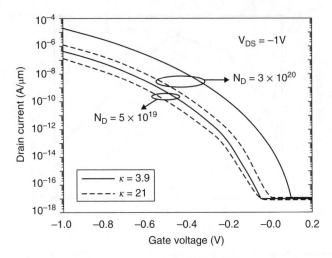

Figure 34.7   Transfer characteristics at $V_{DS} = -1\,\text{V}$ showing the impact of varying the source doping concentration on the gate-dielectric dependence of p-TFET characteristics. No spacer is used in this case. The EOT of the gate dielectric is kept constant at 2 nm. Copyright © 2011 IEEE. Reprinted, with permission, from A. Mallik and A. Chattopadhyay, "The impact of fringing field on the device performance of a p-channel tunnel field-effect transistor with a high-κ gate dielectric," IEEE Trans. Electron Devices, vol. 59, no. 2, pp. 277–282, 2012.

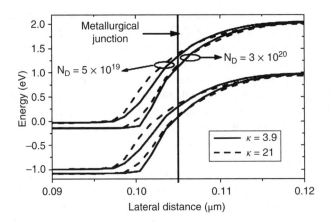

Figure 34.8   Impact of varying the source doping concentration on the simulated energy band diagram near the tunneling junction of the p-TFETs with different gate dielectrics biased at $V_{GS} = V_{DS} = -1\,\text{V}$. Copyright © 2011 IEEE. Reprinted, with permission, from A. Mallik and A. Chattopadhyay, "The impact of fringing field on the device performance of a p-channel tunnel field-effect transistor with a high-κ gate dielectric," IEEE Trans. Electron Devices, vol. 59, no. 2, pp. 277–282, 2012.

34.3.1.3   Impact of Device Scaling

The impact of device scaling on the gate-dielectric dependence of p-TFET characteristics is now investigated. For this purpose, device simulation is done for two different p-TFETs: one with $L_g = 100\,\text{nm}$ and EOT = 1 nm and the other with $L_G = 30\,\text{nm}$ and EOT = 1 nm. All other

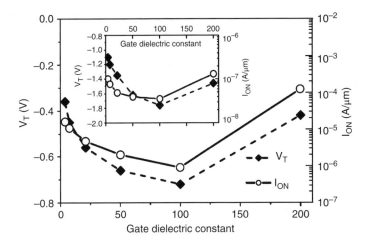

Figure 34.9 Plot showing the variations of V_T and I_{ON} with the κ value of the gate dielectric for a p-TFET with $L_G = 30$ nm and EOT $= 1$ nm. The inset shows the plot for a p-TFET with $L_G = 100$ nm and EOT $= 4$ nm. Copyright © 2011 IEEE. Reprinted, with permission, from A. Mallik and A. Chattopadhyay, "The impact of fringing field on the device performance of a p channel tunnel field-effect transistor with a high-κ gate dielectric," IEEE Trans. Electron Devices, vol. 59, no. 2, pp. 277–282, 2012.

device dimensions and parameters are kept the same as in the previous case. Figure 34.9 shows the plot of V_T and I_{ON} as a function of the κ value of the gate dielectric for the device with $L_G = 30$ nm and EOT $= 1$ nm, whereas the inset of Figure 34.9 shows the plot for the device with $L_G = 100$ nm and EOT $= 4$ nm. It is evident in Figure 34.9 that the qualitative nature of the impact of fringing field does not change with dimension scaling. An increase in κ value, except from $\kappa = 100$ to 200, results in corresponding degradation of the device performance in terms of both V_T and I_{ON} for both the devices, similar to that observed in Figure 34.3b for a device with $L_G = 50$ nm and EOT $= 2$ nm. In addition, an improvement in both V_T and I_{ON} is observed for $\kappa = 200$, compared with that for $\kappa = 100$ in Figure 34.9 for both the devices, similar to that observed in Figure 34.3b.

The TFET characteristics are known to be independent of the channel length down to about 20 nm [7–9]. A careful examination of I_{ON} values in Figure 34.9, however, reveals that I_{ON} increases with decreasing channel length for a given κ value of the gate dielectric. It has been verified that such an increase in I_{ON} is due to the scaling of the EOT in the devices. I_{ON} has been found to be less sensitive to channel length scaling when the EOT is kept at the same value, which is consistent with the reported results.

34.3.1.4   Impact of Higher Electric Field (Constant Physical Thickness)

The impact of the higher electric field at the surface of a p-TFET, when a high-κ gate dielectric is used in it, by keeping the physical thickness of the gate dielectric constant was investigated. For this purpose, a device simulation is carried out for a p-TFET, as shown in Figure 34.1 with $L_G = 50$ nm, for varying κ values of the gate dielectric with a physical thickness of 3 nm. The κ values used in this case are 3.9, 7.5, 21, and 29. No spacer is used in this case also, as earlier.

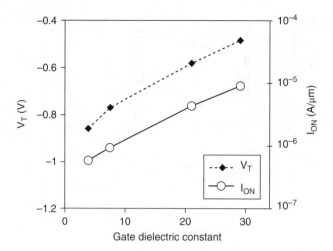

Figure 34.10 V_T and I_{ON} as a function of the κ value of the gate dielectric of 3 nm physical thickness for a p-TFET with $L_G = 50$ nm but without a spacer. Copyright © 2011 IEEE. Reprinted, with permission, from A. Mallik and A. Chattopadhyay, "The impact of fringing field on the device performance of a p-channel tunnel field-effect transistor with a high-κ gate dielectric," IEEE Trans. Electron Devices, vol. 59, no. 2, pp. 277–282, 2012.

The values of V_T and I_{ON} are extracted for different κ values of the gate dielectric from the simulated transfer characteristics. Figure 34.10 shows the plot of V_T and I_{ON} as a function of κ value of the gate dielectric. It is observed in Figure 34.10 that the device performance is improved both in terms of I_{ON} and V_T, when a high-κ gate dielectric of the same physical thickness is used in a p-TFET. A high-κ gate dielectric of the same physical thickness has a larger oxide capacitance, which is expected to increase the tunneling probability as per equation 4 in reference [5], which was originally derived in [10] for silicon-on-insulator MOSFETs. To verify it, the simulated electron energy band diagram around the tunneling junction for the p-TFET biased at $V_{GS} = -0.6$ V and $V_{DS} = -1$ V at a distance of 1 nm below the top oxide-semiconductor interface for different κ values of the gate dielectric of the same physical thickness is shown in Figure 34.11. Increasing the κ value of the gate dielectric, of the same physical thickness, results in a higher surface electric field under the gate. As a result, a larger impact of the high-κ gate dielectric is observed in Figure 34.11 for the tunnel source, compared with that for the tunnel destination. This results in better device performance for a higher κ value of the gate dielectric of the same physical thickness, as observed in Figure 34.10. It may also be noted that, when the κ value of the gate dielectric is increased while keeping its physical thickness constant, the impact of the higher electric field due to increased oxide capacitance, which results in improved device performance, dominates over that of the fringing field, which results in degraded performance, as observed earlier.

34.3.2   *Effects of Variation in the Spacer Dielectric Constant*

The impact of a spacer on the device performance of a p-TFET is studied by varying the κ value of the spacer while keeping its width fixed at 50 nm for the device structure, as shown

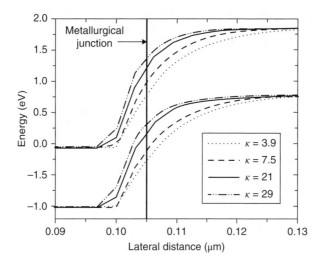

Figure 34.11 Simulated energy band diagram near the tunneling junction at a depth of 1 nm from the oxide-semiconductor interface as a function of the κ value of the gate dielectric of 3 nm physical thickness at $V_{GS}=-0.6$ V and $V_{DS}=1$ V. Copyright © 2011 IEEE. Reprinted, with permission, from A. Mallik and A. Chattopadhyay, "The impact of fringing field on the device performance of a p-channel tunnel field-effect transistor with a high-κ gate dielectric," IEEE Trans. Electron Devices, vol. 59, no. 2, pp. 277–282, 2012.

in Figure 34.1, with $L_G=50$ nm. A silicon dioxide of 1 nm thickness is used as the gate insulator in this case. The κ values of the spacer used are 3.9, 7.5, 21, and 50. The effects of varying the κ value of the spacer on the transfer characteristics of the p-TFET are shown in Figure 34.12. The device characteristics for a p-TFET without a spacer are also shown in Figure 34.12 for comparison. The values of V_T and I_{ON} are extracted from the transfer characteristics in Figure 34.12, and plotted in the inset of Figure 34.12 as a function of the κ value of the spacer. It is evident in Figure 34.12 that an increase in the κ value of the spacer results in a corresponding degradation of the device performance, in terms of both V_T and I_{ON}, of a p-TFET. This is very similar to that reported in [4] for an n-TFET of similar structure. The impact of the fringing field arising out of the spacer is expected to be larger for the source (tunnel destination) than that for the channel (tunnel source) for such a p-TFET structure that causes degradation of the device performance when the κ value of the spacer is increased.

Virani *et al.* [11] proposed an advanced dual-κ spacer technology for an n-TFET consisting of a 2 nm wide high-κ inner spacer adjacent to the gate and an 8 nm wide low-κ outer spacer for a 20 nm gate-length device, which has been found to improve the device performance of such devices. Moreover, it has also been reported in [11] that the fringing field out of a high-κ gate dielectric causes the performance of an n-TFET to deteriorate in the presence of such a dual-κ spacer, which is in contrast to the finding reported in [2] for a relatively simple n-TFET structure. In this chapter, the impact of a high-κ gate dielectric and a spacer on the device performance is investigated for a relatively simple p-TFET structure, similar to that used in [2–5] for an n-TFET, to develop a basic understanding of this subject. It is now open for researchers to investigate the impact of similar dual-κ spacer technology for a p-TFET.

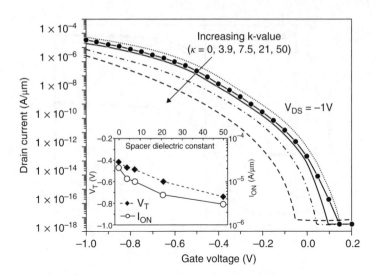

Figure 34.12 Impact of varying the κ value of the spacer on the transfer characteristics for the device in Figure 34.1 with $L_G = 50\,\text{nm}$. The inset shows the variations of V_T and I_{ON} with the κ value of the spacer. A silicon dioxide of 1 nm thickness is used as the gate dielectric in this case. Copyright © 2011 IEEE. Reprinted, with permission, from A. Mallik and A. Chattopadhyay, "The impact of fringing field on the device performance of a p-channel tunnel field-effect transistor with a high-κ gate dielectric," IEEE Trans. Electron Devices, vol. 59, no. 2, pp. 277–282, 2012.

34.4 Summary

The impact of the fringing field arising out of a high-κ gate dielectric on the device performance of a p-TFET was systematically investigated. The fringing field arising out of a high-κ gate dielectric was found to cause the performance of a p-TFET to deteriorate. It was also found that the fringing field arising out of a high-κ gate dielectric had greater impact for a p-TFET with a higher source doping concentration. In addition, the qualitative nature of the impact of fringing field was found to be independent of dimension scaling. The impact of a spacer on a p-TFET, however, was found to be very similar to that on an n-TFET of similar structure, as reported earlier. The findings can be a guide for the design of a p-TFET for complementary TFET applications.

References

[1] G. C.-F. Yeap, S. Krishnan, and M.-R. Lin, "Fringing-induced barrier lowering (FIBL) in sub-100 nm MOSFETs with high-κ gate dielectrics," *Electron. Lett*, vol. 34, no. 11, pp. 1150–1152, 1998.

[2] M. Schlosser, K. K. Bhuwalka, M. Sauter, T. Zilbauer, T. Sulima, and I. Eisele, "Fringing-induced drain current improvement in the tunnel field-effect transistor with high-κ gate dielectrics," *IEEE Trans. Electron Devices*, vol. 56, no. 1, pp. 100–108, 2009.

[3] C. Anghel, P. Chilagani, A. Amara, and A. Vladimirescu, "Tunnel field effect transistor with increased ON current, low-κ spacer and high-κ dielectric," *Appl. Phys. Lett.*, vol. 96, no. 12, pp. 122104-1–122104-3, 2010.

[4] A. Chattopadhyay and A. Mallik, "Impact of a spacer dielectric and a gate overlap/underlap on the device performance of a tunnel field-effect transistor," *IEEE Trans. Electron Devices*, vol. 58, no. 3, pp. 677–683, 2011.

[5] K. Boucart and A. M. Ionescu, "Double-gate tunnel FET with high-κ gate dielectric," *IEEE Trans. Electron Devices*, vol. 54, no. 7, pp. 1725–1733, 2007,

[6] B. Cheng, M. Cao, R. Rao, A. Inani, P. V. Voorde, W. M. Greene, J. M. C. Stork, Z. Yu, P. M. Zeitzoff, and J. C. S. Woo, "The impact of high-κ gate dielectrics and metal gate electrodes on sub-100 nm MOSFETs," *IEEE Trans. Electron Devices*, vol. 46, no. 7, pp. 1537–1544, 1999.

[7] K. Bhuwalka, J. Schulze, and I. Eisele, "Scaling the vertical tunnel FET with tunnel bandgap modulation and gate workfunction engineering," *IEEE Trans. Electron Devices*, vol. 52, no. 5, pp. 909–917, 2005.

[8] K. Boucart and A. M. Ionescu, "A new definition of threshold voltage in tunnel FETs," *Solid State Electron.*, vol. 52, no. 9, pp. 1318–1323, 2008.

[9] C. Sandow, J. Knoch, C. Urban, Q.-T. Zhao, and S. Mantl, "Impact of electrostatics and doping concentration on the performance of silicon tunnel field-effect transistors," *Solid State Electron.*, vol. 53, no. 10, pp. 1126–1129, 2009.

[10] R.-H. Yan, A. Ourmazd, and K. F. Lee, "Scaling the Si MOSFET: from bulk to SOI to bulk," *IEEE Trans. Electron Devices*, vol. 39, no. 7, pp. 1704–1710, 1992.

[11] H. G. Virani, R. B. R. Adari, and A. Kottantharayil, "Dual-κ spacer device architecture for the improvement of performance of silicon n-channel tunnel FETs," *IEEE Trans. Electron Devices*, vol. 57, no. 10, pp. 2410–2417, 2010.

35

Impact of a Spacer-Drain Overlap on the Characteristics of a Silicon Tunnel Field-Effect Transistor Based on Vertical Tunneling

35.1 Introduction

The main challenge to a tunnel field-effect transistor (TFET), particularly where the dominant carrier tunneling takes place from the source to the channel in a direction perpendicular to the gate field (henceforth referred to as *lateral carrier tunneling*), is to improve its ON-state current I_{ON} although some improvement in it is reported by the use of a lower band gap material such as SiGe [1, 2] in the tunneling region, a double-gate architecture [2, 3], a high-κ gate dielectric [3, 4], a thin silicon body [5], and so on.

On the other hand, both experimental and simulation results demonstrate [6–12] that a TFET, in which the dominant carrier tunneling occurs in a direction that is in line with the gate field (henceforth referred to as vertical carrier tunneling), can meet the I_{ON} requirements for sub-0.6 V operation, due to a larger area of tunneling in such devices. Despite producing a large I_{ON}, the compound semiconductor-based TFETs have a few major challenges, in addition to the processing issues, to make them really qualify for sub-0.6 V operation. These include the following: (i) large subthreshold swing (*SS*) and, hence, large OFF-state current I_{OFF} [7, 10] and (ii) large ambipolar current [7, 10]. I_{OFF} and *SS* for the L-shaped Ge source device, as proposed in [11], are also substantially higher in the absence of a technologically challenging SiO_2 isolation between the source and the drain under the channel. A silicon TFET structure based on vertical tunneling, termed a sandwich tunnel barrier field-effect transistor (STBFET), was reported by Asra *et al.* [9]; this suffers from processing complexity due to the requirements of a bottom source contact, a dummy-gate process, and two epitaxial processes. Although a promising complementary metal oxide-semiconductor (CMOS)-compatible silicon TFET structure was proposed by Leonelli *et al.* [12], a detailed understanding of the

MOS Devices for Low-Voltage and Low-Energy Applications, First Edition.
Yasuhisa Omura, Abhijit Mallik, and Naoto Matsuo.
© 2017 John Wiley & Sons Singapore Pte. Ltd. Published 2017 by John Wiley & Sons Singapore Pte. Ltd.

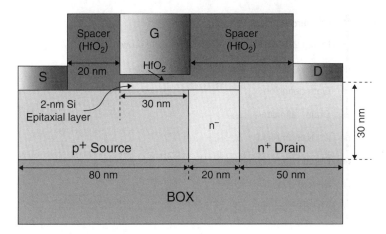

Figure 35.1   Device structure for n-channel silicon TFET. Copyright © 2013 IEEE. Reprinted, with permission, from A. Mallik, A. Chattopadhyay, S. Guin, and A. Karmakar, "Impact of a spacer-drain overlap on the characteristics of a silicon tunnel field-effect transistor based on vertical tunneling," IEEE Trans. Electron Devices, vol. 60, no. 3, pp. 935–943, 2013.

impact of different parameters including the drain potential on the device characteristics as well as the scaling properties of such a device is lacking. In this chapter, a detailed investigation of the impact of variation in the spacer-drain overlap on the device characteristics of such a TFET is made. The influence of the drain potential on the performance of such a device is also investigated, and an investigation of its scaling properties is also made.

35.2   Device Structure and Process Steps

A cross-sectional view of the silicon n-channel tunnel field-effect transistor (n-TFET) structure, used in this study, is shown in Figure 35.1. The highly doped p-region, the epitaxial region, and the highly doped n-region act as the source, the channel, and the drain, respectively. The fabrication steps to implement such a structure on a silicon-on-insulator substrate are summarized in Figure 35.2. After source definition using a photoresist mask and subsequent p^+ implant, the area on which epitaxial growth of silicon is to be made is defined using a hard mask. Following epitaxial growth of 2–3 nm silicon (doping type and concentration may be similar to that of n^--region), the gate consisting of a high-κ gate insulator and a metal of suitable work function φ_m is formed. The detail of the epitaxial growth may be found in [9] and the references therein. This is followed by drain definition using a photoresist mask, subsequent n^+ implant, and rapid thermal annealing for dopant activation. The spacer on both sides of the gate is defined next, which is followed by metallization/silicidation of the source and drain. The details of asymmetric spacer formation, which requires one additional step for selective deposition of the spacer material on the drain side with a mask on the source side, may be found in [13]. It may be noted that it is not easy to align the gate edges to both the epitaxial Si on the source side and the p^+-n^- junction on the drain side. Variability analysis may be performed in the future to find out the impact of such technological constraints on the device performance.

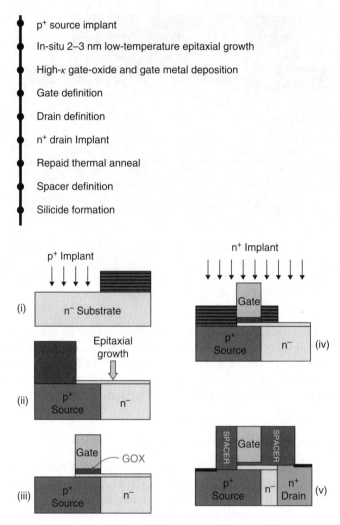

- p⁺ source implant
- In-situ 2–3 nm low-temperature epitaxial growth
- High-κ gate-oxide and gate metal deposition
- Gate definition
- Drain definition
- n⁺ drain Implant
- Repaid thermal anneal
- Spacer definition
- Silicide formation

Figure 35.2 Process flow for the silicon n-TFET.Copyright © 2013 IEEE. Reprinted, with permission, from A. Mallik, A. Chattopadhyay, S. Guin, and A. Karmakar, "Impact of a spacer-drain overlap on the characteristics of a silicon tunnel field-effect transistor based on vertical tunneling," IEEE Trans. Electron Devices, vol. 60, no. 3, pp. 935–943, 2013.

35.3 Simulation Setup

Two-dimensional device simulations are carried out for the TFET structure, as shown in Figure 35.1, using the Version 5.16.3.R Silvaco ATLAS device simulator. Unless otherwise mentioned, the device dimensions, as shown in Figure 35.1, are used for the simulations. The doping concentrations for the source, the drain, and the intermediate n⁻-regions are 2×10^{20}, 1×10^{20}, and 1×10^{17} atoms/cm³, respectively. HfO$_2$ of 0.4-nm equivalent oxide thickness (EOT) is used as the gate dielectric, as used in [9]. HfO$_2$ is also used as the high κ spacer,

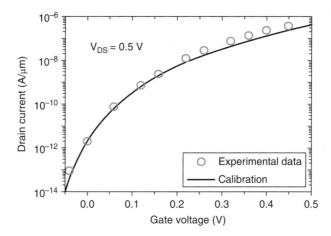

Figure 35.3 Model calibration against the experimental data in [6]. Copyright © 2013 IEEE. Reprinted, with permission, from A. Mallik, A. Chattopadhyay, S. Guin, and A. Karmakar, "Impact of a spacer-drain overlap on the characteristics of a silicon tunnel field-effect transistor based on vertical tunneling," IEEE Trans. Electron Devices, vol. 60, no. 3, pp. 935–943, 2013.

which is required to reduce the channel resistance and, hence, improve I_{ON} [9]. The source and drain contacts are made of aluminum, and the gate electrode is made of a metal of a given φ_m. Gate leakage is neglected in the simulations. The results presented here are obtained by using a nonlocal band-to-band tunneling (BTBT) model combined with a field-dependent mobility model, band-gap-narrowing model, and Shockley–Read–Hall recombination model. A nonlocal BTBT model accounts for the actual spatial charge transfer across a tunnel barrier by considering the actual potential profile along the entire path connected by tunneling. This results in the tunneling current taking into account the local generation rate, which depends on the local Fermi level and the spatial profile of energy bands, instead of electric-field value at each node, as in the case of the local BTBT model. A region consisting of several polygons is constructed around the p$^+$-n$^-$ junction where the quantum-tunnel meshes are carefully defined in both the vertical and horizontal directions, in order to account for tunneling in all directions. It may be mentioned here that no quantization model is used for the quantum well formed by conduction band of the 2 nm thick epitaxial layer with the gate insulator. It has, however, been verified that the energy of at least the first sub-band in the potential well is smaller than that of the top of the valance band of the source region.

The model is first calibrated against the experimental data as reported in [6]. The calibration is made for a device with Ge source, as experimental data on Si TFET based on vertical tunneling are not available. Figure 35.3 shows the plot of such a model calibration for which the device structure, along with its dimensions and parameters, is kept the same, as in [6]. Default material parameters pertaining to Ge, as available in Silvaco ATLAS, are used for the source region to generate the simulated curves in Figure 35.3. A good agreement between the simulated and experimental characteristics is evident in Figure 35.3. It may be noted further that the focus of this work is not on the exact values of currents but more on the relative results based on the electrostatics of the device.

35.4   Results and Discussion

35.4.1   Impact of Variation in the Spacer-Drain Overlap

The impact of variation in the spacer-drain overlap on the device characteristics of the Si-TFET structure, as shown in Figure 35.1, is first studied. In order to do this, six different values of the spacer-drain overlap from −5 (i.e., 5 nm underlap of the spacer with the n^--n^+ junction) to 20 nm in steps of 5 nm are used. The simulated transfer characteristics for such different values of the spacer-drain overlap at a drain-to-source voltage V_{DS}=0.6 V are shown in Figure 35.4. The value of φ_m for the gate metal used in this case is 4.0 eV. It is observed in Figure 35.4 that the subthreshold characteristics for V_{GS}<0.1 V or so strongly depend upon the spacer-drain overlap. In order to compare the device performance, the following are used: (i) I_D=0.1 pA/µm as I_{OFF} and the corresponding V_{GS} as V_{OFF} [9, 11]; (ii) the value of I_D at V_{GS}=V_{OFF}+0.6 V as I_{ON}; (iii) the value of V_{GS} corresponding to I_D=0.1 µA/µm as the threshold voltage V_T; and (iv) average SS=$(V_T-V_{OFF})/6$ [11, 14]. A plot of I_{ON}/I_{OFF} and average SS for different spacer-drain overlaps is shown in the inset of Figure 35.4. It is evident from the inset in Figure 35.4 that an increase in the spacer-drain overlaps results in significant improvement in both the average SS and I_{ON}/I_{OFF}.

In order to gain an insight into it, the simulated BTBT rates are shown in Figure 35.5 for the 0, 10, and 20 nm spacer-drain overlap devices at V_{GS}=0.0 V and V_{DS}=0.6 V. A comparison of the BTBT rates between such devices in Figure 35.5 reveals that a spacer-drain overlap has a significant impact on the BTBT rate at the gate edge. Although an appreciable BTBT occurs at the gate edge for the 0 nm overlap device, it is reduced for increasing values of the spacer-drain overlap. The BTBT becomes negligible for the 20 nm overlap device. It may also be observed in Figure 35.5 that the lateral tunneling dominates over the vertical tunneling at such low values of V_{GS}. It is well known that fringing field arising out of a high-κ spacer causes

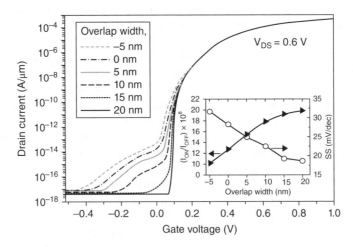

Figure 35.4   Impact of variation in the spacer-drain overlap on the transfer characteristics for the n-TFET with EOT=0.4 nm. The inset shows variation in I_{ON}/I_{OFF} and average SS with spacer-drain overlap. Copyright © 2013 IEEE. Reprinted, with permission, from A. Mallik, A. Chattopadhyay, S. Guin, and A. Karmakar, "Impact of a spacer-drain overlap on the characteristics of a silicon tunnel field-effect transistor based on vertical tunneling," IEEE Trans. Electron Devices, vol. 60, no. 3, pp. 935–943, 2013.

fringing-induced barrier lowering (FIBL) [15]. It has been reported earlier that such FIBL has maximum impact near the gate edge [16, 17]. It is believed that an overlap of the spacer with the drain, by extending it beyond the n^--n^+ junction for the TFET structure, as shown in Figure 35.1, results in reduced fringing field, which is responsible for reduced FIBL at the gate edge of the device. To verify it, the fringing field in the device is shown in Figure 35.6 for the 0, 10, and 20 nm spacer-drain overlap devices at $V_{GS}=0.0$ V and $V_{DS}=0.6$ V. It is evident in Figure 35.6 that the fringing field is larger for the device with 0 nm overlap, which is reduced for increasing spacer-drain overlaps. The simulated energy band diagram in the

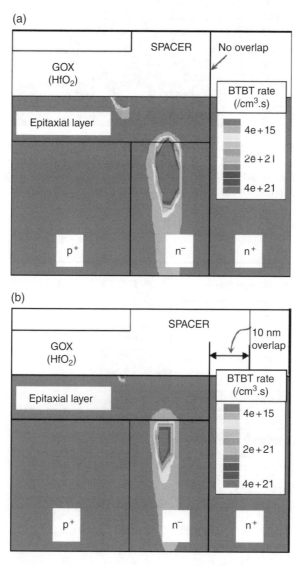

Figure 35.5 Plots showing the BTBT rates for the n-TFET biased at $V_{GS}=0.0$ V and $V_{DS}=0.6$ V for different values of the spacer-drain overlap. (a) No overlap, (b) 10 nm overlap, and

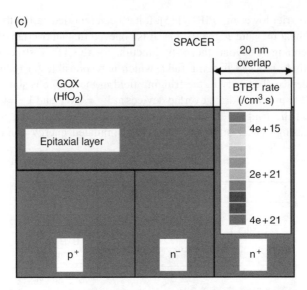

Figure 35.5 (Continued) (c) 20 nm overlap. Copyright © 2013 IEEE. Reprinted, with permission, from A. Mallik, A. Chattopadhyay, S. Guin, and A. Karmakar, "Impact of a spacer-drain overlap on the characteristics of a silicon tunnel field-effect transistor based on vertical tunneling," IEEE Trans. Electron Devices, vol. 60, no. 3, pp. 935–943, 2013.

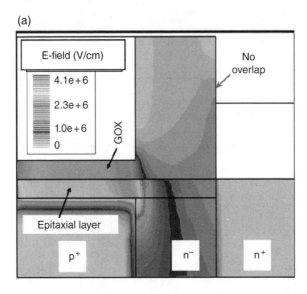

Figure 35.6 Fringing field plots for the n-TFET biased at $V_{GS}=0.0\,\text{V}$ and $V_{DS}=0.6\,\text{V}$ for different values of the spacer-drain overlap. (a) No overlap,

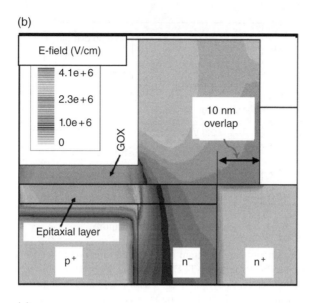

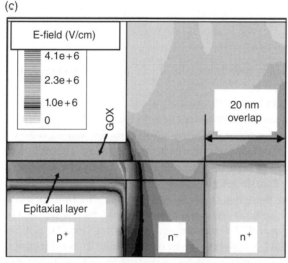

Figure 35.6 (Continued) (b) 10 nm overlap, and (c) 20 nm overlap. Copyright © 2013 IEEE. Reprinted, with permission, from A. Mallik, A. Chattopadhyay, S. Guin, and A. Karmakar, "Impact of a spacer-drain overlap on the characteristics of a silicon tunnel field-effect transistor based on vertical tunneling," IEEE Trans. Electron Devices, vol. 60, no. 3, pp. 935–943, 2013.

lateral direction at 3 nm below the oxide-semiconductor interface at $V_{GS}=0.0$ V and $V_{DS}=0.6$ V is shown in Figure 35.7 for the 0, 10, and 20 nm overlap devices. Reduction in fringing field with an increase in the spacer-drain overlap, as observed in Figure 35.6, results in less FIBL of the channel side of the tunneling region, as can be verified in Figure 35.7. This, in turn, results in increasing tunnel width for increasing values of the spacer-drain overlap, yielding a

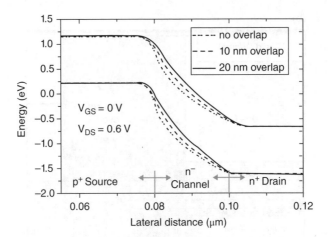

Figure 35.7 Simulated energy band diagram in the lateral direction at a distance of 3 nm below the oxide-semiconductor interface for the n-TFET biased at $V_{GS}=0.0$ V and $V_{DS}=0.6$ V for different values of the spacer-drain overlap. Copyright © 2013 IEEE. Reprinted, with permission, from A. Mallik, A. Chattopadhyay, S. Guin, and A. Karmakar, "Impact of a spacer-drain overlap on the characteristics of a silicon tunnel field-effect transistor based on vertical tunneling," IEEE Trans. Electron Devices, vol. 60, no. 3, pp. 935–943, 2013.

supersteep subthreshold characteristic for the 20 nm overlap device, in comparison with the others, as observed in Figure 35.4.

It is also observed in Figure 35.4 that a variation in the spacer-drain overlap does not have any significant influence on the transfer characteristics for $V_{GS} \geq 0.1$ V or so. In order to provide an insight into it, the simulated BTBT rate at $V_{GS}=0.25$ V and $V_{DS}=0.6$ V for the 0 and 20 nm spacer-drain overlap devices is shown in Figure 35.8. The following are observed in Figure 35.8. First, at such relatively large value of V_{GS}, the vertical BTBT rate is several orders of magnitude higher than the lateral BTBT rate for both the devices. Second, a variation in the spacer-drain overlap does not make any significant variation in the vertical BTBT rate, although it may have some impact on the lateral BTBT rate, which is, however, negligible as compared to the vertical BTBT rate at such value of V_{GS}. Finally, again for both the devices, the vertical BTBT rate is of the same order of magnitude for most part of the tunneling region. As a result, the transfer characteristics, as plotted in Figure 35.4, do not show much dependence on variation in the value of spacer-drain overlap for $V_{GS} \geq 0.1$ V or so.

Figure 35.9 shows a comparison of the simulated transfer characteristics at $V_{DS}=0.6$ V between the STBFET structure in [9] and the device, as shown in Figure 35.1, with different values of spacer-drain overlap and φ_m. A gate length L_G of 30 nm is used for all the devices in Figure 35.9. The value of φ_m for the gate metal used for both the STBFET and the device with no spacer-drain overlap is 4.0 eV. Unless otherwise mentioned, all other dimensions and parameters are kept the same for all the devices in Figure 35.9. A comparison of the transfer characteristics between the STBFET and the structure with no spacer-drain overlap reveals that the subthreshold characteristic is slightly degraded for the STBFET. It is believed that an increase in the lateral BTBT rate, due to the presence of two drains on either side of the gate in the STBFET, causes such degradation in the subthreshold characteristics. As evident in

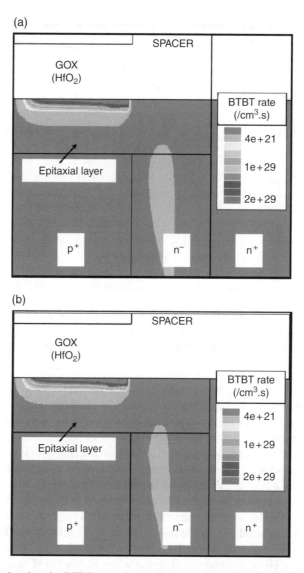

Figure 35.8 Plots showing the BTBT rates for the TFET for two different values of the spacer-drain overlap at $V_{GS}=0.25\,\text{V}$ and $V_{DS}=0.6\,\text{V}$. (a) No overlap and (b) 20 nm overlap. Copyright © 2013 IEEE. Reprinted, with permission, from A. Mallik, A. Chattopadhyay, S. Guin, and A. Karmakar, "Impact of a spacer-drain overlap on the characteristics of a silicon tunnel field-effect transistor based on vertical tunneling," IEEE Trans. Electron Devices, vol. 60, no. 3, pp. 935–943, 2013.

Figure 35.9, SS for the 20 nm overlap device with $\varphi_m=4.0\,\text{eV}$ is significantly less than that for both STBFET and the structure with no spacer-drain overlap. This allows one to reduce both V_{OFF} and V_T by 0.08 V for the 20 nm overlap device by reducing the value of φ_m from 4.0 to 3.92 eV, as can be verified in Figure 35.9. This is significant as it would allow more room to scale V_{DD} even further for such devices. Although the vertical BTBT rate, as observed in

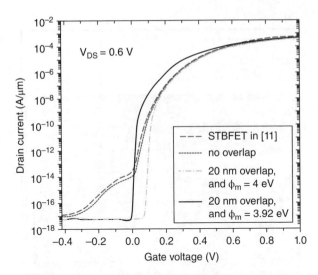

Figure 35.9   Comparison of simulated transfer characteristics at $V_{DS}=0.6$ V between the STBTFET structure in [9] and the device, as shown in Figure 35.1, with different values of the spacer-drain overlap and φ_m. Copyright © 2013 IEEE. Reprinted, with permission, from A. Mallik, A. Chattopadhyay, S. Guin, and A. Karmakar, "Impact of a spacer-drain overlap on the characteristics of a silicon tunnel field-effect transistor based on vertical tunneling," IEEE Trans. Electron Devices, vol. 60, no. 3, pp. 935–943, 2013.

Figure 35.8, is of the same order of magnitude for most part of the tunneling region, a slightly higher rate is observed for the drain side of it, which is probably due to the influence of the drain potential. The presence of two drains on either side of the gate in STBFET results in slightly higher I_D at larger V_{GS}, as observed in Figure 35.9.

In order to find out the impact of variation in EOT on the spacer-drain overlap dependence of the device characteristic, the simulated transfer characteristics for different values of the spacer-drain overlap at $V_{DS}=0.6$ V are shown in Figure 35.10 for EOT=0.8 nm. The qualitative nature of the spacer-drain overlap dependence, as observed in Figure 35.10 for EOT=0.8 nm, is found to be very similar to that observed in Figure 35.4 for EOT=0.4 nm. The values of I_{ON}/I_{OFF} and average SS are shown in the inset of Figure 35.10 both for EOT=0.8 and 0.4 nm, which also look very similar in nature. Degradation in both I_{ON}/I_{OFF} and average SS, for any given spacer-drain overlap, is, however, observed when a higher EOT is used, which is due to the reduced gate control for thicker gate insulator.

Figure 35.11 shows the impact of variation in the doping concentration of the epitaxial layer. It also shows that varying the doping concentration of the epitaxial layer does not have any significant impact on the device characteristics for relatively low doping level ($\leq 10^{17}$ atoms/cm^3). Larger doping levels (up to about 10^{19} atoms/cm^3 or so) reduce the value of V_T. A very large doping level, such as 10^{20} atoms/cm^3 or so, however, degrades the subthreshold performance due to significant carrier tunneling from the source region to the epitaxial region above the n$^-$-layer, as the drain field cannot penetrate easily through the epitaxial region under the gate, as explained in the next section. Not shown in this chapter is the impact of variation in the thickness of the epitaxial layer, which has been found to be very similar to that reported in [12] for a similar structure.

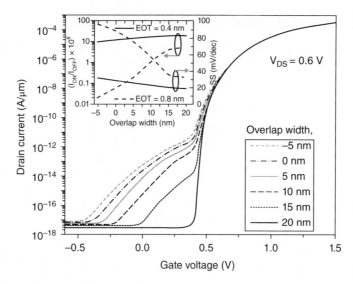

Figure 35.10 Impact of variation in the spacer-drain overlap on the transfer characteristics for the n-TFET with EOT=0.8 nm. The inset shows variation in I_{ON}/I_{OFF} and average SS with spacer-drain overlap both for EOT=0.8 and 0.4 nm. Copyright © 2013 IEEE. Reprinted, with permission, from A. Mallik, A. Chattopadhyay, S. Guin, and A. Karmakar, "Impact of a spacer-drain overlap on the characteristics of a silicon tunnel field-effect transistor based on vertical tunneling," IEEE Trans. Electron Devices, vol. 60, no. 3, pp. 935–943, 2013.

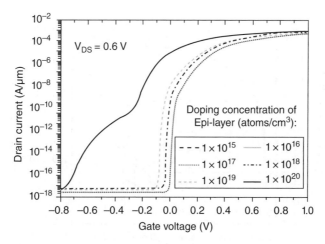

Figure 35.11 Impact of variation in the doping concentration of the epitaxial layer. Copyright © 2013 IEEE. Reprinted, with permission, from A. Mallik, A. Chattopadhyay, S. Guin, and A. Karmakar, "Impact of a spacer-drain overlap on the characteristics of a silicon tunnel field-effect transistor based on vertical tunneling," IEEE Trans. Electron Devices, vol. 60, no. 3, pp. 935–943, 2013.

35.4.2 *Influence of Drain on the Device Characteristics*

The transfer characteristics at different values of V_{DS} and the output characteristics at different values of the gate over-drive voltage V_{GT} $(= V_{GS} - V_T)$, for the device under study, are shown in Figure 35.12 and Figure 35.13, respectively. Unless otherwise mentioned, a 20 nm spacer-drain overlap and $\varphi_m = 3.92\,eV$ are used for the devices in this section, as it produces the best characteristics, as observed in Figure 35.9. It is evident in Figure 35.12 that the transfer characteristics do not show any significant V_{DS}-dependence in the subthreshold region, except in the p-i-n leakage current. Also, long-channel MOSFET-like output characteristics are observed in Figure 35.13 for the device, which look very similar to those for the STBFET and the Ge-source TFET, as reported in [6] and [9], respectively. The impact of short-channel effects, such as drain-induced barrier lowering (DIBL), on the device is greatly reduced both below and above V_T, as observed in Figure 35.12 and Figure 35.13.

The output characteristics of a typical long-channel MOSFET are characterized by a channel-resistance limited region (comprising a linear and a nonlinear subregions), and a saturation region. On the other hand, a TFET, based on lateral tunneling, is characterized by a tunnel-resistance limited region and a channel-resistance limited region [14, 18]. In addition, depending upon the construction of the channel, it may also have a perfect saturation region [19]. It may be clearly observed in Figure 35.13 that the output characteristics for the device do not show any tunnel-resistance limited region, and, hence, look very similar to those of a long-channel MOSFET.

To gain an insight into these observations, the electron concentration and the simulated energy band diagram at $V_{GT} = 0.5\,V$ in the lateral direction at a distance 1 nm below the oxide-semiconductor interface for different values of V_{DS} are shown in Figure 35.14. A similar energy

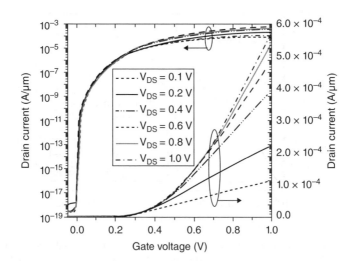

Figure 35.12 Simulated transfer characteristics for different values of V_{DS} for the n-TFET with 20 nm spacer-drain overlap and $\varphi_m = 3.92\,eV$. Copyright © 2013 IEEE. Reprinted, with permission, from A. Mallik, A. Chattopadhyay, S. Guin, and A. Karmakar, "Impact of a spacer-drain overlap on the characteristics of a silicon tunnel field-effect transistor based on vertical tunneling," IEEE Trans. Electron Devices, vol. 60, no. 3, pp. 935–943, 2013.

band diagram for different values of V_{DS} is reported in [9] for the STBFET. A physical explanation for such a band diagram, as observed in Figure 35.14, and the MOSFET-like output characteristics, as observed in Figure 35.13, is given hereinafter. When $V_{DS}=0$ V, a relatively high electron concentration exists in the n⁻-region under the spacer, as observed in Figure 35.14, which is due to accumulation caused by the high-κ insulator. As a result, the

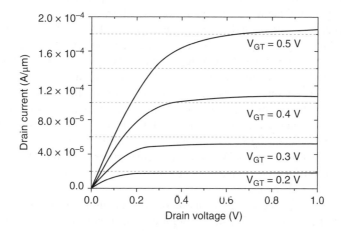

Figure 35.13 Simulated output characteristics at different values of the gate overdrive voltage V_{GT} for the n-TFET with 20 nm spacer-drain overlap and $\varphi_m = 3.92$ eV. Copyright © 2013 IEEE. Reprinted, with permission, from A. Mallik, A. Chattopadhyay, S. Guin, and A. Karmakar, "Impact of a spacer-drain overlap on the characteristics of a silicon tunnel field-effect transistor based on vertical tunneling," IEEE Trans. Electron Devices, vol. 60, no. 3, pp. 935–943, 2013.

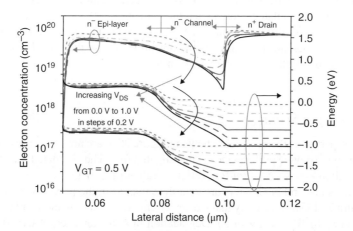

Figure 35.14 Electron concentration and simulated energy band diagram at $V_{GT}=0.5$ V in the lateral direction at a distance 1 nm below the oxide-semiconductor interface for different values of V_{DS} for the n-TFET with 20 nm spacer-drain overlap and $\varphi_m = 3.92$ eV. Copyright © 2013 IEEE. Reprinted, with permission, from A. Mallik, A. Chattopadhyay, S. Guin, and A. Karmakar, "Impact of a spacer-drain overlap on the characteristics of a silicon tunnel field-effect transistor based on vertical tunneling," IEEE Trans. Electron Devices, vol. 60, no. 3, pp. 935–943, 2013.

n⁻-region under the spacer acts like a resistor of relatively low value. On the other hand, although the electron concentration in the epitaxial region under the gate is high at such large value of V_{GT}, as observed in Figure 35.14, the resistance of this region is also relatively high due to its small thickness. Therefore, when a small V_{DS} is applied (up to 0.2 V or so), although the drain field can penetrate through the n⁻-region, it cannot do so easily through the epitaxial region. Hence, the applied V_{DS} mainly appears across the junction comprising the epitaxial layer under the gate and the n⁻-region, with a very small fraction of it causing band lowering of the epitaxial channel region under the gate, as evident in Figure 35.14, which results in the MOSFET-like linear I_D-V_{DS} region in its output characteristics, as observed in Figure 35.13. This also increases the potential of the n⁻-region under the spacer, which, in turn, reduces the electron concentration in this region, as can be observed in Figure 35.14, thereby increasing the resistance of this region. As a result, when V_{DS} is increased further (in the range between 0.2 and 0.4 V or so), a significant fraction of it also appears as resistive drop across the n⁻-region, due to its enhanced resistance, as evident in Figure 35.14. As a result, I_D increases at a retarded pace with increasing V_{DS} resulting in the nonlinear region of the output characteristics, as observed in Figure 35.13. When V_{DS} is increased further (above 0.4 V or so), due to reduced electron concentration in the n⁻-region, particularly near the drain end of it, the resistance of this becomes such that the drain field finds it difficult to penetrate this region, resulting in the additional V_{DS} appearing mainly across the junction comprising the n⁻-region and the drain, as evident in Figure 35.14. A careful examination of the output characteristics in Figure 35.13 reveals that a perfect saturation in I_D does not occur even at higher V_{DS}, although the rate of increase in I_D with V_{DS} is very small, and may often be interpreted as a perfect saturation. A perfect saturation occurs for a long-channel MOSFET due to the pinchoff of the channel near the drain. On the other hand, such saturation also occurs in a TFET with a relatively thin channel due to complete pinchoff of the channel [19]. In the absence of such a complete pinchoff, as evident in Figure 35.14, a perfect saturation does not occur for the device under study. A simulated energy band diagram at V_{GT}=0.5 V in the vertical direction at the middle of the gate for different values of V_{DS} is shown in Figure 35.15, which confirms that an increase in V_{DS} does not have any significant influence on the tunnel parameters in the vertical direction, particularly at larger values of V_{DS}.

35.4.3   Impact of Scaling

The scaling properties of the device, as shown in Figure 35.1, are now investigated. The impact of L_G scaling on the device performance is investigated first. The simulated transfer characteristics are plotted in Figure 35.16 for four different values of L_G, namely, 30, 20, 15, and 10 nm. All other dimensions and parameters are kept the same, as in the previous section, for all the devices. Figure 35.16 shows clearly that the device characteristics are more or less independent of variation in L_G for V_{GS}<0.1 V or so. Degradation in I_D is, however, observed for large V_{GS} values, particularly for the devices with lower L_G, which is due to the reduction in the area of tunneling. However, the reduction in I_D with L_G is not proportional, as the BTBT rate is higher toward the drain side, due to the influence of drain potential, as observed earlier in Figure 35.8.

The impact of scaling of the width of the n⁻-layer between the source and the drain L_{n^-} is investigated next for a device of L_G=30 nm. All other dimensions and parameters are kept the same, as before. The simulated transfer characteristics are plotted in Figure 35.17 for three

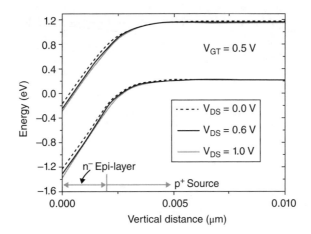

Figure 35.15 Simulated energy band diagram at $V_{GT}=0.5\,V$ in the vertical direction at the middle of the gate for different values of V_{DS} for the n-TFET with 20 nm spacer-drain overlap and $\varphi_m=3.92\,eV$. Copyright © 2013 IEEE. Reprinted, with permission, from A. Mallik, A. Chattopadhyay, S. Guin, and A. Karmakar, "Impact of a spacer-drain overlap on the characteristics of a silicon tunnel field-effect transistor based on vertical tunneling," IEEE Trans. Electron Devices, vol. 60, no. 3, pp. 935–943, 2013.

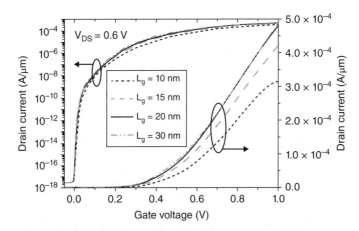

Figure 35.16 Impact of gate-length scaling on the device characteristics for a TFET with 20 nm overlap spacer and $\varphi_m=3.92\,eV$. Copyright © 2013 IEEE. Reprinted, with permission, from A. Mallik, A. Chattopadhyay, S. Guin, and A. Karmakar, "Impact of a spacer-drain overlap on the characteristics of a silicon tunnel field-effect transistor based on vertical tunneling," IEEE Trans. Electron Devices, vol. 60, no. 3, pp. 935–943, 2013.

different values of L_{n^-}, namely, 20, 15, and 10 nm. It is observed in Figure 35.17 that I_D is improved with L_{n^-} scaling for large V_{GS} values, which is due to the reduction in series resistance of the n⁻ layer with L_{n^-} scaling. Degradation in SS and I_{OFF} with L_{n^-} scaling is, however, evident in Figure 35.17, which may be attributed to increased lateral tunneling, very similar to that observed in Figure 35.4 for reduced spacer-drain overlaps.

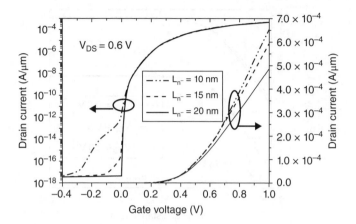

Figure 35.17 Impact of scaling of the n^--layer between the source and the drain on the device characteristics for a TFET of $L_G = 30$ nm with 20 nm overlap spacer and $\varphi_m = 3.92$ eV. Copyright © 2013 IEEE. Reprinted, with permission, from A. Mallik, A. Chattopadhyay, S. Guin, and A. Karmakar, "Impact of a spacer-drain overlap on the characteristics of a silicon tunnel field-effect transistor based on vertical tunneling," IEEE Trans. Electron Devices, vol. 60, no. 3, pp. 935–943, 2013.

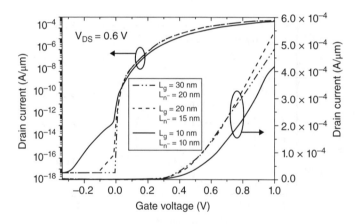

Figure 35.18 Impact of scaling of both L_G and L_{n^-} on the device characteristics for a TFET with 20 nm overlap spacer and $\varphi_m = 3.92$ eV. Copyright © 2013 IEEE. Reprinted, with permission, from A. Mallik, A. Chattopadhyay, S. Guin, and A. Karmakar, "Impact of a spacer-drain overlap on the characteristics of a silicon tunnel field-effect transistor based on vertical tunneling," IEEE Trans. Electron Devices, vol. 60, no. 3, pp. 935–943, 2013.

The impact of combined L_G and L_{n^-} scaling is studied next for area scaling. All other dimensions and parameters are kept the same, as before. The simulated transfer characteristics are plotted in Figure 35.18 for three different combinations of L_G and L_{n^-}: (i) $L_G = 30$ nm and L_{n^-} $L_n- = 20$ nm; (ii) $L_G = 20$ nm, and $L_{n^-} = 15$ nm; and (iii) $L_G = 10$ nm and $L_{n^-} = 10$ nm. Degradation in SS and I_{OFF} with area scaling, as observed in Figure 35.18, is due to l_n scaling.

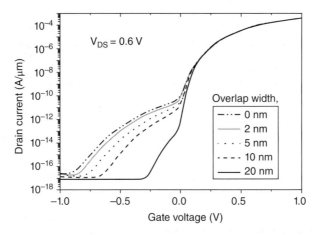

Figure 35.19 Impact of variation in the spacer-drain overlap on the transfer characteristics for an n-TFET with $L_G = 10$ nm and $L_{n^-} = 10$ nm. Copyright © 2013 IEEE. Reprinted, with permission, from A. Mallik, A. Chattopadhyay, S. Guin, and A. Karmakar, "Impact of a spacer-drain overlap on the characteristics of a silicon tunnel field-effect transistor based on vertical tunneling," IEEE Trans. Electron Devices, vol. 60, no. 3, pp. 935–943, 2013.

similar to that observed in Figure 35.17. I_D at relatively large values of V_{GS} initially increases and then decreases with area scaling, which is due to combined influences of the reduction in series resistance due to L_{n^-} scaling and the reduction in the area of tunneling, consistent with the earlier observations.

The impact of variation in the spacer-drain overlap on the device characteristics of a scaled device with $L_G = 10$ nm and $L_{n^-} = 10$ nm is shown in Figure 35.19. The qualitative nature of the impact of variation in the spacer-drain overlap in Figure 35.19 is very similar to that observed in Figure 35.4. Larger degradation in the subthreshold characteristics in Figure 35.19 is due to the additional influence of L_{n^-} scaling, as observed in Figure 35.17.

A larger spacer-drain overlap would also increase the overlap capacitance, which is likely to increase the delay time and, hence, reduce the switching speed. A future investigation may examine this.

35.5 Summary

An extensive investigation was made of the impact of a spacer-drain overlap on the device characteristics of CMOS-compatible Si-TFET structure based on vertical tunneling. It was demonstrated that, by careful design of the drain-side spacer, the impact of FIBL at the gate edge can be reduced at relatively low values of V_{GS}, where lateral tunneling dominated over the vertical tunneling, resulting in supersteep SS and greatly reduced I_{OFF}. This is significant as it allows more room to reduce further both V_T and V_{DD}. The absence of a tunnel-resistance limited region and greatly reduced short-channel effects, such as DIBL, in the device resulted in long-channel MOSFET-like output characteristics for such device structure. It was also found that, although I_{ON} is reduced with L_G scaling, the reduction is not proportional.

References

[1] K. K. Bhuwalka, J. Schulze, and I. Eisele, "Performance enhancement of vertical tunnel field-effect transistor with SiGe in the δp+ layer," *Jpn. J. Appl. Phys.*, 1, vol. 43, no. 7A, pp. 4073–4078, 2004.

[2] E.-H. Toh, G. H. Wang, L. Chan, D. Sylvester, C.-H. Heng, G.S. Samudra, and Y.-C. Lee, "Device design and scalability of a double-gate tunneling field-effect transistor with silicon-germanium source," *Jpn. J. Appl. Phys.*, vol. 47, no. 4, pp. 2593–2597, 2008.

[3] K. Boucart and A. M. Ionescu, "Double-gate tunnel FET with high-κ gate dielectric," *IEEE Trans. Electron Devices*, vol. 54, no. 7, pp. 1725–1733, 2007.

[4] M. Schlosser, K. K. Bhuwalka, M. Sauter, T. Zilbauer, T. Sulima, and I. Eisele, "Fringing-induced drain current improvement in the tunnel field-effect transistor with high-κ gate dielectrics," *IEEE Trans. Electron Devices*, vol. 56, no. 1, pp. 100–108, 2009.

[5] E.-H. Toh, G. H. Wang, G. Samudra, and Y.-C. Yeo, "Device physics and design of double-gate tunneling field-effect transistor by silicon film thickness optimization," *Appl. Phys. Lett.*, vol. 90, no. 26, pp. 263507-1–263507-3, 2007.

[6] S. H. Kim, H. Kam, C. Hu, and T. J. K. Liu, "Germanium-Source Tunnel Field Effect Transistors with Record High I_{ON}/I_{OFF}," Proc. Int. Symp. VLSI Technol., Jun. 2009, pp. 178–179, 2009.

[7] R. Li, Y. Lu, G. Zhou, Q. Liu, S. D. Chae, T. Vasen, W. S. Hwang, Q. Zhang, P. Fay, T. Kosel, M. Wistey, H. Xing, and A. Seabaugh, "AlGaSb/InAs tunnel field-effect transistor with on-current of 78 μA/μm at 0.5 V," *IEEE Electron Device Lett.*, vol. 33, no. 3, pp. 363–365, 2012.

[8] L. Lattanzio, L. D. Michielis, and A. M. Ionescu, "Complementary germanium electron–hole bilayer tunnel FET for sub-0.5-V operation," *IEEE Electron Device Lett.*, vol. 33, no. 2, pp. 167–169, 2012.

[9] R. Asra, M. Shrivastava, K. V. R. M. Murali, R. K. Pandey, H. Gossner, and V. R. Rao, "A tunnel FET for V_{DD} scaling below 0.6 V with a CMOS-comparable performance," *IEEE Trans. Electron Devices*, vol. 58, no. 7, pp. 1855–1863, 2011.

[10] Y. Lu, G. Zhou, R. Li, Q. Liu, Q. Zhang, T. Vasen, S. D. Chae, T. Kosel, M. Wistey, H. Xing, A. Seabaugh, and P. Fay, "Performance of AlGaSb/InAs TFETs with gate electric field and tunneling direction aligned," *IEEE Electron Device Lett.*, vol. 33, no. 5, pp. 655–657, 2012.

[11] K. L. Low, C. Zhan, G. Han, Y. Yang, K.-H. Goh, P. Guo, E.-H. Toh, and Y.-C. Yeo, "Device physics and design of a L-shaped germanium source tunneling transistor," *Jpn. J. Appl. Phys.*, vol. 51, no. 2, pp. 02BC04-1–02BC04-6, 2012.

[12] D. Leonelli, A. Vandooren, R. Rooyackers, A. S. Verhulst, C. Huyghebaert, S. De Gendt, M. M. Heyns, and G. Groeseneken, "Novel Architecture to Boost the Vertical Tunneling in Tunnel Field Effect Transistors," Proc. IEEE Int. SOI Conf., pp. 1–2, 2011.

[13] A. Goel, S. K. Gupta, and K. Roy, "Asymmetric drain spacer extension (ADSE) FinFETs for low-power and robust SRAMs," *IEEE Trans. Electron Devices*, vol. 58, no. 2, pp. 296–308, 2011.

[14] K. K. Bhuwalka, J. Schulze, and I. Eisele, "A simulation approach to optimize the electrical parameters of a vertical tunnel FET," *IEEE Trans. Electron Devices*, vol. 52, no. 7, pp. 1541–1547, 2005.

[15] G. C.-F. Yeap, S. Krishnan, and M.-R. Lin, "Fringing-induced barrier lowering (FIBL) in sub-100 nm MOSFETs with high-κ gate dielectrics," *Electron. Lett.*, vol. 34, no. 11, pp. 1150–1152, 1998.

[16] A. Chattopadhyay and A. Mallik, "Impact of a spacer dielectric and a gate overlap/underlap on the device performance of a tunnel field-effect transistor," *IEEE Trans. Electron Devices*, vol. 58, no. 3, pp. 677–683, 2011.

[17] A. Mallik and A. Chattopadhyay, "The impact of fringing field on the device performance of a P-channel tunnel field-effect transistor with a high-κ gate dielectric," *IEEE Trans. Electron Devices*, vol. 59, no. 2, pp. 277–282, 2012.

[18] K. K. Bhuwalka, S. Sedlmaier, A. Ludsteck, C. Tolksdorf, J. Schulze, and I. Eisele, "Vertical tunnel field-effect transistor," *IEEE Trans. Electron Devices*, vol. 51, no. 2, pp. 279–282, 2004.

[19] A. Mallik and A. Chattopadhyay, "Drain-dependence of tunnel field-effect transistor characteristics: the role of the channel," *IEEE Trans. Electron Devices*, vol. 58, no. 12, pp. 4250–4257, 2011.

36

Gate-on-Germanium Source Tunnel Field-Effect Transistor Enabling Sub-0.5-V Operation

36.1 Introduction

A tunnel field-effect transistor (TFET), in which the dominant carrier tunneling occurs in a direction that is in line with the gate electric field (henceforth referred to as *vertical carrier tunneling*), shows great promise for sub-0.6 V operation. A Si TFET structure based on vertical carrier tunneling has already been discussed in the previous chapter. The use of a lower band gap material such as Ge in the device, particularly in its source region, has also evinced a lot of interest. A Ge-source TFET has been demonstrated experimentally in [1] to produce a high value of I_{ON}/I_{OFF} ratio (>10^6). The relatively low subthreshold swing (*SS*) and I_{ON} for sub-0.5 V operation for the Ge-source TFET in [1] are due to the occurrence of lateral carrier tunneling [2]. Improvement in device performance, by suppressing lateral carrier tunneling, has been reported for an elevated Ge-source TFET in [2]. The stringent requirements of uniformity and quality of the ultrathin spacer layer in such elevated TFETs make the fabrication of such devices technologically challenging. Moreover, the gate capacitance for such a device in [2] as well as for the U-shape channel device with a SiGe source, as proposed in [3], is very large owing to the large effective channel length that would degrade switching characteristics. In this chapter, a novel device structure for a gate-on-germanium source (GoGeS) TFET is proposed, which not only eliminates the above difficulties of an elevated Ge-source TFET but also achieve both supersteep *SS* and high I_{ON} enabling sub-0.5 V operation.

36.2 Proposed Device Structure

A cross-sectional view of the proposed n-channel gate-on germanium source tunnel field-effect transistor (GoGeS TFET) is shown in Figure 36.1. The highly doped p-type Ge region, the p-type Si substrate, and the highly doped n-type Si region act, respectively, as

MOS Devices for Low-Voltage and Low-Energy Applications, First Edition.
Yasuhisa Omura, Abhijit Mallik, and Naoto Matsuo.
© 2017 John Wiley & Sons Singapore Pte. Ltd. Published 2017 by John Wiley & Sons Singapore Pte. Ltd.

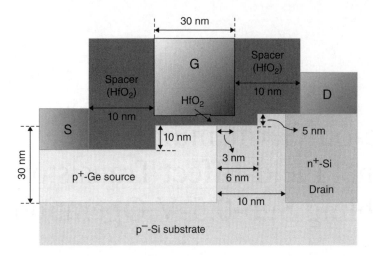

Figure 36.1   Device structure of the proposed n-channel GoGeS TFET. Copyright © 2014 The Japan Society of Applied Physics. Reprinted, with permission, from A. Mallik, A. Chattopadhyay, and Y. Omura, "Gate-on-germanium source tunnel field-effect transistor enabling sub-0.5-V operation," Jpn. J. Appl. Phys., vol. 53, no. 10, pp. 104201-1–104201-7, 2014.

the source, channel, and drain. The device may be implemented on a bulk Si substrate following the process steps shown in Figure 36.2. After shallow-trench isolation, a dummy gate consisting of SiO_2 and polysilicon may be defined over the intended drain region, which is followed by the formation of the first spacer and the first silicon recessing (Figure 36.2a) to form a step in the silicon channel region. The first silicon recessing may be carried out by either anisotropic dry etching or oxidation followed by wet etching [4]. Both the lateral diffusion during drain implantation (to be performed later) and undercutting of the silicon during first silicon recessing should be considered in determining the width of the first spacer layer. The second spacer may be defined next, which is followed by the second silicon recessing and the growth/deposition of the p^+ Ge layer for the source (Figure 36.2b). A process similar to that described in [1] for experimentally demonstrating the Ge-source TFET may be used for such silicon recessing and Ge growth. The undercutting of the silicon during the second silicon recessing should be considered in determining the width of the second spacer layer. Next, a low-temperature oxide (LTO) layer may be deposited, followed by chemical-mechanical polishing (CMP) to expose polysilicon in the dummy gate, followed by dummy gate etching and drain implantation (Figure 36.2c). After the removal of the LTO layer and spacer dielectric, the gate may be defined, which is followed by self-aligned Ge recessing to form a step in the Ge source region (Figure 36.2d). The issue related to ensuring a small gate-to-channel overlap will be discussed in section 36.4.3. The gate side-wall spacer is defined next, followed by silicidation/metallization.

36.3   Simulation Setup

Two-dimensional device simulations are performed for the proposed structure, as shown in Figure 36.1, using the Version 5.16.3.R Silvaco ATLAS device simulator. Unless otherwise mentioned, the device dimensions shown in Figure 36.1 are used for the simulations.

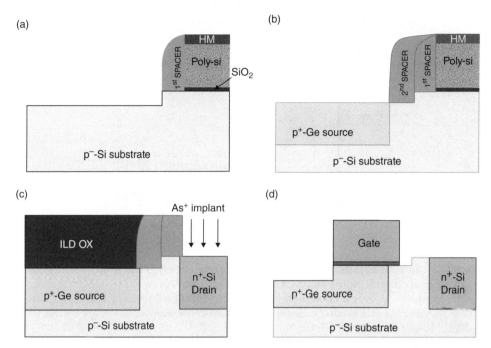

Figure 36.2 Process flow for the proposed GoGeS TFET. Copyright © 2014 The Japan Society of Applied Physics. Reprinted, with permission, from A. Mallik, A. Chattopadhyay, and Y. Omura, "Gate-on-germanium source tunnel field-effect transistor enabling sub-0.5-V operation," Jpn. J. Appl. Phys., vol. 53, no. 10, pp. 104201-1–104201-7, 2014.

The doping concentrations for the source, drain, and bulk Si substrate are 1×10^{19}, 1×10^{19}, and 1×10^{17} atoms/cm^3, respectively, with an abrupt doping profile at both the source-channel and channel-drain junctions. The effect of using a practical doping profile is discussed in section 36.4. A gate length L_G of 30 nm and a HfO$_2$ film of 0.6 nm equivalent oxide thickness (EOT) as the gate dielectric are used. The HfO$_2$ film is also used as the high-κ spacer, which is required to reduce the channel resistance and, hence, improve I_{ON} [5]. The source and drain contacts are made of aluminum, and the gate electrode is made of a metal with a work function φ_m. Unless otherwise mentioned, a φ_m of 3.7 eV is used. Such a low φ_m is used to adjust I_{OFF} at $V_{GS} = 0$ V. In a practical device, the work function to achieve I_{OFF} at $V_{GS} = 0$ V would depend on various other parameters such as fixed-oxide charge, interface-trap density, and EOT. The effects of varying the fixed-oxide charge, the EOT of the gate dielectric, and φ_m of the gate metal on device characteristics will be shown later in section 36.4. Gate leakage is neglected in the simulations. It may be mentioned here that strained Ge is not used for device simulation.

The results presented here are obtained using a nonlocal band-to-band tunneling (BTBT) model combined with a field-dependent mobility model, a band-gap narrowing model, and the Shockley–Read–Hall recombination model. The effects of quantum confinement are taken into account following the procedure described in [6]. Under each bias condition, the self-consistent solution of two-dimensional Schrodinger and Poisson equations is first obtained, followed by current calculation by solving drift-diffusion equations in which nonlocal BTBT

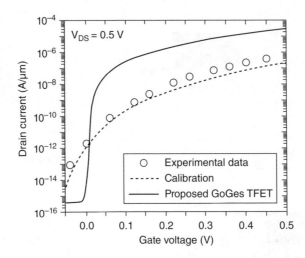

Figure 36.3 Simulated transfer characteristic of the proposed GoGeS TFET. Model calibration against the experimental data in [1] is also shown for reference. Copyright © 2014 The Japan Society of Applied Physics. Reprinted, with permission, from A. Mallik, A. Chattopadhyay, and Y. Omura, "Gate-on-germanium source tunnel field-effect transistor enabling sub-0.5-V operation," Jpn. J. Appl. Phys., vol. 53, no. 10, pp. 104201-1–104201-7, 2014.

injection is used as a generation term. The same values of BTBT model parameters ($A = 1.46 \times 10^{17}\,\text{cm}^{-3}\cdot\text{s}^{-1}$, $B = 3.59 \times 10^6\,\text{V}\cdot\text{cm}^{-1}$) and fixed oxide charge Q_f ($= 10^{11}\,\text{q/cm}^2$) are used as in [2]. A polygon consisting of several quadrilaterals is constructed around the tunneling region, where quantum-tunnel meshes are carefully defined in both the vertical and horizontal directions, to account for tunneling in all directions.

36.4   Results and Discussion

36.4.1   Device Characteristics

The simulated transfer characteristic at a drain-to-source voltage $V_{DS} = 0.5\,\text{V}$ for the proposed GoGeS TFET is shown in Figure 36.3. Also shown in Figure 36.3 is a plot of the model calibration against the experimental data in [1], which is similar to that, as shown in Figure 2 of Ref. 2. A good agreement between the simulated and experimental characteristics is observed in Figure 36.3. Significantly improved device characteristics with supersteep SS and high I_{ON} are evident in Figure 36.3 for the proposed GoGeS TFET, in comparison with those for the Ge-source TFET, as reported in [1]. The following are assumed: (i) the gate-to-source voltage V_{GS} corresponding to $I_D = 0.1\,\mu\text{A}/\mu\text{m}$ as the threshold voltage V_{T}; (ii) the I_D corresponding to the onset of vertical carrier tunneling as I_{OFF} and the corresponding V_{GS} as V_{OFF}, and (iii) the I_D at $V_{GS} = V_{OFF} + 0.5\,\text{V}$ as I_{ON}.

Figure 36.4 shows the energy band diagram at $V_{GS} = 0.4\,\text{V}$ and $V_{DS} = 0.5\,\text{V}$ in the vertical direction of the source region under the gate for the proposed device. The energy quantization is evident in the triangular potential well formed by the insulator and conduction band of the Ge-source region

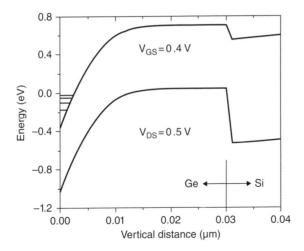

Figure 36.4 Simulated energy band diagram at V_{GS}=0.4 V and V_{DS}=0.5 V in the vertical direction of the source region under the gate for the proposed device. Copyright © 2014 The Japan Society of Applied Physics. Reprinted, with permission, from A. Mallik, A. Chattopadhyay, and Y. Omura, "Gate-on-germanium source tunnel field-effect transistor enabling sub-0.5-V operation," Jpn. J. Appl. Phys., vol. 53, no. 10, pp. 104201-1–104201-7, 2014.

36.4.2 Effects of Different Structural Parameters

To explain the improved device characteristics for the proposed GoGeS structure, the simulated transfer characteristics at V_{DS}=0.5 V are shown in Figure 36.5 for a device that is very similar to the GoGeS TFET, as shown in Figure 36.1, but without one of the following: (i) the step in the Ge source that is defined by Ge etching following gate definition (Dev. 1), (ii) the gate-channel overlap (Dev. 2), and (iii) the step in the Si channel region that is defined by the first silicon recessing (Dev. 3). The transfer characteristic for the proposed GoGeS TFET is also shown in Figure 36.5 for comparison. It is observed in Figure 36.5 that the subthreshold characteristics are degraded for all the devices in comparison with the GoGeS TFET. I_{ON} is also degraded in the cases of Dev. 2 and Dev. 3. To explain such observations, BTBT rates corresponding to I_D=1 pA/µm are shown in Figure 36.6 for all the devices. The structural differences between the different devices can also be clearly seen in Figure 36.6. An early occurrence of lateral carrier tunneling at relatively low gate voltages in the case of Dev. 1, as observed in Figure 36.6, degrades the subthreshold characteristics of Dev. 1, as observed in Figure 36.5. Selective etching of Ge for the proposed GoGeS TFET facilitates the removal of such lateral tunneling, as can be verified from the corresponding plot in Figure 36.6. Fringing-induced barrier lowering (FIBL) has maximum impact at the gate edges [7, 8]. Hence, gate alignment with the channel, in the case of Dev. 2, results in the occurrence of nonuniform BTBT with a high rate at the channel side of the source, as observed in Figure 36.6. The occurrence of such nonuniform BTBT degrades not only the subthreshold characteristics but also I_{ON}, owing to the reduced effective area of tunneling. The absence of a step in the channel region in the case of Dev. 3 results in the spreading of the fringing electric field in the gate insulator, as can be verified in Figure 36.7, where the fringing electric field at V_{GS}=0.0 V and V_{DS}=0.5 V is shown for Dev. 3 and the proposed GoGeS TFET. This fringing field spreading

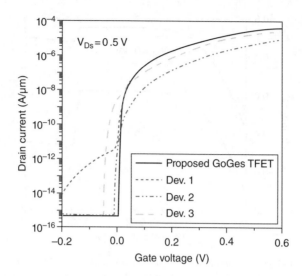

Figure 36.5 Simulated transfer characteristics of the GoGeS TFET and a device that is very similar to the GoGeS TFET, but without one of the following: (i) the Ge etching step following gate definition (Dev. 1), (ii) the gate-channel overlap (Dev. 2), and (iii) the first Si etching step following drain definition (Dev. 3). Copyright © 2014 The Japan Society of Applied Physics. Reprinted, with permission, from A. Mallik, A. Chattopadhyay, and Y. Omura, "Gate-on-germanium source tunnel field-effect transistor enabling sub-0.5-V operation," Jpn. J. Appl. Phys., vol. 53, no. 10, pp. 104201-1–104201-7, 2014.

in the case of Dev. 3, as observed in Figure 36.7, results in nonuniform BTBT in the source, as can be verified from the corresponding plot in Figure 36.6. This is responsible for the degradation of device performance in terms of both SS and I_{ON}, as observed in Figure 36.5, for reasons explained earlier. Figure 36.8 shows the BTBT generation contour plot corresponding to I_{ON}, that is, $V_{GS} = V_{OFF} + 0.5\,V$ and $V_{DS} = 0.5\,V$, for the proposed GoGeS TFET. It is evident in Figure 36.8 that vertical carrier tunneling is the dominant tunneling mechanism for the proposed device.

It may also be noted that the use of the HfO_2 film as the spacer would result in increased parasitic capacitance, which is likely to decrease the switching speed to some extent.

36.4.3   Optimization of Different Structural Parameters

In the previous section, it was explained that the following three device structural parameters have significant effects on the performance of a GoGeS TFET: (i) the step in the Ge source region formed by selective etching of Ge following gate definition, (ii) the step in the silicon channel region formed by the first silicon recessing, and (iii) gate-channel overlap. In this section, the effect of the variation in the following parameters on the performance of a GoGeS TFET are first investigated: (i) the height of the step in the Ge source region, (ii) the height of the step in the silicon channel region, and (iii) the gate-channel overlap length L_{OV}.

Figure 36.9 shows the simulated transfer characteristics at $V_{DS} = 0.5\,V$ for different step heights in the Ge source region for the GoGeS TFET. Four different step heights, 0, 3, 7, and 10 nm, were used to generate the plots in Figure 36.9. All other device parameters were kept

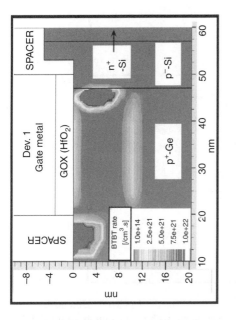
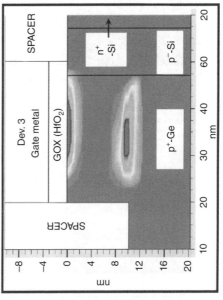
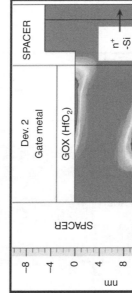
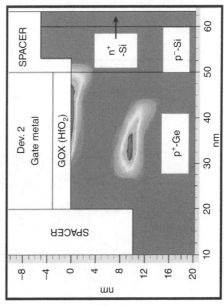

Figure 36.6 BTBT rate at $I_D = 1\,pA/\mu m$ for the proposed GoGeS TFET, Dev. 1, Dev. 2, and Dev. 3. Copyright © 2014 The Japan Society of Applied Physics. Reprinted, with permission, from A. Mallik, A. Chattopadhyay, and Y. Omura, "Gate-on-germanium source tunnel field-effect transistor enabling sub-0.5-V operation," Jpn. J. Appl. Phys., vol. 53, no. 10, pp. 104201-1–104201-7, 2014.

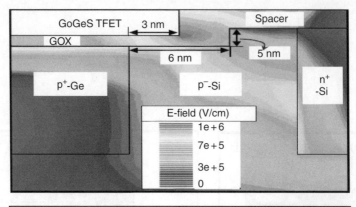

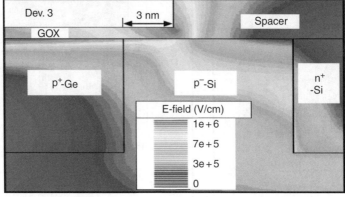

Figure 36.7 Fringing electric field at $V_{GS} = 0.0\,V$ and $V_{DS} = 0.5\,V$ for the GoGeS TFET and Dev. 3. Copyright © 2014 The Japan Society of Applied Physics. Reprinted, with permission, from A. Mallik, A. Chattopadhyay, and Y. Omura, "Gate-on-germanium source tunnel field-effect transistor enabling sub-0.5-V operation," Jpn. J. Appl. Phys., vol. 53, no. 10, pp. 104201-1–104201-7, 2014.

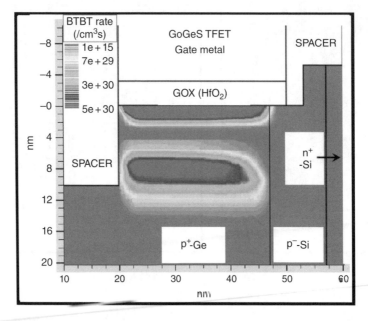

Figure 36.8 BTBT generation contour at $I_D = I_{ON}$ (i.e., $V_{GS} = V_{OFF} + 0.5\,V$ and $V_{DS} = 0.5\,V$) for the proposed GoGeS TFET. Copyright © 2014 The Japan Society of Applied Physics. Reprinted, with permission, from A. Mallik, A. Chattopadhyay, and Y. Omura, "Gate-on-germanium source tunnel field-effect transistor enabling sub-0.5-V operation," Jpn. J. Appl. Phys., vol. 53, no. 10, pp. 104201-1–104201-7, 2014.

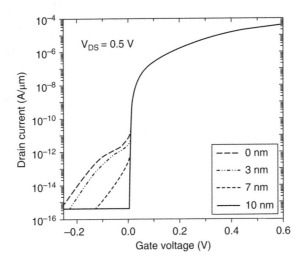

Figure 36.9 Simulated transfer characteristics at $V_{DS}=0.5\,\text{V}$ showing the effect of the step height in the Ge source region. Copyright © 2014 The Japan Society of Applied Physics. Reprinted, with permission, from A. Mallik, A. Chattopadhyay, and Y. Omura, "Gate-on-germanium source tunnel field-effect transistor enabling sub-0.5-V operation," Jpn. J. Appl. Phys., vol. 53, no. 10, pp. 104201-1–104201-7, 2014.

the same. Figure 36.9 shows that the current due to lateral carrier tunneling, which degrades the subthreshold characteristics of a GoGeS TFET, decreases with increasing step height. The use of a step height of 10 nm completely eliminates the current due to lateral carrier tunneling in the device. Note that the minimum step height that can completely eliminate lateral carrier tunneling is related to the band bending of the source region under the gate, a typical case of which is shown in Figure 36.4 for the proposed GoGeS TFET under a particular biasing condition. This would, therefore, be different for different source materials as well as for different source doping concentrations. When the source doping concentration increases, the minimum step height decreases because the gate-induced field increases. Similarly, the minimum step height also decreases when a source material with either a low band gap or a low permittivity is used.

The simulated transfer characteristics at $V_{DS}=0.5\,\text{V}$ for various step heights in the silicon channel region are shown in Figure 36.10. Four different step heights – 0, 3, 5, and 8 nm – were used in this case. I_{ON} and SS are extracted for different step heights from the corresponding plot in Figure 36.10 and plotted against the step heights in the inset in Figure 36.10. It is evident in the inset in Figure 36.10 that an increase in step height up to 5 nm results in a corresponding improvement in the device performance in terms of both SS and I_{ON}. Marginal or no improvement is observed when the step height is increased beyond 5 nm. The optimum step height may, however, be different for different channel materials as well as for different channel doping concentrations.

The effects of varying the gate-channel overlap length L_{OV} on both I-V and C-V characteristics of a GoGeS TFET are shown in Figure 36.11. Four different values of L_{OV}, 0, 3, 5, and 7 nm, were used to generate the plots in Figure 36.11. It is evident from the transfer characteristics in Figure 36.11a that an increase in L_{OV} results in the corresponding improvement in the device performance in terms of both SS and I_{ON}. The improvement is, however, marginal when L_{OV} is increased beyond 3 nm. This makes the gate definition less critical as a longer L_{OV}

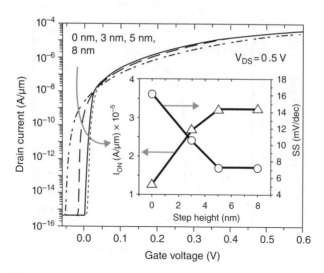

Figure 36.10 Simulated transfer characteristics at $V_{DS}=0.5$ V showing the effect of the step height in the Si channel region. Copyright © 2014 The Japan Society of Applied Physics. Reprinted, with permission, from A. Mallik, A. Chattopadhyay, and Y. Omura, "Gate-on-germanium source tunnel field-effect transistor enabling sub-0.5-V operation," Jpn. J. Appl. Phys., vol. 53, no. 10, pp. 104201-1–104201-7, 2014.

than 3 nm does not degrade the device performance. Hence, a longer silicon channel region may also be used to ensure $L_{OV} \geq 3$ nm, if necessary. On the other hand, the total gate capacitance C_{GG} is found to increase with increasing L_{OV} in Figure 36.11b, which is simply due to the increased area of the gate with increasing L_{OV}. The average overlap-capacitance component in C_{GG}, as computed for a V_{GS} swing of 0.5 V corresponding to $L_{OV}=3$ nm, is less than 10%. The negative effect of overlap capacitance on dynamic operation is, therefore, limited. Switching speed is, however, expected to deteriorate owing to the increased overlap capacitance for $L_{OV}>3$ nm.

Next, the effect of the EOT of the gate dielectric on the device performance of the proposed GoGeS TFET is investigated. The transfer characteristics at $V_{DS}=0.5$ V are shown in Figure 36.12 for five different EOTs, 0.6, 0.8, 1.0, 1.4, and 2.0 nm. V_T is found to decrease with decreasing EOT, as can be verified in the inset of Figure 36.12, simply because of the increased gate capacitance, as expected.

A high-κ gate insulator on Ge gives rise to a relatively poor interface [9]. To study the effect of Q_f on the device characteristics of a GoGeS TFET, device simulations are carried out for five different values of Q_f, 1×10^{10}, 1×10^{11}, 5×10^{11}, 1×10^{12}, and 5×10^{12} q/cm^2. The transfer characteristics at $V_{DS}=0.5$ V are shown in Figure 36.13 for different values of Q_f. It is clear in Figure 36.13 that a variation in Q_f results in a corresponding shift in V_T, as expected. I_{ON} and SS are extracted for different Q_f values from the transfer characteristics in Figure 36.13 and plotted against Q_f in the inset of Figure 36.13. It is evident in the inset of Figure 36.13 that an increase in Q_f degrades both SS and I_{ON}, as expected.

The effect of the thickness of the Ge layer in the source region is investigated next. The simulated transfer characteristics at $V_{DS}=0.5$ V are shown in Figure 36.14 for four different thicknesses of the Ge layer, 12, 15, 20, and 30 nm. Figure 36.14 shows that the characteristics

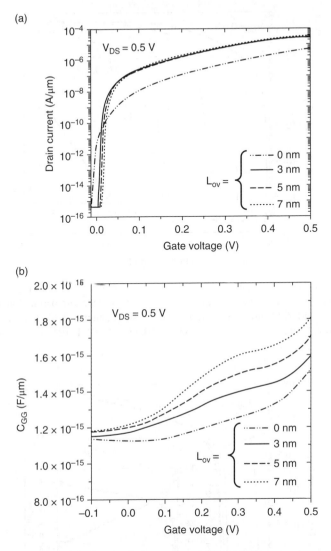

Figure 36.11 Simulated device characteristics at $V_{DS}=0.5\,\mathrm{V}$ for different values of gate-to-channel overlap L_{OV}: (a) I_D–V_{GS} and (b) C_{GG}–V_{GS}. Copyright © 2014 The Japan Society of Applied Physics. Reprinted, with permission, from A. Mallik, A. Chattopadhyay, and Y. Omura, "Gate-on-germanium source tunnel field-effect transistor enabling sub-0.5-V operation," Jpn. J. Appl. Phys., vol. 53, no. 10, pp. 104201-1–104201-7, 2014.

are independent of a relatively thick Ge layer. For a relatively small thickness, I_{ON} is degraded, which is due to the higher resistance of the Ge layer between the source contact and the tunneling region. The increase in both V_T and leakage current for a relatively small thickness of the Ge layer, as observed in Figure 36.14, can be attributed to the increased potential of the source region owing to the increased resistance between the source contact and the tunneling region. It is also worth noting that the optimum thickness of the source region is related to the

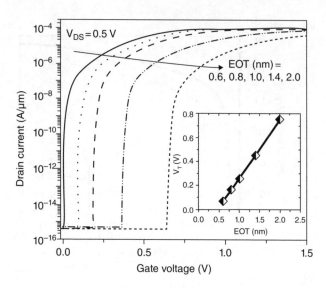

Figure 36.12   Plot showing the effect of EOT scaling. The inset shows the variation in V_T with EOT. Copyright © 2014 The Japan Society of Applied Physics. Reprinted, with permission, from A. Mallik, A. Chattopadhyay, and Y. Omura, "Gate-on-germanium source tunnel field-effect transistor enabling sub-0.5-V operation," Jpn. J. Appl. Phys., vol. 53, no. 10, pp. 104201-1–104201-7, 2014.

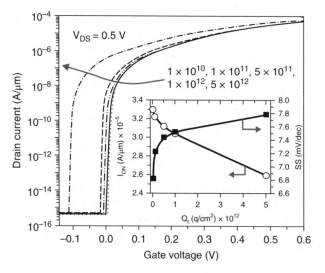

Figure 36.13   Transfer characteristics at $V_{DS}=0.5\,\text{V}$ showing the effect of Q_f. The inset shows the variations in I_{ON} and SS with Q_f. Copyright © 2014 The Japan Society of Applied Physics. Reprinted, with permission, from A. Mallik, A. Chattopadhyay, and Y. Omura, "Gate-on-germanium source tunnel field-effect transistor enabling sub-0.5-V operation," Jpn. J. Appl. Phys., vol. 53, no. 10, pp. 104201-1–104201-7, 2014.

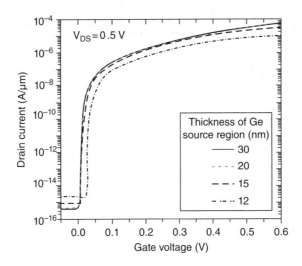

Figure 36.14 Transfer characteristics at $V_{DS}=0.5\,V$ showing the effect of the thickness of the Ge layer in the source region. Copyright © 2014 The Japan Society of Applied Physics. Reprinted, with permission, from A. Mallik, A. Chattopadhyay, and Y. Omura, "Gate on-germanium source tunnel field-effect transistor enabling sub-0.5-V operation," Jpn. J. Appl. Phys., vol. 53, no. 10, pp. 104201-1–104201-7, 2014.

band bending of this region under the gate as well as to BTBT path length. This would, therefore, be different for different source materials as well as for different source doping concentrations. The use of a source material with either a low permittivity or a low band gap would reduce the optimum thickness of the source layer. Similarly, when the source doping concentration is increased, the optimum thickness of the source layer also decreases.

The effect of φ_m on the device characteristics is shown in Figure 36.15. Four different values of φ_m, 3.7 (Mg), 3.9 (Hf), 4.1 (Zr), and 4.5 (Cr, Mo, W), were used in this case. It is observed in Figure 36.15 that a change in φ_m results in a linear shift of the transfer characteristics, including a linear shift of the threshold voltage, which is consistent with the report in [10]. The inset shows no variation in either I_{ON} or SS with φ_m.

To investigate the effect of using a practical doping profile in the proposed device, a Gaussian doping profile with a doping gradient of 2 nm/dec is used for both the source and drain regions of the device. The transfer characteristics of the device are shown in Figure 36.16 with those of a device having an abrupt doping profile at the source-channel and channel-drain junctions. It is evident in Figure 36.16 that the proposed GoGeS TFET is insensitive to the doping profiles of the source-channel and channel-drain junctions. This is expected, because tunneling occurs within the Ge source region in a GoGeS TFET, not across the source-to-channel junction.

It was mentioned above that the only critical process step in implementing the proposed GoGeS TFET is to ensure $L_{OV} \geq 3\,nm$. To do so, a longer channel may be used. The transfer characteristics of two GoGeS TFETs with different dimensions are shown in Figure 36.17. For GoGeS TFET-1, the dimensions shown in Figure 36.1 were used. For GoGeS TFET-2, $L_{OV}=5\,nm$, and the silicon step was 10 and 5 nm away from the source-channel and channel-drain junctions, respectively. It is evident in Figure 36.17 that the use of such large dimensions has no effect on the device's performance.

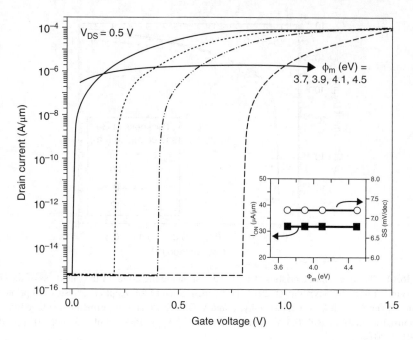

Figure 36.15 Plot showing the effect of φ_m. The inset shows no variation either in I_{ON} or SS with φ_m. Copyright © 2014 The Japan Society of Applied Physics. Reprinted, with permission, from A. Mallik, A. Chattopadhyay, and Y. Omura, "Gate-on-germanium source tunnel field-effect transistor enabling sub-0.5-V operation," Jpn. J. Appl. Phys., vol. 53, no. 10, pp. 104201-1–104201-7, 2014.

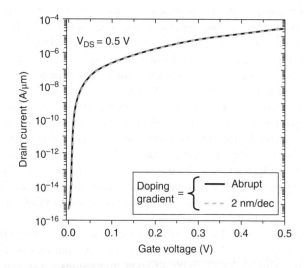

Figure 36.16 Plot showing the effects of the doping profiles of the source-channel and channel-drain junctions. Copyright © 2014 The Japan Society of Applied Physics. Reprinted, with permission, from A. Mallik, A. Chattopadhyay, and Y. Omura, "Gate-on-germanium source tunnel field-effect transistor enabling sub-0.5-V operation," Jpn. J. Appl. Phys., vol. 53, no. 10, pp. 104201-1–104201-7, 2014.

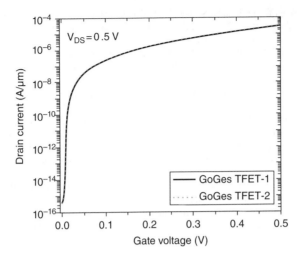

Figure 36.17 Plot showing the impact of using a longer channel to ensure $L_{ov} \geq 3$ nm. For GoGeS TFET-1, the dimensions shown in Figure 36.1 are used. For GoGeS TFET-2, $L_{ov} = 5$ nm and the silicon step is 10 and 5 nm away from the source-channel and channel-drain junctions, respectively (i.e., a channel length of 15 nm). Copyright © 2014 The Japan Society of Applied Physics. Reprinted, with permission, from A. Mallik, A. Chattopadhyay, and Y. Omura, "Gate-on-germanium source tunnel field-effect transistor enabling sub-0.5-V operation," Jpn. J. Appl. Phys., vol. 53, no. 10, pp. 104201-1–104201-7, 2014.

36.5 Summary

A novel GoGeS TFET structure that can be implemented using a standard complementary metal oxide semiconductor (CMOS) technology was presented. To improve I_{ON}, the concept of carrier tunneling in line with the gate electric field was employed by constructing the gate over the germanium source region of the device. A supersteep subthreshold swing was achieved by employing the following: (i) a step in the Ge source region that eliminates lateral carrier tunneling, (ii) a small gate-to-channel overlap that reduces the FIBL, and (iii) a step in the silicon channel region that reduces the spreading of the fringing electric field. Both the reduced FIBL resulting from the small gate-to-channel overlap and the reduced fringing electric field spreading due to the step in the silicon channel region further improve the I_{ON} of the device. Simultaneously achieving both supersteep SS and high I_{ON} makes the device very attractive for sub-0.5 V operation.

References

[1] S. H. Kim, H. Kam, C. Hu, and T. J. K. Liu, "Germanium-Source Tunnel Field Effect Transistors with Record High ION/IOFF," Proc. Int. Symp. VLSI Technol., June 2009, pp. 178–179, 2009.

[2] S. H. Kim, S. Agarwal, Z. A. Jacobson, P. Matheu, C. Hu, and T.-J. K. Liu, "Tunnel field effect transistor with raised germanium source," IEEE Electron Device Lett., vol. 31, no. 10, pp. 1107–1109, 2010.

[3] W. Wang, P.-F. Wang, C.-M. Zhang, X. Lin, X.-Y. Liu, Q.-Q. Sun, P. Zhou, and D. W. Zhang, "Design of U-shape channel tunnel FETs with SiGe source regions," IEEE Trans. Electron Devices, vol. 61, no. 1, pp. 193–197, 2014.

[4] S. Yamakawa, J. Wang, Y. Tateshita, K. Nagano, M. Tsukamoto, H. Ohri, N. Nagashima, and H. Ansai, "Analysis of Novel Stress Enhancement Effect Based on Damascene Gate Process with eSiGe S/D for pFETs," Proc. SISPAD, Sept., 2007, pp. 109–112, 2007.

[5] R. Asra, M. Shrivastava, K. V. R. M. Murali, R. K. Pandey, H. Gossner, and V. R. Rao, "A tunnel FET for VDD scaling below 0.6 V with a CMOS-comparable performance," *IEEE Trans. Electron Devices*, vol. 58, no. 7, pp. 1855–1863, 2011.

[6] J. L. Padilla, F. Gámiz, and A. Godoy, "A simple approach to quantum confinement in tunneling field-effect transistors," *IEEE Electron Device Lett.*, vol. 33, no. 10, pp. 1342–1344, 2012.

[7] A. Chattopadhyay and A. Mallik, "Impact of a spacer dielectric and a gate overlap/underlap on the device performance of a tunnel field-effect transistor," *IEEE Trans. Electron Devices*, vol. 58, no. 3, pp. 677–683, 2011.

[8] A. Mallik and A. Chattopadhyay, "The impact of fringing field on the device performance of a P-channel tunnel field-effect transistor with a high-κ gate dielectric," *IEEE Trans. Electron Devices*, vol. 59, no. 2, pp. 277–282, 2012.

[9] D. Kuzum, A. J. Pethe, T. Krishnamohan, and K. C. Saraswat, "Ge (100) and (111) N- and P-FETs with high mobility and low-T mobility characterization," *IEEE Trans. Electron Devices*, vol. 56, no. 4, pp. 648–655, 2009.

[10] A. Mallik and A. Chattopadhyay, "Drain-dependence of tunnel field-effect transistor characteristics: the role of the channel," *IEEE Trans. Electron Devices*, vol. 58, no. 12, pp. 4250–4257, 2011.

Part VIII

PROSPECTS OF LOW-ENERGY DEVICE TECHNOLOLGY AND APPLICATIONS

37

Performance Comparison of Modern Devices

We discussed the performance of cross-current tetrode silicon-on-insulator complementary metal oxide semiconductors (XCT-SOI CMOSs) based on the partially depleted silicon-on-insulator metal oxide-semiconductor field-effect transistor (SOI MOSFET) in Part V, where we analyzed the impact of the source-potential floating effect (SPFE) on the switching energy of the device. It was shown that the SPFE can reduce the switching energy by approximately one order of magnitude [1]. In order to understand comprehensively the position of XCT-SOI CMOS, a summary of the expected switching performance of various devices based on the International Technology Roadmap for Semiconductors (ITRS) of 2011 , as shown in Part V, is shown again in Figure 37.1. In Figure 37.1, the switching performances of the silicon-on insulator complementary metal oxide semiconductor (SOI CMOS), XCT-SOI CMOS, tunnel FET, tunnel barrier junction metal oxide semiconductor (TBJ-MOS), and gate-all-around (GAA) wire MOSFET are compared. The thermal energy is $\sim 4 \times 10^{-21}$ J at 300 K. Switching performance estimation assumed that the gate-induced drain leakage (GIDL) current (band-to-band tunnel current) of the SOI CMOS is well suppressed in the "OFF" state. Figure 37.1 shows that the tunnel FET is the best choice for reducing dynamic dissipation energy. Other than the tunnel FET, XCT-SOI CMOS seems attractive when the "SPFE" [1] can contribute significantly. Although the TBJ MOSFET offers fairly low energy, the switching delay time appears to be sensitive to the material of source and drain electrodes [2]. It is anticipated that the tunneling dielectric thin-film transistor (TDTFT) will also show behavior similar to the TBJ MOSFET [3]. In contrast, the gate-all-around (GAA) wire MOSFET is superior to the others in terms of switching speed, but not in terms of switching energy [4]. As all data shown in Figure 37.1 are still higher than the thermal energy, the thermal energy is not yet a crucial concern for such devices [5].

The switching delay time of the tunnel FET shown in Figure 37.1 is the smallest theoretically predicted value, but has yet to be verified because of many technical issues. The primary

MOS Devices for Low-Voltage and Low-Energy Applications, First Edition.
Yasuhisa Omura, Abhijit Mallik, and Naoto Matsuo.
© 2017 John Wiley & Sons Singapore Pte. Ltd. Published 2017 by John Wiley & Sons Singapore Pte. Ltd.

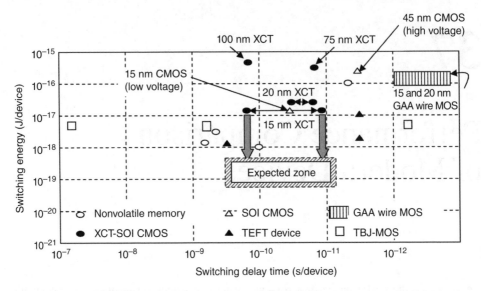

Figure 37.1 Comprehensive examination of scaled XCT-SOI MOSFET based on the ITRS roadmap, 2011. Yasuhisa Omura and Daiki Sato, "Mechanisms of low-energy operation of XCT-SOI CMOS devices – prospect of sub-20-nm regime," J. Low Power Electron. Appl. vol. 4, no. 1, pp. 15–25, 2014; doi:10.3390/jlpea4010015.

performance advantage of the tunnel FET is the steep subthreshold swing (see Part VII) [6–13]. If realized, the tunnel FET will be widely used in mobile electronics and sensor-network field because of its extremely low standby power dissipation, even though its switching speed is not so outstanding [14].

The big issue that needs to be resolved soon is how to reduce the dissipation energy in the "OFF" state or in standby mode, rather than how to suppress the dynamic dissipation energy in the "ON" state (see Parts I, III, IV, and VII). From the description of device operations in Part II and the many discussions in Parts III–VII, it is not difficult to understand why the issue has remained unresolved until now. An understanding of the various aspects of semiconductor devices will help to elucidate the different approaches that may resolve the issue. Such approaches include (but are not limited to): (i) new devices, (ii) overcoming technical barriers, (iii) advances in conventional technologies for the realization of promising devices, and so on [15, 16]. The authors expect that, after reading this book, readers will develop the requisite background to undertake the challenge of resolving this issue.

In the following chapters, we review emerging device technologies, circuit technologies, and future prospects on chip design strategies.

References

[1] Y. Omura and D. Sato, "Mechanisms of low-energy operation of XCT-SOI CMOS devices – prospect of sub-20-nm regime," *J. Low-Power Electron. Appl.*, vol. 4, pp. 14–25, 2014.

[2] H. Nakajima, A. Kawamura, K. Komiya, and Y. Omura, "Simulation models for silicon on-insulator tunnelling-barrier-junction metal oxide-semiconductor field-effect transistor and performance perspective," *Jpn. J. Appl. Phys.*, vol. 42, pp. 1206–1211, 2003.

[3] N. Matsuo, A. Fukushima, K. Ohkura, A. Heya, and S. Yokoyama, "Fabrication of tunneling dielectric thin-film transistor with very thin SiNx films onto source and drain," *IEICE Electron. Express*, vol. 4, pp. 442–447, 2007.

[4] Y. Omura, S. Nakano, and O. Hayashi, "Gate field engineering and source/drain diffusion engineering for high-performance Si wire GAA MOSFET and low-power strategy in sub-30-nm-channel regime," *IEEE Trans. Nanotechnol.*, vol. 10, pp. 715–726, 2011.

[5] R. Landauer, "Irreversibility and heat generation in the computing process," *IBM J. Res. Dev.*, vol. 5, pp. 183–191, 1961.

[6] K. Bhuwalka, J. Schulze, and I. Eisele, "Performance enhancement of vertical tunnel field-effect transistor with SiGe in the δp+ layer," *Jpn. J. Appl. Phys.*, vol. 43, no. 7A, pp. 4073–4078, 2004.

[7] F. Wang, K. Hilsenbeck, T. Nirschl, M. Oswald, C. Stepper, M. Weiss, D. Schmitt-Landsiedel, and W. Hansch, "Complementary tunneling transistor for low power applications," *Solid State Electron.*, vol. 48, no. 12, pp. 2281–2286, 2004.

[8] Y. Choi, B.-G. Park, J. D. Lee, and T.-J. K. Liu, "Tunneling field-effect transistors (TFETs) with subthreshold swing (SS) less than 60 mV/dec," *IEEE Electron Device Lett.*, vol. 28, no. 8, pp. 743–745, 2007.

[9] S. H. Kim, H. Kam, C. Hu, and T. J. K. Liu, Germanium-Source Tunnel Field Effect Transistors with Record High I_{ON}/I_{OFF}," Proc. Int. Symp. VLSI Technol., Jun. 2009, pp. 178–179, 2009.

[10] A. P. Chandrakasan, D. C. Daly, D. F. Finchelstein, J. Kwong, Y. K. Ramadas, M. E. Sinangil, V. Sze, and N. Verma, "Technologies for ultradynamic voltage scaling," *Proc. IEEE*, vol. 98, pp. 191–214, 2010.

[11] A. Chattopadhyay and A. Mallik, "Impact of a spacer dielectric and a gate overlap/underlap on the device performance of a tunnel field-effect transistor," *IEEE Trans. Electron Devices*, vol. 58, no. 3, pp. 677–683, 2011.

[12] A. Mallik and A. Chattopadhyay, "Drain-dependence of tunnel field-effect transistor characteristics: the role of the channel," *IEEE Trans. Electron Devices*, vol. 58, no. 12, pp. 4250–4257, 2011.

[13] A. Mallik and A. Chattopadhyay, "The impact of fringing field on the device performance of a P-channel tunnel field-effect transistor with a high-κ gate dielectric," *IEEE Trans. Electron Devices*, vol. 59, no. 2, pp. 277–282, 2012.

[14] A. Mallik, A. Chattopadhyay, and Y. Omura, "A gate-on-germanium source (GoGeS) tunnel field-effect transistor enabling sub-0.5-V operation," *Jpn. J. Appl. Phys.*, vol. 53, pp. 104201–104208, 2014.

[15] H. Ota, S. Migita, K. Fukuda, and A. Toriumi, "Steep Subthreshold Swing Metal oxide-Semiconductor Field-Effect Transistors Utilizing Nonlinear Gate Dielectric Insulators," Ext. Abstr., 2015 Int. Conf. Solid State Devices and Materials (Sapporo, 2015) pp. 980–981, 2015.

[16] K. Furukawa, R. Kuroda, T. Suwa, K. Hashimoto, A. Teramoto, and S. Sugawa, "Proposal of Tunneling and Diffusion Current Hybrid MOSFET," Ext. Abstr., 2015 Int. Conf. Solid State Devices and Materials (Sapporo, 2015) pp. 86–87, 2015.

38

Emerging Device Technology and the Future of MOSFET

Although the Si metal oxide semiconductor field effect transistor (MOSFET) may be physically scaled to the atomic scale [1], Dr. Landauer has predicted an information processing limit based on Von Neumann's ideas [2]. Dr. Landauer anticipated that the stable operation of atomic-scale devices becomes impossible at room temperature when the supply voltage is lower than the thermal voltage, which is a substantial restriction created by the environment. This may induce some pessimism as regards the future of the Si-MOSFET-based electronics industry, but such concerns are unwarranted because conventional physical scaling approach is not the only doctrine available to us [3]. One interesting alternative is the new instructive guide of "advancing device architectures by aiming at the targets desired." "Technical singularity" is an important keyword for a new business opportunity. This is an unconventional proposal because it differs from the conventional approach of the "application-specific integrated circuit."

38.1 Studies to Realize High-Performance MOSFETs based on Unconventional Materials

Scaling the conventional MOSFET is expected to be hindered by several performance limitations; the limitation of cutoff frequency (determined by carrier mobility), the limitation of drivability (associated with carrier scattering events), and the limitation of threshold-voltage control (due to local doping fluctuations). The primary mechanisms restricting the cutoff frequency are the channel length and carrier mobility [4]. Overcoming this issue by improving carrier mobility [5] is the goal of many scientists and engineers. Mobility enhancement of carriers by using SiGe [6] and strain technology [7] are well-known examples. Recently, carbon nanotube (CNT) transistor [8], graphene field effect transistor (FET) [9], and MoS_2-based FET [10] have also been examined for this purpose.

MOS Devices for Low-Voltage and Low-Energy Applications, First Edition.
Yasuhisa Omura, Abhijit Mallik, and Naoto Matsuo
© 2017 John Wiley & Sons Singapore Pte. Ltd. Published 2017 by John Wiley & Sons Singapore Pte. Ltd.

Use of compound semiconductors was also considered early on but the cost of large-diameter monocrystalline wafers of compound semiconductors is very high. Thus the electronics industry is unlikely to use such expensive compound semiconductor wafers for mass production, at least in the near future. This background emphasizes the importance of semiconductor-on-insulator wafers, not silicon-on-insulator wafers, in the world of commercial electronics. For example, the application of epitaxial transfers to the surface of the silicon-on-insulator wafer makes it possible to reduce the consumption of expensive compound semiconductor material per wafer [11]. In addition, the direct epitaxial growth of a compound semiconductor layer on the silicon-on-insulator substrate with a buffer layer has been proposed [12]. These methods may eliminate the physical limitations of Si and the costs of compound semiconductor materials in the future, and look promising for low-cost RF devices.

The use of ballistic conduction in Si MOSFET [13, 14] is another possible way to improve device drivability. It was predicted in [15] that extremely thin films and narrow wire structures will offer higher carrier mobility than the bulk material. Advanced considerations on ballistic conduction have been described in recent studies [13, 14]. In another trend, the Seebeck effect in Si nano-wires is also attracting attention [16, 17] because they have smaller thermal conductivity. Establishing the relationship between the atomic surface roughness of the narrow wire and thermal conductivity has been of primary interest [18]. Here it is assumed that significant roughness reduces the thermal conductivity. Establishing such a relationship is important to exploit new applications of Si nano-wires, even though many scientists think that it is difficult, at least now, to fabricate fine Si nano-wires. Based on extensive studies of Si nano-wires, it has been predicted that it is not easy to fabricate the high performance gate-all-around (GAA) wire MOSFET regardless of the semiconductor material used. It has been predicted [19, 20] that the diffusion coefficient of a Si wire significantly decreases with its cross-section area. We therefore need an advanced proposal that can overcome the above technical issues so that, for example, non-Si wire MOSFETs with suppressed short-channel effects and better transport characteristics, can be designed.

38.2 Challenging Studies to Realize High-Performance MOSFETs based on the Nonconventional Doctrine

In the previous section, we discussed how technical issues could be overcome after discarding the doctrine of "scaling for high cutoff frequency." Power dissipation, which is another performance parameter of the scaled MOSFET that is independent of wafer materials such as bulk or SOI substrate, has also been discussed widely; both active dissipation power and standby dissipation power in circuits have been thoroughly reconsidered.

As these issues have already been discussed in Part I, we will not address them again here. The nonconventional doctrine stems from a new idea aiming to maximize the efficient usage of energy [3], which is different from the past doctrine. Papers published in IEEE Spectrum in 2010 discussed the following topics:

- ultradynamic voltage scaling [21];
- near-threshold operation [22];
- subthreshold operation [23].

These topics offer the electronics industry possible technical opportunities beyond the conventional applications of MOSFETs and integrated circuits, the symbols of previous mainstream thinking [3]. "Low-energy operation" is the fundamental requirement of future devices; new devices, changes in usage, and the reconsideration and review of circuit technologies are inevitable. Local security systems, security supervising systems for cars, and medical devices such as pacemakers, are possible examples that should be developed around low-energy devices. When such device technology is available, we will have new business opportunities.

References

[1] F.-Q. Xie, L. Nittler, Ch. Obermair, and Th. Schimmel, "Gate-controlled atomic quantum switch," *Phys. Rev. Lett.*, vol. 93, pp. 128303-1–128303-3, 2004.

[2] R. Landauer, "Irreversibility and heat generation in the computing process," *IBM J. Res. Dev.*, vol. 5, pp. 183–191, 1961.

[3] J. Koomey and S. Naffziger, "Moore's Law Might Be Slowing Down, But Not Energy Efficiency," IEEE Spectrum, April Issue, Special Report: 50 years of Moore's law (Posted 31 Mar 2015), 2015.

[4] S. M. Sze and K. K. Ng, "Physics of Semiconductor Devices," 3rd ed. (John Wiley & Sons, Inc., 2007), pp. 347–349.

[5] Y. Song, H. Zhou, Q. Xu, J. Luo, H. Yin, J. Yan, and H. Zhong, "Mobility enhancement technology for scaling of CMOS devices: overview and status," *J. Electron. Mater.*, vol. 40, pp. 1584–1612, 2011.

[6] T. Tezuka, S. Nakaharai, Y. Moriyama, N. Sugiyama, and S. Takagi, "High-mobility strained SiGe-on-insulator pMOSFETs with Ge-rich surface channels fabricated by local condensation technique," *IEEE Electron Device Lett.*, vol. 26, pp. 243–245, 2005.

[7] Z. Y. Cheng, M. T. Currie, C. W. Leitz, G. Taraschi, E. A. Fitzgerald, J. L. Hoyt, and D. A. Antoniadis, "Electron mobility enhancement in strained-Si n-MOSFETs fabricated on SiGe-on-insulator (SGOI) substrates," *IEEE Electron Device Lett.*, vol. 22, pp. 321–323, 2001.

[8] S. J. Trans, A. R. M. Verschueren, and C. Dekker, "Room-temperature transistor based on a single carbon nanotube," *Nature*, vol. 393, pp. 49–51, 1998.

[9] F. Schwierz, "Graphene transistors," *Nat. Nanotechnol.*, vol. 5, pp. 487–496, 2010.

[10] B. Radisavljevic, A. Radenovic, J. Brivio, V. Giacometti, and A. Kis, "Single-layer MoS$_2$ transistors," *Nat. Nanotechnol.*, vol. 6, pp. 147–150, 2011.

[11] H. Ko, K. Takei, R. Kapadia, S. Chuang, H. Fang, P. W. Leu, K. Ganapathi, E. Plis, H. S. Kim, S.-Y. Chen, M. Madsen, A. C. Ford, Y.-L. Chueh, S. Krishna, S. Salahuddin, and A. Javey, "Ultrathin compound semiconductor on insulator layers for high-performance nanoscale transistors," *Nature*, vol. 468, pp. 286–289, 2010.

[12] M. J. Rosker, V. Greanya, and T.-H. Chang, "The DARPA COmpound Semiconductor Materials On Silicon (COSMOS) Program," IEEE Compound Semicond. Integ. Circ. Symp. (Monterey, CA, 2008), pp. 1–4, 2008.

[13] K. Natori, "Ballistic metal-oxide-semiconductor field effect transistor," *J. Appl. Phys.*, vol. 76, pp. 4879–4890, 1994.

[14] R. Clerc, P. Palestri, and L. Selmi, "On the physical understanding of the kT-layer concept in quasi-ballistic regime of transport in nanoscale devices," *IEEE Trans. Electron Devices*, vol. 53, pp. 1634–1640, 2006.

[15] H. Sakaki, "Scattering suppression and high-mobility effect of size-quantized electrons in ultrafine semiconductor wire structures," *Jpn. J. Appl. Phys.*, vol. 19, pp. L735–L738, 1980.

[16] E. Pop, "Energy dissipation and transport in nanoscale devices," *Nano Res.*, vol. 3, pp. 147–169, 2010.

[17] T. Feng and X. Ruan, "Prediction of spectral phonon mean free path and thermal conductivity with applications to thermoelectrics and thermal management: a review," *Hindawi, J. Nanomater.*, vol. 2014, Article ID206370, 2014.

[18] A. I. Hochbaum, P. Chen, R. D. Delgado, W. Liang, E. C. Garnett, M. Najarian, A. Majumdar, and P. Yang, "Enhanced thermoelectric performance of rough silicon nanowires," *Nature*, vol. 451, pp. 163–168, 2008.

[19] Y. Omura and S. Sato, "Theoretical Modeling for Carrier Diffusion Coefficient in One-Dimensional Si Wires around Room Temperature," IEEE Nanoelectron. Conf. (Sapporo, 2014), No. INEC-0135, 2014.

[20] S. Sato and Y. Omura, "Possible theoretical models for carrier diffusion coefficient of one-dimensional Si wire devices," *Jpn. J. Appl. Phys.*, vol. 54, pp. 054001-1–054001-7, 2015.

[21] A. P. Chandrakasan, D. C. Daly, D. F. Finchelstein, J. Kwong, Y. K. Ramadas, M. E. Sinangil, V. Sze, and N. Verma, "Technologies for ultradynamic voltage scaling," *Proc. IEEE*, vol. 98, pp. 191–214, 2010.

[22] D. Markovic, C. C. Wang, L. P. Alarcon, T.-T. Liu, and J. M. Rabaey, "Ultralow-power design in near-threshold region," *Proc. IEEE*, vol. 98, pp. 237–252, 2010.

[23] S. A. Vitale, P. W. Wyatt, N. Checka, J. Kedzierski, and C. L. Keast, "FDSOI process technology for subthreshold-operation ultralow-power electronics," *Proc. IEEE*, vol. 98, pp. 333–342, 2010.

39

How Devices Are and Should Be Applied to Circuits

39.1 Past Approach

In the 1990s, when bulk complementary metal oxide semiconductor (CMOS) technology emerged as the dominant technology for semiconductor products, circuit engineers explored the following techniques to reduce integrated circuit (IC) power dissipation:

1. Using multithreshold devices [1] in the same IC.
2. Using power gating technique [2] to supply power to different circuit blocks of an IC.

Technique 1. is simple but its fabrication cost is so high that it has not been commercialized widely. In contrast to 1., technique 2. does not have a high fabrication cost and so has been used in many commercial ICs, except extremely high-speed ones because of the signal-transfer delay created by the power switching.

Scaled CMOS technology, on the other hand, yields MOSFETs with the stack architecture that allows circuit blocks to have different threshold voltage values, which suppresses the energy dissipated by circuits [3].

39.2 Latest Studies

As described in the previous chapter, the latest methods for designing low-energy circuits are as follows:

- ultradynamic voltage scaling [4];
- use of near-threshold operation [5];
- use of subthreshold operation of FD-SOI MOSFET [6].

MOS Devices for Low-Voltage and Low-Energy Applications, First Edition.
Yasuhisa Omura, Abhijit Mallik, and Naoto Matsuo.
© 2017 John Wiley & Sons Singapore Pte. Ltd. Published 2017 by John Wiley & Sons Singapore Pte. Ltd.

To replace the IC design methodology for conventional devices, such as the MOSFETs described in Part III, leading scientists are investigating how emerging devices can advance circuit performance. For example, adders composed of carbon nanotube (CNT) transistors [7] and microelectromechanical system (MEMS) sensors [8] have been demonstrated. It is anticipated that more such studies will be performed to confirm the feasibility of using such emerging devices in future low-energy ICs [9].

References

[1] T. Douseki, J. Yamada, and H. Kyuragi, "Ultra Low-power CMOS/SOI LSI Design for Future Mobile Systems," Symp. VLSI Circuits Digest of Technical Papers, June 2002, pp. 6–9, 2002.

[2] S. Mutoh, T. Douseki, Y. Matsuya, T. Aoki, S. Shigematsu, and J. Yamada, "1-V power supply high-speed digital circuit technology with multithreshold-voltage CMOS," *IEEE J. Solid-State Circuits*, vol. 30, pp. 847–854, 1995.

[3] S. Narendra, V. De, S. Borkar, D. A. Antoniadis, and A. P. Chandrakasan, "Full-chip subthreshold leakage power prediction and reduction techniques for sub-0.18-μm CMOS", *IEEE J. Solid-State Circuits*, vol. 39, pp. 501–510, 2004.

[4] A. P. Chandrakasan, D. C. Daly, D. F. Finchelstein, J. Kwong, Y. K. Ramadas, M. E. Sinangil, V. Sze, and N, Verma, "Technologies for ultradynamic voltage scaling," *Proc. IEEE*, vol. 98, pp. 191–214, 2010.

[5] D. Markovic, C. C. Wang, L. P. Alarcon, T.-T. Liu, and J. M. Rabaey, "Ultralow-power design in near-threshold region," *Proc. IEEE*, vol. 98, pp. 237–252, 2010.

[6] S. A. Vitale, P. W. Wyatt, N. Checka, J. Kedzierski, and C. L. Keast, "FESOI process technology for subthreshold-operation ultralow-power electronics," *Proc. IEEE*, vol. 98, pp. 333–342, 2010.

[7] M. Masoudi, M. Mazaheri, A. Rezaei, and K. Navi, "Designing high-speed, low-power full adder cells based on carbon nanotube technology," *Int. J. VLSI Design and Comm. Systems*, vol. 5, pp. 31–43, 2014.

[8] "MEMS digital output motion sensor ultra low-power high performance 3-axes 'nano' accelerometer," Product ID: LIS331DLH (ST Micronics Crop.), Doc. ID 15094 (Rev. 3), 2009.

[9] R. Courtland, "Soggy Computing: Liquid Devices Might Match the Brain's Efficiency," IEEE Spectrum, April 23th, 2015, 2015.

40

Prospects for Low-Energy Device Technology and Applications

If humanoids were to be developed in the near future, and if they were fabricated using conventional devices, we would mainly need the following three types of integrated circuits (ICs) (their applications are shown in brackets):

1. ultra-high-speed ICs (artificial intelligence or AI);
2. ordinary high-speed ICs (motor control);
3. low-power ICs (data acquisition from various sensors).

High-level AI architecture needs hardware with a self-control function based on a high-speed CPU and huge-scale memory. DC–DC converters are frequently used for robot motor control. Analog-to-digital converters are generally used to implement sensor functions and require more devices than DC–DC converters.

In order to minimize the energy consumed by the humanoid system, reducing the power dissipation of analog-to-digital converters should be the primary target because they are always used for sensing units. Please refer to Chapter 39 for practical guidance and instructions for circuit design technology.

Now, let us return to academic and scientific considerations. One crucial issue of information processing based on von Neumann's ideas is the fact that increasing the bit length of a signal processor increases its energy dissipation. Although strategies such as multicore processors and parallel signal processing partly contribute to energy efficiencies, they may not be sufficient. Some leading scientists have put forward an information technology (so-called "non-von-Neumann information technology") that may be able to overcome the information processing issues raised by von Neumann's ideas. At the same time, we must develop an innovative electronic device that offers unconventional signal processing. Quantum computing is a good example of the innovation needed. Although the quantum computing is still under

MOS Devices for Low-Voltage and Low-Energy Applications, First Edition.
Yasuhisa Omura, Abhijit Mallik, and Naoto Matsuo.
© 2017 John Wiley & Sons Singapore Pte. Ltd. Published 2017 by John Wiley & Sons Singapore Pte. Ltd.

mathematical investigation and electronic device development despite the recent announcement of commercialization [1], its technical origin lies in the pursuit of minimization of volume density of energy dissipation achieved by downsizing computers.

It is known that the human brain does not offer high-speed signal processing, but its comprehensive decision function (so-called "intelligence") reveals very high performance with a low-energy dissipation for its physical volume. Some researchers are trying to simulate the human brain's activity with electronic devices that offer low energy dissipation density in relation to physical volume. Although scientists and engineers are still investigating how the human brain works, studies of circuits to simulate some brain functions have been performed widely, as well as research on advanced control technologies and fuzzy logic circuits. As we still have no devices that can realize the brain function, we must pay attention to how we can reproduce some brain functions with devices that have the lowest energy dissipation.

We must continue to try to propose new devices that will offer advanced functions with low dissipation energy following the prediction of Dr. R. Landauer, who suggested that the energy dissipation per function of a circuit cannot be suppressed as far as simply binary computation is concerned [2].

References

[1] J. Hsu, "How D-Wave Built Quantum Computing Hardware for the Next Generation," IEEE Spectrum, posted 11 July, 2014.
[2] R. Landauer, "Irreversibility and heat generation in the computing process," *IBM J. Res. and Dev.*, vol. 5, pp. 183–191, 1961.

Bibliography

Part II

[1] "Fully-Depleted SOI CMOS Circuits and Technology for Ultralow-Power Applications," edited by Y. Sakurai, A. Matsuzawa, T. Dozeki (Springer., Feb. 2006), pp. 48–58.

[2] Y. Omura and D. Sato, "Theoretical Modeling of Double-Gate Lateral Tunnel FET", IEICE Technical Report, SDM2013-109, pp. 55–62, 2013.

[3] Y. Omura and A. Mallik, "Physics-Based Analytical Model for Gate-on-Germanium Source (GoGeS)", IEICE Technical Report, SDM2014-97, pp. 7–12, 2014.

Part III

[1] S. Chakraborty, S. Baishya, A. Mallik and C. K. Sarkar, "Performance evaluation of analog circuits with deep submicrometer MOSFETs in the subthreshold regime of operations," in *IEEE International Conference on Industrial and Information Systems (ICIIS)*, Aug. 8-11, 2006, Peradeniya, Sri Lanka, pp. 99–102 (2006).

[2] S. Chakraborty, A. Mallik, C. K. Sarkar, and V. R. Rao, "Impact of halo doping on the subthreshold performance of deep-submicrometer CMOS devices and circuits for ultralow power analog/mixed-signal applications," *IEEE Trans. Electron Devices*, vol. 54, no. 2, pp. 241–248, Feb. 2007.

[3] A. Debnath, S. Chakraborty, C. K. Sarkar and A. Mallik, "Study of the subthreshold performance and the effect of channel engineering on deep submicron single stage CMOS smplifiers," in *IEEE TENCON 2007*, Oct. 30- Nov. 2, pp. 140–143, 2007, Taipei.

[4] S. Chakraborty, A. Mallik, and C. K. Sarkar, "Subthreshold performance of dual-material gate CMOS devices and circuits for ultralow power analog/mixed-signal applications," *IEEE Trans. Electron Devices*, vol. 55, no. 3, pp. 827–832, Mar. 2008.

MOS Devices for Low-Voltage and Low-Energy Applications, First Edition.
Yasuhisa Omura, Abhijit Mallik, and Naoto Matsuo.
© 2017 John Wiley & Sons Singapore Pte. Ltd. Published 2017 by John Wiley & Sons Singapore Pte. Ltd.

Part IV

[1] Y. Omura, S. Nakashima, K. Izumi and T. Ishii, "0.1-μm-Gate, Ultrathin-Film CMOS Devices Using SIMOX Substrate with 80-nm-Thick Buried Oxide Layer," 1991 IEEE Int. Electron Devices Meeting, Tech. Dig., pp. 675–678, 1991.

[2] Y. Omura, S. Nakashima, K. Izumi and T. Ishii,"0.1-μm-Gate, Ultrathin-Film CMOS Devices Using SIMOX Substrate with 80-nm-Thick Buried Oxide Layer," *IEEE Trans. Electron Devices*, vol. 40, No. 5, pp. 1019–1022, 1993

[3] Y. Omura and Y. Iida, "Performance Prospects of Fully-Depleted SOI MOSFET-Based Diodes Applied to Schenkel Circuit for RF-ID Chips", *Sci. Res., J. Cir. and Syst.*, vol. 4, No. 2, pp. 173–180, 2013.

[4] Y. Omura, S. Nakano, and O. Hayashi, "Gate Field Engineering and Source/Drain Diffusion Engineering for High-Performance Si Wire GAA MOSFET and Low-Power Strategy in sub-30-nm-Channel Regime," *IEEE Trans. Nanotechnol.*, vol. 10, pp. 715–726, 2011.

[5] Y. Omura, O. Hayashi, and S. Nakano, "Impact of Local High-κ Insulator on Drivability and Standby Power of Gate-All-Around Silicon-on-Insulator Metal-Oxide-Semiconductor Field-Effect Transistor," *Jpn. J. Appl. Phys.*, vol. 49, pp. 044303–044308, 2010.

Part V

[1] Y. Omura, K. Uchimura, and K. Izumi, "A 60 dB Single-Stage CMOS Amplifier Using High-Gain Cross-Current Tetrode MOSFET/SIMOX", Ext. Abstr., Int. Conf. Solid State Devices and Materials (Tokyo, 1986), pp. 715–716.

[2] Y. Omura, Y. Azuma, Y. Yoshioka, K. Fukuchi, and D. Ino, "Proposal of Preliminary Device Model and Scaling Scheme of Cross-Current Tetrode Silicon-on-Insulator Metal-Oxide-Semiconductor Field-Effect Transistor Aiming at Low-Energy Circuit Applications", *Solid-State Electron.*, vol. 64, pp. 18–27, 2011.

[3] Y. Omura, "Cross-Current Silicon-on-Insulator Metal-Oxide-Semiconductor Field-Effect Transistor and Application to Multiple Voltage Reference Circuits," *Jpn. J. Appl. Phys.*, vol. 48, pp. 04C07-1-04C07-5, 2009.

[4] Y. Omura and D. Sato, "Mechanisms of Low-Energy Operation of XCT-SOI CMOS Devices - Prospect of sub-20-nm Regime -", *J. Low-Power Electron. and Appl.*, vol. 4, pp. 14–25, 2014.

Part VI

[1] Y. Omura, "A Tunneling-Barrier Junction MOSFET on SOI Substrates with a Suppressed Short-Channel Effect for the Ultimate Device Structure," Proc. of 10th Int. Symp. on Silicon-on-Insulator Technology and Devices (The Electrochem. Soc., Washington, D. C., 2001) PV2001-3, pp. 451–456, 2001.

[2] Y. Omura,"A Tunneling-Barrier Junction SOI MOSFET with a Suppressed Short-Channel Effect for the Ultimate Device Structure,"Abstract of IEEE 2001 Silicon Nanoelectronics Workshop (Kyoto, June, 2001), pp. 58–59, 2001.

[3] H. Nakajima, A. Kawamura, K. Komiya and Y. Omura, "Simulation Models for Silicon-on-Insulator Tunneling-Barrier-Junction Metal-Oxide-Semiconductor Field-Effect Transistor and Performance Perspective," *Jpn. J. Appl. Phys.*, vol. 42, No. 3, pp. 1206–1211, 2003.

[4] Y. Omura, "Proposal of High-Temperature-Operation Tolerant (HTOT) SOI MOSFET and Preliminary Study on Device Performance Evaluation", *J. Active and Passive Electronic Components –Field-Effect Transistor- (Hindawi Pub.)*, vol. 2011, pp. 1–8, 2011.

[5] N. Matsuo, T. Miura, J. Yamauchi, H. Hamada, and T. Miyoshi, "Simulation of Si Resonant Tunneling MOS Transistor – Transconductance and Subthreshold Swing," *IEICE Trans. C*, vol. 82, pp. 131–133, 1999.

[6] N. Matsuo, Y. Kitagawa, Y. Takami, and H. Hamada, "Study of Barrier-Height of Si Resonant Tunneling MOS Transistor (SRTMOST) – Relationship between Barrier Height and Gate Length-," *IEICE Trans. C*, vol. 84, pp. 326–327, 2001.

[7] N. Matsuo, Y. Takami, T. Nozaki, and H. Hamada, "Silicon Resonant Tunneling Metal-Oxide-Semiconductor Transistor for Sub-0.1um Era," *IEICE Trans. Electron.*, Vol. E85-C, pp. 1086–1090, 2002.

[8] N. Matsuo, Y. Kitagawa, T. Takami, J. Yamauchi, H. Hamada, and T. Miyoshi, "Influence of Off-set Energy on the Electrical Characteristics of Si Resonant Tunneling MOST (SRTMOST)," *Superlattices and Microstructures*, vol. 28, pp. 407–412, 2000.

[9] N. Matsuo, A. Fukushima, K. Ohkura, A. Heya, and S. Yokoyama, "Fabrication of Tunneling Dielectric Thin-Film Transistor with Very Thin SiNx Films onto Source and Drain," *IEICE Electron. Express (on-line journal)*, vol. 4, pp. 442–447, 2007.

[10] T. Kobayashi, A. Fukushima, N. Matsuo, A. Heya, K. Ohkura, S. Yokoyama and Y. Omura,"Electrical Characteristic in the Low Temperature for Thin-Film Transistor with Very Thin SiNx Film Formed at Source and Drain Region," IEEE International Meeting for Future of Electron Devices, Kansai (April, 2009), pp. 88–89.

[11] T. Kobayashi, N. Matsuo, A. Heya, Y. Omura and S. Yokoyama, "Electrical Characteristic in the High Temperatures for Thin-Film Transistor with Very Thin SiNx Film Formed at Source and Drain Region," IEEE International Meeting for Future of Electron Devices, Kansai (May, 2010), pp. 78–79.

[12] T. Kobayashi, N. Matsuo, T. Tochio, K. Ohkura, Y. Omura, S. Yokoyama and A. Heya, "Temperature Dependence of Drain Currents of Thin-Film Transistor with Very Thin SiNx Film Formed at Source and Drain Region," *IEICE Trans. Electronics (Japanese edition)*,vol. 94, no. 3, pp. 79–87, 2011.

[13] T. Kobayashi, N. Matsuo, Y. Omura, S. Yokoyama and A.H eya, "Electrical Conduction Mechanism in The High Temperatures For Thin-Film Transistor Utilizing Tunnel Effects," *J. Electron Devices*, vol. 20, pp. 1733–1739, 2014.

[14] T. Kobayashi, N. Matsuo, A. Heya and S. Yokoyama, "Improvement of Hump Phenomenon of Thin-Film Transistor by SiNx Film," *IEICE Trans. Electron.*, vol. E97, no. 11, pp. 1112–1116, 2014.

Part VII

[1] A. Chattopadhyay and A. Mallik, "Impact of a spacer dielectric and a gate overlap/underlap on the device performance of a tunnel field-effect transistor," *IEEE Trans. Electron Devices*, vol. 58, no. 3, pp. 677–683, Mar. 2011.

[2] A. Mallik and A. Chattopadhyay, "The impact of fringing field on the device performance of a p-channel tunnel field-effect transistor with a high-κ gate dielectric," *IEEE Trans. Electron Devices*, vol. 59, no. 2, pp. 277–282, Feb. 2012.

[3] A. Mallik, A. Chattopadhyay, S. Guin, and A. Karmakar, "Impact of a spacer-drain overlap on the characteristics of a silicon tunnel field-effect transistor based on vertical tunneling," *IEEE Trans. Electron Devices*, vol. 60, no. 3, pp. 935–943, Mar. 2013.

[4] A. Mallik, A. Chattopadhyay, and Y. Omura, "A gate-on-germanium source tunnel field-effect transistor enabling sub-0.5-V operation," *Jpn. J. Appl. Phys.*, vol. 53, no. 10, pp. 104201-1–104201-7, Oct. 2014.

Part VIII

[1] Y. Omura and D. Sato, "Mechanisms of Low-Energy Operation of XCT-SOI CMOS Devices - Prospect of sub-20-nm Regime -", *J. Low-Power Electron. and Appl.*, vol. 4, pp. 14–25, 2014.

Index

MOS Devices for Low-Voltage and Low-Energy Applications, First Edition.
Yasuhisa Omura, Abhijit Mallik, and Naoto Matsuo.
© 2017 John Wiley & Sons Singapore Pte. Ltd. Published 2017 by John Wiley & Sons Singapore Pte. Ltd.